MW00464253

CONTINUUM MECHANICS FOR ENGINEERS

THIRD EDITION

CRC Series in
COMPUTATIONAL MECHANICS and APPLIED ANALYSIS

Series Editor: J.N. Reddy
Texas A&M University

Published Titles

ADVANCED THERMODYNAMICS ENGINEERING
Kalyan Annamalai and Ishwar K. Puri

APPLIED FUNCTIONAL ANALYSIS
J. Tinsley Oden and Leszek F. Demkowicz

COMBUSTION SCIENCE AND ENGINEERING
Kalyan Annamalai and Ishwar K. Puri

CONTINUUM MECHANICS FOR ENGINEERS, Third Edition
Thomas Mase, Ronald E. Smelser, and George E. Mase

EXACT SOLUTIONS FOR BUCKLING OF STRUCTURAL MEMBERS
C.M. Wang, C.Y. Wang, and J.N. Reddy

THE FINITE ELEMENT METHOD IN HEAT TRANSFER AND FLUID DYNAMICS, Second Edition
J.N. Reddy and D.K. Gartling

MECHANICS OF LAMINATED COMPOSITE PLATES AND SHELLS: THEORY AND ANALYSIS, Second Edition
J.N. Reddy

PRACTICAL ANALYSIS OF COMPOSITE LAMINATES
J.N. Reddy and Antonio Miravete

SOLVING ORDINARY and PARTIAL BOUNDARY VALUE PROBLEMS in SCIENCE and ENGINEERING
Karel Rektorys

CONTINUUM MECHANICS FOR ENGINEERS

THIRD EDITION

G. THOMAS MASE
RONALD E. SMELSER
GEORGE E. MASE

CRC Press is an imprint of the
Taylor & Francis Group, an **informa** business

CRC Press
Taylor & Francis Group
6000 Broken Sound Parkway NW, Suite 300
Boca Raton, FL 33487-2742

Printed in the United States of America on acid-free paper
10 9 8 7 6 5 4 3 2

International Standard Book Number: 978-1-4200-8538-9 (Hardback)

Library of Congress Cataloging-in-Publication Data

Mase, George Thomas.
Continuum mechanics for engineers / G. Thomas Mase, George E. Mase. -- 3rd ed. / Ronald E. Smelser.
p. cm. -- (CRC series in computational mechanics and applied analysis)
Includes bibliographical references and index.
ISBN 978-1-4200-8538-9 (hardcover : alk. paper)
1. Continuum mechanics. I. Mase, George E. II. Smelser, Ronald M., 1942- III. Title. IV. Series.

QA808.2.M364 2009
531--dc22 2009022575

Visit the Taylor & Francis Web site at
http://www.taylorandfrancis.com

and the CRC Press Web site at
http://www.crcpress.com

Contents

List of Figures

List of Tables

Preface to the Third Edition

First, a thank you to all the users of the second edition over the past years. We hope that you will find the updates made in the text make it a more valuable introduction for students to continuum mechanics. The changes made in this edition were substantial but we did not change the basic concept of the book. We seek to provide engineering students with a complete, concise introduction to continuum mechanics that is not intimidating.

Just like previous editions, the third edition is an outgrowth of course notes and problems used to teach the topic to senior undergraduate or first year graduate students. The impetus to do the third edition was to expand it into a text suitable for a two quarter graduate course sequence at Cal Poly. This course sequence introduces continuum mechanics and subsequently covers linear elasticity, nonlinear elastcity, and viscoelasticity. At Cal Poly the terminal degree is a masters degree so the combination of these topics is essential.

One of the things that students struggle with in continuum mechanics and subsequent topics is notation. In the third edition, we have made some changes in notation making the book more consistent with modern continuum mechanics literature. Minor additions were made in many places in the text. The chapter on elasticity was rearranged and expanded to give Saint-Venant's solutions more complete coverage. The extension, torsion, pure bending and flexure subsections give the student a good foundation for posing and solving basic elasticity problems. We have also added some new applications applying continuum mechanics to biological materials in light of their current importance. Finally, a limited amount of material using MATLAB® has been introduced in this edition. We did not want to minimize the fundamental principles of continuum mechanics by making the topic seem like it can be mastered by learning mathematical software. Yet at the same time, these tools can provide valuable help allowing one to stay focused on fundamentals. In addition, most current graduate students are quite proficient at using tools such as MATLAB®, so we did not feel we had to emphasize that topic.

There are many people to acknowledge in the writing of this edition, and we ask the reader to see the Acknowledgments so these people receive their well deserved recognition.

G. Thomas Mase
San Luis Obispo, California, USA

Ronald E. Smelser
Charlotte, North Carolina, USA

George E. Mase
East Lansing, Michigan, USA

Preface to the Third Edition

Preface to the Second Edition

(Note: Some chapter reference information has changed in the Third Edition.)

It is fitting to start this, the preface to our second editions, by thanking all of those who used the text over the last six years. Thanks also to those of you who have inquired about this revised and expanded version. We hope that you find this edition as helpful as the first to introduce seniors or graduate students to continuum mechanics.

The second edition, like its predecessor, is an outgrowth of teaching continuum mechanics to first- or second-year graduate students. Since my father is now fully retired, the course is being taught to students whose final degree will most likely be a Masters at Kettering University. A substantial percentage of these students are working in industry, or have worked in industry, when they take this class. Because of this, the course has to provide the students with the fundamentals of continuum mechanics and demonstrate its applications.

Very often, students are interested in using sophisticated simulation programs that use nonlinear kinematics and a variety of constitutive relationships. Additions to the second edition have been made with these needs in mind. A student who masters its contents should have the mechanics foundation necessary to be a skilled user of today's advanced design tools such as nonlinear, explicit finite elements. Of course, students need to augment the mechanics foundation provided herein with rigorous finite element training.

Major highlights of the second edition include two new chapters, as well as significant expansion of two other chapters. First, Chapter Five, Fundamental Laws and Equations, was expanded to add materials regarding constitutive equation development. This includes material on the second law of thermodynamics and invariance with respect to restrictions on constitutive equations. The first edition applications chapter covering elasticity and fluids has been split into two separate chapters. Elasticity coverage has been expanded by adding sections on Airy stress functions, torsion of non-circular cross sections, and three dimensional solutions. A chapter on nonlinear elasticity has been added to give students a molecular and phenomenological introduction to rubber-like materials. Finally, a chapter introducing students to linear viscoelasticity is given since many important modern polymer applications involve some sort of rate dependent material response.

It is not easy singling out certain people in order to acknowledge their help while not citing others; however, a few individuals should be thanked. Ms. Sheri Burton was instrumental in preparation of the second edition manuscript. We wish to acknowledge the many useful suggestions by users of the previous edition, especially Prof. Morteza M. Mehrabadi, Tulane University, for his detailed comments. Thanks also go to Prof. Charles Davis, Kettering University, for helpful comments on the molecular approach to rubber and thermoplastic elastomers. Finally, our families deserve sincerest thanks for their encouragement.

It has been a great thrill to be able to work as a father-son team in publishing this text, so again we thank you, the reader, for your interest.

G. Thomas Mase
Flint, Michigan, USA

George E. Mase
East Lansing, Michigan, USA

Preface to the First Edition

(Note: Some chapter reference information has changed in the Third Edition.)

Continuum mechanics is the fundamental basis upon which several graduate courses in engineering science such as elasticity, plasticity, viscoelasticity and fluid mechanics are founded. With that in mind, this introductory treatment of the principles of continuum mechanics is written as a text suitable for a first course that provides the student with the necessary background in continuum theory to pursue a formal course in any of the aforementioned subjects. We believe that first-year graduate students, or upper-level undergraduates, in engineering or applied mathematics with a working knowledge of calculus and vector analysis, and a reasonable competency in elementary mechanics will be attracted to such a course.

This text evolved from the course notes of an introductory graduate continuum mechanics course at Michigan State University, which was being taught on a quarter basis. We feel that this text is well suited for either a quarter or semester course in continuum mechanics. Under a semester system, more time can be devoted to later chapters dealing with elasticity and fluid mechanics. For either a quarter or a semester system, the text is intended to be used in conjunction with a lecture course.

The mathematics employed in developing the continuum concepts in the text is the algebra and calculus of Cartesian tensors; these are introduced and discussed in some detail in Chapter Two, along with a review of matrix methods, which are useful for computational purposes in problem solving. Because of the introductory nature of the text, curvilinear coordinates are not introduced and so no effort has been made to involve general tensors in this work. There are several books listed in the Reference Section that a student may refer to for a discussion of continuum mechanics in terms of general tensors. Both indicial and symbolic notations are used in deriving the various equations and formula of importance.

Aside from the essential mathematics presented in Chapter Two, the book can be seen as divided into two parts. The first part develops the principles of stress, strain and motion in Chapters Three and Four, followed by the derivation of the fundamental physical laws relating to continuity, energy and momentum in Chapter Five. The second portion, Chapter Six, presents some elementary applications of continuum mechanics to linear elasticity and classic fluids behavior. Since this text is meant to be a first text in continuum mechanics, these topics are presented as constitutive models without any discussion as to the theory of how the specific constitutive equations was derived. Interested readers should pursue more advanced texts listed in the Reference Section for constitutive equation development. At the end of each chapter (with the exception of Chapter One) there appears a collection of problems, with answers to most, by which the student may reinforce her/his understanding of the material presented in the text. In all, 186 such practice problems are provided, along with numerous worked examples in the text itself.

Like most authors, we are indebted to may people who have assisted in the preparation of this book. Although we are unable to cite each of them individually, we are pleased to acknowledge the contributions of all. In addition, sincere thanks must go to the students who have given feedback for the classroom notes which served as the forerunner to the book. Finally, and most sincerely of all, we express thanks to our family for their encouragement from beginning to end of this work.

G. Thomas Mase
Flint, Michigan, USA

George E. Mase
East Lansing, Michigan, USA

Acknowledgments

There are too many people to thank for their help in preparing this third edition. We can only mention the key contributors. Ryan Miller was a superb help in moving the early manuscript into LaTeX 2ε before his masters research redirected his focus. We were fortunate that one of us (GTM) was teaching ME 501 and 503 in the fall and winter quarters at Cal Poly while preparing the manuscript. The class was quite helpful in proofreading the manuscript. Specifically, Nickolai Volkoff-Shoemaker, Peter Brennen, Roger Sharpe, John Wildharbor, Kevin Ng, and Jason Luther found many typographical errors and suggested helpful corrections and clarifications. Nickolai Volkoff-Shoemaker, Peter Brennen and Roger Sharpe helped in creating some of the figures.

One author (GTM) is very appreciative of Don Bently's generous gift to Cal Poly allowing for partial release time during the Fall 2009 quarter. In addition, many thanks to the devoted teachers that shaped him as a student including George E. Mase, George C. Johnson, Paul M. Naghdi, Michael M. Carroll and David B. Bogy. The other (RES) was privileged to benefit from interactions with several outstanding colleagues and teachers including Ronald Huston, University of Cincinnati, William J. Shack, MIT and Argonne National Laboratories, Morton E. Gurtin, Carnegie Mellon University and the late Owen Richmond, US Steel Research Laboratories and Alcoa Technical Center.

Of course, our greatest thanks go to our families who very patiently kept asking if the book was done. Now it is done; so we can spend more time with the ones we love.

Authors

G. Thomas Mase, Ph.D., is Associate Professor of Mechanical Engineering at California Polytechnic State University, San Luis Obispo, California. Dr. Mase received his B.S. degree from Michigan State University in 1980 from the Department of Metallurgy, Mechanics and Materials Science. He obtained his M.S. and Ph.D. degrees in 1982 and 1985, respectively, from the Department of Mechanical Engineering at the University of California, Berkeley. After graduate school, he has worked at several positions in industry and academia. Industrial companies Dr. Mase has worked full time for include General Motors Research Laboratories, Callaway Golf and Acushnet Golf Company. He has taught or held research positions at the University of Wyoming, Kettering University, Michigan State University and California Polytechnic State University. Dr. Mase is a member of numerous professional societies including the American Society of Mechanical Engineers, American Society for Engineering Education, International Sports Engineering Association, Society of Experimental Mechanics, Pi Tau Sigma and Sigma Xi. He received an ASEE/NASA Summer Faculty Fellowship in 1990 and 1991 to work at NASA Lewis Research Center (currently NASA Glenn Research Center). While at the University of California, he twice received a distinguished teaching assistant award in the Department of Mechanical Engineering. His research interests include mechanics, design and applications of explicit finite element simulation. Specific areas include golf equipment design and performance and vehicle crashworthiness.

Ronald E. Smelser, Ph.D., P.E., is Professor and Associate Dean for Academic Affairs in the William States Lee College of Engineering at the University of North Carolina at Charlotte. Dr. Smelser received his B.S.M.E. from the University of Cincinnati in 1971. He was awarded the S.M.M.E. in 1972 from M.I.T. and completed his Ph.D. (1978) in mechanical engineering at Carnegie Mellon University. He gained industrial experience working for the United States Steel Research Laboratory, the Alcoa Technical Center, and Concurrent Technologies Corporation. Dr. Smelser served as a fulltime or adjunct faculty member at the University of Pittsburgh, Carnegie Mellon University, and the University of Idaho and was a visiting research scientist at Colorado State University. Dr. Smelser is a member of the American Academy of Mechanics, the American Society for Engineering Education, Pi Tau Sigma, Sigma Xi, and Tau Beta Pi. He is also a member and Fellow of the American Society of Mechanical Engineers. Dr. Smelser's research interests are in the areas of process modeling including rolling, casting, drawing and extrusion of single and multi-phase materials, the micromechanics of material behavior and the inclusion of material structure into process models, and the failure of materials.

George E. Mase (1920-2007), Ph.D., was Emeritus Professor, Department of Metallurgy, Mechanics and Materials Science (MMM), College of Engineering, at Michigan State University. Dr. Mase received a B.M.E in Mechanical Engineering (1948) from the Ohio State University, Columbus. He completed his Ph.D. in Mechanics at Virginia Polytechnic Institute and State University (VPI), Blacksburg, Virginia (1958). Previous to his initial appointment as Assistant Professor in the Department of Applied Mechanics at Michigan State University in 1955, Dr. Mase taught at Pennsylvania State University (instructor),

1950-1951, and at Washington University, St. Louis, Missouri (assistant professor), 1951-1954. He was appointed associate professor in 1959 and professor in 1965, and served as acting chairperson of the MMM Department 1965-1966 and again in 1978 to 1979. He taught as visiting assistant professor at VPI during the summer terms, 1953 through 1956. Dr. Mase held membership in Tau Beta Pi and Sigma Xi. His research interests and publications were in the areas of continuum mechanics, viscoelasticity and biomechanics.

Nomenclature

$\boldsymbol{x}$ or x_i	Spatial or current coordinates
$\mathbf{X}$ or X_A	Material or referential coordinates
$\boldsymbol{u}$ or u_i	Displacement components or displacement vector
$\boldsymbol{v}$ or v_i	Velocity or general vector
$\boldsymbol{a}$ or a_i	Acceleration or general vector
x_1^*, x_2^*, x_3^*	Principal axes
$\hat{e}_i$	Unit vectors along coordinate axes
$\hat{\mathbf{I}}_A$	Unit vectors along coordinate axes in reference configuration
δ_{ij}	Kronecker delta
$\mathcal{A}$ or a_{ij}	Transformation or general matrix
$\mathbf{I}$	Identity matrix
ε_{ijk}	Permutation symbol
∂_t	Partial derivative with respect to time
$\dot{\overline{(\cdot)}}$	Derivative with respect to time
$\boldsymbol{\nabla}$ or ∂_x	Spatial gradient operator
$\boldsymbol{\nabla}\phi = \text{grad } \phi = \phi_{,j}$	Scalar gradient
$\boldsymbol{\nabla v} = \partial_j v_i = v_{i,j}$	Vector gradient
$\boldsymbol{\nabla} \cdot v$ or $v_{i,i}$	Divergence of a vector $\boldsymbol{v}$
$\boldsymbol{\nabla} \times \boldsymbol{v}$ or $\varepsilon_{ijk} v_{i,j}$	Curl of a vector $\boldsymbol{v}$
$d/dt = \partial/\partial t + v_k \partial/\partial x_k$	Material derivative operator

$\mathbf{b}$ or b_i	Body force (force per unit mass)
$\mathbf{p}$ or p_i	Body force (force per unit volume)
$\mathbf{f}$ or f_i	Surface force (force per unit area)
$\mathcal{V}$, $\mathcal{V}^0$	Current and referential total volumes
$\Delta\mathcal{V}$, $d\mathcal{V}$	Small and infinitesimal element of volumes
$\mathcal{S}$, $\mathcal{S}^0$	Current and referential total surfaces
$\Delta\mathcal{S}$, $d\mathcal{S}$	Small and infinitesimal elements of surface
ρ	Density
$\hat{\mathbf{n}}$ or n_i	Unit normal in current configuration
$\hat{\mathbf{N}}$ or N_A	Unit normal in reference configuration
$\mathbf{t}^{(\hat{\mathbf{n}})}$ or $t_i^{(\hat{\mathbf{n}})}$	Traction vector
σ_N, σ_S	Normal and shear components of traction vector
$\mathbf{T}$ or t_{ij}	Cauchy stress or general tensor
$\mathbf{T}^*$ or t^*_{ij}	Cauchy stress referred to principal axes
$\mathbf{p}^{0(\hat{\mathbf{N}})}$ or $p_i^{0(\hat{\mathbf{N}})}$	Piola-Kirchhoff stress vector referred to referential area
$\mathbf{P}$ or P_{iA}	First Piola-Kirchhoff stress
$\mathbf{s}$ or s_{AB}	Second Piola-kirchhoff stress
$\sigma_{(1)}$, $\sigma_{(2)}$, $\sigma_{(3)}$	Principal stress values
I_T, II_T, III_T	First, second and third stress invariants
$\sigma_M = \frac{1}{3}t_{ii}$	Mean normal stress
S_{ij}	Deviatoric stress components
η_{ij}	Deviatoric strain components
$J_1 = 0$, J_2, J_3	Deviatoric stress invariants
σ_{oct}	Octahedral shear stress
$\mathbf{F}$ or F_{iA}	Deformation gradient tensor

$\mathbf{C}$ or C_{AB}	Right Cauchy-Green deformation tensor
$\mathbf{E}$ or E_{AB}	Lagrangian finite strain tensor
$\mathbf{c}$ or c_{ij}	Cauchy deformation tensor
$\mathbf{e}$ or e_{ij}	Eulerian finite strain tensor
$\boldsymbol{\epsilon}$ or ϵ_{ij}	Infinitesimal strain tensor
$\epsilon_{(1)}, \epsilon_{(2)}, \epsilon_{(3)}$	Principal strain values
$I_{\boldsymbol{\epsilon}}, II_{\boldsymbol{\epsilon}}, III_{\boldsymbol{\epsilon}}$	First, second and third infinitesimal strain invariants
$\mathbf{B}$ or B_{ij}	Left Cauchy-Green deformation tensor
$I_{\mathbf{B}}, II_{\mathbf{B}}, III_{\mathbf{B}}$, or I_1, I_2, I_3	Invariants of right deformation tensor
W	Strain energy per unit volume or strain energy density
$e_{\hat{\mathbf{N}}}$	Normal strain in the $\hat{\mathbf{N}}$ direction
γ_{ij}	Engineering shear strain
$e = \Delta\mathcal{V}/\mathcal{V} = \epsilon_{ii} = \boldsymbol{\epsilon}_I$	Cubical dilatation
$\boldsymbol{\omega}$ or ω_{ij}	Infinitesimal rotation tensor
$\boldsymbol{\omega}$ or ω_j	Rotation vector
$\Lambda_{\hat{\mathbf{N}}} = dx/dX$	Stretch ratio or stretch in the direction of $\hat{\mathbf{N}}$
$\lambda_{\hat{\mathbf{n}}} = dX/dx$	Stretch ratio in the direction of $\hat{\mathbf{n}}$
$\mathbf{R}$ or R_{ij}	Rotation tensor
$\mathbf{U}$ or U_{AB}	Right stretch tensor
$\mathbf{V}$ or V_{ij}	Left stretch tensor
$\mathbf{L}$ or L_{ij}	Spatial velocity gradient
$\mathbf{D}$ or d_{ij}	Rate of deformation tensor
$\mathbf{W}$ or w_{ij}	Vorticity, or spin tensor
$J = \det \mathbf{F}$	Jacobian
$\mathbf{P}(t)$ or P_i	Linear momentum vector

$K(t)$	Kinetic energy
$P(t)$	Mechanical power or rate of work done by forces
$S(t)$	Stress work
Q	Heat input rate
r	Heat supply per unit mass
$\mathbf{q}$ or q_i	Heat flux vector
θ	Temperature or angle
$\mathbf{g} = \text{grad}\,\theta$ or $g_i = \theta_{,i}$	Temperature gradient
u	Specific internal energy
η	Specific entropy or viscoelastic viscosity
ψ	Gibbs' free energy
ζ	Free enthalpy
χ	Enthalpy
γ	Specific entropy production
E	modulus of elasticity or Young's modulus
G or μ	shear modulus
K	Bulk modulus
ν	Poisson's ratio
λ, μ	Lamé constants
C_{ijkl}	General elastic constants
τ_{ij}	Viscous stress tensor
β_{ij}	Deviatoric rate of deformation
λ^*, μ^*	Viscosity coefficients
κ^*	Bulk viscosity coefficient

1

Continuum Theory

The atomic/molecular composition of matter is well established. On a small enough scale, a body of aluminum, is really a collection of discrete aluminum atoms stacked on one another in a particular repetitive lattice. And on an even smaller scale, the atoms consist of a core of protons and neutrons around which electrons orbit. Thus matter is not continuous. At the same time, the physical space in which we live is truly a continuum, for mathematics teaches us that between any two points in space we can always find another point regardless of how close together we choose the original pair. Clearly then, although we may speak of a material body as "occupying" a region of physical space, it is evident that the body does not totally "fill" the space it occupies. It is this "occupying" of space that will be the basis of our study of continuum mechanics.

1.1 Continuum Mechanics

If we accept the continuum concept of matter, we agree to ignore the discrete composition of material bodies, and to assume that the substance of such bodies is distributed uniformly throughout, and completely fills the space it occupies. In keeping with this continuum model, we assert that matter may be divided indefinitely into smaller and smaller portions, each of which retains all of the physical properties of the parent body. Accordingly, we are able to ascribe field quantities such as density and velocity to each and every point of the region of space which the body occupies.

The continuum model for material bodies is important to engineers for two very good reasons. On the scale by which we consider bodies of steel, aluminum, concrete, etc., the characteristic dimensions are extremely large compared to molecular distances so that the continuum model provides a very useful and reliable representation. Additionally, our knowledge of the mechanical behavior of materials is based almost entirely upon experimental data gathered by tests on relatively large specimens.

The analysis of the kinematic and mechanical behavior of materials modeled on the continuum assumption is what we know as Continuum Mechanics. There are two main themes into which the topics of continuum mechanics are divided. In the first, emphasis is on the derivation of fundamental equations which are valid for all continuous media. These equations are based upon universal laws of physics such as the conservation of mass, the principles of energy and momentum, etc. In the second, the focus of attention is on the development of the *constitutive equations* characterizing the behavior of specific idealized materials; the perfectly elastic solid and the viscous fluid being the best known examples. These equations provide the focal points around which studies in elasticity, plasticity, viscoelasticity and fluid mechanics proceed.

Mathematically, the fundamental equations of continuum mechanics mentioned above may be developed in two separate but essentially equivalent formulations. One, the integral, or global form, derives from a consideration of the basic principles being applied to a finite volume of the material. The other, a differential, or field approach, leads to equations resulting from the basic principles being applied to a very small (infinitesimal) element of volume. In practice, it is often useful and convenient to deduce the field equations from their global counterparts.

As a result of the continuum assumption, field quantities such as density and velocity which reflect the mechanical or kinematic properties of continuum bodies are expressed mathematically as continuous functions, or at worst as piecewise continuous functions, of the space and time variables. Moreover, the derivatives of such functions, if they enter into the theory at all, will be likewise continuous.

Inasmuch as this is an introductory textbook, we shall make two further assumptions on the materials we discuss in addition to the principal one of continuity. First, we require the materials to be *homogeneous*, that is to have identical properties at all locations. And secondly, that the materials be *isotropic* with respect to certain mechanical properties, meaning that those properties are the same in all directions at a given point. Later, we will relax this isotropy restriction to discuss briefly anisotropic materials which have important meaning in the study of composite materials.

1.2 Starting Over

The topic of continuum mechanics typically comes at the end of an undergraduate or at the beginning of a graduate program. Continuum mechanics has a reputation of being a theoretical course without many applications (during the course). The first of part of continuum mechanics' reputation is correct: it is based on fundamental mathematics and mechanics. However, the second part is not founded. There are many, many applications for continuum mechanics, but it is hard to cover the basics and develop the applications in a single quarter or semester. Continuum mechanics takes all the mathematical, physical and engineering principles and casts them in a single structure from which the student is prepared to pursue advanced engineering topics. After having a course in continuum mechanics many applications become accessible to the student: elasticity, nonlinear elasticity, plasticity, crashworthiness, biomechanics, polymers and more. Many of the sophisticated simulation programs such as LS-DYNA® become a playground for advanced design and analysis once continuum mechanics has been mastered.

Some students find continuum mechanics a difficult subject. However, outside of a new notation, the topics studied should be very familiar to the student. Vectors have to be written in component form, and we need to be able to use "dot" and "cross" products. These are skills from the sophomore level statics course. Also needed will be a description for conservation of linear and angular momentum. Taking the time rate of change of these quantities is really no different than what was done in an undergraduate course in dynamics. Just like dynamics, a description of the energy equation will be examined. Rather than study only rigid bodies as done in undergraduate statics and dynamics, deformation is allowed. This requires defining stress and strain that were first introduced in a mechanics of solids class. Stress and strain are tensors which are an order more complex than vectors. When looking at strain and the resulting stress one needs to have a material model. At the undergraduate level students have studied linear elastic, fluid and gas behavior. But the topics generally are taught in separate courses, and often the common, underlying theory is not noticed. Also, the methods used to determine the relationship between stress and strain are not considered.

So continuum mechanics is not a new or challenging topic. Rather it is a chance to start over and put all that was studied previously under a single umbrella. A course in continuum mechanics is a chance to synthesize what was learned during an undergraduate education into a coherent structure. One of the challenges of doing this is developing a common notation, but no new physics is presented. Continuum mechanics is just the process of confirming the foundation for all that was done in undergraduate studies.

1.3 Notation

As one would imagine, building a theoretical foundation for the study of continuum mechanics creates notational difficulties. This is especially true for the student just learning continuum mechanics. In the pages that follow there are many different symbols used for all the quantities of interest. There are more symbols than a student experienced as an undergraduate because a general, nonlinear theory is being contructed. For instance, consider stress. There are different measures of stress that are indistinguishable in the linear theory: Cauchy, first Piola-Kirchhoff, and second Piola-Kirchhoff. In addition, von Mises and octahedral stresses are defined to help analyze yield and failure theory. Finally, it is often advantageous to subdivide stress into deviatoric and spherical parts because of the different role the two have in deformed bodies.

With all these quantities it is hard to come up with symbols for each of them. Often, one symbol is very close to another symbol, and the context has to be used to fully understand the meaning. This is one of the things that makes continuum mechanics difficult for the beginner.

Finally, as students continue beyond this course, they find the disheartening reality that not everybody uses the same notation. Often notational marks will have to be made in margins when reading the literature. The reason there is not a single notation comes from the fact that people were not flying in airplanes to technical conferences when this material was being developed. Even when reading a single author's works spanning a decade the notation can change. For example, Table 1.1 [1] shows some historical notation for stress.

[1] Adapted from *Nonlinear Theory of Continuous Media*, A. Cemal Eringen, McGraw-Hill Inc., (1962)

TABLE 1.1
Historical notation for stress.

Naghdi, Eringen, Clebsch, Truesdell	$t_{11}\ t_{22}\ t_{33}\ t_{12}\ t_{23}\ t_{31}$
Cauchy (early work)	A B C D E F
Cauchy (later work), St. Venant, Maxwell	$p_{xx}\ p_{yy}\ p_{zz}\ p_{xy}\ p_{yz}\ p_{zx}$
F. Neumann, Kirchhoff, Love	$X_x\ Y_y\ Z_z\ X_y\ Y_z\ Z_x$
Green and Zerna, Russian and German writers	$\tau_{11}\ \tau_{22}\ \tau_{33}\ \tau_{12}\ \tau_{23}\ \tau_{31}$
Karman, Timoshenko	$\sigma_x\ \sigma_y\ \sigma_z\ \tau_{xy}\ \tau_{yz}\ \tau_{zx}$
Some English and American writers	$\sigma_{11}\ \sigma_{22}\ \sigma_{33}\ \sigma_{12}\ \sigma_{23}\ \sigma_{31}$
	$\sigma_{xx}\ \sigma_{yy}\ \sigma_{zz}\ \sigma_{xy}\ \sigma_{yz}\ \sigma_{zx}$

2

Essential Mathematics

Learning a discipline's language is the first step a student takes towards becoming competent in that discipline. The language of continuum mechanics is the algebra and calculus of *tensors*. Here, tensor is the generic name for those mathematical entities which are used to represent the important physical quantities of continuum mechanics. A tensor, or linear transformation, assigns any vector $\mathbf{v}$ to another vector $\mathbf{T}\mathbf{v}$ such that

$$\mathbf{T}(\mathbf{v}+\mathbf{w}) = \mathbf{T}\mathbf{v} + \mathbf{T}\mathbf{w}\,, \tag{2.1a}$$

and

$$\mathbf{T}(\alpha\mathbf{v}) = \alpha\mathbf{T}\mathbf{v} \tag{2.1b}$$

for all $\mathbf{v}$ and $\mathbf{w}$. Furthermore, the sum of two tensors and the scalar multiple of a tensor is defined by

$$(\mathbf{T}+\mathbf{S})\,\mathbf{v} = \mathbf{T}\mathbf{v} + \mathbf{S}\mathbf{v}\,, \tag{2.1c}$$

and

$$(\alpha\mathbf{T})\,\mathbf{v} = \alpha\,(\mathbf{T}\mathbf{v})\;. \tag{2.1d}$$

Because of these properties, tensors constitute a vector space.

Tensors have a most useful property in the way that they transform from one basis (reference frame) to another. Having the tensor defined with respect to one reference frame, the tensor quantity (components) can be written in any admissible reference frame. An example of this would be stress defined in principal and non-principal components. Both representations are of the same stress tensor even though the individual components may be different. As long as the relationship between the reference frames is known, the components with respect to one frame may be found from the other.

Only that category of tensors known as *Cartesian tensors* is used in this text, and definitions of these will be given in the pages that follow. General tensor notation is presented in the Appendix for completeness, but it is not necessary for the main text. The tensor equations used to develop the fundamental theory of continuum mechanics may be written in either of two distinct notations; the *symbolic notation*, or the *indicial notation*. We shall make use of both notations, employing whichever is more convenient for the derivation or analysis at hand but taking care to establish the inter-relationships between the two. However, an effort to emphasize indicial notation in most of the text has been made. An introductory course must teach indicial notation to the student who may have little prior exposure to the topic.

2.1 Scalars, Vectors and Cartesian Tensors

A considerable variety of physical and geometrical quantities have important roles in continuum mechanics, and fortunately, each of these may be represented by some form of tensor. For example, such quantities as *density* and *temperature* may be specified completely by giving their magnitude, i.e., by stating a numerical value. These quantities are represented mathematically by scalars, which are referred to as *zero-order tensors*. It should be emphasized that scalars are not constants, but may actually be functions of position and/or time. Also, the exact numerical value of a scalar will depend upon the units in which it is expressed. Thus, the temperature may be given by either 68 °F, or 20 °C at a certain location. As a general rule, lower-case Greek letters in italic print such as α, β, λ, etc. will be used as symbols for scalars in both the indicial and symbolic notations.

Several physical quantities of mechanics such as force and velocity require not only an assignment of magnitude, but also a specification of direction for their complete characterization. As a trivial example, a 20 N force acting vertically at a point is substantially different than a 20 N force acting horizontally at the point. Quantities possessing such directional properties are represented by vectors, which are *first-order tensors*. Geometrically, vectors are generally displayed as *arrows*, having a definite length (the magnitude), a specified orientation (the direction), and also a sense of action as indicated by the head and the tail of the arrow. In this text arrow lengths are not to scale with vector magnitude. Certain quantities in mechanics which are not truly vectors are also portrayed by arrows, for example, finite rotations.

Consequently, in addition to the magnitude and direction characterization, the complete definition of a vector requires the further statement: vectors add (and subtract) in accordance with the triangle rule by which the arrow representing the vector sum of two vectors extends from the tail of the first component arrow to the head of the second when the component arrows are arranged "head-to-tail".

Although vectors are independent of any particular coordinate system, it is often useful to define a vector in terms of its coordinate components, and in this respect it is necessary to reference the vector to an appropriate set of axes. In view of our restriction to Cartesian tensors, we limit ourselves to consideration of Cartesian coordinate systems for designating the components of a vector.

A significant number of physical quantities having important status in continuum mechanics require mathematical entities of higher order than vectors for their representation in the hierarchy of tensors. As we shall see, among the best known of these are the stress and the strain tensors. These particular tensors are *second-order tensors*, and are said to have a rank of two. Third-order and fourth-order tensors are not uncommon in continuum mechanics but they are not nearly as plentiful as second-order tensors. Accordingly, the unqualified use of the word *tensor* in this text will be interpreted to mean *second-order tensor*. With only a few exceptions, primarily those representing the stress and strain tensors, we shall denote second-order tensors by upper-case sans serif Latin letters in bold-faced print, a typical example being the tensor $\mathbf{T}$. The components of the said tensor will, in general, be denoted by lower-case Latin letters with appropriate indices: t_{ij}.

Tensors, like vectors, are independent of any coordinate system, but just as with vectors, when we wish to specify a tensor by its components we are obliged to refer to a suitable set of reference axes. The precise definitions of tensors of various order will be given subsequently in terms of the transformation properties of their components between two related sets of Cartesian coordinate axes.

As a quick notation summary, the International Standards Organization (ISO) conventions for typesetting mathematics are summarized below:

1. Scalar variables are written as italic letters. The letters may be either Roman or Greek style fonts depending on the physical quantity they represent. The following examples are a partial list of scalar notation:

   (a) a – magnitude of acceleration

(b) ν – magnitude of velocity
(c) r – radius
(d) θ – temperature or angle depending on context
(e) α – coefficient of thermal expansion
(f) σ – principal value of stress
(g) λ – eigenvalue or stretch

2. Vectors are written as boldface italic. Examples are as follows:

   (a) $\boldsymbol{x}$ – position
   (b) $\boldsymbol{v}$ – velocity
   (c) $\boldsymbol{a}$ – acceleration
   (d) $\hat{\boldsymbol{e}}_1$ – base vector in x_1 direction

3. Second- and higher-order tensors are designated by uppercase fonts. Additionally, matrices are shown in the calligraphic form to differentiate them from tensors. Tensors can be represented by matrices, but not all matrices are tensors. In the case of several well known engineering quantities this convention will not be accommodated. For example, linear strain has been chosen to be represented by ϵ. Here are some samples of tensor and matrix symbols:

   (a) $\mathcal{Q}$ – orthogonal matrix
   (b) E – finite strain
   (c) T – Cauchy stress tensor
   (d) $\boldsymbol{\epsilon}$ – infinitesimal strain tensor
   (e) $\mathcal{R}$ – rotation matrix

2.2 Tensor Algebra in Symbolic Notation - Summation Convention

The three-dimensional physical space of everyday life is the space in which many of the events of continuum mechanics occur. Mathematically, this space is known as a Euclidean three-space , and its geometry can be referenced to a system of Cartesian coordinate axes. In some instances, higher order dimension spaces play integral roles in continuum topics. Because a scalar has only a single component, it will have the same value in every system of axes, but the components of vectors and tensors will have different component values, in general, for each set of axes.

In order to represent vectors and tensors in component form we introduce in our physical space a right-handed system of rectangular Cartesian axes $Ox_1x_2x_3$, and identify with these axes the triad of unit base vectors, $\hat{\boldsymbol{e}}_1$, $\hat{\boldsymbol{e}}_2$, $\hat{\boldsymbol{e}}_3$, shown in Fig. 2.1(a). All unit vectors in this text will be written with a caret placed above the bold-faced symbol. Due to the mutual perpendicularity of these base vectors they form an orthogonal basis, and furthermore, because they are unit vectors, the basis is said to be orthonormal. In terms of this basis an arbitrary vector $\boldsymbol{v}$ is given in component form by

$$\boldsymbol{v} = v_1\hat{\boldsymbol{e}}_1 + v_2\hat{\boldsymbol{e}}_2 + v_3\hat{\boldsymbol{e}}_3 = \sum_{i=1}^{3} v_i\hat{\boldsymbol{e}}_i \ . \tag{2.2}$$

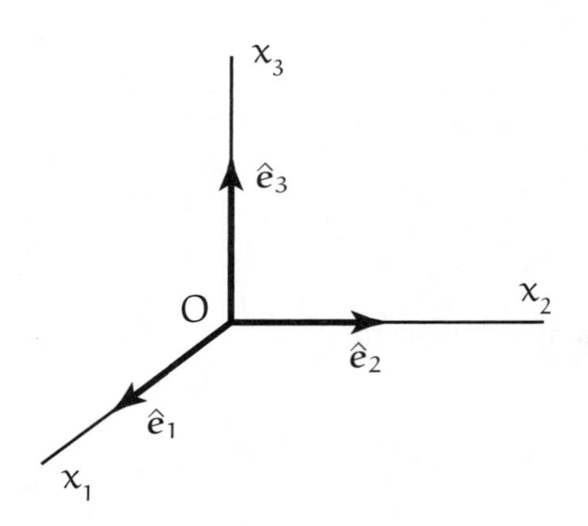

(a) Unit vectors in the coordinate directions x_1, x_2 and x_3.

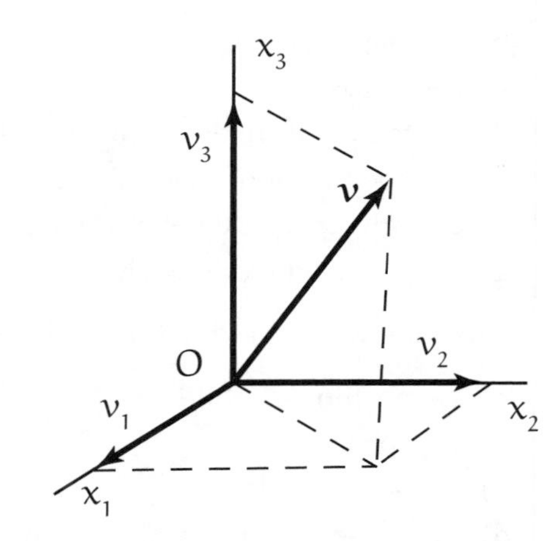

(b) Rectangular Cartesian components of the vector $\boldsymbol{v}$.

FIGURE 2.1
Base vectors and components of a Cartesian vector.

This vector and its coordinate components are pictured in Fig. 2.1(b). For the symbolic description, vectors will usually be given by lower-case Latin letters in bold-faced print, with the vector magnitude denoted by the same letter. Thus v is the magnitude of $\boldsymbol{v}$.

At this juncture of our discussion it is helpful to introduce a notational device called the *summation convention* that will greatly simplify the writing of the equations of continuum mechanics. Stated briefly, we agree that whenever a subscript appears exactly *twice* in a given term, that subscript will take on the values 1, 2, 3 successively, and the resulting terms summed. For example, using this scheme, we may now write Eq 2.2 in the simple form

$$\boldsymbol{v} = v_i \hat{\mathbf{e}}_i \ , \qquad (2.3)$$

and delete entirely the summation symbol $\sum$. For Cartesian tensors, only subscripts are required on the components; for general tensors both subscripts and superscripts are used. The summed subscripts are called *dummy indices* since it is immaterial which particular letter is used. Thus $v_i \hat{\mathbf{e}}_i$ is completely equivalent to $v_j \hat{\mathbf{e}}_j$, or to $v_k \hat{\mathbf{e}}_k$, when the summation convention is used. A word of caution, however; no subscript may appear more than twice in a singe term. But as we shall soon see, more than one pair of dummy indices may appear in a given expression that is the summation of terms (see Example 2.2). Note also that the summation convention may involve subscripts from both the unit vectors and the scalar coefficients.

Example 2.1

Without regard for meaning as far as mechanics is concerned, expand the following expressions according to the summation convention:

(a) $u_i v_i w_j \hat{\mathbf{e}}_j$ (b) $t_{ij} v_i \hat{\mathbf{e}}_j$ (c) $t_{ii} v_j \hat{\mathbf{e}}_j$

Solution

(a) Summing first on i, and then on j

$$u_i v_i w_j \hat{\mathbf{e}}_j = (u_1 v_1 + u_2 v_2 + u_3 v_3)(w_1 \hat{\mathbf{e}}_1 + w_2 \hat{\mathbf{e}}_2 + w_3 \hat{\mathbf{e}}_3)$$

(b) Summing on i, then on j and collecting terms on the unit vectors.

$$\begin{aligned} t_{ij} v_j \hat{\mathbf{e}}_i &= t_{1j} v_j \hat{\mathbf{e}}_1 + t_{2j} v_j \hat{\mathbf{e}}_2 + t_{3j} v_j \hat{\mathbf{e}}_3 \\ &= (t_{11} v_1 + t_{12} v_2 + t_{13} v_3)\, \hat{\mathbf{e}}_1 + (t_{21} v_1 + t_{22} v_2 + t_{23} v_3)\, \hat{\mathbf{e}}_2 \\ &\quad + (t_{31} v_1 + t_{32} v_2 + t_{33} v_3)\, \hat{\mathbf{e}}_3 \end{aligned}$$

(c) Summing on i, then on j,

$$t_{ii} v_j \hat{\mathbf{e}}_j = (t_{11} + t_{22} + t_{33})\,(v_1 \hat{\mathbf{e}}_1 + v_2 \hat{\mathbf{e}}_2 + v_3 \hat{\mathbf{e}}_3)$$

Note the similarity between (a) and (c).

With the above background in place it is now possible, using symbolic notation, to present many useful definitions from vector/tensor algebra. There are two symbols needed prior to writing out all of the vector and tensor algebra necessary. These two symbols are the Kronecker delta and the permutation symbol. Additionally, there are several useful relationships between the Kronecker delta and permutation symbol that are used throughout continuum mechanics. The following three subsections introduce the Kronecker delta, permutation symbol and their relationships. Following that, vector/tensor algebra is presented.

The Kronecker delta is similar to the identity matrix, so the reader should quickly embrace this new entity. However, the permutation symbol is a little more abstract than the Kronecker delta since it cannot be represented by a matrix. In subsequent chapters the Kronecker delta and the permutation symbol play integral roles in describing how forces are carried by continuum bodies and how the position of a particle is described.

2.2.1 Kronecker Delta

Since the base vectors $\hat{\mathbf{e}}_i$ (i = 1,2,3) are unit vectors and orthogonal

$$\hat{\mathbf{e}}_i \cdot \hat{\mathbf{e}}_j = \begin{cases} 1 & \text{if numerical value of } i = \text{numerical value of } j \\ 0 & \text{if numerical value of } i \neq \text{numerical value of } j \end{cases} .$$

Therefore, if we introduce the Kronecker delta defined by

$$\delta_{ij} = \begin{cases} 1 & \text{if numerical value of } i = \text{numerical value of } j \\ 0 & \text{if numerical value of } i \neq \text{numerical value of } j \end{cases}$$

we see that

$$\hat{\mathbf{e}}_i \cdot \hat{\mathbf{e}}_j = \delta_{ij} \qquad (i, j = 1, 2, 3) \ . \tag{2.4}$$

Also, note that by the summation convention

$$\delta_{ii} = \delta_{jj} = \delta_{11} + \delta_{22} + \delta_{33} = 1 + 1 + 1 = 3 \,,$$

and furthermore, we call attention to the substitution property of the Kronecker delta by expanding (summing on j) the expression

$$\delta_{ij}\hat{\mathbf{e}}_j = \delta_{i1}\hat{\mathbf{e}}_1 + \delta_{i2}\hat{\mathbf{e}}_2 + \delta_{i3}\hat{\mathbf{e}}_3 \,.$$

But for a given value of i in this equation, only one of the Kronecker deltas on the right hand side is non-zero, and it has the value one. Therefore,

$$\delta_{ij}\hat{\mathbf{e}}_j = \hat{\mathbf{e}}_i \,,$$

and the Kronecker delta in $\delta_{ij}\hat{\mathbf{e}}_j$ causes the summed subscript j of $\hat{\mathbf{e}}_j$ to be replaced by i reducing the expression to simply $\hat{\mathbf{e}}_i$.

2.2.2 Permutation Symbol

By introducing the permutation symbol ε_{ijk} defined by

$$\varepsilon_{ijk} = \begin{cases} 1 & \text{if numerical values of } ijk \text{ appear as in the sequence } 12312 \\ -1 & \text{if numerical values of } ijk \text{ appear as in the sequence } 32132 \\ 0 & \text{if numerical values of } ijk \text{ appear in any other sequence} \end{cases} \qquad (2.5)$$

we may express the cross products of the base vectors (i=1,2,3) by the use of Eq 2.5 as

$$\hat{\mathbf{e}}_i \times \hat{\mathbf{e}}_j = \varepsilon_{ijk}\hat{\mathbf{e}}_k \qquad (i, j, k = 1, 2, 3) \,. \qquad (2.6)$$

Also, note from its definition that the interchange of any two subscripts in ε_{ijk} causes a sign change so that for example,

$$\varepsilon_{ijk} = -\varepsilon_{kji} = \varepsilon_{kij} = -\varepsilon_{ikj} \,,$$

and, furthermore, that for repeated subscripts is zero as in

$$\varepsilon_{113} = \varepsilon_{212} = \varepsilon_{133} = \varepsilon_{222} = 0 \,.$$

2.2.3 ε - δ Identity

The product of permutation symbols $\varepsilon_{miq}\varepsilon_{jkq}$ may be expressed in terms of Kronecker deltas by the ε - δ identity

$$\varepsilon_{miq}\varepsilon_{jkq} = \delta_{mj}\delta_{ik} - \delta_{mk}\delta_{ij} \qquad (2.7a)$$

as may be proven by direct expansion. This is a *most important formula* used throughout this text and is well worth memorizing. Also, by the sign-change property of ε_{ijk},

$$\varepsilon_{miq}\varepsilon_{jkq} = \varepsilon_{miq}\varepsilon_{qjk} = \varepsilon_{qmi}\varepsilon_{qjk} = \varepsilon_{qmi}\varepsilon_{jkq} \,.$$

Additionally, it is easy to show from Eq 2.7a that

$$\varepsilon_{jkq}\varepsilon_{mkq} = 2\delta_{jm} \,, \qquad (2.7b)$$

by setting $i = k$, and

$$\varepsilon_{jkq}\varepsilon_{jkq} = 6 \,. \qquad (2.7c)$$

2.2.4 Tensor/Vector Algebra

To begin with, vector addition is easily written in indicial form

$$\mathbf{w} = \mathbf{u} + \mathbf{v} \qquad \text{or} \qquad w_i \hat{\mathbf{e}}_i = (u_i + v_i)\,\hat{\mathbf{e}}_i \tag{2.8}$$

where the components simply add together.

Simple vector multiplication can take one of several forms. The specific form depends on the type of entity multiplying the vector. For now, two forms of vector multiplication can be defined in symbolic form. Multiplication of a vector by a scalar is written as

$$\lambda \mathbf{v} = \lambda v_i \hat{\mathbf{e}}_i \ , \tag{2.9}$$

and the *dot (scalar) product* between of two vectors is

$$\mathbf{u} \cdot \mathbf{v} = \mathbf{v} \cdot \mathbf{u} = uv\cos\theta \tag{2.10}$$

where θ is the smaller angle between the two vectors when drawn from a common origin.

From the definition of δ_{ij} and its substitution property the *dot product* $\mathbf{u} \cdot \mathbf{v}$ may be written as

$$\mathbf{u} \cdot \mathbf{v} = u_i \hat{\mathbf{e}}_i \cdot v_j \hat{\mathbf{e}}_j = u_i v_j \hat{\mathbf{e}}_i \cdot \hat{\mathbf{e}}_j = u_i v_j \delta_{ij} = u_i v_i \ . \tag{2.11}$$

Note that scalar components pass through the dot product since it is a vector operator.

The *vector cross (vector) product* of two vectors is defined by

$$\mathbf{u} \times \mathbf{v} = -\mathbf{v} \times \mathbf{u} = (uv\sin\theta)\,\hat{\mathbf{e}}$$

where $0 \leqslant \theta \leqslant \pi$, is the angle between the two vectors when drawn from a common origin, and where $\hat{\mathbf{e}}$ is a unit vector perpendicular to their plane such that a right-handed rotation about $\hat{\mathbf{e}}$ through the angle θ carries $\mathbf{u}$ into $\mathbf{v}$.

The *vector cross product* may be written in terms of the permutation symbol (Eq 2.5) as follows:

$$\mathbf{u} \times \mathbf{v} = u_i \hat{\mathbf{e}}_i \times v_j \hat{\mathbf{e}}_j = u_i v_j \left(\hat{\mathbf{e}}_i \times \hat{\mathbf{e}}_j\right) = \varepsilon_{ijk} u_i v_j \hat{\mathbf{e}}_k \ . \tag{2.12}$$

Again, notice how the scalar components pass through the vector cross product operator.

There are a couple of useful ways three vectors can be multiplied. The *triple scalar product* (*box product*) is

$$\mathbf{u} \cdot \mathbf{v} \times \mathbf{w} = \mathbf{u} \times \mathbf{v} \cdot \mathbf{w} = [\mathbf{u}, \mathbf{v}, \mathbf{w}] \ ,$$

or

$$\begin{aligned} [\mathbf{u}, \mathbf{v}, \mathbf{w}] &= u_i \hat{\mathbf{e}}_i \cdot \left(v_j \hat{\mathbf{e}}_j \times w_k \hat{\mathbf{e}}_k\right) = u_i \hat{\mathbf{e}}_i \cdot \varepsilon_{jkq} v_j w_k \hat{\mathbf{e}}_q \\ &= \varepsilon_{jkq} u_i v_j w_k \delta_{iq} = \varepsilon_{ijk} u_i v_j w_k \end{aligned} \tag{2.13}$$

where in the final step we have used both the substitution property of δ_{iq} and the sign-change property of ε_{ijk}. The *triple cross product* is similar to the triple scalar product

$$\begin{aligned} \mathbf{u} \times (\mathbf{v} \times \mathbf{w}) &= u_i \hat{\mathbf{e}}_i \times \left(v_j \hat{\mathbf{e}}_j \times w_k \hat{\mathbf{e}}_k\right) = u_i \hat{\mathbf{e}}_i \times \left(\varepsilon_{jkq} v_j w_k \hat{\mathbf{e}}_q\right) \\ &= \varepsilon_{iqm} \varepsilon_{jkq} u_i v_j w_k \hat{\mathbf{e}}_m = \varepsilon_{miq} \varepsilon_{jkq} u_i v_j w_k \hat{\mathbf{e}}_m \end{aligned} \tag{2.14}$$

which may be written as

$$\begin{aligned} \mathbf{u} \times (\mathbf{v} \times \mathbf{w}) &= \left(\delta_{mj}\delta_{ik} - \delta_{mk}\delta_{ij}\right) u_i v_j w_k \hat{\mathbf{e}}_m \\ &= \left(u_i v_m w_i - u_i v_i w_m\right) \hat{\mathbf{e}}_m = u_i w_i v_m \hat{\mathbf{e}}_m - u_i v_i w_m \hat{\mathbf{e}}_m \ . \end{aligned} \tag{2.15}$$

Observation of the indices in Eq 2.15 admits

$$\mathbf{u} \times (\mathbf{v} \times \mathbf{w}) = (\mathbf{u} \cdot \mathbf{w})\,\mathbf{v} - (\mathbf{u} \cdot \mathbf{v})\,\mathbf{w}$$

a well-known identity from vector algebra.

In addition to the common vector products above, two vectors can be multiplied together to yield a tensor. The tensor product of two vectors creates a *dyad*

$$\mathbf{u}\mathbf{v} = u_i \hat{\mathbf{e}}_i v_j \hat{\mathbf{e}}_j = u_i v_j \hat{\mathbf{e}}_i \hat{\mathbf{e}}_j \tag{2.16}$$

which in expanded form, summing first on i, yields

$$u_i v_j \hat{\mathbf{e}}_i \hat{\mathbf{e}}_j = u_1 v_j \hat{\mathbf{e}}_1 \hat{\mathbf{e}}_j + u_2 v_j \hat{\mathbf{e}}_2 \hat{\mathbf{e}}_j + u_3 v_j \hat{\mathbf{e}}_3 \hat{\mathbf{e}}_j \ ,$$

and then summing on j

$$\begin{aligned} u_i v_j \hat{\mathbf{e}}_i \hat{\mathbf{e}}_j &= u_1 v_1 \hat{\mathbf{e}}_1 \hat{\mathbf{e}}_1 + u_1 v_2 \hat{\mathbf{e}}_1 \hat{\mathbf{e}}_2 + u_1 v_3 \hat{\mathbf{e}}_1 \hat{\mathbf{e}}_3 \\ &+ u_2 v_1 \hat{\mathbf{e}}_2 \hat{\mathbf{e}}_1 + u_2 v_2 \hat{\mathbf{e}}_2 \hat{\mathbf{e}}_2 + u_2 v_3 \hat{\mathbf{e}}_2 \hat{\mathbf{e}}_3 \\ &+ u_3 v_1 \hat{\mathbf{e}}_3 \hat{\mathbf{e}}_1 + u_3 v_2 \hat{\mathbf{e}}_3 \hat{\mathbf{e}}_2 + u_3 v_3 \hat{\mathbf{e}}_3 \hat{\mathbf{e}}_3 \ . \end{aligned} \tag{2.17}$$

This nine-term sum is called the *nonion* form of the *dyad*, $\mathbf{uv}$. A sum of dyads such as

$$\mathbf{u}_1 \mathbf{v}_1 + \mathbf{u}_2 \mathbf{v}_2 + \cdots + \mathbf{u}_N \mathbf{v}_N \tag{2.18}$$

is called a *dyadic*.

A common, alternative notation frequently used for the dyad product is

$$\mathbf{a} \otimes \mathbf{b} = a_i \hat{\mathbf{e}}_i \otimes b_j \hat{\mathbf{e}}_j = a_i b_j \hat{\mathbf{e}}_i \otimes \hat{\mathbf{e}}_j \tag{2.19}$$

which is called a *tensor product*. The tensor product of vectors $\mathbf{a}$ and $\mathbf{b}$ is defined by how $\mathbf{a} \otimes \mathbf{b}$ maps all vectors $\mathbf{u}$:

$$(\mathbf{a} \otimes \mathbf{b})\,\mathbf{u} = \mathbf{a}\,(\mathbf{b} \cdot \mathbf{u}) \ . \tag{2.20}$$

If one takes vectors $\mathbf{a}$, $\mathbf{b}$ and $\mathbf{u}$ to be vectors from a Euclidian 3-space, the expanded form for the tensor product may be written as

$$\begin{aligned} (\mathbf{a} \otimes \mathbf{b})\,\mathbf{u} = \mathbf{a}\,(\mathbf{b} \cdot \mathbf{u}) &= \begin{bmatrix} a_1 \\ a_2 \\ a_3 \end{bmatrix} (b_1 u_1 + b_2 u_2 + b_3 u_3) \\ &= \begin{bmatrix} a_1 b_1 u_1 + a_1 b_2 u_2 + a_1 b_3 u_3 \\ a_2 b_1 u_1 + a_2 b_2 u_2 + a_2 b_3 u_3 \\ a_3 b_1 u_1 + a_3 b_2 u_2 + a_3 b_3 u_3 \end{bmatrix} \\ &= \begin{bmatrix} a_1 b_1 & a_1 b_2 & a_1 b_3 \\ a_2 b_1 & a_2 b_2 & a_2 b_3 \\ a_3 b_1 & a_3 b_2 & a_3 b_3 \end{bmatrix} \begin{bmatrix} u_1 \\ u_2 \\ u_3 \end{bmatrix} \\ &= \left(\begin{bmatrix} a_1 \\ a_2 \\ a_3 \end{bmatrix} \begin{bmatrix} b_1 & b_2 & b_3 \end{bmatrix} \right) \begin{bmatrix} u_1 \\ u_2 \\ u_3 \end{bmatrix} \ . \end{aligned} \tag{2.21}$$

The last line of Eq 2.21 shows why the tensor product is sometimes called the outer product. The inner product, of course, is defined by

$$\mathbf{a}\cdot\mathbf{b}=\begin{bmatrix}a_1 & a_2 & a_3\end{bmatrix}\begin{bmatrix}b_1\\ b_2\\ b_3\end{bmatrix}=a_1b_1+a_2b_2+a_3b_3\ .$$

The second line of Eq 2.21 shows the tensor product to be equivalent to the dyad product of two vectors. Both notations will be used in the book.

A dyad can be multiplied by a vector giving a *vector-dyad product*:

1. $$\mathbf{u}\cdot(\mathbf{v}\mathbf{w})=u_i\hat{\mathbf{e}}_i\cdot(v_j\hat{\mathbf{e}}_j w_k\hat{\mathbf{e}}_k)=u_iv_iw_k\hat{\mathbf{e}}_k \tag{2.22}$$

2. $$(\mathbf{u}\mathbf{v})\cdot\mathbf{w}=(u_i\hat{\mathbf{e}}_i v_j\hat{\mathbf{e}}_j)\cdot w_k\hat{\mathbf{e}}_k=u_iv_jw_j\hat{\mathbf{e}}_i \tag{2.23}$$

3. $$\mathbf{u}\times(\mathbf{v}\mathbf{w})=(u_i\hat{\mathbf{e}}_i\times v_j\hat{\mathbf{e}}_j)\,w_k\hat{\mathbf{e}}_k=\varepsilon_{ijq}u_iv_jw_k\hat{\mathbf{e}}_q\hat{\mathbf{e}}_k \tag{2.24}$$

4. $$(\mathbf{u}\mathbf{v})\times\mathbf{w}=u_i\hat{\mathbf{e}}_i\,(v_j\hat{\mathbf{e}}_j\times w_k\hat{\mathbf{e}}_k)=\varepsilon_{jkq}u_iv_jw_k\hat{\mathbf{e}}_i\hat{\mathbf{e}}_q \tag{2.25}$$

(Note that in products 3 and 4 the order of the base vectors $\hat{\mathbf{e}}_i$ is important.)

Dyads can be multiplied by each other to yield another dyad

$$(\mathbf{u}\mathbf{v})\cdot(\mathbf{w}\mathbf{s})=u_i\hat{\mathbf{e}}_i\,(v_j\hat{\mathbf{e}}_j\cdot w_k\hat{\mathbf{e}}_k)\,s_q\hat{\mathbf{e}}_q=u_iv_jw_js_q\hat{\mathbf{e}}_i\hat{\mathbf{e}}_q\ . \tag{2.26}$$

Vectors can be multiplied by a tensor to give a vector. The reduction in the order of the tensor is why the "dot" is used in direct notation.

1. $$\mathbf{v}\cdot\mathbf{T}=v_i\hat{\mathbf{e}}_i\cdot t_{jk}\hat{\mathbf{e}}_j\hat{\mathbf{e}}_k=v_it_{jk}\delta_{ij}\hat{\mathbf{e}}_k=v_it_{ik}\hat{\mathbf{e}}_k \tag{2.27}$$

2. $$\mathbf{T}\cdot\mathbf{v}=t_{ij}\hat{\mathbf{e}}_i\hat{\mathbf{e}}_j\cdot v_k\hat{\mathbf{e}}_k=t_{ij}\hat{\mathbf{e}}_i\delta_{jk}v_k=t_{ij}v_j\hat{\mathbf{e}}_i \tag{2.28}$$

(Note that these products are also written as simply $\mathbf{v}\mathbf{T}$ and $\mathbf{T}\mathbf{v}$.)

Finally, two tensors can be multiplied resulting in a tensor

$$\mathbf{T}\cdot\mathbf{S}=t_{ij}\hat{\mathbf{e}}_i\hat{\mathbf{e}}_j\cdot s_{pq}\hat{\mathbf{e}}_p\hat{\mathbf{e}}_q=t_{ij}s_{pq}\delta_{jp}\hat{\mathbf{e}}_i\hat{\mathbf{e}}_q=t_{ij}s_{jp}\hat{\mathbf{e}}_i\hat{\mathbf{e}}_q\ . \tag{2.29}$$

Example 2.2

Let the vector $\mathbf{v}$ be given by $\mathbf{v}=(\mathbf{a}\cdot\hat{\mathbf{n}})\,\hat{\mathbf{n}}+\hat{\mathbf{n}}\times(\mathbf{a}\times\hat{\mathbf{n}})$ where $\mathbf{a}$ is an arbitrary vector, and $\hat{\mathbf{n}}$ is a unit vector. Express $\mathbf{v}$ in terms of the base vectors $\hat{\mathbf{e}}_i$, expand, and simplify. (Note that $\hat{\mathbf{n}}\cdot\hat{\mathbf{n}}=n_i\hat{\mathbf{e}}_i\cdot n_j\hat{\mathbf{e}}_j=n_in_j\delta_{ij}=n_in_i=1$.)

Solution

In terms of the base vectors $\hat{\mathbf{e}}_i$, the given vector $\mathbf{v}$ is expressed by the equation

$$\mathbf{v}=(a_i\hat{\mathbf{e}}_i\cdot n_j\hat{\mathbf{e}}_j)\,n_k\hat{\mathbf{e}}_k+n_i\hat{\mathbf{e}}_i\times(a_j\hat{\mathbf{e}}_j\times n_k\hat{\mathbf{e}}_k)\ .$$

We note here that indices i, j, and k appear four times in this line, however, the summation convention has not been violated. Terms that are separated by a plus or a minus sign are considered different terms each having summation convention rules applicable within them. Vectors joined by a dot or cross product are not distinct terms, and the summation convention must be adhered to in that case. Carrying out the indicated multiplications, we see that

$$\begin{aligned} \mathbf{v} &= (a_i n_j \delta_{ij}) n_k \hat{\mathbf{e}}_k + n_i \hat{\mathbf{e}}_i \times (\varepsilon_{jkq} a_j n_k \hat{\mathbf{e}}_q) \\ &= a_i n_i n_k \hat{\mathbf{e}}_k + \varepsilon_{iqm} \varepsilon_{jkq} n_i a_j n_k \hat{\mathbf{e}}_m \\ &= a_i n_i n_k \hat{\mathbf{e}}_k + \varepsilon_{miq} \varepsilon_{jkq} n_i a_j n_k \hat{\mathbf{e}}_m \\ &= a_i n_i n_k \hat{\mathbf{e}}_k + (\delta_{mj} \delta_{ik} - \delta_{mk} \delta_{ij}) n_i a_j n_k \hat{\mathbf{e}}_m \\ &= a_i n_i n_k \hat{\mathbf{e}}_k + n_i a_j n_i \hat{\mathbf{e}}_j - n_i a_i n_k \hat{\mathbf{e}}_k \\ &= n_i n_i a_j \hat{\mathbf{e}}_j = a_j \hat{\mathbf{e}}_j = \mathbf{a} \, . \end{aligned}$$

Since $\mathbf{a}$ must equal $\mathbf{v}$, this example demonstrates that the vector $\mathbf{v}$ may be resolved into a component $(\mathbf{v} \cdot \hat{\mathbf{n}}) \hat{\mathbf{n}}$ in the direction of $\hat{\mathbf{n}}$, and a component $\hat{\mathbf{n}} \times (\mathbf{v} \times \hat{\mathbf{n}})$ perpendicular to $\hat{\mathbf{n}}$.

Example 2.3

Using Eq 2.7 show that (a) $\varepsilon_{mkq}\varepsilon_{jkq} = 2\delta_{mj}$ and that (b) $\varepsilon_{jkq}\varepsilon_{jkq} = 6$. (Recall that $\delta_{kk} = 3$ and $\delta_{mk}\delta_{kj} = \delta_{mj}$.)

Solution

(a) Write out Eq 2.7a with index i replaced by k to get

$$\begin{aligned} \varepsilon_{mkq}\varepsilon_{jkq} &= \delta_{mj}\delta_{kk} - \delta_{mk}\delta_{kj} \\ &= 3\delta_{mj} - \delta_{mj} = 2\delta_{mj} \, . \end{aligned}$$

(b) Start with the first equation in Part (a) and replace the index m with j, giving

$$\begin{aligned} \varepsilon_{jkq}\varepsilon_{jkq} &= \delta_{jj}\delta_{kk} - \delta_{jk}\delta_{jk} \\ &= (3)\,(3) - \delta_{jj} = 9 - 3 = 6 \, . \end{aligned}$$

Example 2.4

Double dot products of dyads are defined by

$$\text{(a)} \quad (\mathbf{u}\mathbf{v}) \cdot\cdot (\mathbf{w}\mathbf{s}) = (\mathbf{v} \cdot \mathbf{w})(\mathbf{u} \cdot \mathbf{s})$$

$$\text{(b)} \quad (\mathbf{u}\mathbf{v}) : (\mathbf{w}\mathbf{s}) = (\mathbf{u} \cdot \mathbf{w})(\mathbf{v} \cdot \mathbf{s}) \, .$$

Expand these products and compare the component forms.

Solution

$$\text{(a)}\ (\mathbf{uv}) \cdot\cdot (\mathbf{ws}) = (v_i\hat{\mathbf{e}}_i \cdot w_j\hat{\mathbf{e}}_j)(u_k\hat{\mathbf{e}}_k \cdot s_q\hat{\mathbf{e}}_q) = v_i w_i u_k s_k$$

$$\text{(b)}\ (\mathbf{uv}) : (\mathbf{ws}) = (u_i\hat{\mathbf{e}}_i \cdot w_j\hat{\mathbf{e}}_j)(v_k\hat{\mathbf{e}}_k \cdot s_q\hat{\mathbf{e}}_q) = u_i w_i v_k s_k$$

Summary of Symbolic Notations

1. *Addition of vectors*, Eq 2.8:

$$\mathbf{w} = \mathbf{u} + \mathbf{v} \qquad \text{or} \qquad w_i\hat{\mathbf{e}}_i = (u_i + v_i)\hat{\mathbf{e}}_i$$

2. *Multiplication:*

(*a*) *of a vector by a scalar*, Eq 2.9:

$$\lambda\mathbf{v} = \lambda v_i\hat{\mathbf{e}}_i$$

(*b*) *dot (scalar) product of two vectors*, Eq 2.11:

$$\mathbf{u} \cdot \mathbf{v} = \mathbf{v} \cdot \mathbf{u} = uv\cos\theta = u_i v_i$$

(*c*) *cross (vector) product of two vectors*, Eq 2.12:

$$\mathbf{u} \times \mathbf{v} = -\mathbf{v} \times \mathbf{u} = (uv\sin\theta)\,\hat{\mathbf{e}} = \varepsilon_{ijk} u_i v_j \hat{\mathbf{e}}_k$$

(*d*) *triple scalar product (box product)*, Eq 2.13:

$$\begin{aligned}[\mathbf{u}, \mathbf{v}, \mathbf{w}] &= u_i\hat{\mathbf{e}}_i \cdot (v_j\hat{\mathbf{e}}_j \times w_k\hat{\mathbf{e}}_k) = u_i\hat{\mathbf{e}}_i \cdot \varepsilon_{jkq} v_j w_k \hat{\mathbf{e}}_q \\ &= \varepsilon_{jkq} u_i v_j w_k \delta_{iq} = \varepsilon_{ijk} u_i v_j w_k\end{aligned}$$

(*e*) *triple cross product*, Eq 2.14:

$$\begin{aligned}\mathbf{u} \times (\mathbf{v} \times \mathbf{w}) &= u_i\hat{\mathbf{e}}_i \times (v_j\hat{\mathbf{e}}_j \times w_k\hat{\mathbf{e}}_k) = u_i\hat{\mathbf{e}}_i \times (\varepsilon_{jkq} v_j w_k \hat{\mathbf{e}}_q) \\ &= \varepsilon_{iqm}\varepsilon_{jkq} u_i v_j w_k \hat{\mathbf{e}}_m = \varepsilon_{miq}\varepsilon_{jkq} u_i v_j w_k \hat{\mathbf{e}}_m\end{aligned}$$

(*f*) *tensor product of two vectors (dyad)*, Eq 2.16:

$$\mathbf{uv} = u_i\hat{\mathbf{e}}_i v_j\hat{\mathbf{e}}_j = u_i v_j \hat{\mathbf{e}}_i\hat{\mathbf{e}}_j = u_i\hat{\mathbf{e}}_i \otimes v_j\hat{\mathbf{e}}_j = u_i v_j \hat{\mathbf{e}}_i \otimes \hat{\mathbf{e}}_j$$

(*g*) *Vector-dyad products*, Eqs 2.22–2.25:

1. $\mathbf{u} \cdot (\mathbf{vw}) = u_i\hat{\mathbf{e}}_i \cdot (v_j\hat{\mathbf{e}}_j w_k\hat{\mathbf{e}}_k) = u_i v_i w_k \hat{\mathbf{e}}_k$

2. $(\mathbf{uv}) \cdot \mathbf{w} = (u_i\hat{\mathbf{e}}_i v_j\hat{\mathbf{e}}_j) \cdot w_k\hat{\mathbf{e}}_k = u_i v_j w_j \hat{\mathbf{e}}_i$

3. $\mathbf{u} \times (\mathbf{vw}) = (u_i\hat{\mathbf{e}}_i \times v_j\hat{\mathbf{e}}_j) w_k\hat{\mathbf{e}}_k = \varepsilon_{ijq} u_i v_j w_k \hat{\mathbf{e}}_q\hat{\mathbf{e}}_k$

4. $(\mathbf{uv}) \times \mathbf{w} = u_i\hat{\mathbf{e}}_i (v_j\hat{\mathbf{e}}_j \times w_k\hat{\mathbf{e}}_k) = \varepsilon_{jkq} u_i v_j w_k \hat{\mathbf{e}}_i\hat{\mathbf{e}}_q$

TABLE 2.1
Indicial form for a variety of tensor quantities.

λ	=	*scalar* (zeroth-order tensor) λ
v_i	=	*vector* (first-order tensor) $\mathbf{v}$, or equivalently, its 3 components
$u_i v_j$	=	*dyad* (second-order tensor) $\mathbf{uv}$, or its 9 components
t_{ij}	=	*dyadic* (second-order tensor) $\mathbf{T}$, or its 9 components
Q_{ijk}	=	*triadic* (third-order tensor) $\mathbf{Q}$, or its 27 components
C_{ijkm}	=	*tetradic* (forth-order tensor) $\mathbf{C}$, or its 81 components

(*h*) *dyad-dyad product*, Eq 2.26:

$$(\mathbf{uv})\cdot(\mathbf{ws}) = u_i\hat{\mathbf{e}}_i\,(v_j\hat{\mathbf{e}}_j\cdot w_k\hat{\mathbf{e}}_k)\,s_q\hat{\mathbf{e}}_q = u_i v_j w_j s_q\hat{\mathbf{e}}_i\hat{\mathbf{e}}_q$$

(*i*) *vector-tensor products*, Eqs 2.27–2.28:

1. $\mathbf{v}\cdot\mathbf{T} = v_i\hat{\mathbf{e}}_i\cdot t_{jk}\hat{\mathbf{e}}_j\hat{\mathbf{e}}_k = v_i t_{jk}\delta_{ij}\hat{\mathbf{e}}_k = v_i t_{ik}\hat{\mathbf{e}}_k$

2. $\mathbf{T}\cdot\mathbf{v} = t_{ij}\hat{\mathbf{e}}_i\hat{\mathbf{e}}_j\cdot v_k\hat{\mathbf{e}}_k = t_{ij}\hat{\mathbf{e}}_i\delta_{jk}v_k = t_{ij}v_j\hat{\mathbf{e}}_i$

(*j*) *tensor-tensor product*, Eq 2.29:

$$\mathbf{T}\cdot\mathbf{S} = t_{ij}\hat{\mathbf{e}}_i\hat{\mathbf{e}}_j\cdot s_{pq}\hat{\mathbf{e}}_p\hat{\mathbf{e}}_q = t_{ij}s_{jp}\hat{\mathbf{e}}_i\hat{\mathbf{e}}_q$$

(*k*) *double dot product:*

1. $(\mathbf{uv})\cdot\cdot(\mathbf{ws}) = (v_i\hat{\mathbf{e}}_i\cdot w_j\hat{\mathbf{e}}_j)\,(u_k\hat{\mathbf{e}}_k\cdot s_q\hat{\mathbf{e}}_q) = v_i w_i u_k s_k$

2. $(\mathbf{uv}):(\mathbf{ws}) = (u_i\hat{\mathbf{e}}_i\cdot w_j\hat{\mathbf{e}}_j)\,(v_k\hat{\mathbf{e}}_k\cdot s_q\hat{\mathbf{e}}_q) = u_i w_i v_k s_k$

2.3 Indicial Notation

By assigning special meaning to the subscripts, *indicial notation* permits us to carry out the tensor operations of addition, multiplication, differentiation, etc. without the use, or even the appearance of the base vectors $\hat{\mathbf{e}}_i$ in the equations. We simply agree that the tensor rank (order) of a term is indicated by the number of "free," that is, unrepeated, subscripts appearing in that term. Accordingly, a term with no free indices represents a scalar, a term with one free index a vector, a term having two free indices a second order tensor, and so on. The specific meaning of these symbols are given in Table 2.1.

For tensors defined in a three-dimensional space, the free indices take on the values 1,2,3 successively, and we say that these indices have a *range* of three. If N is the number of free indices in a tensor, that tensor has 3^N components in three space.

We must emphasize that in the indicial notation exactly *two types* of subscripts appear:

TABLE 2.2
Forms for inner and outer products.

Outer Products:	Contraction(s):	Inner Products:
$u_i v_j$	$i = j$	$u_i v_i$ (vector dot product)
$\varepsilon_{ijk} u_q v_m$	$j = q, k = m$	$\varepsilon_{ijk} u_j v_k$ (vector cross product)
$\varepsilon_{ijk} u_q v_m w_n$	$i = q, j = m, k = n$	$\varepsilon_{ijk} u_i v_j w_k$ (box product)

1. "free" indices, which are represented by letters that occur only once in a given term,
2. "summed," or "dummy" indices which are represented by letters that appear only *twice* in a given term.

Furthermore, every term in a valid equation must have the same letter subscripts for the free indices. No letter subscript may appear more than twice in any given term.

Mathematical operations among tensors are readily carried out using the indicial notation. Thus addition (and subtraction) among tensors of equal rank follows according to the typical equations; $u_i + v_i - w_i = s_i$ for vectors, and $t_{ij} - v_{ij} + s_{ij} = q_{ij}$ for second-order tensors. Multiplication of two tensors to produce an *outer tensor* product is accomplished by simply setting down the tensor symbols side by side with no dummy indices appearing in the expression. As a typical example, the outer product of the vector v_i and tensor t_{jk} is the third-order tensor $v_i t_{jk}$. *Contraction* is the process of *identifying* (that is, setting equal to one another) any two indices of a tensor term. An *inner tensor product* is formed from an outer tensor product by one or more contractions involving indices from separate tensors in the outer product. We note that the rank of a given tensor is reduced by *two* for each contraction. Some outer products, which contract, form well-known inner products listed in Table 2.2.

A tensor is *symmetric* in any two indices if interchange of those indices leaves the tensor value unchanged. For example, if $s_{ij} = s_{ji}$ and $c_{ijm} = c_{jim}$, both of these tensors are said to be symmetric in the indices i and j. A tensor is *anti-symmetric* (or *skew-symmetric*) in any two indices if interchange of those indices causes a sign change in the value of the tensor. Thus if $a_{ij} = -a_{ji}$, it is anti-symmetric in i and j. Also, recall that by definition, $\varepsilon_{ijk} = -\varepsilon_{jik} = \varepsilon_{jki}$, etc., and hence the permutation symbol is anti-symmetric in all indices.

Example 2.5

Show that the inner product $s_{ij} a_{ij}$ of a symmetric tensor $s_{ij} = s_{ji}$, and an anti-symmetric tensor $a_{ij} = -a_{ji}$ is zero.

Solution
By definition of symmetric tensor s_{ij} and skew-symmetric tensor a_{ij}, we have

$$s_{ij} a_{ij} = -s_{ji} a_{ji} = -s_{mn} a_{mn} = -s_{ij} a_{ij}$$

where the last two steps are the result of all indices being dummy indices. Therefore, $2 s_{ij} a_{ij} = 0$, or $s_{ij} a_{ij} = 0$.

One of the most important advantages of the indicial notation is the compactness it provides in expressing equations in three dimensions. A brief listing of typical equations of continuum mechanics is presented below to illustrate this feature.

1. $\phi = s_{ij}t_{ij} - s_{ii}t_{jj}$ (1 equation, 18 terms on RHS)
2. $t_i = q_{ij}n_j$ (3 equations, 3 terms on RHS of each)
3. $t_{ij} = \lambda\delta_{ij}E_{kk} + 2\mu E_{ij}$ (9 equations, 4 terms on RHS of each)

Example 2.6

By direct expansion of the expression $v_i = \varepsilon_{ijk}w_{jk}$ determine the components of the vector v_i in terms of the components of the tensor w_{jk}.

Solution

By summing first on j and then on k and then omitting the zero terms, we find that

$$\begin{aligned} v_i &= \varepsilon_{i1k}w_{1k} + \varepsilon_{i2k}w_{2k} + \varepsilon_{i3k}w_{3k} \\ &= \varepsilon_{i12}w_{12} + \varepsilon_{i13}w_{13} + \varepsilon_{i21}w_{21} + \varepsilon_{i23}w_{23} + \varepsilon_{i31}w_{31} + \varepsilon_{i32}w_{32} \ . \end{aligned}$$

Therefore,

$$v_1 = \varepsilon_{123}w_{23} + \varepsilon_{132}w_{32} = w_{23} - w_{32} \ ,$$

$$v_2 = \varepsilon_{213}w_{13} + \varepsilon_{231}w_{31} = w_{31} - w_{13} \ ,$$

$$v_3 = \varepsilon_{312}w_{12} + \varepsilon_{321}w_{21} = w_{12} - w_{21} \ .$$

Note that if the tensor w_{jk} were symmetric, the vector v_i would be a null (zero) vector.

To end this section, consider a skew–symmetric second order tensor $\mathbf{W} = w_{ij}\hat{\mathbf{e}}_i\hat{\mathbf{e}}_j$. All skew–symmetric tensors can be represented in terms of an *axial vector* by using the permutation symbol. Let the axial vector for w_{ij} be ω_i defined by

$$\omega_i = -\frac{1}{2}\varepsilon_{ijk}w_{jk} \ . \tag{2.30}$$

It is an easy exercise in indicial manipulation to find the inverse of Eq 2.30. We can write w_{jk} in terms of ω_i as follows:

$$\begin{aligned} \varepsilon_{imn}\omega_i &= -\tfrac{1}{2}\varepsilon_{imn}\varepsilon_{ijk}w_{jk} \\ &= -\tfrac{1}{2}\left(\delta_{mj}\delta_{nk} - \delta_{mk}\delta_{nj}\right)w_{jk} \\ &= -\tfrac{1}{2}\left(w_{mn} - w_{nm}\right) \\ &= -\tfrac{1}{2}\left(2w_{mn}\right) \\ &= -w_{mn} \end{aligned} \tag{2.31}$$

where Eq 2.7a and the skew–symmetric property of w_{ij} were used.

2.4 Matrices and Determinants

For computational purposes it is often expedient to use the *matrix representation* of vectors and tensors. Accordingly, we review here several definitions and operations of elementary matrix theory.

A *matrix* is an ordered rectangular array of elements enclosed by square brackets and subjected to certain operational rules. The typical element $\mathcal{A}_{ij}$ of the matrix is located in the ith (horizontal) row, and in the jth (vertical) column of the array. A matrix having elements $\mathcal{A}_{ij}$, which may be numbers, variables, functions, or any of several mathematical entities, is designated by $[\mathcal{A}_{ij}]$, or symbolically by the kernel letter $\boldsymbol{\mathcal{A}}$. An M by N matrix (written $M \times N$) has M rows and N columns, and may be displayed as

$$\boldsymbol{\mathcal{A}} = [\mathcal{A}_{ij}] = \begin{bmatrix} \mathcal{A}_{11} & \mathcal{A}_{12} & \cdots & \mathcal{A}_{1N} \\ \mathcal{A}_{21} & \mathcal{A}_{22} & \cdots & \mathcal{A}_{2N} \\ \vdots & \vdots & & \vdots \\ \mathcal{A}_{M1} & \mathcal{A}_{M2} & \cdots & \mathcal{A}_{MN} \end{bmatrix} . \tag{2.32}$$

If M = N, the matrix is a *square matrix*. A $1 \times N$ matrix $[\mathcal{A}_{1N}]$ is an *row matrix*, and an $M \times 1$ matrix $[\mathcal{A}_{M1}]$ is a *column matrix*. Row and column matrices represent vectors, whereas a 3×3 square matrix represents a second-order tensor. A scalar is represented by a 1×1 matrix (a single element). The unqualified use of the word *matrix* in this text is understood to mean a 3×3 square matrix, that is, the matrix representation of a second-order tensor or a matrix that is not a tensor.

A *zero*, or *null* matrix has all elements equal to zero. A *diagonal matrix* is a *square matrix* whose elements not on the *principal diagonal*, which extends from $\mathcal{A}_{11}$ to $\mathcal{A}_{NN}$, are all zeros. Thus for a diagonal matrix, $\mathcal{A}_{ij} = 0$ for $i \neq j$. The unit or *identity matrix* **I**, which, incidentally, is the matrix representation of the Kronecker delta, is a diagonal matrix whose diagonal elements all have the value one.

The $N \times M$ matrix formed by interchanging the rows and columns of the $M \times N$ matrix $\boldsymbol{\mathcal{A}}$ is called the *transpose* of $\boldsymbol{\mathcal{A}}$, and is written as $\boldsymbol{\mathcal{A}}^T$, or $[\mathcal{A}_{ij}]^T$. By definition, the elements of a matrix $\boldsymbol{\mathcal{A}}$ and its transpose are related by the equation $\mathcal{A}^T_{ij} = \mathcal{A}_{ji}$. A square matrix for which $\boldsymbol{\mathcal{A}} = \boldsymbol{\mathcal{A}}^T$, or in element form, $\mathcal{A}_{ij} = \mathcal{A}_{ji}$, is called a symmetric matrix; one for which $\boldsymbol{\mathcal{A}} = -\boldsymbol{\mathcal{A}}$, or, $\mathcal{A}_{ij} = -\mathcal{A}_{ji}$, is called an anti-symmetric, or skew-symmetric matrix. The elements of the principal diagonal of a skew-symmetric matrix are all zeros. Two matrices are *equal* if they are identical element by element. Matrices having the same number of rows and columns may be added (or subtracted) element by element. Thus if $\boldsymbol{\mathcal{A}} = \boldsymbol{\mathcal{B}} + \boldsymbol{\mathcal{C}}$, the elements of $\boldsymbol{\mathcal{A}}$ are given by

$$\mathcal{A}_{ij} = \mathcal{B}_{ij} + \mathcal{C}_{ij} \ . \tag{2.33}$$

Addition of matrices is commutative, $\boldsymbol{\mathcal{A}} + \boldsymbol{\mathcal{B}} = \boldsymbol{\mathcal{B}} + \boldsymbol{\mathcal{A}}$, and associative, $\boldsymbol{\mathcal{A}} + (\boldsymbol{\mathcal{B}} + \boldsymbol{\mathcal{C}}) = (\boldsymbol{\mathcal{A}} + \boldsymbol{\mathcal{B}}) + \boldsymbol{\mathcal{C}}$.

Example 2.7

Show that the square matrix $\mathcal{A}$ can be expressed as the sum of a symmetric and a skew-symmetric matrix by the decomposition

$$\mathcal{A} = \frac{\mathcal{A} + \mathcal{A}^T}{2} + \frac{\mathcal{A} - \mathcal{A}^T}{2} .$$

Solution

Let the decomposition be written as $\mathcal{A} = \mathcal{B} + \mathcal{C}$ where $\mathcal{B} = \frac{1}{2}\left(\mathcal{A} + \mathcal{A}^T\right)$ and $\mathcal{C} = \frac{1}{2}\left(\mathcal{A} - \mathcal{A}^T\right)$. Then writing $\mathcal{B}$ and $\mathcal{C}$ in element form,

$$\mathcal{B}_{ij} = \frac{\mathcal{A}_{ij} + \mathcal{A}^T_{ij}}{2} = \frac{\mathcal{A}_{ij} + \mathcal{A}_{ji}}{2} = \frac{\mathcal{A}^T_{ji} + \mathcal{A}_{ji}}{2} = \mathcal{B}_{ji} = \mathcal{B}^T_{ij} \qquad \text{(symmetric)} ,$$

$$\mathcal{C}_{ij} = \frac{\mathcal{A}_{ij} - \mathcal{A}^T_{ij}}{2} = \frac{\mathcal{A}_{ij} - \mathcal{A}_{ji}}{2} = -\frac{\mathcal{A}^T_{ji} - \mathcal{A}_{ji}}{2} = -\mathcal{C}_{ji} = -\mathcal{C}^T_{ij} \qquad \text{(skew-symmetric)} .$$

Therefore $\mathcal{B}$ is symmetric, and $\mathcal{C}$ skew-symmetric.

Multiplication of the matrix $\mathcal{A}$ by the scalar λ results in the matrix $\lambda\mathcal{A}$, or $[\lambda\mathcal{A}_{ij}] = \lambda[\mathcal{A}_{ij}]$. The product of two matrices $\mathcal{A}$ and $\mathcal{B}$, denoted by $\mathcal{A}\mathcal{B}$, is defined only if the matrices are *conformable*, that is, if the *prefactor* matrix $\mathcal{A}$ has the same number of columns as the *postfactor* matrix $\mathcal{B}$ has rows. Thus the product of an $M \times Q$ matrix multiplied by a $Q \times N$ matrix is an $M \times N$ matrix. The product matrix $\mathcal{C} = \mathcal{A}\mathcal{B}$ has elements given by

$$\mathcal{C}_{ij} = \mathcal{A}_{ik}\mathcal{B}_{kj} \tag{2.34}$$

in which k is, of course, a summed index. Therefore each element $\mathcal{C}_{ij}$ of the product matrix is an inner product of the ith row of the prefactor matrix with the jth column of the postfactor matrix. In general, matrix multiplication is not commutative, $\mathcal{A}\mathcal{B} \neq \mathcal{B}\mathcal{A}$, but the associative and distributive laws of multiplication do hold for matrices. The product of a matrix with itself is the square of the matrix, and is written $\mathcal{A}\mathcal{A} = \mathcal{A}^2$. Likewise the cube of the matrix is $\mathcal{A}\mathcal{A}\mathcal{A} = \mathcal{A}^3$, and in general, matrix products obey the exponent rule

$$\mathcal{A}^m\mathcal{A}^n = \mathcal{A}^n\mathcal{A}^m = \mathcal{A}^{m+n} \tag{2.35}$$

where m and n are positive integers, or zero. Also, we note that

$$\left(\mathcal{A}^n\right)^T = \left(\mathcal{A}^T\right)^n , \tag{2.36}$$

and if $\mathcal{B}\mathcal{B} = \mathcal{A}$ then

$$\mathcal{B} = \sqrt{\mathcal{A}} = \mathcal{A}^{\frac{1}{2}} \tag{2.37}$$

but the square root is not unique.

Example 2.8

Use indicial notation to show that for arbitrary matrices $\mathcal{A}$ and $\mathcal{B}$:

(a) $(\mathcal{A}+\mathcal{B})^{\mathsf{T}} = \mathcal{A}^{\mathsf{T}} + \mathcal{B}^{\mathsf{T}}$

(b) $(\mathcal{A}\mathcal{B})^{\mathsf{T}} = \mathcal{B}^{\mathsf{T}}\mathcal{A}^{\mathsf{T}}$

(c) $\mathbf{I}\mathcal{B} = \mathcal{B}\mathbf{I} = \mathcal{B}$ where $\mathbf{I}$ is the identity matrix.

Solution

(a) Let $\mathcal{A}+\mathcal{B} = \mathcal{C}$, then in element form $\mathcal{C}_{ij} = \mathcal{A}_{ij} + \mathcal{B}_{ij}$ and therefore $\mathcal{C}^{\mathsf{T}}$ is given by

$$\mathcal{C}^{\mathsf{T}}_{ij} = \mathcal{C}_{ji} = \mathcal{A}_{ji} + \mathcal{B}_{ji} = \mathcal{A}^{\mathsf{T}}_{ij} + \mathcal{B}^{\mathsf{T}}_{ji} \,,$$

or

$$\mathcal{C}^{\mathsf{T}} = (\mathcal{A}+\mathcal{B})^{\mathsf{T}} = \mathcal{A}^{\mathsf{T}} + \mathcal{B}^{\mathsf{T}} \,.$$

(b) Let $\mathcal{A}\mathcal{B} = \mathcal{C}$, then in element form

$$\mathcal{C}_{ij} = \mathcal{A}_{ik}\mathcal{B}_{kj} = \mathcal{A}^{\mathsf{T}}_{ki}\mathcal{B}^{\mathsf{T}}_{jk} = \mathcal{B}^{\mathsf{T}}_{jk}\mathcal{A}^{\mathsf{T}}_{ki} = \mathcal{C}^{\mathsf{T}}_{ji} \,.$$

Hence $(\mathcal{A}\mathcal{B})^{\mathsf{T}} = \mathcal{B}^{\mathsf{T}}\mathcal{A}^{\mathsf{T}}$. Note that exchanging the order $\mathcal{A}^{\mathsf{T}}_{ki}\mathcal{B}^{\mathsf{T}}_{jk} = \mathcal{B}^{\mathsf{T}}_{jk}\mathcal{A}^{\mathsf{T}}_{ki}$ is not necessary when using indicial notation. It is done for clarity. In direct or matrix notation the order of the terms is critical.

(c) Let $\mathbf{I}\mathcal{B} = \mathcal{C}$, then in element form

$$\mathcal{C}_{ij} = \delta_{ik}\mathcal{B}_{kj} = \mathcal{B}_{ij} = \mathcal{B}_{ik}\delta_{kj}$$

by the substitution property. Thus $\mathbf{I}\mathcal{B} = \mathcal{B}\mathbf{I} = \mathcal{B}$.

The *determinant* of a square matrix is formed from the square array of elements of the matrix and this array evaluated according to established mathematical rules. The determinant of the matrix $\mathcal{A}$ is designated by either $\det\mathcal{A}$, or by $|\mathcal{A}_{ij}|$, and for a 3×3 matrix $\mathcal{A}$,

$$\det\mathcal{A} = |\mathcal{A}_{ij}| = \begin{vmatrix} \mathcal{A}_{11} & \mathcal{A}_{12} & \mathcal{A}_{13} \\ \mathcal{A}_{21} & \mathcal{A}_{22} & \mathcal{A}_{23} \\ \mathcal{A}_{31} & \mathcal{A}_{32} & \mathcal{A}_{33} \end{vmatrix} \,. \tag{2.38}$$

A minor of $\det\mathcal{A}$ is another determinant $|\mathcal{M}_{ij}|$ formed by deleting the ith row and j^{th} column of $|A_{ij}|$. The cofactor of the element $\mathcal{A}_{ij}$ (sometimes referred to as the *signed minor*) is defined by

$$\mathcal{A}^{(c)}_{ij} = (-1)^{i+j}\,|\mathcal{M}_{ij}| \tag{2.39}$$

where superscript (c) denotes cofactor of matrix $\mathcal{A}$.

Evaluation of a determinant may be carried out by a standard method called *expansion by cofactors*. In this method, any row (or column) of the determinant is chosen, and each element in that row (or column) is multiplied by its cofactor. The sum of these products gives the value of the determinant. For example, expansion of the determinant of Eq 2.38 by the first row becomes

$$\det \mathcal{A} = \mathcal{A}_{11} \begin{vmatrix} \mathcal{A}_{22} & \mathcal{A}_{23} \\ \mathcal{A}_{32} & \mathcal{A}_{33} \end{vmatrix} - \mathcal{A}_{12} \begin{vmatrix} \mathcal{A}_{21} & \mathcal{A}_{23} \\ \mathcal{A}_{31} & \mathcal{A}_{33} \end{vmatrix} + \mathcal{A}_{13} \begin{vmatrix} \mathcal{A}_{21} & \mathcal{A}_{22} \\ \mathcal{A}_{31} & \mathcal{A}_{32} \end{vmatrix} \tag{2.40}$$

which upon complete expansion gives

$$\begin{aligned} \det \mathcal{A} &= \mathcal{A}_{11} (\mathcal{A}_{22}\mathcal{A}_{33} - \mathcal{A}_{23}\mathcal{A}_{32}) - \mathcal{A}_{12} (\mathcal{A}_{21}\mathcal{A}_{33} - \mathcal{A}_{23}\mathcal{A}_{31}) \\ &\quad + \mathcal{A}_{13} (\mathcal{A}_{21}\mathcal{A}_{32} - \mathcal{A}_{22}\mathcal{A}_{31}) \ . \end{aligned} \tag{2.41}$$

Several interesting properties of determinants are worth mentioning at this point. To begin with, the interchange of any two rows (or columns) of a determinant causes a sign change in its value. Because of this property and because of the sign-change property of the permutation symbol, the $\det \mathcal{A}$ of Eq 2.38 may be expressed in the indicial notation by the alternative forms (see Prob. 2.13)

$$\det \mathcal{A} = \varepsilon_{ijk} \mathcal{A}_{i1} \mathcal{A}_{j2} \mathcal{A}_{k3} = \varepsilon_{ijk} \mathcal{A}_{1i} \mathcal{A}_{2j} \mathcal{A}_{3k} \ . \tag{2.42}$$

Furthermore, following an arbitrary number of column interchanges with the accompanying sign change of the determinant for each, it can be shown from the first form of Eq 2.42 that, (see Prob 2.14)

$$\varepsilon_{qmn} \det \mathcal{A} = \varepsilon_{ijk} \mathcal{A}_{iq} \mathcal{A}_{jm} \mathcal{A}_{kn} \ . \tag{2.43}$$

Finally, we note that if the $\det \mathcal{A} = 0$, the matrix is said to be *singular*. It may be easily shown that every 3×3 skew-symmetric matrix is singular. Also, the determinant of the diagonal matrix, $\mathcal{D}$, is simply the product of its diagonal elements: $\det \mathcal{D} = \mathcal{D}_{11}\mathcal{D}_{22} \cdots \mathcal{D}_{NN}$.

Example 2.9

Show that for matrices $\mathcal{A}$ and $\mathcal{B}$, $\det(\mathcal{A}\mathcal{B}) = \det(\mathcal{B}\mathcal{A}) = \det(\mathcal{A}) \det(\mathcal{B})$.

Solution

Let $\mathcal{C} = \mathcal{A}\mathcal{B}$, then $\mathcal{C}_{ij} = \mathcal{A}_{ik}\mathcal{B}_{kj}$ and from Eq 2.42

$$\begin{aligned} \det \mathcal{C} &= \varepsilon_{ijk} \mathcal{C}_{i1} \mathcal{C}_{j2} \mathcal{C}_{k3} \\ &= \varepsilon_{ijk} \mathcal{A}_{iq} \mathcal{B}_{q1} \mathcal{A}_{jm} \mathcal{B}_{m2} \mathcal{A}_{kn} \mathcal{B}_{n3} \\ &= \varepsilon_{ijk} \mathcal{A}_{iq} \mathcal{A}_{jm} \mathcal{A}_{kn} \mathcal{B}_{q1} \mathcal{B}_{m2} \mathcal{B}_{n3} \ . \end{aligned}$$

But from Eq 2.43

$$\varepsilon_{ijk} \mathcal{A}_{iq} \mathcal{A}_{jm} \mathcal{A}_{kn} = \varepsilon_{qmn} \det \mathcal{A} \ ,$$

so now

$$\det \mathcal{C} = \det \mathcal{AB} = \varepsilon_{qmn}\mathcal{B}_{q1}\mathcal{B}_{m2}\mathcal{B}_{n3}\det \mathcal{A} = \det \mathcal{A}\det \mathcal{B} \ .$$

By a direct interchange of $\mathcal{A}$ and $\mathcal{B}$, $\det \mathcal{AB} = \det \mathcal{BA}$.

Example 2.10

Use Eq 2.40 and Eq 2.41 to show that $\det \mathcal{A} = \det \mathcal{A}^{\mathsf{T}}$.

Solution
Since

$$|\mathcal{A}^{\mathsf{T}}| = \begin{vmatrix} \mathcal{A}_{11} & \mathcal{A}_{21} & \mathcal{A}_{31} \\ \mathcal{A}_{12} & \mathcal{A}_{22} & \mathcal{A}_{32} \\ \mathcal{A}_{13} & \mathcal{A}_{23} & \mathcal{A}_{33} \end{vmatrix}$$

cofactor expansion by the first column here yields

$$\det \mathcal{A} = \mathcal{A}_{11}\begin{vmatrix} \mathcal{A}_{22} & \mathcal{A}_{32} \\ \mathcal{A}_{23} & \mathcal{A}_{33} \end{vmatrix} - \mathcal{A}_{12}\begin{vmatrix} \mathcal{A}_{21} & \mathcal{A}_{31} \\ \mathcal{A}_{23} & \mathcal{A}_{33} \end{vmatrix} + \mathcal{A}_{13}\begin{vmatrix} \mathcal{A}_{21} & \mathcal{A}_{31} \\ \mathcal{A}_{22} & \mathcal{A}_{32} \end{vmatrix}$$

which is equal to Eq 2.41.

The *inverse* of the matrix $\mathcal{A}$ is written $\mathcal{A}^{-1}$, and is defined by

$$\mathcal{A}\mathcal{A}^{-1} = \mathcal{A}^{-1}\mathcal{A} = \mathbf{I} \tag{2.44}$$

where $\mathbf{I}$ is the identity matrix. Thus if $\mathcal{AB} = \mathbf{I}$, then $\mathcal{B} = \mathcal{A}^{-1}$, and $\mathcal{A} = \mathcal{B}^{-1}$. The *adjoint matrix* $\mathcal{A}^*$ is defined as the transpose of the cofactor matrix

$$\mathcal{A}^* = \left[\mathcal{A}^{(c)}\right]^{\mathsf{T}} \ . \tag{2.45}$$

In terms of the adjoint matrix the inverse matrix is expressed by

$$\mathcal{A}^{-1} = \frac{\mathcal{A}^*}{\det \mathcal{A}} \tag{2.46}$$

which is actually a working formula by which an inverse matrix may be calculated. This formula shows that the inverse matrix exists only if $\det \mathcal{A} \neq 0$, i.e., only if the matrix $\mathcal{A}$ is non-singular. In particular, a 3×3 skew-symmetric matrix has no inverse.

Example 2.11

Show from the definition of the inverse, Eq 2.44 that

(a) $(\mathcal{A}\mathcal{B})^{-1} = \mathcal{B}^{-1}\mathcal{A}^{-1}$,

(b) $\left(\mathcal{A}^{\mathsf{T}}\right)^{-1} = \left(\mathcal{A}^{-1}\right)^{\mathsf{T}}$.

Solution

(a) By pre-multiplying the matrix product $\mathcal{A}\mathcal{B}$ by $\mathcal{B}^{-1}\mathcal{A}^{-1}$, we have (using Eq 2.44),

$$\mathcal{B}^{-1}\mathcal{A}^{-1}\mathcal{A}\mathcal{B} = \mathcal{B}^{-1}\mathbf{I}\mathcal{B} = \mathcal{B}^{-1}\mathcal{B} = \mathbf{I}$$

and therefore $\mathcal{B}^{-1}\mathcal{A}^{-1} = (\mathcal{A}\mathcal{B})^{-1}$.

(b) Taking the transpose of both sides of Eq 2.44 and using the result of Example 2.8 we have

$$\left(\mathcal{A}\mathcal{A}^{-1}\right)^{\mathsf{T}} = \left(\mathcal{A}^{-1}\right)^{\mathsf{T}}\mathcal{A}^{\mathsf{T}} = \mathbf{I}^{\mathsf{T}} = \mathbf{I} .$$

Hence $\left(\mathcal{A}^{-1}\right)^{\mathsf{T}}$ must be the inverse of $\mathcal{A}^{\mathsf{T}}$, or $\left(\mathcal{A}^{-1}\right)^{\mathsf{T}} = \left(\mathcal{A}^{\mathsf{T}}\right)^{-1}$.

An *orthogonal* matrix, call it $\mathcal{Q}$, is a square matrix for which $\mathcal{Q}^{-1} = \mathcal{Q}^{\mathsf{T}}$. From this definition we note that a *symmetric orthogonal* matrix is its own inverse since in this case

$$\mathcal{Q}^{-1} = \mathcal{Q}^{\mathsf{T}} = \mathcal{Q} . \tag{2.47}$$

Also, if $\mathcal{A}$ and $\mathcal{B}$ are orthogonal matrices.

$$(\mathcal{A}\mathcal{B})^{-1} = \mathcal{B}^{-1}\mathcal{A}^{-1} = \mathcal{B}^{\mathsf{T}}\mathcal{A}^{\mathsf{T}} = (\mathcal{A}\mathcal{B})^{\mathsf{T}} , \tag{2.48}$$

so that the product matrix is likewise orthogonal. Furthermore, if $\mathcal{A}$ is orthogonal it may shown (see Prob. 2.18) that

$$\det \mathcal{A} = \pm 1 . \tag{2.49}$$

As mentioned near the beginning of this section, a vector may be represented by a row or column matrix, a second-order tensor by a square 3×3 matrix. For computational purposes, it is frequently advantageous to transcribe vector/tensor equations into their matrix form. As a very simple example, the vector-tensor product, $\mathbf{u} = \mathbf{T}\mathbf{v}$ (symbolic notation) or $u_i = T_{ij}v_j$ (indicial notation) appears in matrix form as

$$[u_{i1}] = [T_{ij}]\,[v_{j1}] \quad \text{or} \quad \begin{bmatrix} u_1 \\ u_2 \\ u_3 \end{bmatrix} = \begin{bmatrix} T_{11} & T_{12} & T_{13} \\ T_{21} & T_{22} & T_{23} \\ T_{31} & T_{32} & T_{33} \end{bmatrix} \begin{bmatrix} v_1 \\ v_2 \\ v_3 \end{bmatrix} . \tag{2.50}$$

TABLE 2.3
Transformation table between $Ox_1x_2x_3$ and $Ox'_1x'_2x'_3$.

	$\hat{e}_1$, x_1	$\hat{e}_2$, x_2	$\hat{e}_3$, x_3
$\hat{e}'_1$, x'_1	a_{11}	a_{12}	a_{13}
$\hat{e}'_2$, x'_2	a_{21}	a_{22}	a_{23}
$\hat{e}'_3$, x'_3	a_{31}	a_{32}	a_{33}

In much the same way the product

$$\mathbf{w} = \mathbf{v} \cdot \mathbf{T} \quad \text{or} \quad w_i = v_j T_{ji}$$

appears as

$$[w_{1i}] = [v_{1j}]\,[T_{ji}] \quad \text{or} \quad \begin{bmatrix} w_1 & w_2 & w_3 \end{bmatrix} = \begin{bmatrix} v_1 & v_2 & v_3 \end{bmatrix} \begin{bmatrix} T_{11} & T_{21} & T_{31} \\ T_{12} & T_{22} & T_{32} \\ T_{13} & T_{23} & T_{33} \end{bmatrix}. \quad (2.51)$$

Note that in Eq 2.51 the order of the subscripts are exchanged since the transpose of the matrix is being written. Equations 2.51 and 2.50 could be written in the direct form the following way:

$$\mathbf{w} = \mathbf{v} \cdot \mathbf{T} = (\mathbf{T} \cdot \mathbf{v})^T = \mathbf{v}^T \cdot \mathbf{T}^T .$$

From this, it is clear that

$$\mathbf{w} = \mathbf{u}^T$$

since $\mathbf{u} = \mathbf{T} \cdot \mathbf{v}$.

2.5 Transformations of Cartesian Tensors

Although, as already mentioned, vectors and tensors have an identity independent of any particular reference or coordinate system, the relative values of their respective components do depend upon the specific axes to which they are referred. The relationships among these various components when given with respect to two separate sets of coordinate axes are known as the *transformation equations*. In developing these transformation equations for Cartesian tensors, we consider two sets of rectangular Cartesian axes, $Ox_1x_2x_3$ and $Ox'_1x'_2x'_3$, sharing a common origin, and oriented relative to one another so that the direction cosines between the primed and unprimed axes are given by $a_{ij} = \cos(x'_i, x_j)$ as shown in Fig. 2.2.

The square array of the nine direction cosines displayed in Table 2.3 is useful in relating the unit base vectors $\hat{e}_i$ and $\hat{e}'_i$ to one another, as well as relating the primed and unprimed coordinates x'_i and x_i of a point. Thus the primed base vectors $\hat{e}'_i$ are given in terms of the unprimed vectors $\hat{e}_i$ by the equations (as is also easily verified from the geometry of the vectors in the diagram of Fig. 2.2),

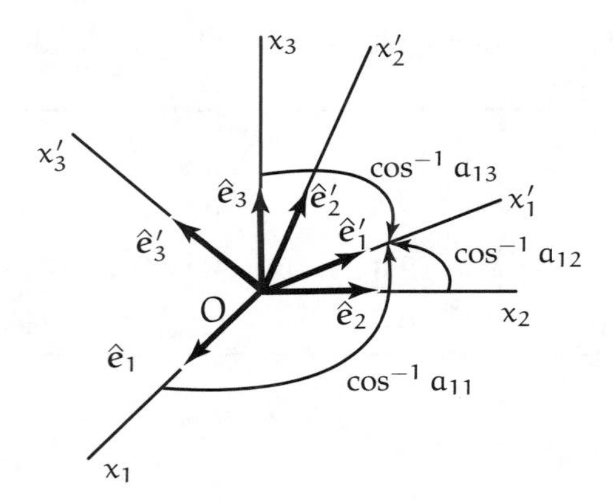

FIGURE 2.2
Rectangular coordinate system $Ox'_1x'_2x'_3$ relative to $Ox_1x_2x_3$. Direction cosines shown for coordinate x'_1 relative to unprimed coordinates. Similar direction cosines are defined for x'_2 and x'_3 coordinates.

$$\hat{\mathbf{e}}'_1 = a_{11}\hat{\mathbf{e}}_1 + a_{12}\hat{\mathbf{e}}_2 + a_{13}\hat{\mathbf{e}}_3 = a_{1j}\hat{\mathbf{e}}_j \ , \tag{2.52a}$$
$$\hat{\mathbf{e}}'_2 = a_{21}\hat{\mathbf{e}}_1 + a_{22}\hat{\mathbf{e}}_2 + a_{23}\hat{\mathbf{e}}_3 = a_{2j}\hat{\mathbf{e}}_j \ , \tag{2.52b}$$
$$\hat{\mathbf{e}}'_3 = a_{31}\hat{\mathbf{e}}_1 + a_{32}\hat{\mathbf{e}}_2 + a_{33}\hat{\mathbf{e}}_3 = a_{3j}\hat{\mathbf{e}}_j \ , \tag{2.52c}$$

or in compact indicial form

$$\hat{\mathbf{e}}'_i = a_{ij}\hat{\mathbf{e}}_j \ . \tag{2.53}$$

By defining the matrix $\mathcal{A}$ whose elements are the direction cosines a_{ij}, Eq 2.53 can be written in matrix form as

$$[\hat{\mathbf{e}}'_{i1}] = [a_{ij}]\,[\hat{\mathbf{e}}_{j1}] \qquad \text{or} \qquad \begin{bmatrix} \hat{\mathbf{e}}'_1 \\ \hat{\mathbf{e}}'_2 \\ \hat{\mathbf{e}}'_3 \end{bmatrix} = \begin{bmatrix} a_{11} & a_{12} & a_{13} \\ a_{21} & a_{22} & a_{23} \\ a_{31} & a_{32} & a_{33} \end{bmatrix} \begin{bmatrix} \hat{\mathbf{e}}_1 \\ \hat{\mathbf{e}}_2 \\ \hat{\mathbf{e}}_3 \end{bmatrix} \tag{2.54}$$

where the elements of the column matrices are unit vectors. The matrix $\mathcal{A}$ is called the *transformation matrix* because, as we shall see, of its role in transforming the components of a vector (or tensor) referred to one set of axes into the components of the same vector (or tensor) in a rotated set.

Because of the perpendicularity of the primed axes, $\hat{\mathbf{e}}'_i \cdot \hat{\mathbf{e}}'_j = \delta_{ij}$. But also, in view of Eq 2.53,

$$\hat{\mathbf{e}}'_i \cdot \hat{\mathbf{e}}'_j = a_{iq}\hat{\mathbf{e}}_q \cdot a_{jm}\hat{\mathbf{e}}_m = a_{iq}a_{jm}\delta_{qm} = a_{iq}a_{jq} = \delta_{ij}$$

from which we extract the *orthogonality condition* on the direction cosines (given here in both indicial and matrix form),

$$a_{iq}a_{jq} = \delta_{ij} \qquad \text{or} \qquad \mathcal{A}\mathcal{A}^T = \mathbf{I} \ . \tag{2.55}$$

Note that this is simply the inner product of the ith row with the jth row of the matrix $\mathcal{A}$. By an analogous derivation to that leading to Eq 2.53, but using the columns of $\mathcal{A}$, we obtain

$$\hat{\mathbf{e}}_i = a_{ji}\hat{\mathbf{e}}'_j \,, \tag{2.56}$$

which in matrix form is

$$[\hat{\mathbf{e}}_{1i}] = [\hat{\mathbf{e}}'_{1j}]\,[a_{ij}] \quad \text{or} \quad \begin{bmatrix} \hat{\mathbf{e}}_1 & \hat{\mathbf{e}}_2 & \hat{\mathbf{e}}_3 \end{bmatrix} = \begin{bmatrix} \hat{\mathbf{e}}'_1 & \hat{\mathbf{e}}'_2 & \hat{\mathbf{e}}'_3 \end{bmatrix} \begin{bmatrix} a_{11} & a_{12} & a_{13} \\ a_{21} & a_{22} & a_{23} \\ a_{31} & a_{32} & a_{33} \end{bmatrix} . \tag{2.57}$$

Note that using the transpose $\mathcal{A}^T$, Eq 2.57 may also be written

$$[\hat{\mathbf{e}}_{1i}] = [a_{ij}]^T\,[\hat{\mathbf{e}}'_{1j}] \quad \text{or} \quad \begin{bmatrix} \hat{\mathbf{e}}_1 \\ \hat{\mathbf{e}}_2 \\ \hat{\mathbf{e}}_3 \end{bmatrix} = \begin{bmatrix} a_{11} & a_{21} & a_{31} \\ a_{12} & a_{22} & a_{32} \\ a_{13} & a_{23} & a_{33} \end{bmatrix} \begin{bmatrix} \hat{\mathbf{e}}'_1 \\ \hat{\mathbf{e}}'_2 \\ \hat{\mathbf{e}}'_3 \end{bmatrix} \tag{2.58}$$

in which column matrices are used for the vectors $\hat{\mathbf{e}}_i$ and $\hat{\mathbf{e}}'_i$. By a consideration of the dot product and Eq 2.56 we obtain a second orthogonality condition

$$a_{ij}a_{ik} = \delta_{jk} \qquad \text{or} \qquad \mathcal{A}^T\mathcal{A} = \mathbf{I} \tag{2.59}$$

which is the inner product of the jth column with the kth column of $\mathcal{A}$.

Consider next an arbitrary vector $\mathbf{v}$ having components v_i in the unprimed system, and v'_i in the primed system. Then using Eq 2.56,

$$\mathbf{v} = v'_j\hat{\mathbf{e}}'_j = v_i\hat{\mathbf{e}}_i = v_i a_{ji}\hat{\mathbf{e}}'_j$$

from which, by matching coefficients on $\hat{\mathbf{e}}'_j$, we have (in both the indicial and matrix forms),

$$v'_j = a_{ji}v_i \qquad \text{or} \qquad \mathbf{v}' = \mathcal{A}\mathbf{v} = \mathbf{v}\mathcal{A}^T \tag{2.60}$$

which is the *transformation law* expressing the primed components of an arbitrary vector in terms of its unprimed components. Although the elements of the transformation matrix are written as a_{ij} we must emphasize that they are not the components of a second-order Cartesian tensor as it might appear. Multiplication of Eq 2.60 by a_{jk} and using the orthogonality condition Eq 2.59 we obtain the inverse law

$$v_k = a_{jk}v'_j \qquad \text{or} \qquad \mathbf{v} = \mathbf{v}'\mathcal{A} = \mathcal{A}^T\mathbf{v}' \tag{2.61}$$

giving the unprimed components in terms of the primed.

By a direct application of Eq 2.61 to the dyad $\mathbf{uv}$ we have

$$u_i v_j = a_{qi}u'_q a_{mj}v'_m = a_{qi}a_{mj}u'_q v'_m \,. \tag{2.62}$$

But a dyad is, after all, one form of a second-order tensor, and so by an obvious adaptation of Eq 2.62 we obtain the transformation law for a second-order tensor, $\mathbf{T}$ as

$$t_{ij} = a_{qi}a_{mj}t'_{qm} \qquad \text{or} \qquad \mathbf{T} = \mathcal{A}^T\mathbf{T}'\mathcal{A} \tag{2.63}$$

which may be readily inverted with the help of the orthogonality conditions to yield

$$t'_{ij} = a_{iq}a_{jm}t_{qm} \qquad \text{or} \qquad \mathbf{T}' = \mathcal{A}\mathbf{T}\mathcal{A}^T \,. \tag{2.64}$$

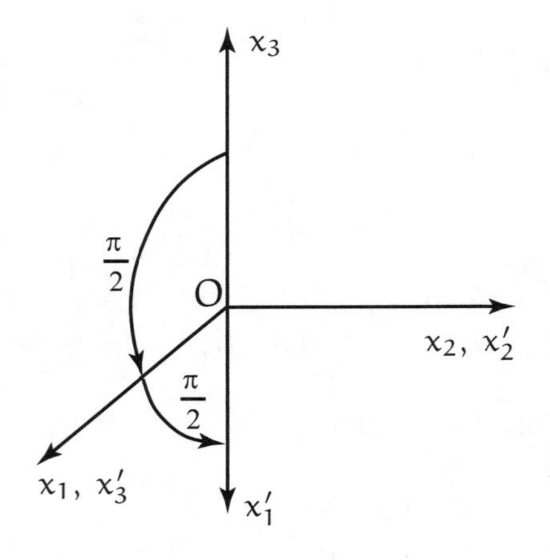

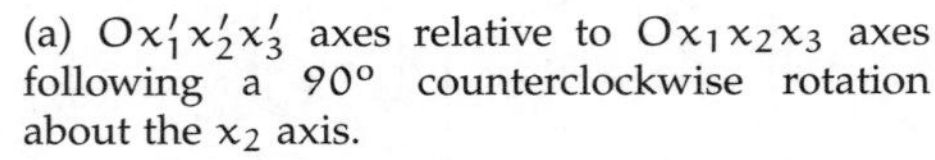
(a) $Ox_1'x_2'x_3'$ axes relative to $Ox_1x_2x_3$ axes following a 90° counterclockwise rotation about the x_2 axis.

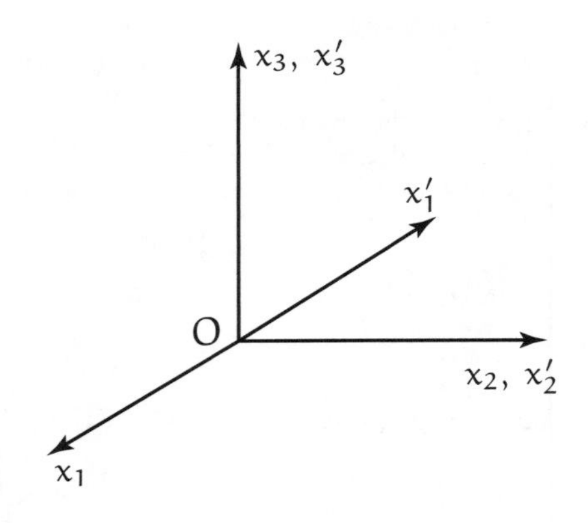

(b) $Ox_1'x_2'x_3'$ axes relative to $Ox_1x_2x_3$ axes following a reflection about the x_2x_3 plane.

FIGURE 2.3
Rotation and reflection of reference axes.

Note carefully the location of the summed indices q and m in Equations 2.63 and 2.64. Finally, by a logical generalization of the pattern of the transformation rules developed thus far we state that for an arbitrary Cartesian tensor of any order

$$R'_{ij\dots k} = a_{iq}a_{jm}\cdots a_{kn}R_{qm\dots n}\,. \tag{2.65}$$

The primed axes may be related to the unprimed axes through either a *rotation* about an axis through the origin, or by a *reflection* of the axes in one of the coordinate planes. (Or by a combination of such changes). As a simple example, consider a 90° (counterclockwise) rotation about the x_2 axis shown in Fig. 2.3(a). The matrix of direction cosines for this rotation is

$$[a_{ij}] = \begin{bmatrix} 0 & 0 & -1 \\ 0 & 1 & 0 \\ 1 & 0 & 0 \end{bmatrix}$$

and det $\mathcal{A} = 1$. The transformation of tensor components in this case is called a *proper orthogonal transformation*. For a reflection of axes in the x_2x_3 plane shown in Fig 2.3(b) the transformation matrix is

$$[a_{ij}] = \begin{bmatrix} -1 & 0 & 0 \\ 0 & 1 & 0 \\ 0 & 0 & 1 \end{bmatrix}$$

where det $\mathcal{A} = -1$, and we have an *improper orthogonal transformation*. It may be shown that *true (polar)* vectors transform by the rules $v_i' = a_{ij}v_j$ and $v_i = a_{ij}v_j'$ regardless of whether the axes transformation is proper or improper. However, *pseudo (axial)* vectors transform correctly only according to $v_i' = (\det \mathcal{A})\, a_{ij}v_j$ and $v_j = (\det \mathcal{A})\, a_{ij}v_i'$ under an improper transformation of axes.

Example 2.12

Let the primed axes $Ox_1'x_2'x_3'$ be given with respect to the unprimed axes by a 45°(counterclockwise) rotation about the x_2 axis as shown. Determine the primed components of the vector given by $\mathbf{v} = \hat{\mathbf{e}}_1 + \hat{\mathbf{e}}_2 + \hat{\mathbf{e}}_3$.

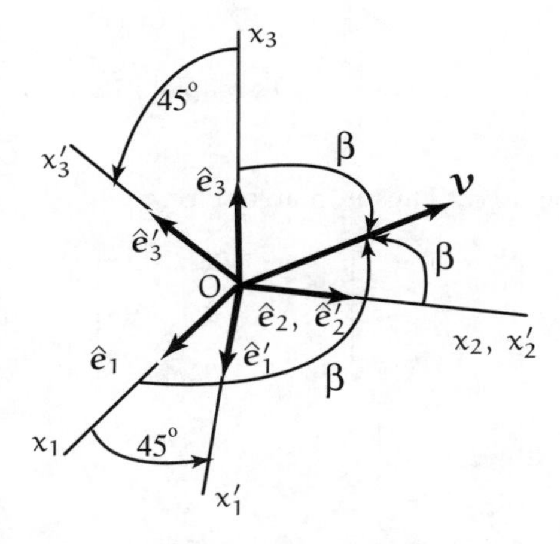

Vector $\mathbf{v}$ with respect to $Ox_1'x_2'x_3'$ and $Ox_1x_2x_3$.

Solution
Here the transformation matrix is

$$[a_{ij}] = \begin{bmatrix} \frac{1}{\sqrt{2}} & 0 & -\frac{1}{\sqrt{2}} \\ 0 & 1 & 0 \\ \frac{1}{\sqrt{2}} & 0 & \frac{1}{\sqrt{2}} \end{bmatrix},$$

and from Eq 2.60 in matrix form

$$\begin{bmatrix} v_1' \\ v_2' \\ v_3' \end{bmatrix} = \begin{bmatrix} \frac{1}{\sqrt{2}} & 0 & -\frac{1}{\sqrt{2}} \\ 0 & 1 & 0 \\ \frac{1}{\sqrt{2}} & 0 & \frac{1}{\sqrt{2}} \end{bmatrix} \begin{bmatrix} 1 \\ 1 \\ 1 \end{bmatrix} = \begin{bmatrix} 0 \\ 1 \\ \sqrt{2} \end{bmatrix}.$$

Example 2.13

Determine the primed components of the tensor

$$[t_{ij}] = \begin{bmatrix} 2 & 6 & 4 \\ 0 & 8 & 0 \\ 4 & 2 & 0 \end{bmatrix}$$

under the rotation of axes described in Example 2.12.

Solution
Here Eq 2.64 may be used. Thus in matrix form

$$\begin{aligned}
\left[t'_{ij}\right] &= \begin{bmatrix} \frac{1}{\sqrt{2}} & 0 & -\frac{1}{\sqrt{2}} \\ 0 & 1 & 0 \\ \frac{1}{\sqrt{2}} & 0 & \frac{1}{\sqrt{2}} \end{bmatrix} \begin{bmatrix} 2 & 6 & 4 \\ 0 & 8 & 0 \\ 4 & 2 & 0 \end{bmatrix} \begin{bmatrix} \frac{1}{\sqrt{2}} & 0 & \frac{1}{\sqrt{2}} \\ 0 & 1 & 0 \\ -\frac{1}{\sqrt{2}} & 0 & \frac{1}{\sqrt{2}} \end{bmatrix} \\
&= \begin{bmatrix} -3 & \frac{4}{\sqrt{2}} & 1 \\ 0 & 8 & 0 \\ 1 & \frac{8}{\sqrt{2}} & 5 \end{bmatrix} .
\end{aligned}$$

2.6 Principal Values and Principal Directions of Symmetric Second-Order Tensors

First, let us note that in view of the form of the inner product of a second-order tensor $\mathbf{T}$ with the arbitrary vector $\mathbf{u}$ (which we write here in both the indicial and symbolic notation),

$$t_{ij}u_j = v_i \qquad \text{or} \qquad \mathbf{T} \cdot \mathbf{u} = \mathbf{v} \tag{2.66}$$

any second-order tensor may be thought of as a *linear transformation* which transforms the *antecedent* vector $\mathbf{u}$ into the *image* vector $\mathbf{v}$ in a Euclidean three-space. In particular, for every symmetric tensor $\mathbf{T}$ having real components t_{ij}, and defined at some point in physical space, there is associated with each direction at that point (identified by the unit vector n_i), an image vector v_i given by

$$t_{ij}n_j = v_i \qquad \text{or} \qquad \mathbf{T} \cdot \hat{\mathbf{n}} = \mathbf{v} . \tag{2.67}$$

If the vector v_i determined by Eq 2.67 happens to be a scalar multiple of n_i, that is, if

$$t_{ij}n_j = \lambda n_i \qquad \text{or} \qquad \mathbf{T} \cdot \hat{\mathbf{n}} = \lambda \hat{\mathbf{n}} , \tag{2.68}$$

the direction defined by n_i is called a *principal direction*, or *eigenvector*, of $\mathbf{T}$, and the scalar λ is called a *principal value*, or *eigenvalue* of $\mathbf{T}$. Using the substitution property of the Kronecker delta, Eq 2.68 may be rewritten as

$$(t_{ij} - \lambda\delta_{ij})\, n_j = 0 \quad \text{or} \quad (\mathbf{T} - \lambda\mathbf{I}) \cdot \hat{\mathbf{n}} = 0 \,, \tag{2.69}$$

or in expanded form

$$(t_{11} - \lambda)\, n_1 + t_{12}n_2 + t_{13}n_3 = 0 \,, \tag{2.70a}$$

$$t_{21}n_1 + (t_{22} - \lambda)\, n_2 + t_{23}n_3 = 0 \,, \tag{2.70b}$$

$$t_{31}n_1 + t_{32}n_2 + (t_{33} - \lambda)\, n_3 = 0 \,. \tag{2.70c}$$

This system of homogeneous equations for the unknown direction n_i and the unknown λ's will have non-trivial solutions only if the determinant of coefficients vanishes. Thus,

$$|t_{ij} - \lambda\delta_{ij}| = 0 \tag{2.71}$$

which upon expansion leads to the cubic in λ (called the *characteristic equation*)

$$\lambda^3 - I_T\lambda^2 + II_T\lambda - III_T = 0 \tag{2.72}$$

where the coefficients here are expressed in terms of the known components of $\mathbf{T}$ by

$$I_T = t_{ii} = \text{tr } \mathbf{T} \,, \tag{2.73a}$$

$$II_T = \frac{1}{2}(t_{ii}t_{jj} - t_{ij}t_{ji}) = \frac{1}{2}\left[(\text{tr } \mathbf{T})^2 - \text{tr}\left(\mathbf{T}^2\right)\right] \,, \tag{2.73b}$$

$$III_T = \varepsilon_{ijk}t_{1i}t_{2j}t_{3k} = \det \mathbf{T} \,, \tag{2.73c}$$

and are known as the *first*, *second*, and *third invariants*, respectively, of the tensor $\mathbf{T}$. [1] The sum of the elements on the principal diagonal of the matrix form of any tensor is called the *trace* of that tensor, and for the tensor $\mathbf{T}$ is written $\text{tr}\,\mathbf{T}$ as in Eq 2.73.

The roots $\lambda_{(1)}$, $\lambda_{(2)}$, and $\lambda_{(3)}$ of Eq 2.72 are all real for a symmetric tensor $\mathbf{T}$ having real components. With each of these roots $\lambda_{(q)}$ $(q = 1, 2, 3)$ we can determine a principal direction $n_i^{(q)}$ $(q = 1, 2, 3)$ by solving Eq 2.69 together with the normalizing condition $n_i n_i = 1$. Thus, Eq 2.69 is satisfied by

$$\left[t_{ij} - \lambda_{(q)}\delta_{ij}\right] n_i^{(q)} = 0 \qquad (q = 1, 2, 3) \tag{2.74}$$

with

$$n_i^{(q)} n_i^{(q)} = 1 \qquad (q = 1, 2, 3). \tag{2.75}$$

If the $\lambda_{(q)}$'s are distinct the principal directions are unique and mutually perpendicular. If, however, there is a pair of equal roots, say $\lambda_{(1)} = \lambda_{(2)}$, then only the direction associated with $\lambda_{(3)}$ will be unique. In this case any other two directions which are orthogonal to $n_i^{(3)}$, and to one another so as to form a right-handed system, may be taken as principal directions. If $\lambda_{(1)} = \lambda_{(2)} = \lambda_{(3)}$, every set of right-handed orthogonal axes qualifies as principal axes, and every direction is said to be a principal direction.

In order to reinforce the concept of principal directions, let the components of the tensor $\mathbf{T}$ be given initially with respect to arbitrary Cartesian axes $Ox_1x_2x_3$, and let the principal axes of $\mathbf{T}$ be designated by $Ox_1^*x_2^*x_3^*$, as shown in Fig. 2.4. The transformation matrix $\mathcal{A}$ between these two sets of axes is established by taking the direction cosines $n_i^{(q)}$ as calculated from Eq 2.74 and Eq 2.75 as the elements of the qth row of $\mathcal{A}$. Therefore, by definition, $a_{ij} \equiv n_j^{(i)}$ as detailed in the table below.

[1]Note that some authors use the negative of the second invariant as defined in Eq 2.73b. This changes the sign on the $II_T\lambda$ term in Eq 2.72 and must be kept consistent throughout the use of the invariants.

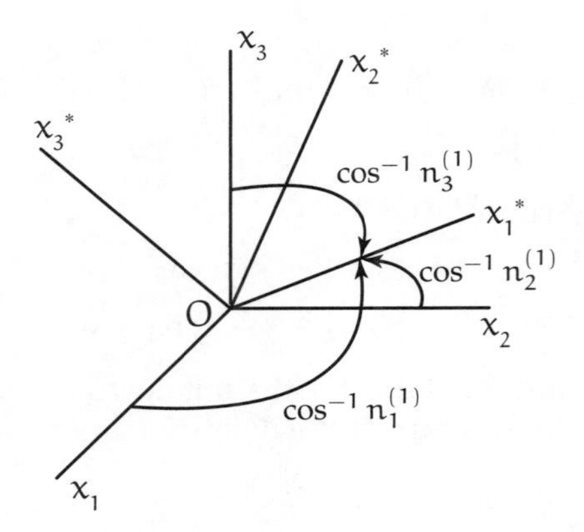

FIGURE 2.4
Principal axes $Ox_1^*x_2^*x_3^*$ relative to axes $Ox_1x_2x_3$.

	x_1 or $\hat{e}_1$	x_2 or $\hat{e}_2$	x_3 or $\hat{e}_3$
x_1^* or $\hat{e}_1^*$	$a_{11} = n_1^{(1)}$	$a_{12} = n_2^{(1)}$	$a_{13} = n_3^{(1)}$
x_2^* or $\hat{e}_2^*$	$a_{21} = n_1^{(2)}$	$a_{22} = n_2^{(2)}$	$a_{23} = n_3^{(2)}$
x_3^* or $\hat{e}_3^*$	$a_{31} = n_1^{(3)}$	$a_{32} = n_2^{(3)}$	$a_{33} = n_3^{(3)}$

(2.76)

The transformation matrix here is orthogonal and in accordance with the transformation law for second-order tensors

$$t_{ij}^* = a_{iq}a_{jm}t_{qm} \quad \text{or} \quad \mathbf{T}^* = \mathcal{A}\mathbf{T}\mathcal{A}^T \tag{2.77}$$

where $\mathbf{T}^*$ is a diagonal matrix whose elements are the principal values $\lambda_{(q)}$.

Example 2.14

Determine the principal values and principal directions of the second-order tensor $\mathbf{T}$ whose matrix representation is

$$[t_{ij}] = \begin{bmatrix} 5 & 2 & 0 \\ 2 & 2 & 0 \\ 0 & 0 & 3 \end{bmatrix} .$$

Solution
Here Eq 2.71 is given by

$$\begin{vmatrix} 5-\lambda & 2 & 0 \\ 2 & 2-\lambda & 0 \\ 0 & 0 & 3-\lambda \end{vmatrix} = 0$$

which upon expansion by the third row becomes

$$(3-\lambda)\left(10 - 7\lambda + \lambda^2 - 4\right) = 0 ,$$

or

$$(3-\lambda)(6-\lambda)(1-\lambda)=0\ .$$

Hence, $\lambda_{(1)}=3$, $\lambda_{(2)}=6$, $\lambda_{(3)}=1$ are the principal values of $\mathbf{T}$. For $\lambda_{(1)}=3$, Eq 2.70 yields the equations

$$\begin{aligned} 2n_1+2n_2 &= 0\ , \\ 2n_1-n_2 &= 0\ , \end{aligned}$$

which are satisfied only if $n_1=n_2=0$, and so from $n_in_i=1$ we have $n_3=\pm1$. For $\lambda_{(2)}=6$, Eq 2.70 yields

$$\begin{aligned} -n_1+2n_2 &= 0\ , \\ 2n_1-4n_2 &= 0\ , \\ -3n_3 &= 0\ , \end{aligned}$$

so that $n_1=2n_2$ and since $n_3=0$, we have $(2n_2)^2+n_2^2=1$, or $n_2=\pm1/\sqrt{5}$, and $n_1=\pm2/\sqrt{5}$. For $\lambda_{(3)}=1$, Eq 2.70 yields

$$\begin{aligned} 4n_1+2n_2 &= 0\ , \\ 2n_1+n_2 &= 0\ , \end{aligned}$$

together with $2n_3=0$. Again $n_3=0$, and here $n_1^2+(-2n_1)^2=1$ so that $n_1=\pm1/\sqrt{5}$ and $n_2=\mp2/\sqrt{5}$. From these results the transformation matrix $\mathcal{A}$ is given by

$$[a_{ij}]=\begin{bmatrix} 0 & 0 & \pm1 \\ \pm\frac{2}{\sqrt{5}} & \pm\frac{1}{\sqrt{5}} & 0 \\ \pm\frac{1}{\sqrt{5}} & \mp\frac{2}{\sqrt{5}} & 0 \end{bmatrix}$$

which identifies two sets of principal direction axes, one a reflection of the other with respect to the origin. Also, it may be easily verified that $\mathcal{A}$ is orthogonal by multiplying it with its transpose $\mathcal{A}^T$ to obtain the identity matrix. Finally, from Eq 2.77 we see that using the upper set of the $\pm$ signs,

$$\begin{bmatrix} 0 & 0 & 1 \\ \frac{2}{\sqrt{5}} & \frac{1}{\sqrt{5}} & 0 \\ \frac{1}{\sqrt{5}} & -\frac{2}{\sqrt{5}} & 0 \end{bmatrix}\begin{bmatrix} 5 & 2 & 0 \\ 2 & 2 & 0 \\ 0 & 0 & 3 \end{bmatrix}\begin{bmatrix} 0 & \frac{2}{\sqrt{5}} & \frac{1}{\sqrt{5}} \\ 0 & \frac{1}{\sqrt{5}} & -\frac{2}{\sqrt{5}} \\ 1 & 0 & 0 \end{bmatrix}=\begin{bmatrix} 3 & 0 & 0 \\ 0 & 6 & 0 \\ 0 & 0 & 1 \end{bmatrix}\ .$$

Example 2.15

Show that the principal values for the tensor having the matrix

$$[t_{ij}] = \begin{bmatrix} 5 & 1 & \sqrt{2} \\ 1 & 5 & \sqrt{2} \\ \sqrt{2} & \sqrt{2} & 6 \end{bmatrix}$$

have a multiplicity of two, and determine the principal directions.

Solution

Here Eq 2.71 is given by

$$\begin{vmatrix} 5-\lambda & 1 & \sqrt{2} \\ 1 & 5-\lambda & \sqrt{2} \\ \sqrt{2} & \sqrt{2} & 6-\lambda \end{vmatrix} = 0$$

for which the characteristic equation becomes

$$\lambda^3 - 16\lambda^2 + 80\lambda - 128 = 0 \ ,$$

or

$$(\lambda - 8)\,(\lambda - 4)^2 = 0 \ .$$

For $\lambda_{(1)} = 8$, Eq 2.70 yields

$$\begin{aligned} -3n_1 + n_2 + \sqrt{2}n_3 &= 0 \ , \\ n_1 - 3n_2 + \sqrt{2}n_3 &= 0 \ , \\ \sqrt{2}n_1 + \sqrt{2}n_2 - 2n_3 &= 0 \ . \end{aligned}$$

From the first two of these equations $n_1 = n_2$, and from the second and third equations $n_3 = \sqrt{2}n_2$. Therefore, using, $n_i n_i = 1$, we have

$$(n_2)^2 + (n_2)^2 + \left(\sqrt{2}n_2\right)^2 = 1 \ ,$$

and so $n_1 = n_2 = \pm 1/2$ and $n_3 = \pm 1/\sqrt{2}$, from which the unit vector in the principal direction associated with $\lambda_{(1)} = 8$ (the so-called normalized eigenvector) is

$$\hat{n}^{(1)} = \frac{1}{2}\left(\hat{e}_1 + \hat{e}_2 + \sqrt{2}\hat{e}_3\right) = \hat{e}_3^* \ .$$

For $\hat{n}^{(2)}$, we choose any unit vector perpendicular to $\hat{n}^{(1)}$; an obvious choice being

$$\hat{n}^{(2)} = \frac{-\hat{e}_1 + \hat{e}_2}{\sqrt{2}} = \hat{e}_2^* \ .$$

Then $\hat{n}^{(3)}$ is constructed from $\hat{n}^{(3)} = \hat{n}^{(1)} \times \hat{n}^{(2)}$, so that

$$\hat{n}^{(3)} = \frac{1}{2}\left(-\hat{e}_1 - \hat{e}_2 + \sqrt{2}\hat{e}_3\right) = \hat{e}_3^* \ .$$

Thus, the transformation matrix $\mathcal{A}$ is given by Eq 2.76 as

$$[a_{ij}] = \begin{bmatrix} \frac{1}{2} & \frac{1}{2} & \frac{1}{\sqrt{2}} \\ -\frac{1}{\sqrt{2}} & \frac{1}{\sqrt{2}} & 0 \\ -\frac{1}{\sqrt{2}} & -\frac{1}{2} & \frac{1}{\sqrt{2}} \end{bmatrix} .$$

It is worthwhile to revisit the invariants of Eq 2.73 within the context of the characteristic equation because of their extensive uses. The first invariant is just the trace of the tensor which is easy to calculate. Similarly, the third invariant may be calculated by a cofactor expansion to evaluate the determinant. Calculating the second invariant may be done two ways. First is the straight forward calculation. Half the trace squared minus the trace of the square may be done using matrix operations. For example, take the generic matrix

$$\mathcal{G} = \begin{bmatrix} a & b & c \\ d & e & f \\ g & h & i \end{bmatrix} . \tag{2.78}$$

Clearly, from Eq 2.73a, the first invariant is

$$I_{\mathcal{G}} = a + e + i . \tag{2.79}$$

Finding the second invariant requires two terms: $\frac{1}{2}I_{\mathcal{G}}^2$ and $-\frac{1}{2}\mathrm{tr}\,(\mathcal{G}^2)$. The first term is expanded as

$$\frac{1}{2}(a + e + i)^2 = \frac{1}{2}(a^2 + e^2 + i^2) + ae + ai + ei .$$

Matrix multiplication of the diagonal terms will evaluate the second term:

$$\frac{1}{2}\mathrm{tr}\left(\begin{bmatrix} a & b & c \\ d & e & f \\ g & h & i \end{bmatrix}\begin{bmatrix} a & b & c \\ d & e & f \\ g & h & i \end{bmatrix}\right) = \frac{1}{2}(\underbrace{a^2 + bd + gc}_{1,1 \text{ term}} + \underbrace{bd + e^2 + fh}_{2,2 \text{ term}} + \underbrace{gc + hf + i^2}_{3,3 \text{ term}}) .$$

Combining these two terms according to Eq 2.73b results in

$$II_{\mathcal{G}} = ei - hf + ai - gc + ae - db . \tag{2.80}$$

A second way to find the second invariant is to sum the cofactors along the diagonal of matrix $\mathcal{G}$. Using this method on Eq 2.78 results in Eq 2.80. This second way is an easier way to calculate the second invariant.

If a problem is given with specific numbers for the matrix, or tensor, there are computational tools that allow for quick invariant calculations. For instance, using MATLAB®, the matrix of Example 2.14 can be entered and the invariants found with the following code:

```
>> T = [5 2 0; 2 2 0; 0 0 3]

T =
     5     2     0
     2     2     0
     0     0     3

>> Tsqrd = T*T

Tsqrd =
    29    14     0
    14     8     0
     0     0     9

>> I_T = trace(T)

I_T = 10

>> II_T = 0.5*(I_T^2 - trace(Tsqrd))

II_T =  27

>> III_T = det(T)

III_T = 18
```

Furthermore, most of these computational tools can use symbolic notation as well as numbers. Consider the analysis done on the matrix defined in Eq 2.78. All of those calculations may be completed with MATLAB® as follows:

```
>> syms a b c d e f g h i
>> A = [a b c; d e f; g h i]

A =
[ a, b, c]
[ d, e, f]
[ g, h, i]

>> Asqrd = A*A

Asqrd =
[ a^2+b*d+c*g, a*b+b*e+c*h, a*c+b*f+c*i]
[ d*a+e*d+f*g, b*d+e^2+f*h, d*c+e*f+f*i]
[ g*a+h*d+i*g, g*b+h*e+i*h, c*g+f*h+i^2]

>> I_A = trace(A)
I_A = a+e+i

>> II_A = 0.5*(I_A^2 - trace(Asqrd))

II_A = 1/2*(a+e+i)^2-1/2*a^2-b*d-c*g-1/2*e^2-f*h-1/2*i^2
```

```
>> simplify(II_A)

ans = a*e+a*i+e*i-b*d-c*g-f*h

>> III_A = det(A)

III_A = a*e*i-a*f*h-d*b*i+d*c*h+g*b*f-g*c*e
```

In concluding this section, we mention several interesting properties of symmetric second-order tensors:

1. The principal values and principal directions of $\mathbf{T}$ and $\mathbf{T}^T$ are the same.
2. The principal values of $\mathbf{T}^{-1}$ are reciprocals of the principal values of $\mathbf{T}$, and both have the same principal directions.
3. The product tensors $\mathbf{TQ}$ and $\mathbf{QT}$ have the same principal values.
4. A symmetric tensor is said to be *positive* (*negative*) *definite* if all of its principal values are positive (negative); and *positive* (*negative*) *semi-definite* if one principal value is zero and the others positive (negative).

2.7 Tensor Fields, Tensor Calculus

A *tensor field* assigns to every location $\mathbf{x}$, at every instant of time t, a tensor $t_{ij\dots k}(\mathbf{x}, t)$, for which $\mathbf{x}$ ranges over a finite region of space, and t varies over some interval of time. The field is continuous and hence differentiable if the components $t_{ij\dots k}(\mathbf{x}, t)$ are continuous functions of $\mathbf{x}$ and t. Tensor fields may be of any order. For example, we may denote typical scalar, vector, and tensor fields by the notations $\phi(\mathbf{x}, t)$, $v_i(\mathbf{x}, t)$, and $t_{ij}(\mathbf{x}, t)$, respectively.

Partial differentiation of a tensor field with respect to the variable t is symbolized by the operator $\partial/\partial t$ and follows the usual rules of calculus. On the other hand, partial differentiation with respect to the coordinate x_q will be indicated by the operator $\partial/\partial x_q$, which may be abbreviated as simply ∂_q. Likewise, the second partial $\partial^2/\partial x_q \partial x_m$ may be written ∂_{qm}, and so on. As an additional measure in notational compactness it is customary in continuum mechanics to introduce the *subscript comma* to denote partial differentiation with respect to the coordinate variables. For example, we write $\phi_{,i}$ for $\partial\phi/\partial x_i$; $v_{i,j}$ for $\partial v_i/\partial x_j$; $t_{ij,k}$ for $\partial t_{ij}/\partial x_k$; and $u_{i,jk}$ for $\partial^2 u_i/\partial x_j \partial x_k$. We note from these examples that differentiation with respect to a coordinate produces a tensor of one order higher. Also, a useful identity results from the derivative $\partial x_i/\partial x_j$, viz.,

$$\frac{\partial x_i}{\partial x_j} = \delta_{ij} \; . \tag{2.81}$$

In the notation adopted here the operator $\boldsymbol{\nabla}$ (del) of vector calculus, which in symbolic notation appears as

$$\boldsymbol{\nabla} = \frac{\partial}{\partial x_1}\hat{\mathbf{e}}_1 + \frac{\partial}{\partial x_2}\hat{\mathbf{e}}_2 + \frac{\partial}{\partial x_3}\hat{\mathbf{e}}_3 = \frac{\partial}{\partial x_i}\hat{\mathbf{e}}_i \tag{2.82}$$

takes on the simple form ∂_i. Therefore, we may write the *scalar gradient* $\nabla\phi = \text{grad}\,\phi$ as

$$\partial_i \phi = \phi_{,i} \,, \tag{2.83}$$

the vector gradient $\nabla \mathbf{v}$ as

$$\partial_i v_j = v_{j,i} \,, \tag{2.84}$$

the divergence of $\mathbf{v}$, $\nabla \cdot \mathbf{v}$ as

$$\partial_i v_i = v_{i,i} \,, \tag{2.85}$$

and the curl of $\mathbf{v}$, $\nabla \times \mathbf{v}$ as

$$\varepsilon_{ijk} \partial_j v_k = \varepsilon_{ijk} v_{k,j} \,. \tag{2.86}$$

Note in passing that many of the identities of vector analysis can be verified with relative ease by manipulations using the indicial notation. For example, to show that $\text{div}(\text{curl}\,\mathbf{v}) = 0$ for any vector $\mathbf{v}$ we write from Eqs 2.86 and Eq 2.85

$$\partial_i \left(\varepsilon_{ijk} v_{k,j} \right) = \varepsilon_{ijk} v_{k,ji} = 0 \,,$$

and because the first term of this inner product is skew–symmetric in i and j, whereas the second term is symmetric in the same indices, (since v_k is assumed to have continuous spatial gradients), their product is zero.

Example 2.16

Use indicial notation to verify the following identities:

$$\text{curl}\,\nabla \mathbf{u} = 0 \tag{2.87a}$$

and

$$\text{curl}\left(\nabla \mathbf{u}^T\right) = \nabla\,\text{curl}\,\mathbf{u} \tag{2.87b}$$

where $\mathbf{u}$ is any continuously differentiable vector field.

Solution

For the first, write $\textbf{curl}\nabla\mathbf{u} = \varepsilon_{ipk} u_{j,kp} \hat{\mathbf{e}}_i \hat{\mathbf{e}}_j$ where the order of partial differentiation of $\mathbf{u}$ does not matter since $\mathbf{u}$ is continuously differentiable. Thus, $u_{j,kp} = u_{j,pk}$. By definition, the permutation symbol is skew–symmetric in kp and the product of a skew–symmetric and symmetric tensor is zero (see Example 2.5).

The second identity is just as easily verified with indicial notation. Write the left-hand side of the expression in indicial notation, exchange the order of partial differentiation on $\mathbf{u}$. Since the permutation symbol is a constant we may write $\varepsilon_{ipk} u_{k,jp} = \left(\varepsilon_{ipk} u_{k,p}\right)_{,j}$

There is also a need for using the divergence and curl of second order tensors in continuum mechanics. Let s_{jk} be a second order tensor that is differentiable at $\mathbf{x}$. The divergence of s_{jk} at $\mathbf{x}$ is defined as

$$\text{div}\,\mathbf{S} = s_{jk,k} \hat{\mathbf{e}}_j \tag{2.88}$$

where

$$s_{ij,k} = \frac{\partial s_{ij}}{\partial x_k} = \hat{\mathbf{e}}_i \cdot (\nabla \mathbf{S} \hat{\mathbf{e}}_k)\, \hat{\mathbf{e}}_j \ . \tag{2.89}$$

Just as the divergence of a vector reduces the tensor order by one to a scalar, the divergence of a second order tensor reduces the tensor order by one to a vector. The curl of a second order tensor s_{jk} is defined as

$$(\text{curl}\, \mathbf{S})_{ij} = \varepsilon_{ipk} s_{jk,p} \tag{2.90}$$

in a manner that parallels the curl of a vector. The curl $\mathbf{u}$ returns a vector and the curl $\mathbf{S}$ results in a second order tensor.

Example 2.17

Consider $s_{ij} = s_{ji}$ to be a continuously differentiable, second order tensor. Show that

$$\text{tr}\,(\text{curl}\, \mathbf{S}) = 0 \ . \tag{2.91}$$

Solution
Write the curl of $\mathbf{S}$ in indicial form:

$$(\text{curl}\, \mathbf{S})_{ij} = \varepsilon_{ipk} s_{jk,p} \ .$$

The trace of this is found by setting $i = j$:

$$(\text{tr}\,(\text{curl}\, \mathbf{S}))_{ii} = \varepsilon_{ipk} s_{ik,p} \ .$$

Since $s_{ik} = s_{ki}$ and $\varepsilon_{ipk} = -\varepsilon_{kpi}$ we see that $\text{tr}\,(\text{curl}\, \mathbf{S}) = 0$.

Example 2.18

Let $\mathbf{W} = w_{ij} \hat{\mathbf{e}}_i \hat{\mathbf{e}}_j = -w_{ji} \hat{\mathbf{e}}_i \hat{\mathbf{e}}_j$ be a skew–symmetric, second order tensor having continuous derivatives. Since $\mathbf{W}$ is skew–symmetric it has an axial vector $\mathbf{u}$ such that $w_{jk} = -\varepsilon_{qjk} u_q$. Show that

$$\text{curl}\, \mathbf{W} = \mathbf{I}\, \text{div}\, \mathbf{u} - \nabla \mathbf{u} \ , \tag{2.92a}$$

or, equivalently

$$\varepsilon_{ipk} w_{jk,p} = \delta_{ij} u_{q,q} - u_{i,j} \ . \tag{2.92b}$$

Solution
Using the definition of an axial vector as well as Eq 2.7a we find that:

$$\begin{aligned} \varepsilon_{ipk} w_{jk,p} &= \varepsilon_{ipk} \left(-\varepsilon_{qjk} u_q\right)_{,p} \\ &= -\left(\delta_{iq}\delta_{pj} - \delta_{ij}\delta_{pq}\right) u_{q,p} \\ &= \delta_{ij} u_{q,q} - u_{i,j} \ . \end{aligned}$$

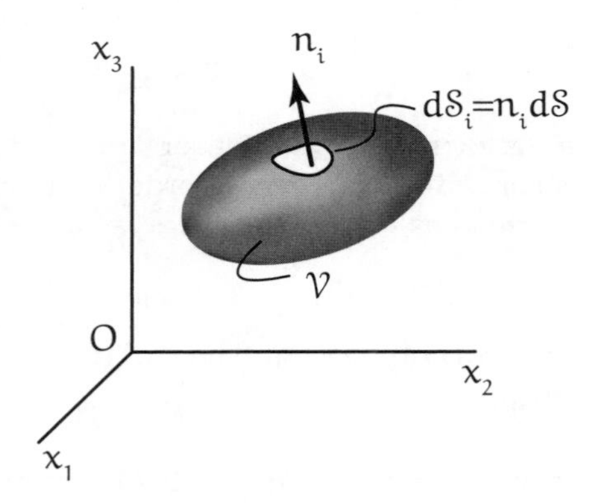

FIGURE 2.5
Volume $\mathcal{V}$ with infinitesimal element $d\mathcal{S}_i$ having a unit normal n_i.

2.8 Integral Theorems of Gauss and Stokes

Consider an arbitrary continuously differentiable tensor field $t_{ij\ldots k}$ defined on some finite region of physical space. Let $\mathcal{V}$ be a volume in this space with a closed surface $\mathcal{S}$ bounding the volume, and let the outward normal to this bounding surface be n_i as shown in Fig. 2.5 so that the element of surface is given by $d\mathcal{S}_i = n_i d\mathcal{S}$. The *divergence theorem of Gauss* establishes a relationship between the surface integral having $t_{ij\ldots k}$ as integrand to the volume integral for which a coordinate derivative of $t_{ij\ldots k}$ is the integrand. Specifically,

$$\int_{\mathcal{S}} t_{ij\ldots k} n_q \, d\mathcal{S} = \int_{\mathcal{V}} t_{ij\ldots k,q} \, d\mathcal{V} \, . \tag{2.93}$$

Several important special cases of this theorem for scalar and vector fields are worth noting, and are given here in both indicial and symbolic notation:

$$\int_{\mathcal{S}} \lambda n_q \, d\mathcal{S} = \int_{\mathcal{V}} \lambda_{,q} \, d\mathcal{V} \quad \text{or} \quad \int_{\mathcal{S}} \lambda \hat{\mathbf{n}} \, d\mathcal{S} = \int_{\mathcal{V}} \operatorname{grad} \lambda \, d\mathcal{V} \, , \tag{2.94}$$

$$\int_{\mathcal{S}} v_q n_q \, d\mathcal{S} = \int_{\mathcal{V}} v_{q,q} \, d\mathcal{V} \quad \text{or} \quad \int_{\mathcal{S}} \mathbf{v} \cdot \hat{\mathbf{n}} \, d\mathcal{S} = \int_{\mathcal{V}} \operatorname{div} \mathbf{v} \, d\mathcal{V} \, , \tag{2.95}$$

$$\int_{\mathcal{S}} \varepsilon_{ijk} n_j v_k \, d\mathcal{S} = \int_{\mathcal{V}} \varepsilon_{ijk} v_{k,j} \, d\mathcal{V} \quad \text{or} \quad \int_{\mathcal{S}} \hat{\mathbf{n}} \times \mathbf{v} \, d\mathcal{S} = \int_{\mathcal{V}} \operatorname{curl} \mathbf{v} \, d\mathcal{V} \, . \tag{2.96}$$

Called *Gauss's divergence theorem*, Eq 2.96 is the one presented in a traditional vector calculus course.

Whereas Gauss's theorem relates an integral over a closed volume to an integral over its bounding surface, *Stokes' theorem* relates an integral over an open surface (a so-called *cap*) to a line integral around the bounding curve of the surface. Therefore, let $\mathcal{C}$ be the

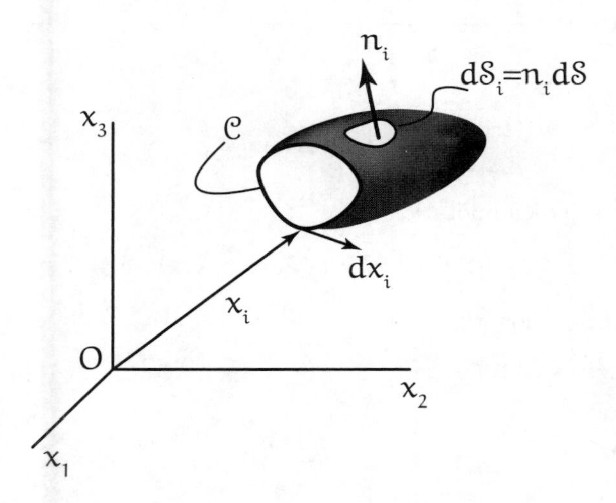

FIGURE 2.6
Bounding space curve $\mathcal{C}$ with tangential vector dx_i and surface element $d\mathcal{S}_i$ for partial volume.

bounding space curve to the surface $\mathcal{S}$, and let dx_i be the *differential tangent vector* to $\mathcal{C}$ as shown in Fig. 2.6. (A hemispherical surface having a circular bounding curve $\mathcal{C}$ is a classic example). If n_i is the outward normal to the surface $\mathcal{S}$, and v_i is any vector field defined on $\mathcal{S}$ and $\mathcal{C}$, Stokes' theorem asserts that

$$\int_{\mathcal{S}} \varepsilon_{ijk} n_i v_{k,j} \, d\mathcal{S} = \int_{\mathcal{C}} v_k \, dx_k \quad \text{or} \quad \int_{\mathcal{S}} \hat{\mathbf{n}} \cdot (\boldsymbol{\nabla} \times \mathbf{v}) \, d\mathcal{S} = \int_{\mathcal{C}} \mathbf{v} \cdot d\mathbf{x} \, . \tag{2.97}$$

The integral on the right-hand side of this equation is often referred to as the *circulation* when the vector $\mathbf{v}$ is the velocity vector.

Problems

Problem 2.1

Let $\mathbf{v} = \mathbf{a} \times \mathbf{b}$, or in indicial notation,

$$v_i \hat{\mathbf{e}}_i = a_j \hat{\mathbf{e}}_j \times b_k \hat{\mathbf{e}}_k = \varepsilon_{ijk} a_j b_k \hat{\mathbf{e}}_i$$

Using indicial notation, show that,

(a) $\mathbf{v} \cdot \mathbf{v} = a^2 b^2 \sin^2 \theta$,

(b) $\mathbf{a} \times \mathbf{b} \cdot \mathbf{a} = 0$,

(c) $\mathbf{a} \times \mathbf{b} \cdot \mathbf{b} = 0$.

Problem 2.2

With respect to the triad of base vectors $\mathbf{u}_1$, $\mathbf{u}_2$, and $\mathbf{u}_3$ (not necessarily unit vectors), the triad $\mathbf{u}^1$,$\mathbf{u}^2$, and $\mathbf{u}^3$ is said to be a *reciprocal basis* if $\mathbf{u}_i \cdot \mathbf{u}^j = \delta_{ij}$ $(i, j = 1, 2, 3)$. Show that to satisfy these conditions,

$$\mathbf{u}^1 = \frac{\mathbf{u}_2 \times \mathbf{u}_3}{[\mathbf{u}_1, \mathbf{u}_2, \mathbf{u}_3]}; \quad \mathbf{u}^2 = \frac{\mathbf{u}_3 \times \mathbf{u}_1}{[\mathbf{u}_1, \mathbf{u}_2, \mathbf{u}_3]}; \quad \mathbf{u}^3 = \frac{\mathbf{u}_1 \times \mathbf{u}_2}{[\mathbf{u}_1, \mathbf{u}_2, \mathbf{u}_3]}$$

and determine the reciprocal basis for the specific base vectors

$$\begin{aligned} \mathbf{u}_1 &= 2\hat{\mathbf{e}}_1 + \hat{\mathbf{e}}_2 , \\ \mathbf{u}_2 &= 2\hat{\mathbf{e}}_2 - \hat{\mathbf{e}}_3 , \\ \mathbf{u}_3 &= \hat{\mathbf{e}}_1 + \hat{\mathbf{e}}_2 + \hat{\mathbf{e}}_3 . \end{aligned}$$

Answer

$$\begin{aligned} \mathbf{u}^1 &= \tfrac{1}{5}(3\hat{\mathbf{e}}_1 - \hat{\mathbf{e}}_2 - 2\hat{\mathbf{e}}_3) \\ \mathbf{u}^2 &= \tfrac{1}{5}(-\hat{\mathbf{e}}_1 + 2\hat{\mathbf{e}}_2 - \hat{\mathbf{e}}_3) \\ \mathbf{u}^3 &= \tfrac{1}{5}(-\hat{\mathbf{e}}_1 + 2\hat{\mathbf{e}}_2 + 4\hat{\mathbf{e}}_3) \end{aligned}$$

Problem 2.3

Let the position vector of an arbitrary point $P(x_1 x_2 x_3)$ be $\mathbf{x} = x_i \hat{\mathbf{e}}_i$, and let $\mathbf{b} = b_i \hat{\mathbf{e}}_i$ be a *constant vector*. Show that $(\mathbf{x} - \mathbf{b}) \cdot \mathbf{x} = 0$ is the vector equation of a spherical surface having its center at $\mathbf{x} = \frac{1}{2}\mathbf{b}$ with a radius of $\frac{1}{2}b$.

Problem 2.4

Using the notations $A_{(ij)} = \frac{1}{2}(A_{ij} + A_{ji})$ and $A_{[ij]} = \frac{1}{2}(A_{ij} - A_{ji})$ show that

(a) the tensor $\mathbf{A}$ having components A_{ij} can always be decomposed into a sum of its symmetric $A_{(ij)}$ and skew-symmetric $A_{[ij]}$ parts, respectively, by the decomposition,

$$A_{ij} = A_{(ij)} + A_{[ij]} ,$$

(b) the trace of $\mathbf{A}$ is expressed in terms of $A_{(ij)}$ by

$$A_{ii} = A_{(ii)} \ ,$$

(c) for arbitrary tensors $\mathbf{A}$ and $\mathbf{B}$,

$$A_{ij}B_{ij} = A_{(ij)}B_{(ij)} + A_{[ij]}B_{[ij]} \ .$$

Problem 2.5

Expand the following expressions involving Kronecker deltas, and simplify where possible.

$$\text{(a) } \delta_{ij}\delta_{ij}, \quad \text{(b) } \delta_{ij}\delta_{jk}\delta_{ki}, \quad \text{(c) } \delta_{ij}\delta_{jk}, \quad \text{(d) } \delta_{ij}A_{ik}$$

Answer

(a) 3, (b) 3, (c) δ_{ik}, (d) A_{jk}

Problem 2.6

If $a_i = \varepsilon_{ijk}b_jc_k$ and $b_i = \varepsilon_{ijk}g_jh_k$, substitute b_j into the expression for a_i to show that

$$a_i = g_kc_kh_i - h_kc_kg_i \ ,$$

or in symbolic notation, $\mathbf{a} = (\mathbf{c} \cdot \mathbf{g})\mathbf{h} - (\mathbf{c} \cdot \mathbf{h})\mathbf{g}$.

Problem 2.7

By summing on the repeated subscripts determine the simplest form of

$$\text{(a) } \varepsilon_{3jk}a_ja_k, \quad \text{(b) } \varepsilon_{ijk}\delta_{kj}, \quad \text{(c) } \varepsilon_{1jk}a_2T_{kj}, \quad \text{(d) } \varepsilon_{1jk}\delta_{3j}v_k \ .$$

Answer

(a) 0, (b) 0, (c) $a_2(T_{32} - T_{23})$, (d) $-v_2$

Problem 2.8

Consider the tensor $B_{ik} = \varepsilon_{ijk}v_j$.

(a) Show that B_{ik} is skew-symmetric.

(b) Let B_{ij} be skew-symmetric, and consider the vector defined by $v_i = \varepsilon_{ijk}B_{jk}$ (often called the *dual vector* of the tensor $\mathbf{B}$). Show that $B_{mq} = \frac{1}{2}\varepsilon_{mqi}v_i$.

Problem 2.9

Use indicial notation to show that

$$A_{mi}\varepsilon_{mjk} + A_{mj}\varepsilon_{imk} + A_{mk}\varepsilon_{ijm} = A_{mm}\varepsilon_{ijk}$$

where $\mathbf{A}$ is any tensor and ε_{ijk} is the permutation symbol.

Problem 2.10

If $A_{ij} = \delta_{ij} B_{kk} + 3B_{ij}$, determine B_{kk} and using that solve for B_{ij} in terms of A_{ij} and its first invariant, A_{ii}.

Answer

$$B_{kk} = \tfrac{1}{6} A_{kk}; \quad B_{ij} = \tfrac{1}{3} A_{ij} - \tfrac{1}{18} \delta_{ij} A_{kk}$$

Problem 2.11

Show that the value of the quadratic form $T_{ij} x_i x_j$ is unchanged if T_{ij} is replaced by its symmetric part, $\frac{1}{2}(T_{ij} + T_{ji})$.

Problem 2.12

With the aid of Eq 2.7, show that any skew–symmetric tensor $\boldsymbol{W}$ may be written in terms of an *axial vector* ω_i given by

$$\omega_i = -\frac{1}{2} \varepsilon_{ijk} w_{jk}$$

where w_{jk} are the components of $\boldsymbol{W}$.

Problem 2.13

Show by direct expansion (or otherwise) that the box product $\lambda = \varepsilon_{ijk} a_i b_j c_k$ is equal to the determinant

$$\begin{vmatrix} a_1 & a_2 & a_3 \\ b_1 & b_2 & b_3 \\ c_1 & c_2 & c_3 \end{vmatrix} .$$

Thus, by substituting $\mathcal{A}_{1i}$ for a_i, $\mathcal{A}_{2j}$ for b_j and $\mathcal{A}_{3k}$ for c_k, derive Eq 2.42 in the form $\det \boldsymbol{\mathcal{A}} = \varepsilon_{ijk} \mathcal{A}_{1i} \mathcal{A}_{2j} \mathcal{A}_{3k}$ where $\mathcal{A}_{ij}$ are the elements of $\boldsymbol{\mathcal{A}}$.

Problem 2.14

Starting with Eq 2.42 of the text in the form

$$\det \boldsymbol{\mathcal{A}} = \varepsilon_{ijk} \mathcal{A}_{i1} \mathcal{A}_{j2} \mathcal{A}_{k3}$$

show that by an arbitrary number of interchanges of columns of $\mathcal{A}_{ij}$ we obtain

$$\varepsilon_{qmn} \det \boldsymbol{\mathcal{A}} = \varepsilon_{ijk} \mathcal{A}_{iq} \mathcal{A}_{jm} \mathcal{A}_{kn}$$

which is Eq 2.43. Further, multiply this equation by the appropriate permutation symbol to derive the formula

$$6 \det \boldsymbol{\mathcal{A}} = \varepsilon_{qmn} \varepsilon_{ijk} \mathcal{A}_{iq} \mathcal{A}_{jm} \mathcal{A}_{kn} .$$

Problem 2.15

Let the determinant of the matrix $\mathcal{A}_{ij}$ be given by

$$\det \boldsymbol{\mathcal{A}} = \begin{vmatrix} \mathcal{A}_{11} & \mathcal{A}_{12} & \mathcal{A}_{13} \\ \mathcal{A}_{21} & \mathcal{A}_{22} & \mathcal{A}_{23} \\ \mathcal{A}_{31} & \mathcal{A}_{32} & \mathcal{A}_{33} \end{vmatrix} .$$

Since the interchange of any two rows or any two columns causes a sign change in the value of the determinant, show that after an arbitrary number of row and column interchanges

$$\begin{vmatrix} \mathcal{A}_{mq} & \mathcal{A}_{mr} & \mathcal{A}_{ms} \\ \mathcal{A}_{nq} & \mathcal{A}_{nr} & \mathcal{A}_{ns} \\ \mathcal{A}_{pq} & \mathcal{A}_{pr} & \mathcal{A}_{ps} \end{vmatrix} = \varepsilon_{mnp}\varepsilon_{qrs} \det \mathcal{A} \,.$$

Now let $\mathcal{A}_{ij} = \delta_{ij}$ in the above determinant which results in $\det \mathcal{A} = 1$ and, upon expansion, yields

$$\varepsilon_{mnp}\varepsilon_{qrs} = \delta_{mq}(\delta_{nr}\delta_{ps} - \delta_{ns}\delta_{pr}) - \delta_{mr}(\delta_{nq}\delta_{ps} - \delta_{ns}\delta_{pq}) + \delta_{ms}(\delta_{nq}\delta_{pr} - \delta_{nr}\delta_{pq}) \,.$$

Thus, by setting $p = q$, establish Eq 2.7 in the form

$$\varepsilon_{mnq}\varepsilon_{qrs} = \delta_{mr}\delta_{ns} - \delta_{ms}\delta_{nr} \,.$$

Problem 2.16

Show that the square matrices

$$[\mathcal{B}_{ij}] = \begin{bmatrix} 1 & 0 & 0 \\ 0 & -1 & 0 \\ 0 & 0 & 1 \end{bmatrix} \quad \text{and} \quad [\mathcal{C}_{ij}] = \begin{bmatrix} 5 & 2 \\ -12 & -5 \end{bmatrix}$$

are both square roots of the identity matrix.

Problem 2.17

Using the square matrices below, demonstrate

(a) that the transpose of the square of a matrix is equal to the square of its transpose (Eq 2.36 with $n = 2$),

(b) that $(\mathcal{A}\mathcal{B})^T = \mathcal{B}^T\mathcal{A}^T$ as was proven in Example 2.8

$$[\mathcal{A}_{ij}] = \begin{bmatrix} 3 & 0 & 1 \\ 0 & 2 & 4 \\ 5 & 1 & 2 \end{bmatrix}, \quad [\mathcal{B}_{ij}] = \begin{bmatrix} 1 & 3 & 1 \\ 2 & 2 & 5 \\ 4 & 0 & 3 \end{bmatrix}.$$

Problem 2.18

Let $\mathcal{A}$ be any orthogonal matrix, i.e., $\mathcal{A}\mathcal{A}^T = \mathcal{A}\mathcal{A}^{-1} = \mathbf{I}$, where $\mathbf{I}$ is the identity matrix. Thus, by using the results in Examples 2.9 and 2.10, show that $\det \mathcal{A} = \pm 1$.

Problem 2.19

A tensor is called *isotropic* if its components have the same set of values in every Cartesian coordinate system at a point. Assume that $\mathbf{T}$ is an isotropic tensor of rank two with components t_{ij} relative to axes $Ox_1x_2x_3$. Let axes $Ox'_1x'_2x'_3$ be obtained with respect to $Ox_1x_2x_3$ by a righthand rotation of 120° about the axis along $\hat{\mathbf{n}} = (\hat{\mathbf{e}} + \hat{\mathbf{e}} + \hat{\mathbf{e}})/\sqrt{3}$. Show by the transformation between these axes that $t_{11} = t_{22} = t_{33}$, as well as other relationships. Further, let axes $Ox''_1x''_2x''_3$ be obtained with respect to $Ox_1x_2x_3$ by a right-hand rotation of 90° about x_3. Thus, show by the additional considerations of this transformation that if $\mathbf{T}$ is any isotropic tensor of second order, it can be written as $\lambda\mathbf{I}$ where λ is a scalar and $\mathbf{I}$ is the identity tensor.

Problem 2.20

For a proper orthogonal transformation between axes $Ox_1x_2x_3$ and $Ox_1'x_2'x_3'$ show the invariance of δ_{ij} and ε_{ijk}. That is, show that

(a) $\delta'_{ij} = \delta_{ij}$,

(b) $\varepsilon'_{ijk} = \varepsilon_{ijk}$.

Hint: For part (b) let $\varepsilon'_{ijk} = a_{iq}a_{jm}a_{kn}\varepsilon_{qmn}$ and make use of Eq 2.43.

Problem 2.21

The angles between the respective axes of the $Ox_1'x_2'x_3'$ and the $Ox_1x_2x_3$ Cartesian systems are given by the table below

	x_1	x_2	x_3
x_1'	45°	90°	45°
x_2'	60°	45°	120°
x_3'	120°	45°	60°

Determine

(a) the transformation matrix between the two sets of axes, and show that it is a proper orthogonal transformation,

(b) the equation of the plane $x_1 + x_2 + x_3 = 1/\sqrt{2}$ in its primed axes form, that is, in the form $b_1x_1' + b_2x_2' + b_3x_3' = b$.

Answer

(a) $$[a_{ij}] = \begin{bmatrix} \frac{1}{\sqrt{2}} & 0 & \frac{1}{\sqrt{2}} \\ \frac{1}{\sqrt{2}} & \frac{1}{\sqrt{2}} & -\frac{1}{\sqrt{2}} \\ -\frac{1}{\sqrt{2}} & \frac{1}{\sqrt{2}} & \frac{1}{\sqrt{2}} \end{bmatrix}$$

(b) $2x_1' + x_2' + x_3' = 1$

Problem 2.22

Making use of Eq 2.42 of the text in the form $\det \mathcal{A} = \varepsilon_{ijk}\mathcal{A}_{1i}\mathcal{A}_{2j}\mathcal{A}_{3k}$ write Eq 2.71 as

$$|t_{ij} - \lambda\delta_{ij}| = \varepsilon_{ijk}(t_{1i} - \delta_{1i})(t_{2j} - \delta_{2j})(t_{3k} - \delta_{3j}) = 0$$

and show by expansion of this equation that

$$\lambda^3 - t_{ii}\lambda^2 + \left[\frac{1}{2}(t_{ii}t_{jj} - t_{ij}t_{ji})\right]\lambda - \varepsilon_{ijk}t_{1i}t_{2j}t_{3k} = 0$$

to verify Eq 2.72 of the text.

Problem 2.23

For the matrix representation of tensor **B** shown below,

$$[b_{ij}] = \begin{bmatrix} 17 & 0 & 0 \\ 0 & -23 & 28 \\ 0 & 28 & 10 \end{bmatrix}$$

determine the principal values (eigenvalues) and the principal directions (eigenvectors) of the tensor.

Answer

$$\lambda_1 = 17, \quad \lambda_2 = 26, \quad \lambda_3 = -39$$
$$\hat{n}^{(1)} = \hat{e}_1, \quad \hat{n}^{(2)} = (4\hat{e}_2 + 7\hat{e}_3)/\sqrt{65}, \quad \hat{n}^{(3)} = (-7\hat{e}_2 + 4\hat{e}_3)/\sqrt{65}$$

Problem 2.24

Consider the symmetrical matrix

$$[\mathcal{B}_{ij}] = \begin{bmatrix} \frac{5}{4} & 0 & \frac{3}{2} \\ 0 & 4 & 0 \\ \frac{3}{2} & 0 & \frac{5}{2} \end{bmatrix}.$$

(a) Show that a multiplicity of two occurs among the principal values of this matrix.

(b) Let λ_1 be the unique principal value and show that the transformation matrix

$$[a_{ij}] = \begin{bmatrix} \frac{1}{\sqrt{2}} & 0 & -\frac{1}{\sqrt{2}} \\ 0 & 1 & 0 \\ \frac{1}{\sqrt{2}} & 0 & \frac{1}{\sqrt{2}} \end{bmatrix}$$

gives $\mathcal{B}^*$ according to $\mathcal{B}^*_{ij} = a_{iq} a_{jm} \mathcal{B}_{qm}$.

(c) Taking the square root of $\left[\mathcal{B}^*_{ij}\right]$ and transforming back to $Ox_1x_2x_3$ axes show that

$$\left[\sqrt{\mathcal{B}_{ij}}\right] = \begin{bmatrix} \frac{3}{2} & 0 & \frac{1}{2} \\ 0 & 2 & 0 \\ \frac{1}{2} & 0 & \frac{3}{2} \end{bmatrix}.$$

(d) Verify that the matrix

$$[\mathcal{C}_{ij}] = \begin{bmatrix} -\frac{1}{2} & 0 & -\frac{3}{2} \\ 0 & 2 & 0 \\ -\frac{3}{2} & 0 & -\frac{1}{2} \end{bmatrix}$$

is also a square root of $[\mathcal{B}_{ij}]$.

Problem 2.25
Determine the principal values of the matrix

$$[\mathcal{K}_{ij}] = \begin{bmatrix} 4 & 0 & 0 \\ 0 & 11 & -\sqrt{3} \\ 0 & -\sqrt{3} & 9 \end{bmatrix},$$

and show that the principal axes $Ox_1^*x_2^*x_3^*$ are obtained from $Ox_1x_2x_3$ by a rotation of 60° about the x_1 axis.

Answer

$$\lambda_1 = 4, \quad \lambda_2 = 8, \quad \lambda_3 = 12.$$

Problem 2.26
Determine the principal values $\lambda_{(q)}$ $(q = 1, 2, 3)$ and principal directions $\hat{\mathbf{n}}^{(q)}$ $(q = 1, 2, 3)$ for the symmetric matrix

$$[\mathcal{T}_{ij}] = \frac{1}{2}\begin{bmatrix} 3 & -\frac{1}{\sqrt{2}} & \frac{1}{\sqrt{2}} \\ -\frac{1}{\sqrt{2}} & \frac{9}{2} & \frac{3}{2} \\ \frac{1}{\sqrt{2}} & \frac{3}{2} & \frac{9}{2} \end{bmatrix}$$

Answer

$$\begin{aligned} \lambda_{(1)} &= 1, \quad \lambda_{(2)} = 2 \quad \lambda_{(3)} = 3 \\ \hat{\mathbf{n}}^{(1)} &= \tfrac{1}{2}\left(\sqrt{2}\hat{\mathbf{e}}_1 + \hat{\mathbf{e}}_2 - \hat{\mathbf{e}}_3\right) \\ \hat{\mathbf{n}}^{(2)} &= \tfrac{1}{2}\left(\sqrt{2}\hat{\mathbf{e}}_1 - \hat{\mathbf{e}}_2 + \hat{\mathbf{e}}_3\right) \\ \hat{\mathbf{n}}^{(3)} &= -\left(\hat{\mathbf{e}}_2 + \hat{\mathbf{e}}_3\right)/\sqrt{2} \end{aligned}$$

Problem 2.27
For the second-order tensor $C_{ij} = u_i v_j$:

(a) Calculate the principal invariants I_C, II_C, and III_C . Reduce your answer to its simplest form, but leave it in index notation. State the reasons for simplification.

(b) If $\mathbf{u} = \begin{bmatrix} 1 & 2 & 3 \end{bmatrix}$ and $\mathbf{v} = \begin{bmatrix} 5 & 10 & 4 \end{bmatrix}$ write out the matrix form for C_{ij}.

(c) Use the numbers of (b) to validate the answer of (a).

Problem 2.28
Let $\mathbf{D}$ be a constant tensor whose components do not depend upon the coordinates. Show that

$$\boldsymbol{\nabla}(\mathbf{x} \cdot \mathbf{D}) = \mathbf{D}$$

where $\mathbf{x} = x_i \hat{\mathbf{e}}_i$ is the position vector.

Problem 2.29

Consider the vector $\mathbf{x} = x_i \hat{\mathbf{e}}_i$ having a magnitude squared $x^2 = x_1^2 + x_2^2 + x_3^3$. Determine

(a) grad x ,

(b) grad (x^{-n}) ,

(c) $\nabla^2\ (1/x)$,

(d) div $(x^n \mathbf{x})$,

(e) curl $(x^n \mathbf{x})$, where n is a positive integer.

Answer

(a) x_i/x, (b) $-nx_i/x^{(n+2)}$, (c) 0, (d) $x^n(n+3)$, (e) 0.

Problem 2.30

If λ and φ are scalar functions of the coordinates x_i, verify the following vector identities. Transcribe the left-hand side of the equations into indicial notation and, following the indicated operations, show that the result is the right-hand side.

(a) $\mathbf{v} \times (\nabla \times \mathbf{v}) = \frac{1}{2}\nabla(\mathbf{v} \cdot \mathbf{v}) - (\mathbf{v} \cdot \nabla)\mathbf{v}$

(b) $\mathbf{v} \cdot \mathbf{u} \times \mathbf{w} = \mathbf{v} \times \mathbf{u} \cdot \mathbf{w}$

(c) $\nabla \times (\nabla \times \mathbf{v}) = \nabla(\nabla \cdot \mathbf{v}) - \nabla^2 \mathbf{v}$

(d) $\nabla \cdot (\lambda \nabla \varphi) = \lambda \nabla^2 \varphi + \nabla \lambda \cdot \nabla \varphi$

(e) $\nabla^2(\lambda\varphi) = \lambda\nabla^2\varphi + 2(\nabla\lambda) \cdot (\nabla\varphi) + \varphi\nabla^2\lambda$

(f) $\nabla \cdot (\mathbf{u} \times \mathbf{v}) = (\nabla \times \mathbf{u}) \cdot \mathbf{v} - \mathbf{u} \cdot (\nabla \times \mathbf{v})$

Problem 2.31

Let the vector $\mathbf{v} = \mathbf{b} \times \mathbf{x}$ be one for which $\mathbf{b}$ does not depend upon the coordinates. Use indicial notation to show that

(a) curl $\mathbf{v} = 2\mathbf{b}$,

(b) div $\mathbf{v} = 0$.

Problem 2.32

Transcribe the left-hand side of the following equations into indicial notation and verify that the indicated operations result in the expressions on the right-hand side of the equations for the scalar φ, and vectors $\mathbf{u}$ and $\mathbf{v}$.

(a) div $(\varphi\mathbf{v}) = \varphi$ div $\mathbf{v} + \mathbf{v} \cdot$ grad φ

(b) $\mathbf{u} \times$ curl $\mathbf{v} + \mathbf{v} \times$ curl $\mathbf{u} = -(\mathbf{u} \cdot \text{grad})\mathbf{v} - (\mathbf{v} \cdot \text{grad})\mathbf{u} + \text{grad}(\mathbf{u} \cdot \mathbf{v})$

(c) div $(\mathbf{u} \times \mathbf{v}) = \mathbf{v} \cdot$ curl $\mathbf{u} - \mathbf{u} \cdot$ curl $\mathbf{v}$

(d) $\text{curl}(\mathbf{u} \times \mathbf{v}) = (\mathbf{v} \cdot \text{grad})\mathbf{u} - (\mathbf{u} \cdot \text{grad})\mathbf{v} + \mathbf{u}\,\text{div}\,\mathbf{v} - \mathbf{v}\,\text{div}\,\mathbf{u}$

(e) $\text{curl}(\text{curl}\,\mathbf{u}) = \text{grad}(\text{div}\,\mathbf{u}) - \nabla^2\mathbf{u}$

Problem 2.33

Let the volume $\mathcal{V}$ have a bounding surface $\mathcal{S}$ with an outward unit normal n_i. Let x_i be the position vector to any point in the volume or on its surface. Show that

$$\text{(a)} \int_{\mathcal{S}} x_i n_j d\mathcal{S} = \delta_{ij} \mathcal{V} \,,$$

$$\text{(b)} \int_{\mathcal{S}} \boldsymbol{\nabla}(\mathbf{x} \cdot \mathbf{x}) \cdot \hat{\mathbf{n}} d\mathcal{S} = 6\mathcal{V} \,,$$

$$\text{(c)} \int_{\mathcal{S}} \lambda \mathbf{w} \cdot \hat{\mathbf{n}} d\mathcal{S} = \int_{\mathcal{V}} \mathbf{w} \cdot \textbf{grad } \lambda d\mathcal{V}, \text{ where } \mathbf{w} = \textbf{curl } \mathbf{v} \text{ and } \lambda = \lambda(\mathbf{x}) \,,$$

$$\text{(d)} \int_{\mathcal{S}} [\hat{\mathbf{e}}_i \times \mathbf{x}, \hat{\mathbf{e}}_j, \hat{\mathbf{n}}] \, d\mathcal{S} = 2\mathcal{V}\delta_{ij} \text{ where } \hat{\mathbf{e}}_i \text{ and } \hat{\mathbf{e}}_j \text{ are coordinate base vectors.}$$

Hint: Write the box product

$$[\hat{\mathbf{e}}_i \times \mathbf{x}, \hat{\mathbf{e}}_j, \hat{\mathbf{n}}] = (\hat{\mathbf{e}}_i \times \mathbf{x}) \cdot (\hat{\mathbf{e}}_j \times \hat{\mathbf{n}})$$

and transcribe into indicial notation.

Problem 2.34

Use Stokes' theorem to show that upon integrating around the space curve $\mathcal{C}$ having a differential tangential vector dx_i that for $\varphi(\mathbf{x})$.

$$\oint_{\mathcal{C}} \varphi_{,i} dx_i = 0$$

Problem 2.35

For the position vector x_i having a magnitude x, show that $x_{,j} = x_j/x$ and therefore,

$$\text{(a)} \; x_{,ij} = \frac{\delta_{ij}}{x} - \frac{x_i x_j}{x^3} \,,$$

$$\text{(b)} \; \left(x^{-1}\right)_{,ij} = \frac{3x_i x_j}{x^5} - \frac{\delta_{ij}}{x^3} \,,$$

$$\text{(c)} \; x_{,ii} = \frac{2}{x} \,.$$

Problem 2.36

Show that for arbitrary tensors $\mathbf{A}$ and $\mathbf{B}$, and arbitrary vectors $\mathbf{a}$ and $\mathbf{b}$,

$$\text{(a)} \; (\mathbf{A} \cdot \mathbf{a}) \cdot (\mathbf{B} \cdot \mathbf{b}) = \mathbf{a} \cdot (\mathbf{A}^T \cdot \mathbf{B}) \cdot \mathbf{b} \,,$$

$$\text{(b)} \; \mathbf{b} \times \mathbf{a} = \tfrac{1}{2}(\mathbf{B} - \mathbf{B}^T) \cdot \mathbf{a}, \text{ if } 2b_i = \varepsilon_{ijk} B_{kj} \,,$$

$$\text{(c)} \; \mathbf{a} \cdot \mathbf{A} \cdot \mathbf{b} = \mathbf{b} \cdot \mathbf{A}^T \cdot \mathbf{a} \,.$$

Problem 2.37

Use Eqs 2.42 and 2.43 as necessary to prove the identities

$$\text{(a)} \; [\mathcal{A}\mathbf{a}, \mathcal{A}\mathbf{b}, \mathcal{A}\mathbf{c}] = (\det \mathcal{A})[\mathbf{a}, \mathbf{b}, \mathbf{c}] \,,$$

(b) $\mathcal{A}^{\mathsf{T}} \cdot (\mathcal{A}\mathbf{a} \times \mathcal{A}\mathbf{b}) = (\det\mathcal{A})(\mathbf{a} \times \mathbf{b})$,

for arbitrary vectors $\mathbf{a}$, $\mathbf{b}$, $\mathbf{c}$, and matrix $\mathcal{A}$.

Problem 2.38

Let $\varphi = \varphi(x_i)$ and $\psi = \psi(x_i)$ be scalar functions of the coordinates. Recall that in the indicial notation $\varphi_{,i}$ represents $\boldsymbol{\nabla}\varphi$ and $\varphi_{,ii}$ represents $\boldsymbol{\nabla}^2\varphi$. Now apply the divergence theorem, Eq 2.93, to the field $\varphi\psi_{,i}$ to obtain

$$\int_{\mathcal{S}} \varphi\psi_{,i}n_i \, d\mathcal{S} = \int_{\mathcal{V}} (\varphi_{,i}\psi_{,i} + \varphi\psi_{,ii}) \, d\mathcal{V} \, .$$

Transcribe this result into symbolic notation as

$$\int_{\mathcal{S}} \varphi\boldsymbol{\nabla}\psi \cdot \hat{\mathbf{n}} \, d\mathcal{S} = \int_{\mathcal{S}} \varphi\frac{\partial\psi}{\partial n} \, d\mathcal{S} = \int_{\mathcal{V}} (\boldsymbol{\nabla}\varphi \cdot \boldsymbol{\nabla}\psi + \varphi\boldsymbol{\nabla}^2\psi) \, d\mathcal{V}$$

which is known as *Green's first identity*. Show also by the divergence theorem that

$$\int_{\mathcal{S}} (\varphi\psi_{,i} - \psi\varphi_{,i})\, n_i \, d\mathcal{S} = \int_{\mathcal{V}} (\varphi\psi_{,ii} - \psi\varphi_{,ii}) \, d\mathcal{V} \, ,$$

and transcribe into symbolic notation as

$$\int_{\mathcal{S}} \left(\varphi\frac{\partial\psi}{\partial n} - \psi\frac{\partial\varphi}{\partial n}\right) d\mathcal{S} = \int_{\mathcal{V}} (\varphi\boldsymbol{\nabla}^2\psi - \psi\boldsymbol{\nabla}^2\varphi) \, d\mathcal{V}$$

which is known as *Green's second identity*.

3

Stress Principles

Stress is a measure of *force intensity*, either within, or on the bounding surface of a body subjected to loads. It should be noted that in continuum mechanics a body is considered stress free if the only forces present are those inter-atomic forces required to hold the body together. And so it follows that the stresses which concern us here are those that result from the application of forces by an external agent.

3.1 Body and Surface Forces, Mass Density

Two basic types of forces are easily distinguished from one another and are defined as follows. First, those forces acting on all volume elements, and are distributed throughout the body, are known as *body forces*. Gravity and inertia forces are the best known examples. We designate body forces by the vector symbol b_i (force per unit mass), or by the symbol p_i (force per unit volume). Secondly, those forces which act upon, and are distributed in some fashion over a surface element of the body, regardless of whether that element is part of the bounding surface, or an arbitrary element of surface within the body, are called *surface forces*. These are denoted by the vector symbol f_i, and have dimensions of force per unit area. Forces which occur on the outer surfaces of two bodies pressed against one another (contact forces), or those which result from the transmission of forces across an internal surface are examples of this type of force.

Next, let us consider a material body $\mathcal{B}$ having a volume $\mathcal{V}$ enclosed by a surface $\mathcal{S}$, and occupying a regular region R_0 of physical space. Let **P** be an interior point of the body located in the small element of volume $\Delta\mathcal{V}$ whose mass is Δm as indicated in Fig. 3.1. Recall that mass is that property of a material body by virtue of which the body possesses inertia, that is, the opposition which the body offers to any change in its motion. We define the average density of this volume element by the ratio

$$\rho_{ave} = \frac{\Delta m}{\Delta \mathcal{V}}, \tag{3.1}$$

and the density ρ at point **P** by the limit of this ratio as the volume shrinks to the point **P**,

$$\rho = \lim_{\Delta\mathcal{V}\to 0} \frac{\Delta m}{\Delta \mathcal{V}} = \frac{dm}{d\mathcal{V}}. \tag{3.2}$$

The units of density are kilograms per cubic meter, (kg/m^3). Notice that the two measures of body forces, b_i having units of Newton per kilogram (N/kg), and p_i having units of Newtons per meter cubed (N/m^3), are related through the density by the equation

$$\rho b_i = p_i \qquad \text{or} \qquad \rho \mathbf{b} = \mathbf{p}. \tag{3.3}$$

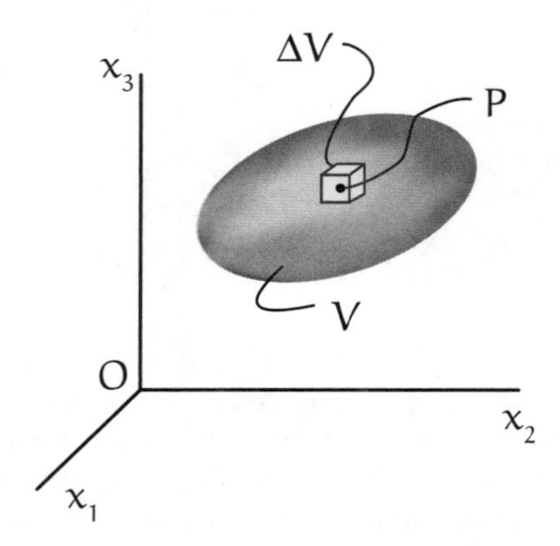

FIGURE 3.1
Typical continuum volume $\mathcal{V}$ with infinitesimal element $\Delta\mathcal{V}$ having mass Δm at point **P**. Point **P** would be in the center of the infinitesimal volume.

Of course, the density is, in general, a scalar function of position and time as indicated by

$$\rho = \rho(x_i, t) \qquad \text{or} \qquad \rho = \rho(\mathbf{x}, t) \tag{3.4}$$

and thus may vary from point to point within a given body.

3.2 Cauchy Stress Principle

Consider a homogeneous, isotropic material body $\mathcal{B}$ having a bounding surface $\mathcal{S}$, and a volume $\mathcal{V}$, which is subjected to arbitrary surface forces f_i and body forces b_i. Let **P** be an interior point of $\mathcal{B}$ and imagine a plane surface $\mathcal{S}^*$ passing through point **P** (sometimes referred to as a *cutting plane*) so as to partition the body into two portions, designated I and II (Fig. 3.2(a)). Point **P** is in the small element of area $\Delta\mathcal{S}^*$ of the cutting plane, which is defined by the unit normal pointing in the direction from portion I into portion II as shown by the free body diagram of portion I in 3.2(b). The internal forces being transmitted across the cutting plane due to the action of portion II upon portion I will give rise to a force distribution on $\mathcal{S}^*$ equivalent to a resultant force Δf_i and a resultant moment ΔM_i at **P**, as is also shown in Fig. 3.2(b). (For simplicity body forces b_i and surface forces f_i acting on the body as a whole are not drawn in Figs. 3.2(a) and 3.2(b).) Notice that Δf_i and ΔM_i are not necessarily in the direction of the unit normal vector n_i at **P**. The *Cauchy stress principle* asserts that in the limit as the area $\mathcal{S}^*$ shrinks to zero with **P** remaining an interior point, we obtain

$$\lim_{\Delta\mathcal{S}^* \to 0} \frac{\Delta f_i}{\Delta\mathcal{S}^*} = \frac{df_i}{d\mathcal{S}^*} = t_i^{(\hat{\mathbf{n}})} \tag{3.5}$$

and

$$\lim_{\Delta\mathcal{S}^* \to 0} \frac{\Delta M_i}{\Delta\mathcal{S}^*} = 0\,. \tag{3.6}$$

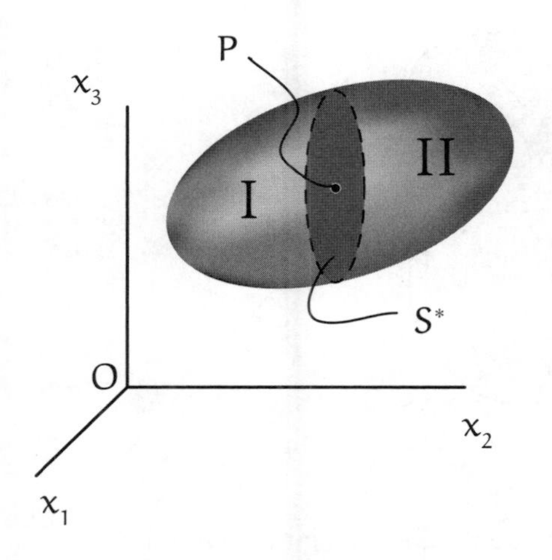

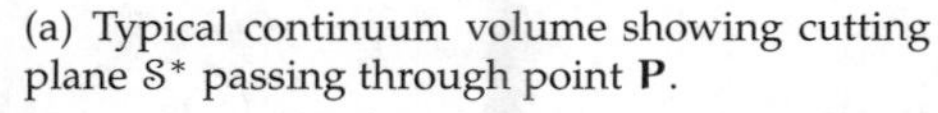
(a) Typical continuum volume showing cutting plane $\mathcal{S}^*$ passing through point **P**.

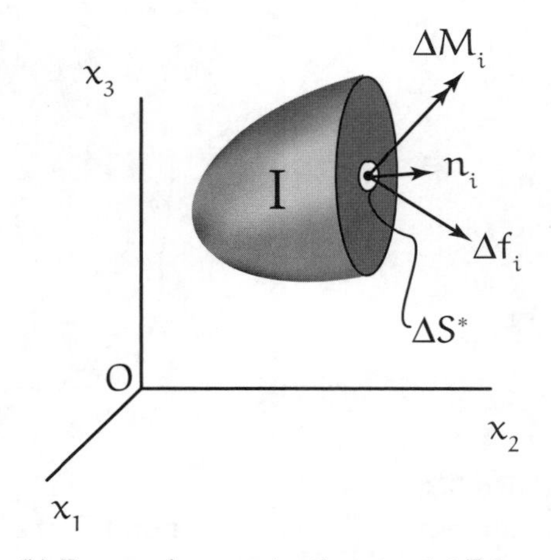

(b) Force and moment acting at point **P** in surface element $\Delta\mathcal{S}^*$.

FIGURE 3.2
Typical continuum volume with cutting plane.

The vector $df_i/d\mathcal{S}^* = t_i^{(\hat{n})}$ is called the *stress vector*, or sometimes the *traction vector*. In Eq 3.6 we have made the assumption that in the limit at **P**, the moment vector vanishes, and there is no remaining *concentrated moment*, or *couple stress* as it is called.

The appearance of $\hat{n}$ in the symbol $t_i^{(\hat{n})}$ for the stress vector serves to remind us that this is a special vector in that it is meaningful only in conjunction with its associated normal vector $\hat{n}$ at **P**. Thus, for the infinity of cutting planes imaginable through point **P**, each identified by a specific $\hat{n}$, there is also an infinity of associated stress vectors $t_i^{(\hat{n})}$ for a given loading of the body. The totality of pairs of the companion vectors $t_i^{(\hat{n})}$ and $\hat{n}$ at **P**, as illustrated by a typical pair in Fig. 3.3, defines the *state of stress* at that point.

By applying Newton's third law of action and reaction across the cutting plane we observe that the force exerted by portion I upon portion II is equal and opposite to the force of portion II upon portion I. Additionally, from the principle of linear momentum (Newton's second law) we know that the time rate of change of the linear momentum of *any* portion of a continuum body is equal to the resultant force acting upon that portion. For portions I and II, this principle may be expressed in integral form by the respective equations (these equations are derived in Section 5.3 from the principle of linear momentum),

$$\int_{\mathcal{S}_I} t_i^{(\hat{n})} d\mathcal{S} + \int_{\mathcal{V}_I} \rho b_i d\mathcal{V} = \frac{d}{dt}\int_{\mathcal{V}_I} \rho v_i d\mathcal{V}\,, \tag{3.7a}$$

$$\int_{\mathcal{S}_{II}} t_i^{(\hat{n})} d\mathcal{S} + \int_{\mathcal{V}_{II}} \rho b_i d\mathcal{V} = \frac{d}{dt}\int_{\mathcal{V}_{II}} \rho v_i d\mathcal{V} \tag{3.7b}$$

where $\mathcal{S}_I$ and $\mathcal{S}_{II}$ are the bounding surfaces and $\mathcal{V}_I$ and $\mathcal{V}_{II}$ are the volumes of portions I and II, respectively. Also, b_i are the body forces, ρ is the density and v_i is the velocity field for the two portions. We note that $\mathcal{S}_I$ and $\mathcal{S}_{II}$ each contain $\mathcal{S}^*$ as part of their total areas.

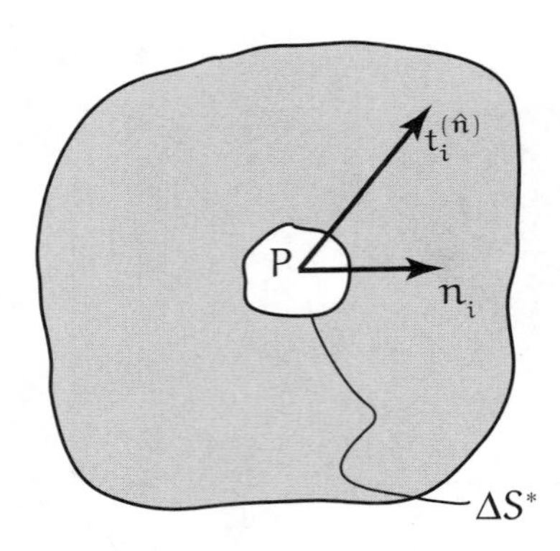

FIGURE 3.3
Traction vector $t_i^{(\hat{n})}$ acting at point **P** of plane element ΔS_i whose normal is n_i.

The linear momentum principle may also be applied to the body $\mathcal{B}$ as a whole, so that

$$\int_{S} t_i^{(\hat{n})} dS + \int_{V} \rho b_i dV = \frac{d}{dt} \int_{V} \rho v_i dV . \tag{3.8}$$

If we add Eq 3.7a and Eq 3.7b and use Eq 3.8, noting that the normal to S^* for portion I is $\hat{n}$, whereas for portion II it is $-\hat{n}$, we arrive at the equation

$$\int_{S^*} \left[t_i^{(\hat{n})} + t_i^{(-\hat{n})} \right] dS = 0 \tag{3.9}$$

since both S_I and S_{II} contain a surface integral over S^*. This equation must hold for arbitrary partitioning of the body (that is, for every imaginable cutting plane through point **P**) which means that the integrand must be identically zero. Hence,

$$t_i^{(\hat{n})} = -t_i^{(-\hat{n})} \tag{3.10}$$

indicating that if portion II had been chosen as the free body in Fig. 3.2(b) instead of portion I, the resulting stress vector would have been $-t_i^{(\hat{n})}$.

3.3 The Stress Tensor

As noted in Section 3.2, the Cauchy stress principle associates with each direction $\hat{n}$ at point **P** a stress vector $t_i^{(\hat{n})}$. In particular, if we introduce a rectangular Cartesian reference frame at **P**, there is associated with each of the area elements $dS_i (i = 1, 2, 3)$ located in the coordinate planes and having unit normals $\hat{e}_i (i = 1, 2, 3)$, respectively, a stress vector $t_i^{(\hat{n})}$ as shown in Fig. 3.4. In terms of their coordinate components these three stress vectors associated with the coordinate planes are expressed by

$$\mathbf{t}^{(\hat{e}_1)} = t_1^{(\hat{e}_1)} \hat{e}_1 + t_2^{(\hat{e}_1)} \hat{e}_2 + t_3^{(\hat{e}_1)} \hat{e}_3 , \tag{3.11a}$$

$$\mathbf{t}^{(\hat{e}_2)} = t_1^{(\hat{e}_2)} \hat{e}_1 + t_2^{(\hat{e}_2)} \hat{e}_2 + t_3^{(\hat{e}_2)} \hat{e}_3 , \tag{3.11b}$$

$$\mathbf{t}^{(\hat{e}_3)} = t_1^{(\hat{e}_3)} \hat{e}_1 + t_2^{(\hat{e}_3)} \hat{e}_2 + t_3^{(\hat{e}_3)} \hat{e}_3 , \tag{3.11c}$$

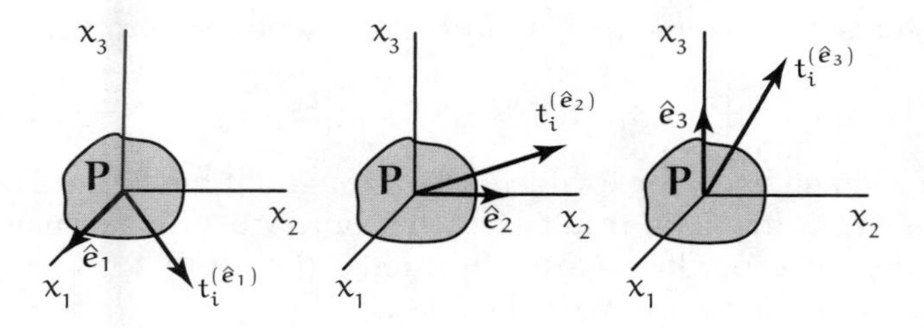

FIGURE 3.4
Traction vectors on the three coordinate planes at point **P**.

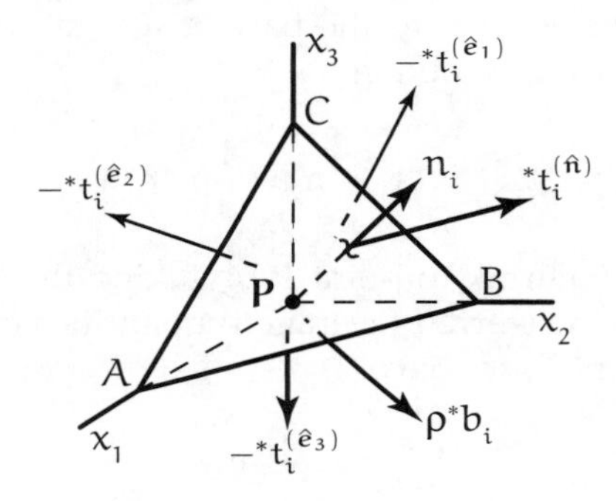

FIGURE 3.5
Free body diagram of tetrahedron element having its vertex at point **P**.

or more compactly, using the summation convention

$$\mathbf{t}^{(\hat{\mathbf{e}}_i)} = t_j^{(\hat{\mathbf{e}}_i)} \hat{\mathbf{e}}_j \ . \tag{3.12}$$

This equation expresses the stress vector at **P** for a given coordinate plane in terms of its rectangular Cartesian components, but what is really needed is an expression for the coordinate components of the stress vector at **P** associated with an arbitrarily oriented plane. For this purpose we consider the equilibrium of a small portion of the body in the shape of a tetrahedron having its vertex at **P**, and its base ABC perpendicular to an arbitrarily oriented normal $\hat{\mathbf{n}} = n_i \hat{\mathbf{e}}_i$ as shown by Fig. 3.5. The coordinate directions are chosen so that the three faces BPC, CPA and APB of the tetrahedron are situated in the coordinate planes. If the area of the base is assigned the value $d\mathcal{S}$, the areas of the respective faces will be the projected areas $d\mathcal{S}_i = d\mathcal{S}\ \cos(\hat{\mathbf{n}}, \hat{\mathbf{e}}_i)$, $(i = 1, 2, 3)$ or specifically,

$$\text{for BPC} \quad d\mathcal{S}_1 = n_1 d\mathcal{S} \ , \tag{3.13a}$$

$$\text{for CPA} \quad d\mathcal{S}_2 = n_2 d\mathcal{S} \ , \tag{3.13b}$$

$$\text{for APB} \quad d\mathcal{S}_3 = n_3 d\mathcal{S} \ . \tag{3.13c}$$

The stress vectors shown on the surfaces of the tetrahedron of Fig. 3.5 represent *average values* over the areas on which they act. This is indicated by a superscript asterisk in front of the stress vector symbols (remember that the stress vector is a point quantity).

Equilibrium requires the vector sum of all forces acting on the tetrahedron to be zero, that is, for,

$$ {}^*t_i^{(\hat{n})} dS - {}^*t_i^{(\hat{e}_1)} dS_1 - {}^*t_i^{(\hat{e}_2)} dS_2 - {}^*t_i^{(\hat{e}_3)} dS_3 + \rho\, {}^*b_i dV = 0 \tag{3.14} $$

where ${}^*b_i$ is an average body force which acts throughout the body. The negative signs on the coordinate-face tractions result from the outward unit normals on those faces pointing in the negative coordinate axes directions. (Recall that $t_i^{(\hat{n})} = -t_i^{(-\hat{n})}$.) Taking into consideration Eq 3.13, we can write Eq 3.14 as

$$ {}^*t_i^{(\hat{n})} dS - {}^*t_i^{(\hat{e}_j)} n_j dS + \rho\, {}^*b_i dV = 0 \,, \tag{3.15} $$

if we permit the indices on the unit vectors of the ${}^*t_i^{(\hat{e}_j)}$ term to participate in the summation process. The volume of the tetrahedron is given by $dV = \frac{1}{3} h dS$, where h is the perpendicular distance from point **P** to the base ABC. Inserting this into Eq 3.15 and canceling the common factor dS we obtain

$$ {}^*t_i^{(\hat{n})} = {}^*t_i^{(\hat{e}_j)} n_j - \frac{1}{3} \rho\, {}^*b_i h \,. \tag{3.16} $$

Now, letting the tetrahedron shrink to point **P** by taking the limit as $h \to 0$ and noting that in this limiting process the starred (averaged) quantities take on the actual values of those same quantities at point **P**, we have

$$ t_i^{(\hat{n})} = t_i^{(\hat{e}_j)} n_j \,, \tag{3.17} $$

or, by defining $t_{ji} \equiv t_i^{(\hat{e}_j)}$,

$$ t_i^{(\hat{n})} = t_{ji} n_j \qquad \text{or} \qquad \mathbf{t}^{(\hat{n})} = \hat{\mathbf{n}} \cdot \mathbf{T} \tag{3.18} $$

which is the *Cauchy stress formula*. We can obtain this same result for bodies which are accelerating by using the conservation of linear momentum instead of a balance of forces on the tetrahedron of Fig. 3.5.

Stress as a Tensor

The quantities $t_{ji} \equiv t_i^{(\hat{e}_j)}$ are the components of a second order tensor **T** known as the stress tensor. This can be shown by considering the transformation of the components of the stress vector $t_i^{(\hat{n})}$ between coordinate systems $Px_1x_2x_3$ and $Px_1'x_2'x_3'$ as given by the transformation matrix having elements (see Section 2.5)

$$ a_{ij} = \hat{e}_i' \cdot \hat{e}_j \,. \tag{3.19} $$

Since $\mathbf{t}^{(\hat{n})}$ can be expressed in terms of its components in either coordinate system,

$$ \mathbf{t}^{(n)} = t_i^{(\hat{n})} \hat{e}_i = t'^{(\hat{n}')}_i \hat{e}_i' \tag{3.20a} $$

or, from Eq 3.18,

$$ \mathbf{t}^{(\hat{n})} = t_{ji} n_j \hat{e}_i = t_{ji}' n_j' \hat{e}_i' \,. \tag{3.20b} $$

But from Eq 2.53, $\hat{e}_i' = a_{ij}\hat{e}_j$ and from Eq 2.60, $n_j' = a_{js} n_s$, so that now Eq 3.20b becomes, after some manipulations of the summed indices,

$$ \left(t_{sr} - a_{js} a_{ir} t_{ji}'\right) n_s \hat{e}_r = 0 \,. \tag{3.21} $$

Because the vectors $\hat{\mathbf{e}}_r$ are linearly independent and since Eq 3.21 must be valid for all vectors n_s, we see that

$$t_{sr} = a_{js} a_{ir} t'_{ji} \ . \tag{3.22}$$

But this is the transformation equation for a second-order tensor, and thus by Eq 2.63 the tensor character of the stress components is clearly established.

The Cauchy stress formula given by Eq 3.18 expresses the stress vector associated with the element of area having an outward normal n_i at point **P** in terms of the stress tensor components t_{ji} at that point. And although the state of stress at **P** has been described as the totality of pairs of the associated normal and traction vectors at that point, we see from the analysis of the tetrahedron element that if we know the stress vectors on the three coordinate planes of any Cartesian system at **P**, or equivalently, the nine stress tensor components t_{ji} at that point, we can determine the stress vector for any plane at that point. For computational purposes it is often convenient to express Eq 3.18 in the matrix form

$$\left[\mathbf{t}_1^{(\hat{\mathbf{n}})} \ \mathbf{t}_2^{(\hat{\mathbf{n}})} \ \mathbf{t}_3^{(\hat{\mathbf{n}})} \right] = [n_1 \ n_2 \ n_3] \begin{bmatrix} t_{11} & t_{12} & t_{13} \\ t_{21} & t_{22} & t_{23} \\ t_{31} & t_{32} & t_{33} \end{bmatrix} . \tag{3.23}$$

The nine components of t_{ji} are often displayed by arrows on the coordinate faces of a rectangular parallelpiped, as shown in Fig. 3.6. We emphasize that this parallelpiped is not a block of material from the continuum body (note that no dimensions are given to the parallelpiped), but is simply a convenient schematic device for displaying the stress tensor components. In an actual physical body $\mathcal{B}$, all nine stress components act at the *single point* **P**. The three stress components shown by arrows acting perpendicular (normal) to the respective coordinate planes and labeled t_{11}, t_{22}, and t_{33} are called *normal stresses*. The six arrows lying in the coordinate planes and pointing in the directions of the coordinate axes, namely, t_{12}, t_{21}, t_{23}, t_{32}, t_{13}, and t_{31} are called *shear stresses*. Note that, for these, the first subscript designates the coordinate plane on which the shear stress acts, and the second subscript identifies the coordinate direction in which it acts. A stress component is positive when its vector arrow points in the positive direction of one of the coordinate axes while acting on a plane whose outward normal also points in a positive coordinate direction. All of the stress components displayed in Fig. 3.6 are positive. In general, positive normal stresses are called *tensile stresses*, and negative normal stresses are referred to as *compressive stresses*. The units of stress are Newtons per square meter (N/m^2) in the SI system, and pounds per square inch (psi) in the English system. One Newton per square meter is called a *Pascal*, but because this is a rather small stress from an engineering point of view, stresses are usually expressed as mega-pascals (MPa) or in English units as kilo-pounds per square inch (ksi).

Example 3.1

Let the components of the stress tensor at **P** be given in matrix form by

$$[t_{ij}] = \begin{bmatrix} 21 & -63 & 42 \\ -63 & 0 & 84 \\ 42 & 84 & -21 \end{bmatrix}$$

in units of mega-Pascals. Determine

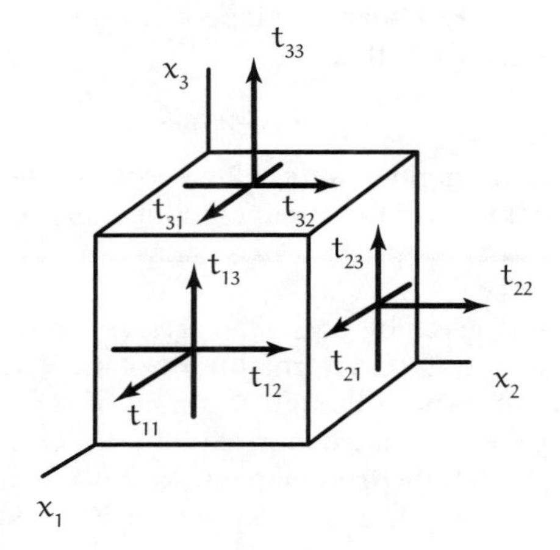

FIGURE 3.6
Cartesian stress components shown in their positive sense.

(a) the stress vector on the plane at **P** having the unit normal

$$\hat{\mathbf{n}} = \frac{1}{7}(2\hat{\mathbf{e}}_1 - 3\hat{\mathbf{e}}_2 + 6\hat{\mathbf{e}}_3) \ ,$$

(b) the stress vector on a plane at **P** parallel to the plane ABC shown in the sketch.

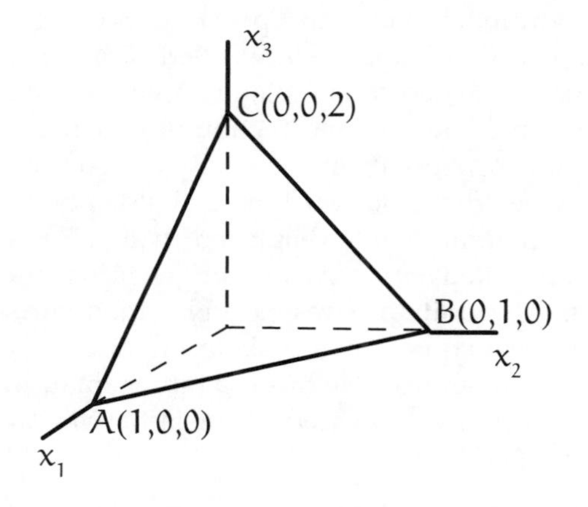

Solution

(a) From Eq 3.23 for the given data,

$$\begin{bmatrix} t_1^{(\hat{\mathbf{n}})} & t_2^{(\hat{\mathbf{n}})} & t_3^{(\hat{\mathbf{n}})} \end{bmatrix} = \begin{bmatrix} \frac{2}{7} & -\frac{3}{7} & \frac{6}{7} \end{bmatrix} \begin{bmatrix} 21 & -63 & 42 \\ -63 & 0 & 84 \\ 42 & 84 & -21 \end{bmatrix} = [69 \quad 54 \quad -42] \ ,$$

or, in vector form, $\mathbf{t}^{(\hat{n})} = 69\hat{\mathbf{e}}_1 + 54\hat{\mathbf{e}}_2 - 42\hat{\mathbf{e}}_3$. This vector represents the components of the force per unit area (traction) on the plane defined by $\frac{1}{7}[2, \quad -3, \quad 6]$.

(b) The equation of the plane ABC in the sketch is easily verified to be $2x_1 + 2x_2 + x_3 = 2$, and the unit outward normal to this plane is $\hat{\mathbf{n}} = \frac{1}{3}(2\hat{\mathbf{e}}_1 + 2\hat{\mathbf{e}}_2 + \hat{\mathbf{e}}_3)$ so that, again from Eq 3.23,

$$\begin{bmatrix} t_1^{(\hat{n})} & t_2^{(\hat{n})} & t_3^{(\hat{n})} \end{bmatrix} = \begin{bmatrix} \frac{2}{3} & \frac{2}{3} & \frac{1}{3} \end{bmatrix} \begin{bmatrix} 21 & -63 & 42 \\ -63 & 0 & 84 \\ 42 & 84 & -21 \end{bmatrix}$$

$$= [-14 \quad -14 \quad 77]$$

or, in vector form, $\mathbf{t}^{(\hat{n})} = -14\hat{\mathbf{e}}_1 - 14\hat{\mathbf{e}}_2 + 77\hat{\mathbf{e}}_3$.

In this example, we clearly see the dependency of the cutting plane and the stress vector. Here, we have considered two different cutting planes at the same point and found that two distinct traction vectors arose from the given stress tensor components.

3.4 Force and Moment Equilibrium; Stress Tensor Symmetry

In the previous section, we used a balance-of-forces condition for a tetrahedron element of a body in equilibrium to define the stress tensor and to develop the Cauchy stress formula. Here, we employ a force balance on the body as a whole to derive what are known as the *local equilibrium equations*. This set of three differential equations must hold for every point in a continuum body that is in equilibrium. As is well known, equilibrium also requires the sum of moments to be zero with respect to any fixed point, and we use this condition, together with the local equilibrium equations, to deduce the fact that the stress tensor is symmetric in the absence of concentrated body moments.

Consider a material body having a volume $\mathcal{V}$ and a bounding surface $\mathcal{S}$. Let the body be subjected to surface tractions $t_i^{(\hat{n})}$ and body forces b_i (force per unit mass), as shown by Fig. 3.7. As before, we exclude concentrated body moments from consideration. Equilibrium requires that the summation of all forces acting on the body be equal to zero. This condition is expressed by the global (integral) equation representing the sum of the total surface and body forces acting on the body,

$$\int_{\mathcal{S}} t_i^{(\hat{n})} d\mathcal{S} + \int_{\mathcal{V}} \rho b_i d\mathcal{V} = 0 \tag{3.24}$$

where $d\mathcal{S}$ is the differential element of the surface $\mathcal{S}$ and $d\mathcal{V}$ that of volume $\mathcal{V}$.

Because $t_i^{(\hat{n})} = t_{ji}n_j$ as a result of Eq 3.18, the divergence theorem Eq 2.93 allows the first term of Eq 3.24 to be written as

$$\int_{\mathcal{S}} t_{ji}n_j d\mathcal{S} = \int_{\mathcal{V}} t_{ji,j} d\mathcal{V} \,, \tag{3.25}$$

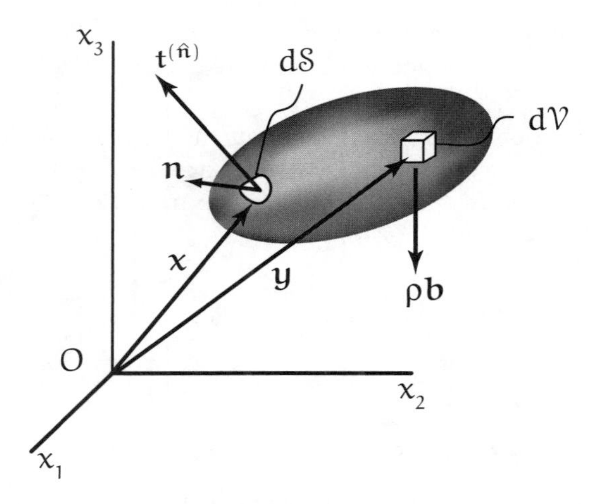

FIGURE 3.7
Material volume showing surface traction vector $t_i^{(\hat{n})}$ on an infinitesimal area element $d\mathcal{S}$ at position x_i, and body force vector b_i acting on an infinitesimal volume element $d\mathcal{V}$ at position y_i. Two positions are taken separately for ease of illustration. When applying equilibrium the traction and body forces are taken at the same point.

so that Eq 3.24 becomes

$$\int_{\mathcal{V}} (t_{ji,j} + \rho b_i)\, d\mathcal{V} = 0 \,. \tag{3.26}$$

This equation must be valid for an arbitrary volume $\mathcal{V}$ (every portion of the body is in equilibrium), which requires the integrand itself to vanish, and we obtain the so-called *local equilibrium equations*

$$t_{ji,j} + \rho b_i = 0 \,. \tag{3.27}$$

In addition to the balance of forces expressed by Eq 3.24, equilibrium requires that the summation of moments with respect to an arbitrary point must also be zero. Recall that the moment of a force about a point is defined by the cross product of its position vector with the force. Therefore, taking the origin of coordinates as the center for moments, and noting that x_i is the position vector for the typical elements of surface and volume (Fig. 3.7), we express the balance of moments for the body as a whole by

$$\int_{\mathcal{S}} \varepsilon_{ijk} x_j t_i^{(\hat{n})} d\mathcal{S} + \int_{\mathcal{V}} \varepsilon_{ijk}\, x_j \rho b_i d\mathcal{V} = 0 \,. \tag{3.28}$$

As before, using the identity $t_k^{(\hat{n})} = t_{qk} n_q$ and Gauss's divergence theorem, we obtain

$$\int_{\mathcal{V}} \varepsilon_{ijk} \left[(x_j t_{qk})_{,q} + x_j \rho b_k \right] d\mathcal{V} = 0 \,,$$

or

$$\int_{\mathcal{V}} \varepsilon_{ijk} \left[x_{j,q} t_{qk} + x_j \left(t_{qk,q} + \rho b_k \right) \right] d\mathcal{V} = 0 \,.$$

But $x_{j,q} = \delta_{jq}$ and by Eq 3.27, $t_{kq,k} + \rho b_k = 0$, so that the latter equation immediately above reduces to

$$\int_{V} \varepsilon_{ijk} t_{jk} d\mathcal{V} = 0 \,.$$

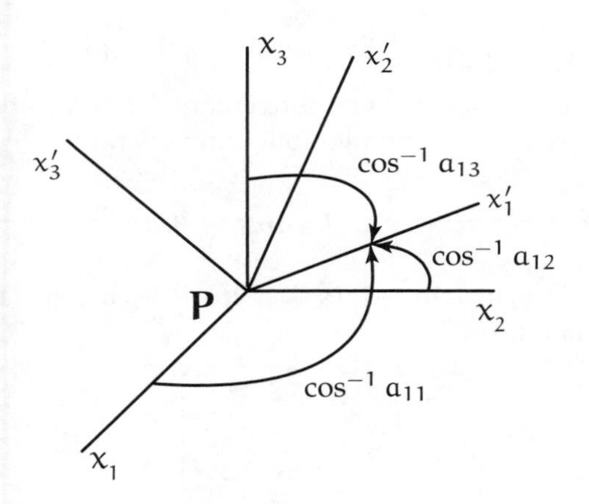

FIGURE 3.8
Rectangular coordinate axes $Px'_1x'_2x'_3$ relative to $Px_1x_2x_3$ at point **P**.

Again, since volume $\mathcal{V}$ is arbitrary, the integrand here must vanish, or

$$\varepsilon_{ijk} t_{jk} = 0 \,. \tag{3.29}$$

By a direct expansion of the left-hand side of this equation, we obtain for the free index $i = 1$, (omitting zero terms) $\varepsilon_{123}t_{23} + \varepsilon_{132}t_{32} = 0$, or $t_{23} - t_{32} = 0$ implying that $t_{23} = t_{32}$. In the same way for $i = 2$ and $i = 3$ we find that $t_{13} = t_{31}$ and $t_{12} = t_{21}$, respectively, so that in general

$$t_{jk} = t_{kj} \,. \tag{3.30}$$

Thus, we conclude from the balance of moments for a body in which concentrated body moments are absent, that the stress tensor is symmetric, and Eq 3.27 may now be written in the form

$$t_{ij,j} + \rho b_i = 0 \qquad \text{or} \qquad \nabla \cdot \mathbf{T} + \rho \mathbf{b} = 0 \,. \tag{3.31}$$

Also, because of this symmetry of the stress tensor, Eq 3.18 may now be expressed in the slightly altered form

$$t_i^{(\hat{n})} = t_{ij} n_j \qquad \text{or} \qquad \mathbf{t}^{(\hat{n})} = \mathbf{T} \cdot \hat{\mathbf{n}} \,. \tag{3.32}$$

In the matrix form of Eq 3.32 the vectors $t_i^{(\hat{n})}$ and n_j are represented by column matrices.

3.5 Stress Transformation Laws

Let the state of stress at point **P** be given with respect to Cartesian axes $Px_1x_2x_3$ shown in Fig. 3.8 by the stress tensor **T** having components t_{ij}. We introduce a second set of axes $Px'_1x'_2x'_3$, which is obtained from $Px_1x_2x_3$ by a rotation of axes so that the transformation matrix $[a_{ij}]$ relating the two is a proper orthogonal matrix. Because **T** is a second-order Cartesian tensor, its components t_{ij} in the primed system are expressed in terms of the unprimed components by Eq 2.64 as

$$t'_{ij} = a_{iq} t_{qm} a_{jm} \qquad \text{or} \qquad T' = \mathcal{A} T \mathcal{A}^T . \tag{3.33}$$

The matrix formulation of Eq 3.33 is very convenient for computing the primed components of stress as demonstrated by the two following examples.

Example 3.2

Let the stress components (in MPa) at point **P** with respect to axes $Px_1x_2x_3$ be expressed by the matrix

$$[t_{ij}] = \begin{bmatrix} 1 & 3 & 2 \\ 3 & 1 & 0 \\ 2 & 0 & -2 \end{bmatrix} ,$$

and let the primed axes $Px'_1x'_2x'_3$ be obtained by a 45° counterclockwise rotation about the x_3 axis. Determine the stress components t'_{ij}.

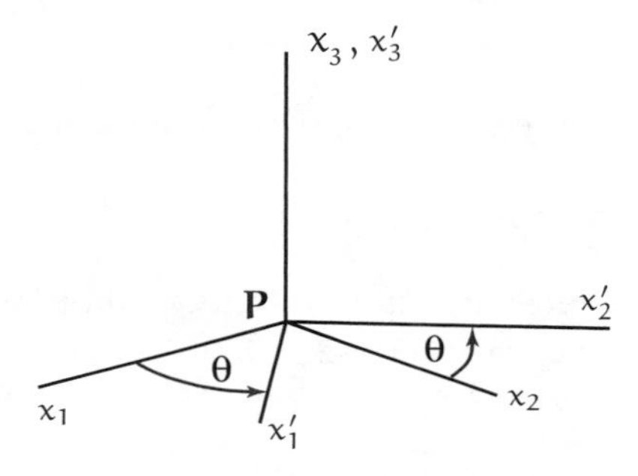

Solution

For a positive rotation θ about x_3 as shown by the sketch, the transformation matrix $[a_{ij}]$ has the general form

$$[a_{ij}] = \begin{bmatrix} \cos\theta & \sin\theta & 0 \\ -\sin\theta & \cos\theta & 0 \\ 0 & 0 & 1 \end{bmatrix} .$$

Thus, from Eq 3.33 expressed in matrix form, a 45° rotation of axes requires

$$[t_{ij}] = \begin{bmatrix} 1/\sqrt{2} & 1/\sqrt{2} & 0 \\ -1/\sqrt{2} & 1/\sqrt{2} & 0 \\ 0 & 0 & 1 \end{bmatrix} \begin{bmatrix} 1 & 3 & 2 \\ 3 & 1 & 0 \\ 2 & 0 & -2 \end{bmatrix} \begin{bmatrix} 1/\sqrt{2} & -1/\sqrt{2} & 0 \\ 1/\sqrt{2} & 1/\sqrt{2} & 0 \\ 0 & 0 & 1 \end{bmatrix}$$

$$= \begin{bmatrix} 4 & 0 & \sqrt{2} \\ 0 & -2 & -\sqrt{2} \\ \sqrt{2} & -\sqrt{2} & -2 \end{bmatrix} .$$

Example 3.3

Assume the stress tensor **T** (in ksi) at **P** with respect to axes $Px_1x_2x_3$ is represented by the matrix

$$[t_{ij}] = \begin{bmatrix} 18 & 0 & -12 \\ 0 & 6 & 0 \\ -12 & 0 & 24 \end{bmatrix} .$$

If the x_1' axis makes equal angles with the three unprimed axes, and the x_2' axis lies in the plane of $x_1'x_3$, as shown by the sketch, determine the primed components of **T** assuming $Px_1'x_2'x_3'$ is a right-handed system.

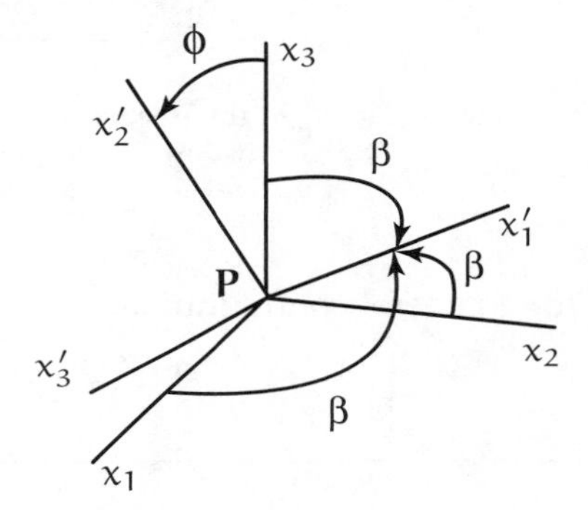

Solution

We must first determine the transformation matrix $[a_{ij}]$. Let β be the common angle which x_1' makes with the unprimed axes, as shown in the figure. Then $a_{11} = a_{12} = a_{13} = \cos\beta$ and from the orthogonality condition Eq 2.55 with $i = j = 1$, $\cos\beta = 1/\sqrt{3}$. Next, let ϕ be the angle between $x_2'$and x_3. Then $a_{23} = \cos\phi = \sin\beta = 2/\sqrt{6}$. As seen from the obvious symmetry of the axes arrangement, x_2' makes equal angles with x_1 and x_2, which means that $a_{21} = a_{22}$. Thus, again from Eq 2.55, with $i = 1$, $j = 2$, we have $a_{21} = a_{22} = -1/\sqrt{6}$ (the minus sign is required because of the positive sign chosen for a_{23}). For the primed axes to be a right-handed system we require $\hat{\mathbf{e}}_3' = \hat{\mathbf{e}}_1' \times \hat{\mathbf{e}}_2'$, with the result that $a_{31} = 1/\sqrt{2}$, $a_{32} = -1/\sqrt{2}$, and $a_{33} = 0$. Finally, from Eq 3.33,

$$\left[t_{ij}'\right] = \begin{bmatrix} \frac{1}{\sqrt{3}} & \frac{1}{\sqrt{3}} & \frac{1}{\sqrt{3}} \\ -\frac{1}{\sqrt{6}} & -\frac{1}{\sqrt{6}} & \frac{2}{\sqrt{6}} \\ \frac{1}{\sqrt{2}} & -\frac{1}{\sqrt{2}} & 0 \end{bmatrix} \begin{bmatrix} 18 & 0 & -12 \\ 0 & 6 & 0 \\ -12 & 0 & 24 \end{bmatrix} \begin{bmatrix} \frac{1}{\sqrt{3}} & -\frac{1}{\sqrt{6}} & \frac{1}{\sqrt{2}} \\ \frac{1}{\sqrt{3}} & -\frac{1}{\sqrt{6}} & -\frac{1}{\sqrt{2}} \\ \frac{1}{\sqrt{3}} & \frac{2}{\sqrt{6}} & 0 \end{bmatrix}$$

$$= \begin{bmatrix} 8 & 2\sqrt{2} & 0 \\ 2\sqrt{2} & 28 & -6\sqrt{3} \\ 0 & -6\sqrt{3} & 12 \end{bmatrix} .$$

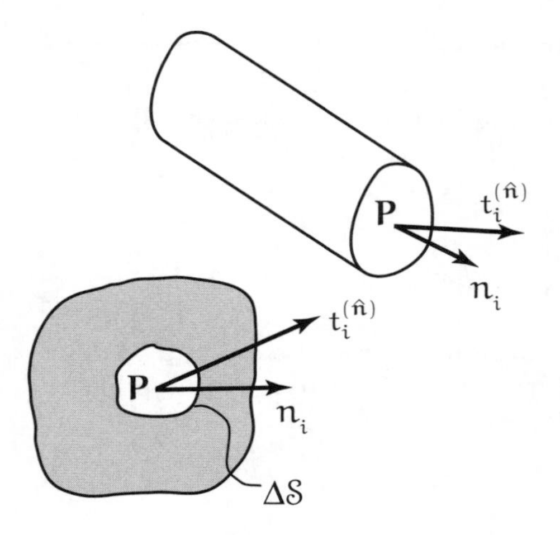

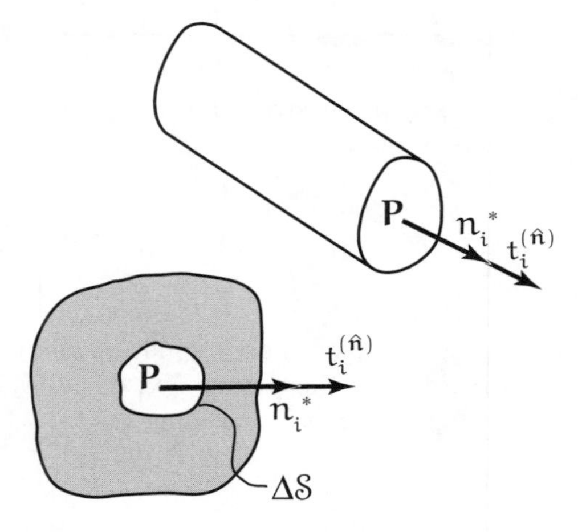

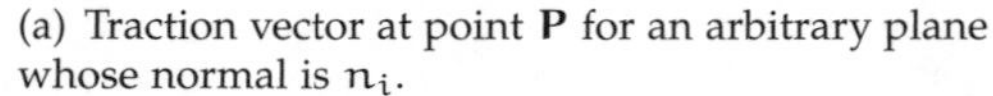
(a) Traction vector at point **P** for an arbitrary plane whose normal is n_i.

(b) Traction vector at point **P** for a principal plane whose normal is n_i^*.

FIGURE 3.9
Traction vector and normal for a general continuum and a prismatic beam.

3.6 Principal Stresses; Principal Stress Directions

Let us turn our attention once more to the state of stress at point **P** and assume it is given with respect to axes $Px_1x_2x_3$ by the stress tensor t_{ij}. As we saw in Example 3.1, for each plane element of area ΔS at **P** having an outward normal n_i, a stress vector $t_i^{(\hat{n})}$ is defined by Eq 3.32. In addition, as indicated by Fig. 3.9(a), this stress vector is not generally in the direction of n_i. However, for certain special directions at **P**, the stress vector does indeed act in the direction of n_i and may therefore be expressed as a scalar multiple of that normal. Thus, as shown in Fig. 3.9(b), for such directions

$$t_i^{(\hat{n})} = \sigma n_i \tag{3.34}$$

where σ is the scalar multiple of n_i. Directions designated by n_i for which Eq 3.34 is valid are called *principal stress directions*, and the scalar σ is called a *principal stress value* of t_{ij}. Also, the plane at **P** perpendicular to n_i is referred to as a principal stress plane. We see from Fig. 3.9(b), that, because of the perpendicularity of $\mathbf{t}^{(\hat{n})}$ to the *principal planes* there are no shear stresses acting in these planes.

The determination of principal stress values and principal stress directions follows precisely the same procedure developed in Section 2.6 for determining principal values and principal directions of any symmetric second-order tensor. In properly formulating the eigenvalue problem for the stress tensor we use the identity $t_i^{(\hat{n})} = t_{ji}n_j$ and the substitution property of the Kronecker delta to rewrite Eq 3.34 as

$$t_{ij}n_j - \sigma\delta_{ij}n_j = 0 \tag{3.35}$$

or, in expanded form, using $t_{ij} = t_{ji}$,

$$(t_{11} - \sigma)n_1 + t_{12}n_2 + t_{13}n_3 = 0\,, \tag{3.36a}$$
$$t_{12}n_1 + (t_{22} - \sigma)n_2 + t_{23}n_3 = 0\,, \tag{3.36b}$$
$$t_{13}n_1 + t_{23}n_2 + (t_{33} - \sigma)n_3 = 0\,. \tag{3.36c}$$

In the three linear homogeneous equations expressed by Eq 3.36, the tensor components t_{ij} are assumed known; the unknowns are the three components of the principal normal n_i, and the corresponding principal stress σ. To complete the system of equations for these four unknowns, we use the normalizing condition on the direction cosines,

$$n_i n_i = 1\,. \tag{3.37}$$

For non-trivial solutions of Eq 3.35 (the solution $n_j = 0$ is not compatible with Eq 3.37), the determinant of coefficients on n_j must vanish. That is,

$$|t_{ij} - \delta_{ij}\sigma| = 0\,, \tag{3.38}$$

which upon expansion yields a cubic equation in σ (called the *characteristic equation* of the stress tensor),

$$\sigma^3 - I_T\sigma^2 + II_T\sigma - III_T = 0 \tag{3.39}$$

whose roots $\sigma_{(1)}, \sigma_{(2)}, \sigma_{(3)}$ are the principal stress values of t_{ij}. The coefficients I_T, II_T and III_T are known as the *first, second, and third invariants*, respectively, of t_{ij} and may be expressed in terms of its components by

$$I_T = t_{ii} = \mathrm{tr}\,\mathbf{T}\,, \tag{3.40a}$$
$$II_T = \frac{1}{2}\left[t_{ii}t_{jj} - t_{ij}t_{ij}\right] = \frac{1}{2}\left[(\mathrm{tr}\,\mathbf{T})^2 - \mathrm{tr}\,\mathbf{T}^2\right]\,, \tag{3.40b}$$
$$III_T = \varepsilon_{ijk}t_{1i}t_{2j}t_{3k} = \det\mathbf{T}\,. \tag{3.40c}$$

Because the stress tensor t_{ij} is a symmetric tensor having real components, the three stress invariants are real, and likewise, the principal stresses being roots of Eq 3.39 are also real. To show this, we recall from the theory of equations that for a cubic with real coefficients at least one root is real, call it $\sigma_{(1)}$, and let the associated principal direction be designated by $n_i^{(1)}$. Introduce a second set of Cartesian axes $Px_1'x_2'x_3'$ so that x_1' is in the direction of $n_i^{(1)}$. In this system the shear stresses, $t_{12}' = t_{13}' = 0$, so that the characteristic equation of t_{ij}' relative to these axes results from the expansion of the determinant

$$\begin{vmatrix} \sigma_{(1)} - \sigma & 0 & 0 \\ 0 & t_{22}' - \sigma & t_{23}' \\ 0 & t_{23}' & t_{33}' - \sigma \end{vmatrix} = 0\,, \tag{3.41}$$

or

$$\left[\sigma_{(1)} - \sigma\right]\left[\sigma^2 - (t_{22}' + t_{33}')\,\sigma + t_{22}'t_{33}' - (t_{23}')^2\right] = 0\,. \tag{3.42}$$

From this equation, the remaining two principal stresses $\sigma_{(2)}$ and $\sigma_{(3)}$ are roots of the quadratic in brackets.

But the discriminant of this quadratic is

$$(t'_{22}+t'_{33})^2 - 4\left[t'_{22}t'_{33} - (t'_{23})^2\right] = (t'_{22}-t'_{33})^2 + 4\,(t'_{23})^2 \ ,$$

which is clearly positive, indicating that both $\sigma_{(2)}$ and $\sigma_{(3)}$ are real.

If the principal stress values $\sigma_{(1)}$, $\sigma_{(2)}$, and $\sigma_{(3)}$ are distinct, the principal directions associated with these stresses are mutually orthogonal. To see why this is true, let $n_i^{(1)}$ and $n_i^{(2)}$ be the normalized principal direction vectors (eigenvectors) corresponding to $\sigma_{(1)}$ and $\sigma_{(2)}$, respectively. Then, from Eq 3.35, $\sigma_{ij}n_j^{(1)} = \sigma_{(1)}n_j$ and $\sigma_{ij}n_j^{(2)} = \sigma_{(2)}n_j$, which, upon forming the inner products, that is, multiplying in turn by $n_i^{(2)}$ and $n_i^{(1)}$, become

$$t_{ij}n_j^{(1)}n_i^{(2)} = \sigma_{(1)}n_i^{(1)}n_i^{(2)} \ , \tag{3.43a}$$

$$t_{ij}n_j^{(2)}n_i^{(1)} = \sigma_{(2)}n_i^{(2)}n_i^{(1)} \ . \tag{3.43b}$$

Furthermore, because the stress tensor is symmetric, and since i and j are dummy indices,

$$t_{ij}n_j^{(1)}n_i^{(2)} = t_{ji}n_i^{(1)}n_j^{(2)} = t_{ij}n_i^{(1)}n_j^{(2)} \ ,$$

so that by the subtraction of Eq 3.43b from Eq 3.43a, the left hand side of the resulting difference is zero, or

$$0 = \left[\sigma_{(1)} - \sigma_{(2)}\right] n_i^{(1)}n_i^{(2)} \ . \tag{3.44}$$

But since we assumed that the principal stresses were distinct, or $\sigma_{(1)} \neq \sigma_{(2)}$, it follows that

$$n_i^{(1)}n_i^{(2)} = 0 \tag{3.45}$$

which expresses orthogonality between $n_i^{(1)}$ and $n_i^{(2)}$. By similar arguments, we may show that $n_i^{(3)}$ is perpendicular to both $n_i^{(1)}$ and $n_i^{(2)}$.

If two principal stress values happen to be equal, say $\sigma_{(1)} = \sigma_{(2)}$, the principal direction $n_i^{(3)}$ associated with $\sigma_{(3)}$ will still be unique, and because of the linearity of Eq 3.35, any direction in the plane perpendicular to $n_i^{(3)}$ may serve as a principal direction. Accordingly, we may determine $n_i^{(3)}$ uniquely and then *choose* $n_i^{(1)}$ and $n_i^{(2)}$ so as to establish a right-handed system of principal axes. If it happens that all three principal stresses are equal, any direction may be taken as a principal direction, and as a result every set of right-handed Cartesian axes at **P** constitutes a set of principal axes in this case.

We give the coordinate axes in the principal stress directions special status by labeling them $Px_1^*x_2^*x_3^*$, as shown in Fig. 3.10. Thus, for example, $\sigma_{(1)}$ acts on the plane perpendicular to x_1^* and is positive (tension) if it acts in the positive direction, negative (compression) if it acts in the negative x_1^* direction. Also, if $n_i^{(q)}$ is the unit normal conjugate to the principal stress $\sigma_{(q)}$ $(q = 1,2,3)$, the transformation matrix relating the principal stress axes to arbitrary axes $Px_1x_2x_3$ has elements defined by $a_{qj} \equiv n_j^{(q)}$, as indicated by Table 3.1. Accordingly, the transformation equation expressing principal stress components in terms of arbitrary stresses at **P** is given by Eq 2.77 in the form

$$t^*_{ij} = a_{iq}a_{jm}t_{qm} \qquad \text{or} \qquad T^* = \mathcal{A}T\mathcal{A}^T \ . \tag{3.46}$$

In addition, notice that Eq 3.35 is satisfied by $n_i^{(q)}$ and $\sigma_{(q)}$ so that

$$t_{ij}n_j^{(q)} = \sigma_{(q)}n_i^{(q)} \tag{3.47}$$

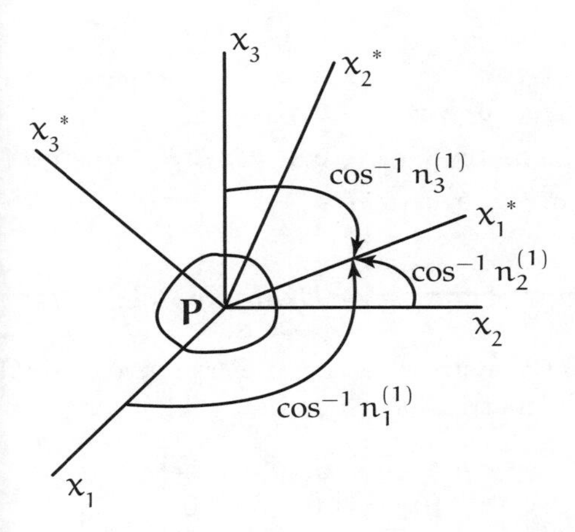

FIGURE 3.10
Principal axes $Px_1^*x_2^*x_3^*$.

TABLE 3.1
Table displaying direction cosines of principal axes $Px_1^*x_2^*x_3^*$ relative to axes $Px_1x_2x_3$.

	x_1	x_2	x_3
x_1^*	$a_{11} = n_1^{(1)}$	$a_{12} = n_2^{(1)}$	$a_{13} = n_3^{(1)}$
x_2^*	$a_{21} = n_1^{(2)}$	$a_{22} = n_2^{(2)}$	$a_{23} = n_3^{(2)}$
x_3^*	$a_{31} = n_1^{(3)}$	$a_{32} = n_2^{(3)}$	$a_{33} = n_3^{(3)}$

for $(q = 1, 2, 3)$, which upon introducing the identity $a_{qi} \equiv n_i^{(q)}$ becomes $t_{ij}a_{qj} = \sigma_{(q)}a_{qi}$. Now, multiplying each side of this equation by a_{mi} and using the symmetry property of the stress tensor, we have

$$t_{ji}a_{qj}a_{mi} = \sigma_{(q)}a_{qi}a_{mi} .$$

The left-hand side of this expression is simply t_{qm}^*, from Eq 3.46. Since, by orthogonality, $a_{qi}a_{mi}$ on the right hand side, the final result is

$$t_{qm}^* = \delta_{qm}\sigma_{(q)} , \tag{3.48}$$

which demonstrates that when referred to principal axes, the stress tensor is a diagonal tensor with principal stress values on the main diagonal. In matrix form, therefore,

$$\left[t_{ij}^*\right] = \begin{bmatrix} \sigma_{(1)} & 0 & 0 \\ 0 & \sigma_{(2)} & 0 \\ 0 & 0 & \sigma_{(3)} \end{bmatrix} \quad \text{or} \quad \left[t_{ij}^*\right] = \begin{bmatrix} \sigma_{I} & 0 & 0 \\ 0 & \sigma_{II} & 0 \\ 0 & 0 & \sigma_{III} \end{bmatrix} \tag{3.49}$$

where in the second equation the notation serves to indicate that the principal stresses are ordered, $\sigma_I \geqslant \sigma_{II} \geqslant \sigma_{III}$, with positive stresses considered greater than negative stresses regardless of numerical values. In terms of the principal stresses, the stress invariants may be written

$$I_T = \sigma_{(1)} + \sigma_{(2)} + \sigma_{(3)} = \sigma_I + \sigma_{II} + \sigma_{III} \,, \tag{3.50a}$$

$$II_T = \sigma_{(1)}\sigma_{(2)} + \sigma_{(2)}\sigma_{(3)} + \sigma_{(3)}\sigma_{(1)} = \sigma_I\sigma_{II} + \sigma_{II}\sigma_{III} + \sigma_{III}\sigma_I \,, \tag{3.50b}$$

$$III_T = \sigma_{(1)}\sigma_{(2)}\sigma_{(3)} = \sigma_I\sigma_{II}\sigma_{III} \,. \tag{3.50c}$$

Example 3.4

The components of the stress tensor at **P** are given in MPa with respect to axes $Px_1x_2x_3$ by the matrix

$$[t_{ij}] = \begin{bmatrix} 57 & 0 & 24 \\ 0 & 50 & 0 \\ 24 & 0 & 43 \end{bmatrix} .$$

Determine the principal stresses and the principal stress directions at **P**.

Solution

For the given stress tensor, Eq 3.38 takes the form of the determinant

$$\begin{vmatrix} 57-\sigma & 0 & 24 \\ 0 & 50-\sigma & 0 \\ 24 & 0 & 43-\sigma \end{vmatrix} = 0$$

which, upon cofactor expansion about the first row, results in the equation

$$(57-\sigma)(50-\sigma)(43-\sigma) - (24)^2(50-\sigma) = 0 \,,$$

or in its readily factored form

$$(50-\sigma)(\sigma-25)(\sigma-75) = 0 \,.$$

Hence, the principal stress values are $\sigma_{(1)} = 25$ MPa, $\sigma_{(2)} = 50$ MPa, and $\sigma_{(3)} = 75$ MPa. Note that, in keeping with Eqs 3.40a and 3.50a, we confirm that the first stress invariant,

$$I_T = 57 + 50 + 43 = 25 + 50 + 75 = 150 \text{ MPa} \,.$$

To determine the principal directions we first consider $\sigma_{(1)} = 25$ MPa, for which Eq 3.36 provides three equations for the direction cosines of the principal direction of $\sigma_{(1)}$, namely,

$$32n_1^{(1)} + 24n_3^{(1)} = 0 \,,$$

$$25n_2^{(1)} = 0 \,,$$

$$24n_1^{(1)} + 18n_3^{(1)} = 0 \,.$$

Obviously, $n_2^{(1)} = 0$ from the second of these equations, and from the other two, $n_3^{(1)} = -\frac{4}{3}n_1^{(1)}$ so that, from the normalizing condition, $n_i n_i = 1$, we see that $\left(n_1^{(1)}\right)^2 = \frac{9}{25}$ which gives $n_1^{(1)} = \pm\frac{3}{5}$ and $n_3^{(1)} = \mp\frac{4}{5}$. The fact that the first and third equations result in the same relationship is the reason the normalizing condition must be used.

Next for $\sigma_{(2)} = 50$ MPa, Eq 3.36 gives

$$7n_1^{(2)} + 24n_3^{(2)} = 0 \ ,$$
$$24n_1^{(2)} - 7n_3^{(2)} = 0$$

which are satisfied only when $n_1^{(2)} = n_3^{(2)} = 0$. Then from the normalizing condition, $n_i n_i = 1$, $n_2^{(2)} = \pm 1$.

Finally, for $\sigma_{(3)} = 75$ MPa, Eq 3.36 gives

$$-18n_1^{(3)} + 24n_3^{(3)} = 0 \ ,$$
$$-25n_2^{(3)} = 0$$

as well as

$$24n_1^{(3)} - 32n_3^{(3)} = 0 \ .$$

Here, from the second equation $n_2^{(3)} = 0$, and from either of the other two equations $4n_3^{(3)} = 3n_1^{(3)}$, so that from $n_i n_i = 1$ we have $n_1^{(3)} = \pm\frac{4}{5}$ and $n_3^{(3)} = \pm\frac{3}{5}$.

From these values of $n_i^{(q)}$, we now construct the transformation matrix $[a_{ij}]$ in accordance with Table 3.1, keeping in mind that to assure a right-handed system of principal axes we must have $\hat{n}^{(3)} = \hat{n}^{(1)} \times \hat{n}^{(2)}$. Thus the transformation matrix has the general form

$$[a_{ij}] = \begin{bmatrix} \pm\frac{3}{5} & 0 & \mp\frac{4}{5} \\ 0 & \pm 1 & 0 \\ \pm\frac{4}{5} & 0 & \pm\frac{3}{5} \end{bmatrix} \ .$$

Therefore, from Eq 3.48, when the upper signs in the above matrix are used,

$$[t_{ij}] = \begin{bmatrix} \frac{3}{5} & 0 & -\frac{4}{5} \\ 0 & 1 & 0 \\ \frac{4}{5} & 0 & \frac{3}{5} \end{bmatrix} \begin{bmatrix} 57 & 0 & 24 \\ 0 & 50 & 0 \\ 24 & 0 & 43 \end{bmatrix} \begin{bmatrix} \frac{3}{5} & 0 & -\frac{4}{5} \\ 0 & 1 & 0 \\ \frac{4}{5} & 0 & \frac{3}{5} \end{bmatrix} = \begin{bmatrix} 25 & 0 & 0 \\ 0 & 50 & 0 \\ 0 & 0 & 75 \end{bmatrix} \text{ MPa} \ .$$

3.7 Maximum and Minimum Stress Values

The stress vector $t_i^{(\hat{n})}$ on an arbitrary plane at **P** may be resolved into a component normal to the plane having a magnitude σ_N, along with a shear component which acts in the plane and has a magnitude σ_S, as shown in Fig. 3.11. (Here, σ_N and σ_S are *not* vectors, but scalar magnitudes of vector components. The subscripts N and S are to be taken as part of the component symbols). Clearly, from Fig. 3.11, it is seen that σ_N is given by the dot product, $\sigma_N = t_i^{(\hat{n})} n_i$, and inasmuch as $t_i^{(\hat{n})} = t_{ij} n_j$, it follows that

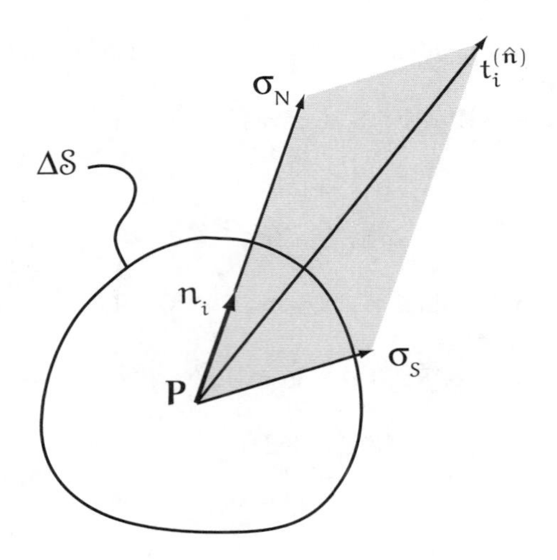

FIGURE 3.11
Traction vector components normal and in-plane (shear) at point **P** on the plane whose normal is n_i.

$$\sigma_N = t_{ij} n_i n_j \qquad \text{or} \qquad \sigma_N = \mathbf{t}^{(\hat{\mathbf{n}})} \cdot \hat{\mathbf{n}} \,. \tag{3.51}$$

Also, from the geometry of the decomposition, we see that

$$\sigma_S^2 = t_i^{(\hat{n})} t_i^{(\hat{n})} - \sigma_N^2 \quad \text{or} \quad \sigma_S^2 = \mathbf{t}^{(\hat{\mathbf{n}})} \cdot \mathbf{t}^{(\hat{\mathbf{n}})} - \sigma_N^2 \,. \tag{3.52}$$

In seeking the maximum and minimum (the so-called extremal) values of the above components, let us consider first σ_N. As the normal n_i assumes all possible orientations at **P**, the values of σ_N will be prescribed by the functional relation in Eq 3.51 subject to the condition that $n_i n_i = 1$. Accordingly, we may use to advantage the *Lagrangian multiplier* method to obtain extremal values of σ_N. To do so we construct the function $f(n_i) = t_{ij} n_i n_j - \sigma(n_i n_i - 1)$, where the scalar σ is called the Lagrangian multiplier. The method requires the derivative of $f(n_i)$ with respect to n_k to vanish; and, noting that $\partial n_i / \partial n_k = \delta_{ik}$, we have

$$\frac{\partial f}{\partial n_k} = t_{ij} \left(\delta_{ik} n_j + \delta_{jk} n_i \right) - \sigma \left(2 n_i \delta_{ik} \right) = 0 \,.$$

But $t_{ij} = t_{ji}$, and $\delta_{kj} n_j = n_k$, so that this equation reduces to

$$\left(t_{ij} - \sigma \delta_{ij} \right) n_j = 0 \tag{3.53}$$

which is identical to Eq 3.35, the eigenvalue formulation for principal stresses. Therefore, we conclude that the Lagrangian multiplier σ assumes the role of a principal stress and, furthermore, that the principal stresses include both the maximum and minimum normal stress values.

With regard to the maximum and minimum values of the shear component σ_S, it is useful to refer the state of stress to principal axes $Px_1^* x_2^* x_3^*$, as shown in Fig. 3.12. Let the principal stresses be ordered in the sequence $\sigma_I > \sigma_{II} > \sigma_{III}$ so that $t_i^{(\hat{n})}$ is expressed in vector form by

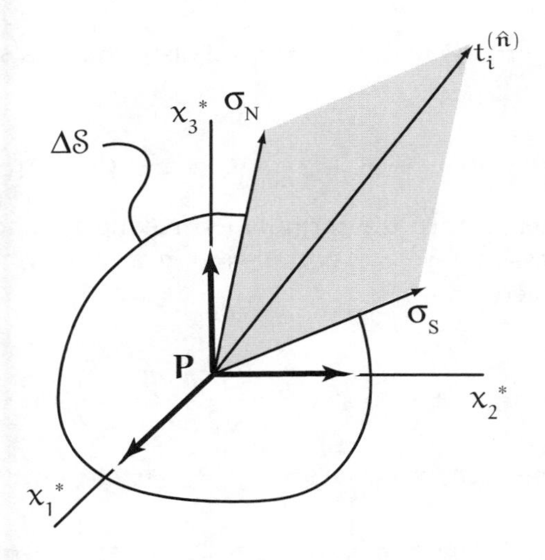

FIGURE 3.12
Normal and shear components at P to plane referred to principal axes.

$$\mathbf{t}^{(\hat{\mathbf{n}})} = \mathbf{T} \cdot \hat{\mathbf{n}} = \sigma_I n_1 \hat{\mathbf{e}}_1^* + \sigma_{II} n_2 \hat{\mathbf{e}}_2^* + \sigma_{III} n_3 \hat{\mathbf{e}}_3^* \,, \tag{3.54}$$

and similarly, $\sigma_N = \mathbf{t}^{(\hat{\mathbf{n}})} \cdot \hat{\mathbf{n}}$ by

$$\sigma_N = \sigma_I n_1^2 + \sigma_{II} n_2^2 + \sigma_{III} n_3^2 \,. \tag{3.55}$$

Then, substituting Eqs 3.54 and 3.55 into Eq 3.52, we have

$$\sigma_S^2 = \sigma_I^2 n_1^2 + \sigma_{II}^2 n_2^2 + \sigma_{III}^2 n_3^2 - (\sigma_I n_1^2 + \sigma_{II} n_2^2 + \sigma_{III} n_3^2) \tag{3.56}$$

which expresses σ_S^2 in terms of the direction cosines n_i. But $n_i n_i = 1$, so that $n_3^2 = 1 - n_1^2 - n_2^2$ and we are able to eliminate n_3 from Eq 3.56, which then becomes a function of n_1 and n_2 only,

$$\begin{aligned} \sigma_S^2 &= (\sigma_I - \sigma_{III})\, n_1^2 + \left(\sigma_{II}^2 - \sigma_{III}^2\right) n_2^2 + \sigma_{III}^2 \\ &\quad - \left[(\sigma_I - \sigma_{III})\, n_1^2 + (\sigma_{II} - \sigma_{III})\, n_2^2 + \sigma_{III}\right]^2 \,. \end{aligned} \tag{3.57}$$

In order to obtain the *stationary*, that is, the *extremal values* of σ_S^2, we must equate the derivatives of the right-hand side of this equation with respect to both n_1 and n_2 to zero, and solve simultaneously. After some algebraic manipulations, we obtain

$$\frac{\partial \sigma_S^2}{\partial n_1} = n_1 \sigma_I - \sigma_{III} \left\{\sigma_I - \sigma_{III} - 2\left[\sigma_I - \sigma_{III} n_1^2 + \sigma_{II} - \sigma_{III} n_2^2\right]\right\} = 0 \,, \tag{3.58a}$$

$$\frac{\partial \sigma_S^2}{\partial n_2} = n_2 \sigma_{II} - \sigma_{III} \left\{\sigma_{II} - \sigma_{III} - 2\left[\sigma_I - \sigma_{III} n_1^2 + \sigma_{II} - \sigma_{III} n_2^2\right]\right\} = 0 \,. \tag{3.58b}$$

An obvious solution to Eq 3.58 is $n_1 = n_2 = 0$ for which $n_3 = \pm 1$, and the corresponding value of σ_S^2 is observed from Eq 3.57 to be zero. This is an expected result since $n_3 = \pm 1$ designates a principal plane upon which the shear is zero. A similar calculation made with n_1 and n_3, or with n_2 and n_3 as the variables, would lead to the other two principal planes as minimum (zero) shear stress planes. It is easily verified that a second solution to Eq 3.58 is obtained by taking $n_1 = 0$ and solving for n_2. The result is

$n_2 = \pm 1/\sqrt{2}$ and, from orthogonality, $n_3 = \pm\sqrt{2}$ also. For this solution, Eq 3.57 yields the results

$$\sigma_S^2 = \frac{1}{4}\left(\sigma_{II} - \sigma_{III}\right)^2 \quad \text{or} \quad \sigma_S = \pm\frac{1}{2}\left(\sigma_{II} - \sigma_{III}\right) . \tag{3.59}$$

As before, if we consider in turn the formulation having n_1 and n_3, or n_2 and n_3 as the variable pairs, and assume $n_3 = 0$, and $n_2 = 0$, respectively, we obtain the complete solution which is presented below,

$$n_1 = 0, \quad n_2 = \pm\frac{1}{\sqrt{2}}, \quad n_3 = \pm\frac{1}{\sqrt{2}}; \quad \sigma_S = \frac{1}{2}\left|\sigma_{II} - \sigma_{III}\right| , \tag{3.60a}$$

$$n_1 = \pm\frac{1}{\sqrt{2}}, \quad n_2 = 0, \quad n_3 = \pm\frac{1}{\sqrt{2}}; \quad \sigma_S = \frac{1}{2}\left|\sigma_{III} - \sigma_{I}\right| , \tag{3.60b}$$

$$n_1 = \pm\frac{1}{\sqrt{2}}, \quad n_2 = \pm\frac{1}{\sqrt{2}}, \quad n_3 = 0; \quad \sigma_S = \frac{1}{2}\left|\sigma_{I} - \sigma_{II}\right| \tag{3.60c}$$

where the vertical bars in the formulas for σ_S indicate absolute values of the enclosed expressions. Because $\sigma_I \geqslant \sigma_{II} \geqslant \sigma_{III}$, it is clear that the largest shear stress value is

$$\sigma_S^{max} = \frac{1}{2}\left|\sigma_{III} - \sigma_{I}\right| . \tag{3.61}$$

It may be shown that, for distinct principal stresses, only the two solutions presented in this section satisfy Eq 3.58.

3.8 Mohr's Circles for Stress

Mohr's circle is a method the student should remember from undergraduate mechanics of materials. It provides a graphic means for the transformation of a second order tensor like that discussed in Section 2.5. Mohr's circle is commonly used to transform stress, area or mass moment of inertia tensors. While this book gives the student the mathematical tools to transform tensors, we feel a discussion of Mohr's circle is worthy of inclusion in this introduction to continuum mechanics. Most of the students prior use of Mohr's circle has likely been with two dimensional problems: plane stress or transformation of the area or mass inertia tensors. Mohr's circle allows for an easy means to derive and compute transformation and principal values in two dimensions.

In three dimensions, the convenience and simplicity of Mohr's circle to transform stress components to an arbitrary reference frame vanishes. When working with three dimensional problems the advantage of using the mathematics of continuum mechanics become evident. This is the case for many problems in mechanics. In two dimensions, intuitive diagrams and relatively simplified equations greet the engineer. Adding the third dimensions leaves most all engineers relying on the proper application of mathematical models. This is the case for mechanics of materials and dynamics; two topics where the student may have experienced this.

Consider again the state of stress at **P** referenced to principal axes (Fig. 3.12) and let the principal stresses be ordered according to $\sigma_I > \sigma_{II} > \sigma_{III}$. As before, we may express σ_N and σ_S on any plane at **P** in terms of the components of the normal to that plane by the

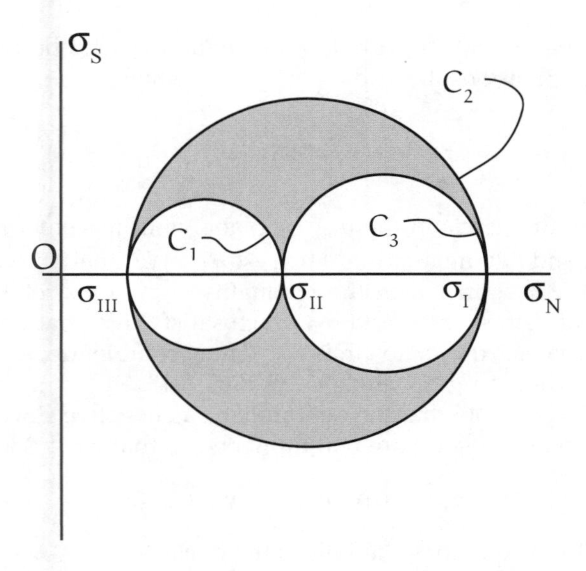

FIGURE 3.13
Typical Mohr's circle for stress.

equations

$$\sigma_N = \sigma_I n_1^2 + \sigma_{II} n_2^2 + \sigma_{III} n_3^2 \,, \tag{3.62a}$$

$$\sigma_N^2 + \sigma_S^2 = \sigma_I^2 n_1^2 + \sigma_{II}^2 n_2^2 + \sigma_{III}^2 n_3^2 \tag{3.62b}$$

which, along with the condition

$$n_1^2 + n_2^2 + n_3^2 = 1 \,,$$

provide us with three equations for the three direction cosines n_1, n_2 and n_3. Solving these equations, we obtain

$$n_1^2 = \frac{(\sigma_N - \sigma_{II})(\sigma_N - \sigma_{III}) + \sigma_S^2}{(\sigma_I - \sigma_{II})(\sigma_I - \sigma_{III})} \,, \tag{3.63a}$$

$$n_2^2 = \frac{(\sigma_N - \sigma_{III})(\sigma_N - \sigma_I) + \sigma_S^2}{(\sigma_{II} - \sigma_{III})(\sigma_{II} - \sigma_I)} \,, \tag{3.63b}$$

$$n_3^2 = \frac{(\sigma_N - \sigma_I)(\sigma_N - \sigma_{II}) + \sigma_S^2}{(\sigma_{III} - \sigma_I)(\sigma_{III} - \sigma_{II})} \,. \tag{3.63c}$$

In these equations, σ_I, σ_{II} and σ_{III} are known; σ_N and σ_S are functions of the direction cosines n_i. Our intention here is to interpret these equations graphically by representing conjugate pairs of σ_N, σ_S values, which satisfy Eq 3.63, as a point in the *stress plane* having σ_N as absicca and σ_S as ordinate (see Fig. 3.13).

To develop this graphical interpretation of the three-dimensional stress state in terms of σ_N and σ_S, we note that the denominator of Eq 3.63a is positive since both $\sigma_I - \sigma_{II} > 0$ and $\sigma_I - \sigma_{III} > 0$, and also that $n_1^2 > 0$, all of which tells us that

$$(\sigma_N - \sigma_{II})(\sigma_N - \sigma_{III}) + \sigma_S^2 \geqslant 0 \,. \tag{3.64}$$

For the case where the equality sign holds, this equation may be rewritten, after some simple algebraic manipulations, to read

$$\left[\sigma_N - \frac{1}{2}(\sigma_{II} + \sigma_{III})\right]^2 + \sigma_S^2 = \left[\frac{1}{2}(\sigma_{II} - \sigma_{III})\right]^2 \tag{3.65}$$

which is the equation of a circle in the σ_N, σ_S plane, with its center at the point $\frac{1}{2}(\sigma_{II} + \sigma_{III})$ on the σ_N axis, and having a radius $\frac{1}{2}(\sigma_{II} - \sigma_{III})$. We label this circle C_1 and display it in Fig. 3.13. For the case in which the inequality sign holds for Eq 3.64, we observe that conjugate pairs of values of σ_N and σ_S which satisfy this relationship result in stress points having coordinates exterior to circle C_1. Thus, combinations of σ_N and σ_S which satisfy Eq 3.63a lie *on*, or *exterior* to, circle C_1 in Fig. 3.13.

Examining Eq 3.63b, we note that the denominator is negative since $\sigma_{II} - \sigma_{III} > 0$, and $\sigma_{II} - \sigma_I < 0$. The direction cosines are real numbers, so that $n_2^2 \geqslant 0$ and we have

$$(\sigma_N - \sigma_{III})(\sigma_N - \sigma_I) + \sigma_S^2 \leqslant 0 \tag{3.66}$$

which for the case of the equality sign defines the circle

$$\left[\sigma_N - \frac{1}{2}(\sigma_I + \sigma_{III})\right]^2 = \left[\frac{1}{2}(\sigma_I - \sigma_{III})\right]^2 \tag{3.67}$$

in the σ_N, σ_S plane. This circle is labeled C_2 in Fig. 3.13, and the stress points which satisfy the inequality of Eq 3.66 lie *interior* to it. Following the same general procedure, we rearrange Eq 3.63c into an expression from which we extract the equation of the third circle, C_3 in Fig. 3.13, namely,

$$\left[\sigma_N - \frac{1}{2}(\sigma_I + \sigma_{II})\right]^2 + \sigma_S^2 = \left[\frac{1}{2}(\sigma_I - \sigma_{II})\right]^2 . \tag{3.68}$$

Admissible stress points in the σ_N, σ_S plane lie *on* or *exterior* to this circle. The three circles defined above, and shown in Fig. 3.13, are called *Mohr's circles for stress*. All possible pairs of values of σ_N and σ_S at **P** which satisfy Eq 3.63 lie on these circles or between the areas enclosed by them. Actually, in conformance with Fig. 3.11 (which is the physical basis for Fig. 3.13), we see that the sign of the shear component is arbitrary so that only the top half of the circle diagram need be drawn, a practice we will occasionally follow hereafter. In addition, it is clear from the Mohr's circles diagram that the maximum shear stress value at **P** is the radius of circle C_2, which confirms the result presented in Eq 3.61.

In order to relate a typical stress point having coordinates σ_N and σ_S in the stress plane of Fig. 3.13 to the orientation of the area element ΔS (denoted by n_i in Fig. 3.12) upon which the stress components σ_N and σ_S act, we consider a small spherical portion of the continuum body centered at **P**. As the unit normal n_i assumes all possible directions at **P**, the point of intersection of its line of action with the sphere will move over the surface of the sphere. However, as seen from Eqs 3.55 and Eq 3.56 the values of σ_N and σ_S are functions of the *squares* of the direction cosines, and hence do not change for n_i reflected in the principal planes. Accordingly, we may restrict our attention to the first octant of the spherical body, as shown in Fig. 3.14(a). Let Q be the point of intersection of the line of action of n_i with the spherical surface ABC in Fig. 3.14(a) and note that

$$\hat{n} = \cos\phi\,\hat{e}_1^* + \cos\beta\,\hat{e}_2^* + \cos\theta\,\hat{e}_3^* . \tag{3.69}$$

If $\hat{n} = \hat{e}_1^*$ so that its intersection point Q coincides with A, $\sigma_N = \sigma_I$. Likewise, when Q coincides with B, $\sigma_N = \sigma_{II}$, and with C, $\sigma_N = \sigma_{III}$. In all three cases, σ_S will be zero.

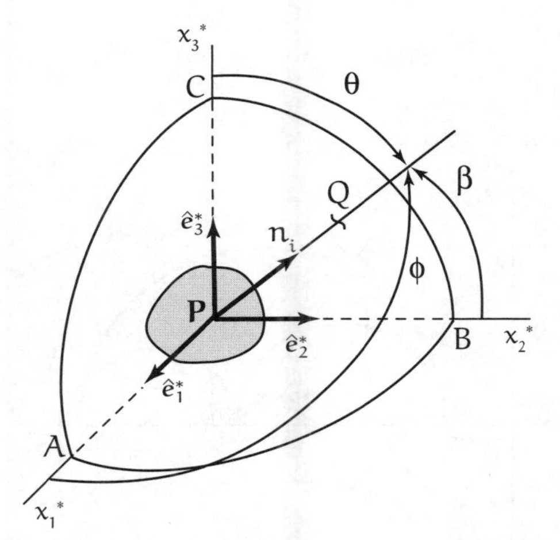

(a) Octant of small spherical portion of body together with plane at **P** with normal n_i referred to principal axes $Ox_1^*x_2^*x_3^*$.

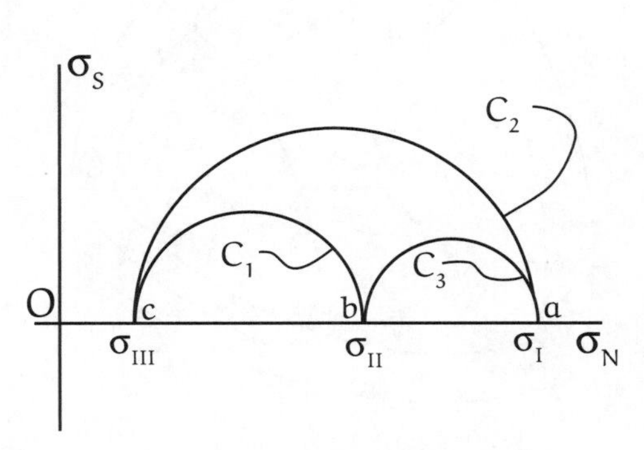

(b) Mohr's stress semicircle for octant of Fig. 3.14(a).

FIGURE 3.14
Typical Mohr's circle representation.

In the Mohr's circle diagram (Fig. 3.14(b)), these stress values are located at points a, b and c, respectively. If now θ is set equal to $\frac{\pi}{2}$ and ϕ allowed to vary from zero to $\frac{\pi}{2}$ (β will concurrently go from $\frac{\pi}{2}$ to zero), Q will move along the quarter-circle arc AB from A to B. In the stress space of Fig. 3.14(b), the stress point q (the image point of Q) having coordinates σ_N and σ_S will simultaneously move along the semicircle of C_3 from a to b. (Note that as Q moves 90° along AB in physical space, q moves 180° along the semicircle, joining a to b in stress space.) Similarly, when Q is located on the quarter circle BC, or CA of Fig. 3.14(a), point q will occupy a corresponding position on the semicircles of bc and ca, respectively, in Fig. 3.14(b).

Now let the angle ϕ be given some fixed value less than $\frac{\pi}{2}$, say $\phi = \phi_1$, and imagine that β and θ take on all values compatible with the movement of Q along the circle arc KD of Fig. 3.15(a). For this case, Eq 3.63a becomes $(\sigma_N - \sigma_{II})(\sigma_N - \sigma_{III}) + \sigma_S^2 = (\sigma_I - \sigma_{II})(\sigma_I - \sigma_{III})\cos^2\phi_1$, which may be cast into the standard form of a circle as

$$\left[\sigma_N - \frac{1}{2}(\sigma_{II} + \sigma_{III})\right]^2 + \sigma_S^2$$
$$= (\sigma_I - \sigma_{II})(\sigma_I - \sigma_{III})\cos^2\phi_1 + \left[\frac{1}{2}(\sigma_{II} - \sigma_{III})\right]^2 = R_1^2 \,. \tag{3.70}$$

This circle is seen to have its center coincident with that of circle C_1 in stress space and to have a radius R_1 indicated by Eq 3.70. Therefore, as Q moves on circle arc KD in Fig. 3.15(a), the stress point q traces the circle arc kd shown in Fig. 3.15(b). (Notice that if $\phi_1 = \frac{\pi}{2}$ so that $\cos\phi_1 = 0$, R_1 reduces to $\frac{1}{2}(\sigma_{II} - \sigma_{III})$, the radius of circle C_1). Next, let $\beta = \beta_1 < \frac{\pi}{2}$ and then, as ϕ and θ range through all admissible values, point Q moves along the circle arc EG of Fig. 3.15(a). For this case Eq 3.63b may be restructured into the

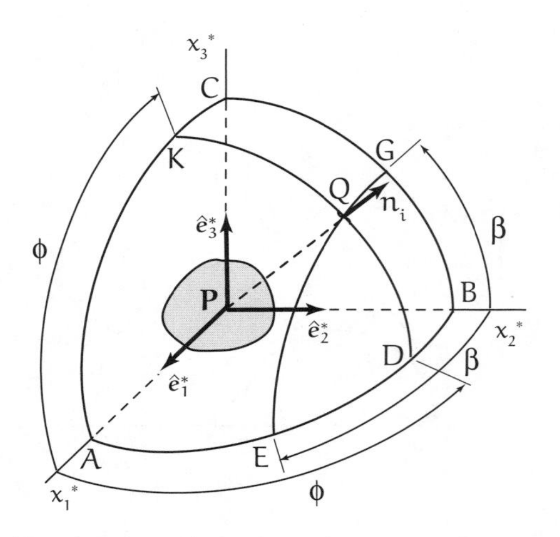

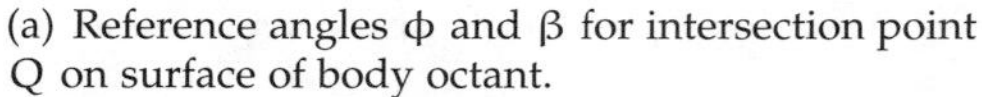
(a) Reference angles φ and β for intersection point Q on surface of body octant.

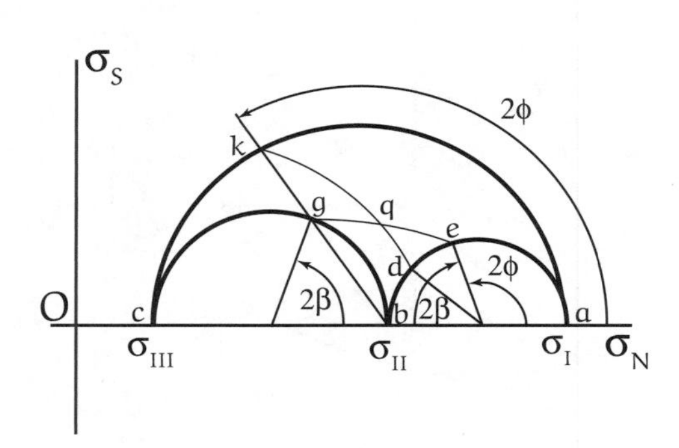

(b) Mohr's stress semicircle for octant of Fig. 3.15(a).

FIGURE 3.15
Typical 3-D Mohr's circle and associated geometry.

form

$$\left[\sigma_N - \frac{1}{2}(\sigma_I + \sigma_{III})\right]^2 + \sigma_S^2$$
$$= (\sigma_{II} - \sigma_{III})(\sigma_{II} - \sigma_I)\cos^2\beta_1 + \left[\frac{1}{2}(\sigma_I - \sigma_{III})\right]^2 = R_2^2 \qquad (3.71)$$

which defines a circle whose center is coincident with that of circle C_2, and having a radius R_2. Here, when $\beta_1 = \frac{\pi}{2}$, the radius R_2 reduces to $\frac{1}{2}(\sigma_I - \sigma_{III})$, which is the radius of circle C_2. As Q moves on the circle arc EG of Fig. 3.15(a), the stress point q traces out the circle arc eg in Fig. 3.15(b).

In summary, for a specific $\hat{\mathbf{n}}$ at point **P** in the body, point Q, where the line of action of $\hat{\mathbf{n}}$ intersects the spherical octant of the body (Fig. 3.15(a)), is located at the common point of circle arcs KD and EG, and at the same time, the corresponding stress point q (having coordinates σ_N and σ_S) is located at the intersection of circle arcs kd and eg in the stress plane of Fig. 3.15(b). The following example provides details of the procedure.

Example 3.5

The state of stress at point **P** is given in MPa with respect to axes $Px_1x_2x_3$ by the matrix

$$[t_{ij}] = \begin{bmatrix} 25 & 0 & 0 \\ 0 & -30 & -60 \\ 0 & -60 & 5 \end{bmatrix}.$$

(a) Determine the stress vector on the plane whose unit normal is $\hat{\mathbf{n}} = \frac{1}{3}(2\hat{\mathbf{e}}_1 + \hat{\mathbf{e}}_2 + 2\hat{\mathbf{e}}_3)$.

(b) Determine the normal stress component σ_N and shear component σ_S on the same plane.

(c) Verify the results of part (b) by the Mohr's circle construction of Fig. 3.15(b).

Solution

(a) Using Eq 3.32 in matrix form gives the stress vector $t_i^{(\hat{\mathbf{n}})}$

$$\begin{bmatrix} t_1^{(\hat{\mathbf{n}})} \\ t_2^{(\hat{\mathbf{n}})} \\ t_3^{(\hat{\mathbf{n}})} \end{bmatrix} = [t_{ij}][n_j] = \begin{bmatrix} 25 & 0 & 0 \\ 0 & -30 & -60 \\ 0 & -60 & 5 \end{bmatrix} \begin{bmatrix} \frac{2}{3} \\ \frac{1}{3} \\ \frac{1}{3} \end{bmatrix} = \frac{1}{3} \begin{bmatrix} 50 \\ -150 \\ -50 \end{bmatrix},$$

or

$$\mathbf{t}^{(\hat{\mathbf{n}})} = \frac{1}{3}(50\hat{\mathbf{e}}_1 - 150\hat{\mathbf{e}}_2 - 50\hat{\mathbf{e}}_3) .$$

(b) Making use of Eq 3.51, we can calculate σ_N conveniently from the matrix product

$$\begin{bmatrix} \frac{2}{3} & \frac{1}{3} & \frac{2}{3} \end{bmatrix} \begin{bmatrix} 25 & 0 & 0 \\ 0 & -30 & -60 \\ 0 & -60 & 5 \end{bmatrix} \begin{bmatrix} \frac{2}{3} \\ \frac{1}{3} \\ \frac{2}{3} \end{bmatrix} = \sigma_N$$

$$\sigma_N = \frac{100}{9} - \frac{150}{9} - \frac{100}{9}$$

so that $\sigma_N = -\frac{150}{9} = -16.67$ MPa. Note that the same result could have been obtained by the dot product

$$\sigma_N = \mathbf{t}^{(\hat{\mathbf{n}})} \cdot \hat{\mathbf{n}} = \frac{1}{3}(50\hat{\mathbf{e}}_1 - 150\hat{\mathbf{e}}_2 - 50\hat{\mathbf{e}}_3) \cdot \frac{1}{3}(2\hat{\mathbf{e}}_1 + \hat{\mathbf{e}}_2 + 2\hat{\mathbf{e}}_3) .$$

The shear component σ_S is given by Eq 3.52, which for the values of σ_N and $t_i^{(\hat{\mathbf{n}})}$ calculated above, results in the equation,

$$\sigma_S^2 = \frac{2\,500 + 22\,500 + 2\,500}{9} - \frac{22\,500}{81} = 2\,777 ,$$

or, finally,

$$\sigma_S = 52.7 \text{ MPa} .$$

(c) Using the procedure of Example 3.4 the student should verify that for the stress tensor t_{ij} given here the principal stress values are $\sigma_I = 50$ MPa, $\sigma_{II} = 25$ MPa, and $\sigma_{III} = -75$ MPa. Also, the transformation matrix from axes $Px_1x_2x_3$ to $Px_1^*x_2^*x_3^*$ is

$$[a_{ij}] = \begin{bmatrix} 0 & -\frac{3}{5} & \frac{4}{5} \\ 1 & 0 & 0 \\ 0 & \frac{4}{5} & \frac{3}{5} \end{bmatrix} ,$$

so that the components of $\hat{\mathbf{n}}$ are given relative to the principal axes by

$$\begin{bmatrix} n_1^* \\ n_2^* \\ n_3^* \end{bmatrix} = \begin{bmatrix} 0 & -\frac{3}{5} & \frac{4}{5} \\ 1 & 0 & 0 \\ 0 & \frac{4}{5} & \frac{3}{5} \end{bmatrix} \begin{bmatrix} \frac{2}{3} \\ \frac{1}{3} \\ \frac{2}{3} \end{bmatrix} = \begin{bmatrix} \frac{1}{3} \\ \frac{2}{3} \\ \frac{2}{3} \end{bmatrix} .$$

Therefore, with respect to Fig. 3.14(a), $\phi = \cos^{-1}(1/3) = 70.53°$; $\beta = \theta = \cos^{-1}(2/3) = 48.19°$, so that—following the procedure outlined for construction of Fig. 3.15(b)—we obtain the Mohr's circle below, from which we may measure the coordinates of the stress point q and confirm the values $\sigma_N = -16.7$ and $\sigma_S = 52.7$, both in MPa.

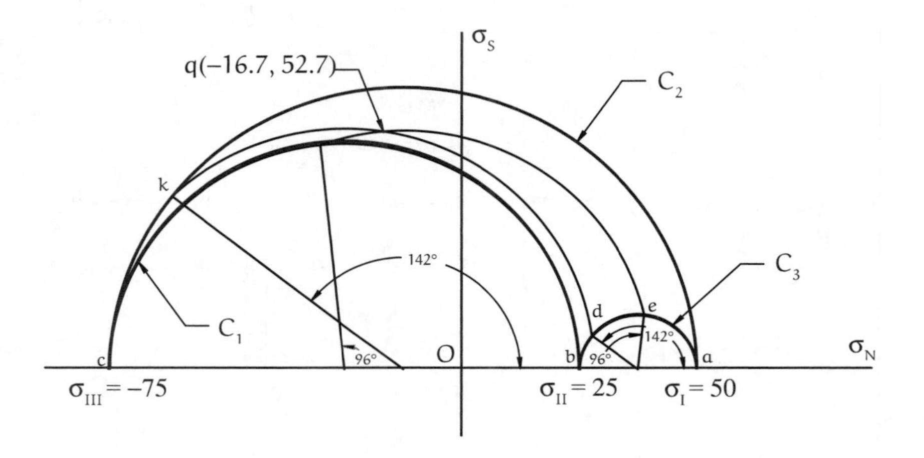

3.9 Plane Stress

When one, and only one principal stress is zero, we have a state of *plane stress* for which the plane of the two nonzero principal stresses is the *designated plane*. This is an important state of stress because it represents the physical situation occurring at an unloaded point on the bounding surface of a body under stress. The zero principal stress may be any one of the three principal stresses as indicated by the corresponding Mohr's circles of Fig. 3.16.

If the principal stresses are not ordered and the direction of the zero principal stress is arbitrarily chosen as x_3, we have plane stress parallel to the x_1x_2-plane and the stress matrix takes the form

$$[t_{ij}] = \begin{bmatrix} t_{11} & t_{12} & 0 \\ t_{12} & t_{22} & 0 \\ 0 & 0 & 0 \end{bmatrix} , \tag{3.72}$$

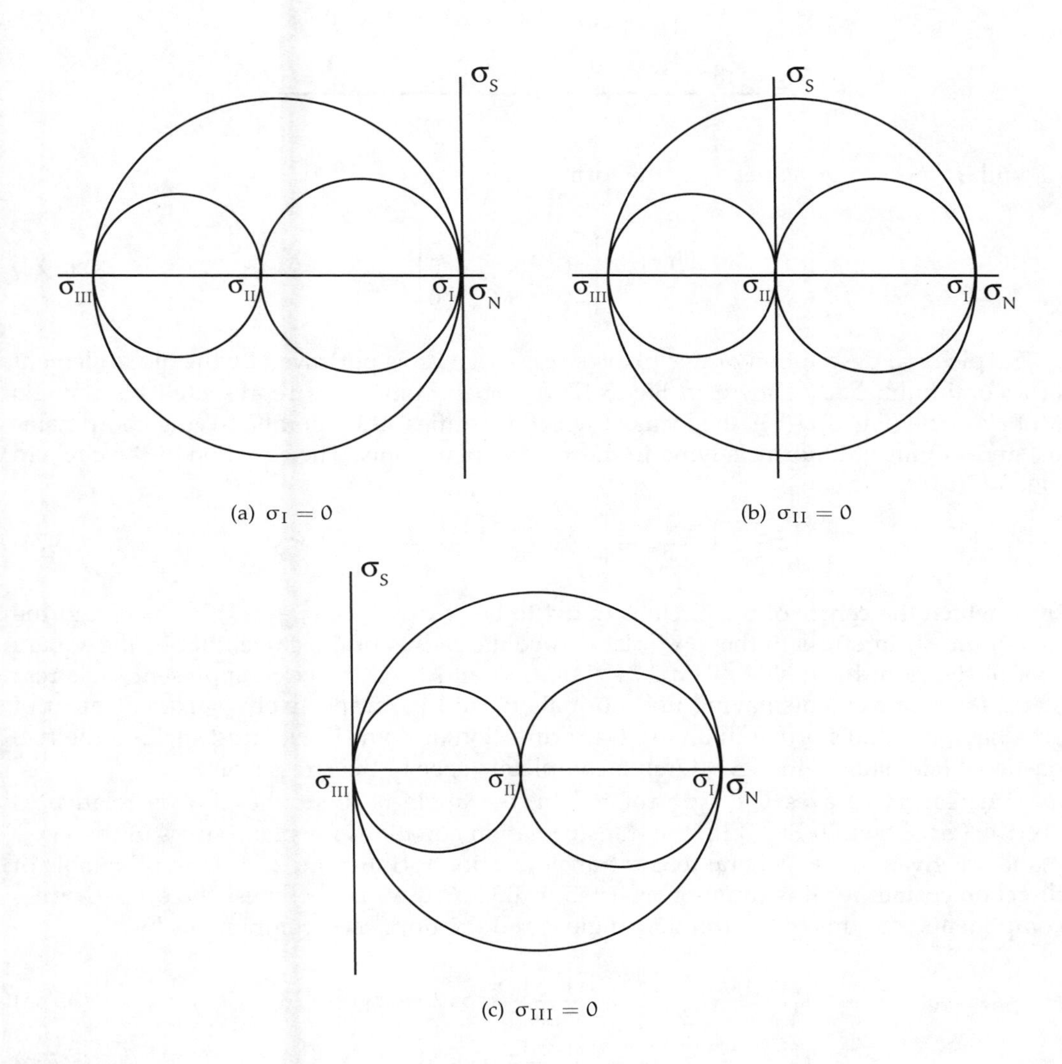

(a) $\sigma_I = 0$

(b) $\sigma_{II} = 0$

(c) $\sigma_{III} = 0$

FIGURE 3.16
Mohr's circle for plane stress.

TABLE 3.2
Transformation table for general plane stress.

	x_1	x_2	x_3
x_1^*	$\cos\theta$	$\sin\theta$	0
x_2^*	$-\sin\theta$	$\cos\theta$	0
x_3^*	0	0	1

or, with respect to principal axes, the form

$$[t_{ij}^*] = \begin{bmatrix} \sigma_{(1)} & 0 & 0 \\ 0 & \sigma_{(2)} & 0 \\ 0 & 0 & 0 \end{bmatrix} . \tag{3.73}$$

The pictorial description of this plane stress situation is portrayed by the block element of a continuum body shown in Fig. 3.17(a), and is sometimes represented by a single Mohr's circle (Fig. 3.17(b)), the locus of which identifies stress points (having coordinates σ_N and σ_S) for unit normals lying in the $x_1 - x_2$ plane only. The equation of the circle in Fig. 3.17(b) is

$$\left[\sigma_N - \frac{t_{11}+t_{22}}{2}\right]^2 + \sigma_S^2 = \left[\frac{t_{11}-t_{22}}{2}\right]^2 + t_{12}^2 \tag{3.74}$$

from which the center of the circle is noted to be at $\sigma_N = \frac{1}{2}(t_{11}+t_{22})$, $\sigma_S = 0$, and the maximum shear stress in the x_1x_2-plane to be the radius of the circle, that is, the square root of the right-hand side of Eq 3.74. Points A and B on the circle represent the stress states for area elements having unit normals $\hat{\mathbf{e}}_1$ and $\hat{\mathbf{e}}_2$, respectively. For an element of area having a unit normal in an arbitrary direction at point **P**, we must include the two dashed circles shown in Fig. 3.17(c) to completely specify the stress state.

With respect to axes $Ox_1'x_2'x_3'$ rotated by the angle θ about the x_3 axis relative to $Ox_1x_2x_3$ as shown in Fig. 3.18, the transformation equations for plane stress in the x_1x_2-plane are given by the general tensor transformation formula, Eq 2.64. Using the table of direction cosines for this situation as listed in Table 3.2, we may express the primed stress components in terms of the rotation angle θ and the unprimed components by

$$t_{11}' = \frac{t_{11}+t_{22}}{2} + \frac{t_{11}-t_{22}}{2}\cos 2\theta + t_{12}\sin 2\theta , \tag{3.75a}$$

$$t_{22}' = \frac{t_{11}+t_{22}}{2} - \frac{t_{11}-t_{22}}{2}\cos 2\theta - t_{12}\sin 2\theta , \tag{3.75b}$$

$$t_{12}' = -\frac{t_{11}-t_{22}}{2}\sin 2\theta + t_{12}\cos 2\theta . \tag{3.75c}$$

In addition, if the principal axes of stress are chosen for the primed directions, it is easily shown that the two nonzero principal stress values are given by

$$\sigma_{(1)}, \sigma_{(2)} = \frac{t_{11}+t_{22}}{2} \pm \sqrt{\left[\frac{t_{11}-t_{22}}{2}\right]^2 + t_{12}^2} \tag{3.76}$$

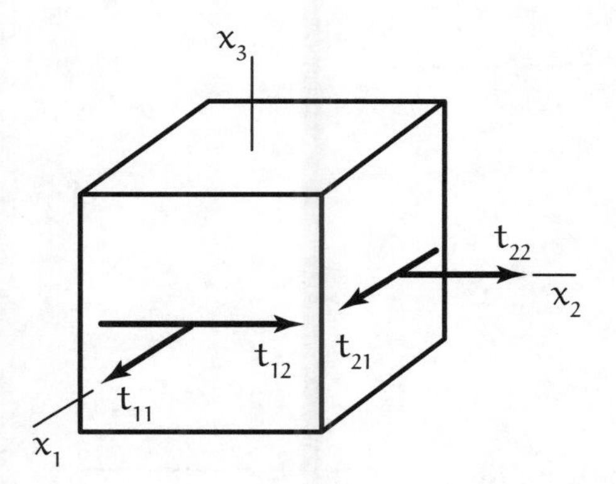

(a) Plane stress element having nonzero x_1 and x_2 components.

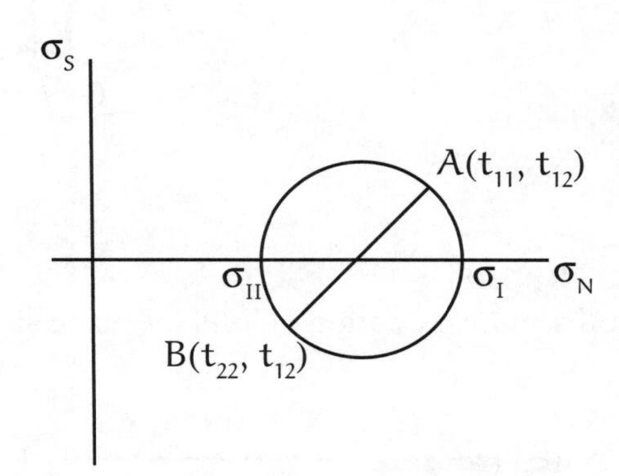

(b) Mohr's circle for in-plane stress components.

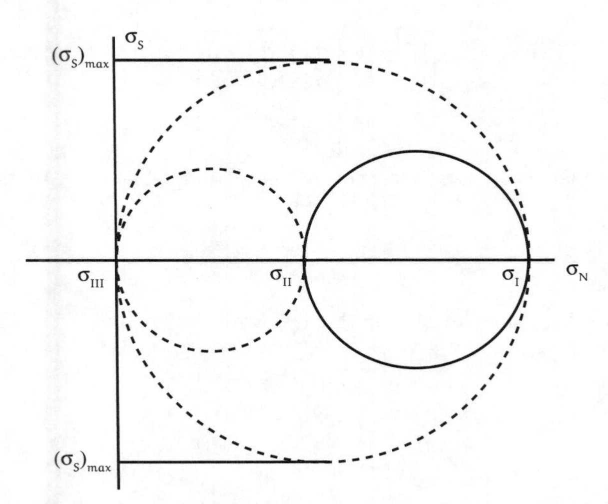

(c) General Mohr's circles for the plane stress element. Dashed lines represent out-of-plane Mohr's circles. Note the maximum shear can occur out-of-plane.

FIGURE 3.17
Mohr's circle for plane stress.

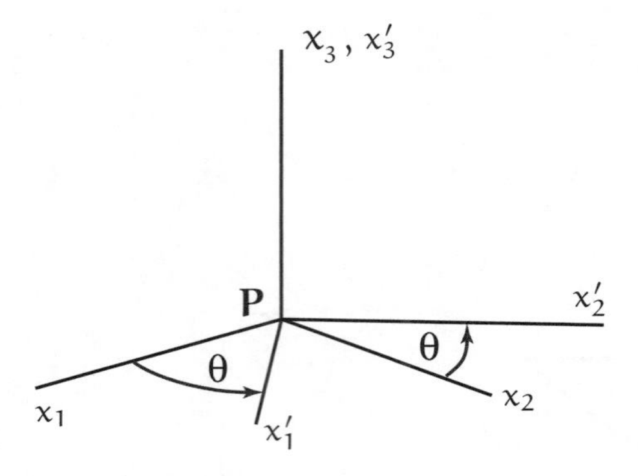

FIGURE 3.18
Representative rotation of axes for plane stress.

Example 3.6

A specimen is loaded with equal tensile and shear stresses. This case of plane stress may be represented by the matrix

$$[t_{ij}] = \begin{bmatrix} \sigma_0 & \sigma_0 & 0 \\ \sigma_0 & \sigma_0 & 0 \\ 0 & 0 & 0 \end{bmatrix}$$

where σ_0 is a constant stress. Determine the principal stress values and plot the Mohr's circles.

Solution
For this stress state, the determinant Eq 3.38 is given by

$$\begin{vmatrix} \sigma_0 - \sigma & \sigma_0 & 0 \\ \sigma_0 & \sigma_0 - \sigma & 0 \\ 0 & 0 & -\sigma \end{vmatrix} = 0$$

which results in a cubic having roots (principal stress values) $\sigma_{(1)} = 2\sigma_0$, $\sigma_{(2)} = \sigma_{(3)} = 0$ (as may be readily verified by Eq 3.76) so that, in principal axes form, the stress matrix is

$$[t^*_{ij}] = \begin{bmatrix} 2\sigma_0 & 0 & 0 \\ 0 & 0 & 0 \\ 0 & 0 & 0 \end{bmatrix}.$$

The Mohr's circle diagram is shown below.

Here, because of the double-zero root, one of the three Mohr's circles degenerates into a point (the origin) and the other two circles coincide. Also, we note that physically this is simply a one-dimensional tension in the x^*_1 direction and that the maximum shear stress values (shown by points A and B) occur on the x_1 and x_2 coordinate planes which make 45° with the principal x^*_1 direction.

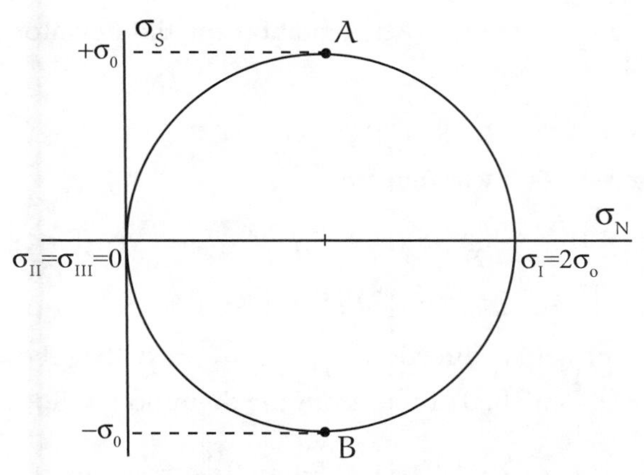

Mohr's circle for principal stresses $\sigma_1 = 2\sigma_o$, $\sigma_2 = \sigma_3 = 0$.

3.10 Deviator and Spherical Stress States

The arithmetic mean of the normal stresses,

$$\sigma_M = \frac{1}{3}(t_{11} + t_{22} + t_{33}) = \frac{1}{3}t_{ii} \tag{3.77}$$

is referred to as the *mean normal stress*. The state of stress having all three principal stresses equal (and therefore equal to σ_M) is called a *spherical state of stress* and is represented by the diagonal matrix

$$[t_{ij}] = \begin{bmatrix} \sigma_M & 0 & 0 \\ 0 & \sigma_M & 0 \\ 0 & 0 & \sigma_M \end{bmatrix} \tag{3.78}$$

for which all directions are principal directions as explained in Section 3.6. The classical physical example for this is the stress in a fluid at rest which is termed *hydrostatic stress*, and for which $\sigma_M = -p_0$, the static pressure.

Every state of stress t_{ij} may be decomposed into a spherical portion and a portion S_{ij} known as the *deviator stress* in accordance with the equation

$$t_{ij} = S_{ij} + \delta_{ij}\sigma_M = S_{ij} + \frac{1}{3}\delta_{ij}t_{kk} \tag{3.79}$$

where δ_{ij} is the Kronecker delta. This equation may be solved for S_{ij}, which then appears in the symmetric matrix form

$$\begin{bmatrix} S_{11} & S_{12} & S_{13} \\ S_{12} & S_{22} & S_{23} \\ S_{13} & S_{23} & S_{33} \end{bmatrix} = \begin{bmatrix} t_{11} - \sigma_M & t_{12} & t_{13} \\ t_{12} & t_{22} - \sigma_M & t_{23} \\ t_{13} & t_{23} & t_{33} - \sigma_M \end{bmatrix}. \tag{3.80}$$

Also from Eq 3.79, we notice immediately that the first invariant of the deviator stress is

$$S_{ii} = t_{ii} - \frac{1}{3}\delta_{ii}t_{kk} = 0 \tag{3.81}$$

(since $\delta_{ii} = 3$), so that the characteristic equation for the deviator stress (analogous to Eq 3.39 for t_{ij}) is

$$S^3 + II_S S - III_S = 0 \tag{3.82}$$

for which the deviator stress invariants are

$$II_S = -\frac{1}{2} S_{ij} S_{ji} = S_I S_{II} + S_{II} S_{III} + S_{III} S_I \, , \tag{3.83a}$$

$$III_S = \varepsilon_{ijk} S_{1i} S_{2j} S_{3k} = S_I S_{II} S_{III} \, . \tag{3.83b}$$

Finally, consider a principal direction $n_j^{(q)}$ of t_{ij} such that the eigenvalue equation $\left[t_{ij} - \sigma_{(q)} \delta_{ij}\right] n_j^{(q)} = 0$ is satisfied. Then, from the definition of S_{ij}, we have

$$[S_{ij} + \sigma_M \delta_{ij} - \sigma_q \delta_{ij}] \, n_j^{(q)} = 0 \, ,$$

or

$$[S_{ij} - (\sigma_q - \sigma_M) \, \delta_{ij}] \, n_j^{(q)} = 0 \tag{3.84}$$

which demonstrates that $n_j^{(q)}$ is also a principal direction of S_{ij}, and furthermore, the principal values of S_{ij} are given in terms of the principal values of t_{ij} by

$$S_q = \sigma_q - \sigma_M \, , \qquad (q = 1, 2, 3) \, . \tag{3.85}$$

Example 3.7

Decompose the stress tensor $\mathbf{T}$ of Example 3.4 into its deviator and spherical portions and determine the principal stress values of the deviator portion.

$$[t_{ij}] = \begin{bmatrix} 57 & 0 & 24 \\ 0 & 50 & 0 \\ 24 & 0 & 43 \end{bmatrix} \quad [\text{MPa}] \, .$$

Solution

By Eq 3.77, σ_M for the given stress is

$$\sigma_M = \frac{1}{3}(57 + 50 + 43) = 50 \quad [\text{MPa}]$$

Thus, decomposition by Eq 3.79 leads to the matrix sum

$$\begin{bmatrix} 57 & 0 & 24 \\ 0 & 50 & 0 \\ 24 & 0 & 43 \end{bmatrix} = \begin{bmatrix} 7 & 0 & 24 \\ 0 & 0 & 0 \\ 24 & 0 & -7 \end{bmatrix} + \begin{bmatrix} 50 & 0 & 0 \\ 0 & 50 & 0 \\ 0 & 0 & 50 \end{bmatrix} \quad [\text{MPa}] \, .$$

Principal stress values of the deviator portion result from the determinant

$$\begin{vmatrix} 7 - S & 0 & 24 \\ 0 & -S & 0 \\ 24 & 0 & -7 - S \end{vmatrix} = -S[(7 - S)(-7 - S) - 24^2] = 0$$

which is readily factored to yield $S_I = 25$ MPa, $S_{II} = 0$ and $S_{III} = -25$ MPa. These results are easily verified using the principal values determined in Example 3.4 together with Eq 3.85.

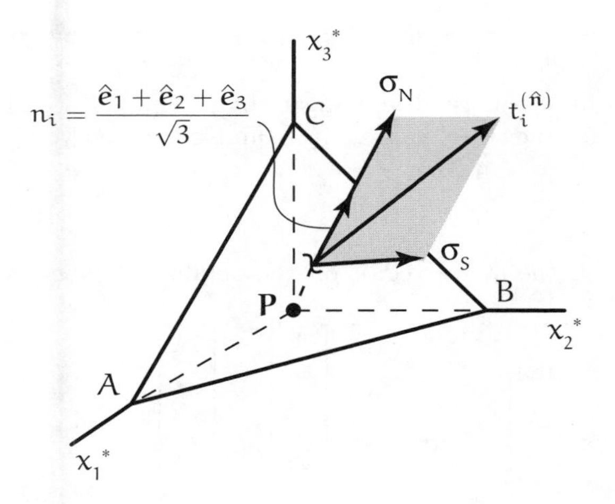

FIGURE 3.19
Octahedral plane (*ABC*) with traction vector $t_i^{(\hat{n})}$, and octahedral normal and shear stresses, σ_N and σ_S.

3.11 Octahedral Shear Stress

Consider the plane at **P** whose unit normal makes equal angles with the principal stress directions. That plane, called the *octahedral plane*, may be pictured as the triangular surface ABC of Fig. 3.19 and imagined to be the face in the first octant of a regular octahedron. The traction vector on this plane is

$$\mathbf{t}^{(\hat{n})} = \mathbf{T}^* \cdot \hat{\mathbf{n}} = \frac{\sigma_{(1)}\hat{\mathbf{e}}_1^* + \sigma_{(2)}\hat{\mathbf{e}}_2^* + \sigma_{(3)}\hat{\mathbf{e}}_3^*}{\sqrt{3}}, \tag{3.86}$$

and its component in the direction of $\hat{\mathbf{n}}$ is

$$\sigma_N = \mathbf{t}^{(\hat{n})} \cdot \hat{\mathbf{n}} = \frac{1}{3}\left[\sigma_{(1)} + \sigma_{(2)} + \sigma_{(3)}\right] = \frac{1}{3}\sigma_{ii}. \tag{3.87}$$

Thus, from Eq 3.52, the square of the shear stress on the octahedral plane, known as the octahedral shear stress, is

$$\sigma_{oct}^2 = \mathbf{t}^{(\hat{n})} \cdot \mathbf{t}^{(\hat{n})} - \sigma_N^2 = \frac{1}{3}[\sigma_{(1)}^2 + \sigma_{(2)}^2 + \sigma_{(3)}^2] - \frac{1}{9}[\sigma_{(1)} + \sigma_{(2)} + \sigma_{(3)}]^2 \tag{3.88}$$

which may be reduced to either the form (see Problem 3.27)

$$\sigma_{oct} = \frac{1}{3}\sqrt{\left(\sigma_{(1)} - \sigma_{(2)}\right)^2 + \left(\sigma_{(2)} - \sigma_{(3)}\right)^2 + \left(\sigma_{(3)} - \sigma_{(1)}\right)^2}, \tag{3.89}$$

or

$$\sigma_{oct} = \frac{\sqrt{S_1^2 + S_2^2 + S_3^2}}{\sqrt{3}} = \sqrt{\frac{-2II_S}{3}}. \tag{3.90}$$

Example 3.8

Determine directly the normal and shear components, σ_N and σ_{oct}, on the octahedral plane for the state of stress in Example 3.4, and verify the result for σ_{oct} by Eq 3.89.

Solution
From Example 3.4, the stress vector on the octahedral plane is given by the matrix product

$$\begin{bmatrix} 75 & 0 & 0 \\ 0 & 50 & 0 \\ 0 & 0 & 25 \end{bmatrix} \begin{bmatrix} \frac{1}{\sqrt{3}} \\ \frac{1}{\sqrt{3}} \\ \frac{1}{\sqrt{3}} \end{bmatrix} = \begin{bmatrix} \frac{75}{\sqrt{3}} \\ \frac{50}{\sqrt{3}} \\ \frac{25}{\sqrt{3}} \end{bmatrix},$$

or

$$\mathbf{t}^{(\hat{\mathbf{n}})} = \frac{75\hat{\mathbf{e}}_1^* + 50\hat{\mathbf{e}}_2^* + 25\hat{\mathbf{e}}_3^*}{\sqrt{3}},$$

so that

$$\sigma_N = \mathbf{t}^{(\hat{\mathbf{n}})} \cdot \hat{\mathbf{n}} = \frac{1}{3}(75 + 50 + 25) = 50 \text{ MPa}.$$

Also, from Eq 3.52,

$$\sigma_{oct}^2 = \mathbf{t}^{(\hat{\mathbf{n}})} \cdot \mathbf{t}^{(\hat{\mathbf{n}})} - \sigma_N^2 = \frac{1}{3}(75^2 + 50^2 + 25^2) - 50^2 = 417 \text{ MPa},$$

and so $\sigma_{oct} = 20.4$ MPa. By Eq 3.89, we verify directly that

$$\sigma_{oct} = \frac{1}{3}\sqrt{(75-50)^2 + (50-25)^2 + (25-75)^2} = 20.41 \text{ MPa}$$

Example 3.9

Show that the octahedral shear stress may be written as

$$\sigma_{oct} = \frac{1}{\sqrt{3}}\left[(\sigma_{(1)} - \sigma_M)^2 + (\sigma_{(2)} - \sigma_M)^2 + (\sigma_{(3)} - \sigma_M)^2\right]^{1/2},$$

and, from this, show that

$$\sigma_{oct} = \frac{\sqrt{2}}{3}\left(I_T^2 + 3II_T\right)^{1/2}$$

is another form of the octahedral shear stress.

Solution
Start by multiplying the squared terms under the radical

$$\sigma_{oct} = \frac{1}{\sqrt{3}}\left[\sigma_{(1)}^2 + \sigma_{(2)}^2 + \sigma_{(3)}^2 + 3\sigma_M^2 - 2\left(\sigma_{(1)} + \sigma_{(2)} + \sigma_{(3)}\right)\sigma_M\right]^{1/2}.$$

Recall Eq 3.77 in the form

$$\sigma_{(1)} + \sigma_{(2)} + \sigma_{(3)} = 3\sigma_M$$

and substitute to get

$$\sigma_{oct}^2 = \left(\frac{\sigma_{(1)}^2 + \sigma_{(2)}^2 + \sigma_{(3)}^2}{3} - \sigma_M^2 \right) .$$

Substitution of Eq 3.77 again for σ_M results in Eq 3.88. Multiply out the mean stress terms and simplify to get

$$\begin{aligned} \sigma_{oct}^2 &= \left[\frac{3(\sigma_{(1)}^2 + \sigma_{(2)}^2 + \sigma_{(3)}^2)}{9} \right. \\ &\quad \left. - \frac{\sigma_{(1)}^2 + \sigma_{(2)}^2 + \sigma_{(3)}^2 + 2(\sigma_{(1)}\sigma_{(2)} + \sigma_{(2)}\sigma_{(3)} + \sigma_{(1)}\sigma_{(3)})}{9} \right] \\ &= \tfrac{2}{9} \left[\sigma_{(1)}^2 + \sigma_{(2)}^2 + \sigma_{(3)}^2 - (\sigma_{(1)}\sigma_{(2)} + \sigma_{(2)}\sigma_{(3)} + \sigma_{(1)}\sigma_{(3)}) \right] . \end{aligned}$$

Finally, add and subtract $2\left(\sigma_{(1)}\sigma_{(2)} + \sigma_{(2)}\sigma_{(3)} + \sigma_{(1)}\sigma_{(3)}\right)$ in the bracket

$$\begin{aligned} \sigma_{oct}^2 &= \tfrac{2}{9} \left[\sigma_{(1)}^2 + \sigma_{(2)}^2 + \sigma_{(3)}^2 + 2(\sigma_{(1)}\sigma_{(2)} + \sigma_{(2)}\sigma_{(3)} + \sigma_{(1)}\sigma_{(3)}) \right. \\ &\quad \left. -3(\sigma_{(1)}\sigma_{(2)} + \sigma_{(2)}\sigma_{(3)} + \sigma_{(1)}\sigma_{(3)}) \right] , \end{aligned}$$

and note that, for principal axes

$$\begin{aligned} I_T^2 &= \left(\sigma_{(1)} + \sigma_{(2)} + \sigma_{(3)}\right)^2 \\ II_T &= \sigma_{(1)}\sigma_{(2)} + \sigma_{(2)}\sigma_{(3)} + \sigma_{(1)}\sigma_{(3)} \end{aligned}$$

leaving

$$\sigma_{oct}^2 = \tfrac{2}{9} \left[I_T^2 + 3II_T \right] ,$$

or

$$\sigma_{oct} = \tfrac{\sqrt{2}}{3} \left[I_T^2 + 3II_T \right]^{1/2} .$$

Problems

Problem 3.1

At a point **P**, the stress tensor relative to axes $Px_1x_2x_3$ has components t_{ij}. On the area element $dS^{(1)}$ having the unit normal $\hat{n}_1$, the stress vector is $\mathbf{t}^{(\hat{n}_1)}$, and on area element $dS^{(2)}$ with normal $\hat{n}_2$ the stress vector is $\mathbf{t}^{(\hat{n}_2)}$. Show that the component of $\mathbf{t}^{(\hat{n}_1)}$ in the direction of $\hat{n}_2$ is equal to the component of $\mathbf{t}^{(\hat{n}_2)}$ in the direction of $\hat{n}_1$.

Problem 3.2

Verify the result established in Problem 3.1 for the area elements having normals

$$\hat{n}_1 = \frac{1}{7}(2\hat{e}_1 + 3\hat{e}_2 + 6\hat{e}_3)$$

$$\hat{n}_2 = \frac{1}{7}(3\hat{e}_1 - 6\hat{e}_2 + 2\hat{e}_3)$$

if the stress matrix at **P** is given with respect to axes $Px_1x_2x_3$ by

$$[t_{ij}] = \begin{bmatrix} 35 & 0 & 21 \\ 0 & 49 & 0 \\ 21 & 0 & 14 \end{bmatrix}.$$

Problem 3.3

The stress tensor at **P** relative to axes $Px_1x_2x_3$ has components in MPa given by the matrix representation

$$[t_{ij}] = \begin{bmatrix} t_{11} & 2 & 1 \\ 2 & 0 & 2 \\ 1 & 2 & 0 \end{bmatrix}$$

where t_{11} is unspecified. Determine a direction $\hat{n}$ at **P** for which the plane perpendicular to $\hat{n}$ will be stress-free, that is, for which $\mathbf{t}^{(\hat{n})} = 0$ on that plane. What is the required value of t_{11} for this condition?

Answer

$$\hat{n} = \frac{1}{3}(2\hat{e}_1 - \hat{e}_2 - 2\hat{e}_3), \quad t_{11} = 2 \text{ MPa}$$

Problem 3.4

The stress tensor has components at point **P** in ksi units as specified by the matrix

$$[t_{ij}] = \begin{bmatrix} -9 & 3 & -6 \\ 3 & 6 & 9 \\ -6 & 9 & -6 \end{bmatrix}.$$

Determine:

(a) the stress vector on the plane at **P** whose normal vector is

$$\hat{n} = \frac{1}{9}(\hat{e}_1 + 4\hat{e}_2 + 8\hat{e}_3) \ ,$$

(b) the magnitude of this stress vector,

(c) the component of the stress vector in the direction of the normal,

(d) the angle in degrees between the stress vector and the normal.

Answer

(a) $\mathbf{t}^{(\hat{n})} = -5\hat{e}_1 + 11\hat{e}_2 - 2\hat{e}_3$

(b) $\mathbf{t}^{(\hat{n})} = \sqrt{150}$

(c) $\dfrac{23}{9}$

(d) $77.96°$

Problem 3.5

Let the stress tensor components at a point be given by $t_{ij} = \pm\sigma_0 n_i n_j$ where σ_0 is a positive constant. Show that this represents a uniaxial state of stress having a magnitude $\pm\sigma_0$ and acting in the direction of n_i.

Problem 3.6

Show that the sum of squares of the magnitudes of the stress vectors on the coordinate planes is independent of the orientation of the coordinate axes, that is, show that the sum

$$t_i^{(\hat{e}_1)} t_i^{(\hat{e}_1)} + t_i^{(\hat{e}_2)} t_i^{(\hat{e}_2)} + t_i^{(\hat{e}_3)} t_i^{(\hat{e}_3)}$$

is an invariant.

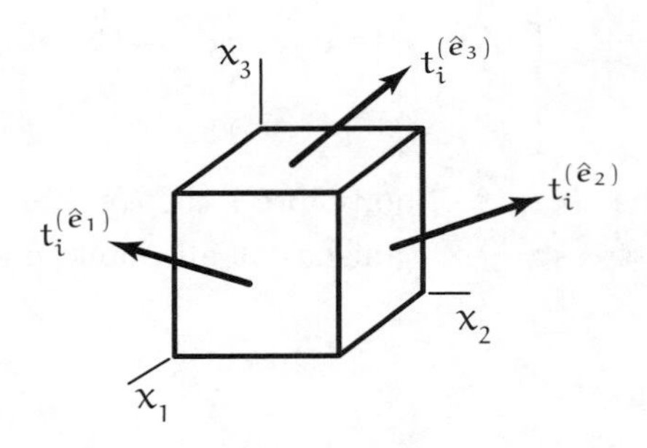

Problem 3.7

With respect to axes $Ox_1x_2x_3$ the stress state is given in terms of the coordinates by the matrix

$$[t_{ij}] = \begin{bmatrix} x_1x_2 & x_2^2 & 0 \\ x_2^2 & x_2x_3 & x_3^2 \\ 0 & x_3^2 & x_3x_1 \end{bmatrix} .$$

Determine

(a) the body force components as functions of the coordinates if the equilibrium equations are to be satisfied everywhere, and
(b) the stress vector at point $\mathbf{P}(1,2,3)$ on the plane whose outward unit normal makes equal angles with the positive coordinate axes.

Answer

$$\text{(a)}\ b_1 = \frac{-3x_2}{\rho},\quad b_2 = \frac{-3x_3}{\rho},\quad b_3 = \frac{-x_1}{\rho}$$

$$\text{(b)}\ \mathbf{t}^{(\hat{\mathbf{n}})} = \frac{(6\hat{\mathbf{e}}_1 + 19\hat{\mathbf{e}}_2 + 12\hat{\mathbf{e}}_3)}{\sqrt{3}}$$

Problem 3.8

Relative to the Cartesian axes $Ox_1x_2x_3$ a stress field is given by the matrix

$$[t_{ij}] = \begin{bmatrix} (1 - x_1^2)\, x_2 + \frac{2}{3}x_2^3 & -(4 - x_2^2)\, x_1 & 0 \\ -(4 - x_2^2)\, x_1 & -\frac{1}{3}(x_2^3 - 12x_2) & 0 \\ 0 & 0 & (3 - x_1^2)\, x_2 \end{bmatrix}.$$

(a) Show that the equilibrium equations are satisfied everywhere for zero body forces.
(b) Determine the stress vector at the point $\mathbf{P}(2,-1,6)$ of the plane whose equation is $3x_1 + 6x_2 + 2x_3 = 12$.

Answer

$$\text{(b)}\quad \mathbf{t}^{(\hat{\mathbf{n}})} = \frac{1}{7}(-29\hat{\mathbf{e}}_1 - 40\hat{\mathbf{e}}_2 + 2\hat{\mathbf{e}}_3)$$

Problem 3.9

The stress components in a circular cylinder of length L and radius r are given by

$$[t_{ij}] = \begin{bmatrix} Ax_2 + Bx_3 & Cx_3 & -Cx_2 \\ Cx_3 & 0 & 0 \\ -C_2 & 0 & 0 \end{bmatrix}.$$

(a) Verify that in the absence of body forces the equilibrium equations are satisfied.
(b) Show that the stress vector vanishes at all points on the curved surface of the cylinder.

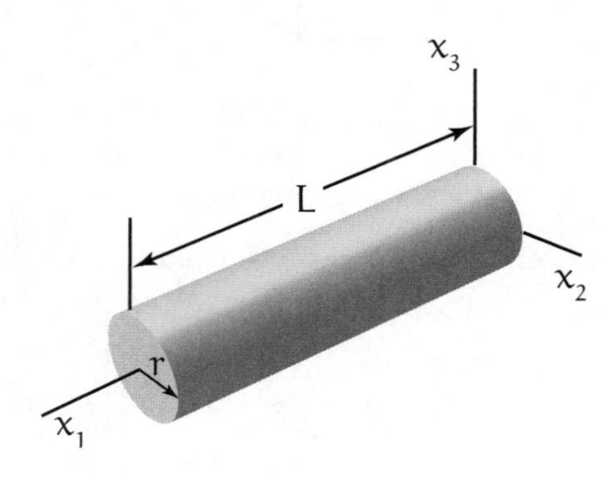

Problem 3.10

Rotated axes $Px'_1x'_2x'_3$ are obtained from axes $Px_1x_2x_3$ by a right-handed rotation about the line PQ that makes equal angles with respect to the $Px_1x_2x_3$ axes (see sketch). Determine the primed stress components for the stress tensor in (MPa)

$$[t_{ij}] = \begin{bmatrix} 3 & 0 & 6 \\ 0 & 0 & 0 \\ 6 & 0 & -3 \end{bmatrix}$$

if the angle of rotation is (a) 120°, or (b) 60°.

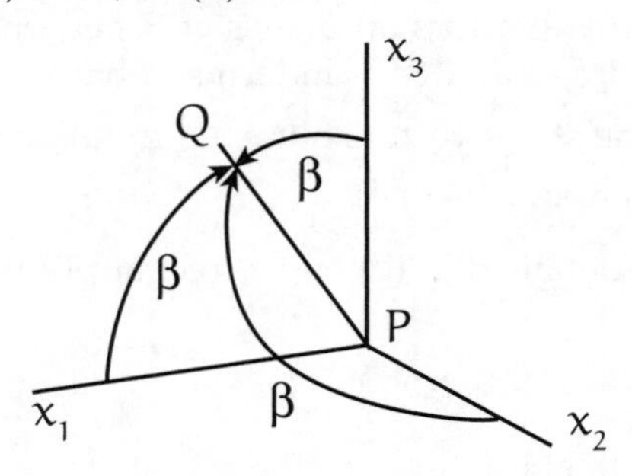

Answer

(a) $\left[t'_{ij}\right] = \begin{bmatrix} 0 & 0 & 0 \\ 0 & -3 & 6 \\ 0 & 6 & 3 \end{bmatrix}$ MPa, (b) $\left[t'_{ij}\right] = \frac{1}{3}\begin{bmatrix} -5 & 10 & 10 \\ 10 & -11 & -2 \\ 10 & -2 & 16 \end{bmatrix}$ MPa

Problem 3.11

At the point **P**, rotated axes $Px'_1x'_2x'_3$ are related to the axes $Px_1x_2x_3$ by the transformation matrix

$$[a_{ij}] = \frac{1}{3}\begin{bmatrix} a & 1-\sqrt{3} & 1+\sqrt{3} \\ 1+\sqrt{3} & b & 1-\sqrt{3} \\ 1-\sqrt{3} & 1+\sqrt{3} & c \end{bmatrix}$$

where a, b, and c are to be determined. Determine $\left[t'_{ij}\right]$ if the stress matrix relative to axes $Px_1x_2x_3$ is given in MPa by

$$[t_{ij}] = \begin{bmatrix} 1 & 0 & 1 \\ 0 & 1 & 0 \\ 1 & 0 & 1 \end{bmatrix}.$$

Answer

$$\left[t'_{ij}\right] = \frac{1}{9}\begin{bmatrix} 11+2\sqrt{3} & 5+\sqrt{3} & -1 \\ 5+\sqrt{3} & 5 & 5-\sqrt{3} \\ -1 & 5-\sqrt{3} & 11-2\sqrt{3} \end{bmatrix} \text{MPa}$$

Problem 3.12

The stress matrix referred to axes $Px_1x_2x_3$ is given in ksi by

$$[t_{ij}] = \begin{bmatrix} 14 & 0 & 21 \\ 0 & 21 & 0 \\ 21 & 0 & 7 \end{bmatrix}$$

Let rotated axes $Px'_1x'_2x'_3$ be defined with respect to axes $Px_1x_2x_3$ by the table of base vectors

	$\hat{e}_1$	$\hat{e}_2$	$\hat{e}_3$
$\hat{e}'_1$	2/7	3/7	6/7
$\hat{e}'_2$	3/7	−6/7	2/7
$\hat{e}'_3$	6/7	2/7	−3/7

(a) Determine the stress vectors on planes at **P** perpendicular to the primed axes; determine $\mathbf{t}^{(\hat{e}'_1)}$, $\mathbf{t}^{(\hat{e}'_2)}$, and $\mathbf{t}^{(\hat{e}'_3)}$ in terms of base vectors $\hat{e}_1$, $\hat{e}_2$, and $\hat{e}_3$.

(b) Project each of the stress vectors obtained in (a) onto the primed axes to determine the nine components of $\left[t'_{ij}\right]$.

(c) Verify the result obtained in (b) by a direct application of Eq 3.33 of the text.

Answer

$$\left[t'_{ij}\right] = \frac{1}{7}\begin{bmatrix} 143 & 36 & 114 \\ 36 & 166 & 3 \\ 114 & 3 & -15 \end{bmatrix} \text{ ksi}$$

Problem 3.13

At point **P**, the stress matrix is given in MPa with respect to axes $Px_1x_2x_3$ by

$$\text{Case 1: } [t_{ij}] = \begin{bmatrix} 6 & 4 & 0 \\ 4 & 6 & 0 \\ 0 & 0 & -2 \end{bmatrix} \quad \text{Case 2: } [t_{ij}] = \begin{bmatrix} 2 & 1 & 1 \\ 1 & 2 & 1 \\ 1 & 1 & 2 \end{bmatrix}$$

Determine for each case

(a) the principal stress values,

(b) the principal stress directions.

Answer

(a) Case 1: $\sigma_{(1)} = 10$ MPa, $\sigma_{(2)} = 2$ MPa, $\sigma_{(3)} = -2$ MPa
Case 2: $\sigma_{(1)} = 4$ MPa, $\sigma_{(2)} = \sigma_{(3)} = 1$ MPa

(b) Case 1: $\hat{n}^{(1)} = \pm\dfrac{\hat{e}_1 + \hat{e}_2}{\sqrt{2}}$, $\hat{n}^{(2)} = \pm\dfrac{\hat{e}_1 - \hat{e}_2}{\sqrt{2}}$, $\hat{n}^{(3)} = \mp\hat{e}_3$

Case 2: $\hat{n}^{(1)} = \dfrac{\hat{e}_1 + \hat{e}_2 + \hat{e}_3}{\sqrt{3}}$, $\hat{n}^{(2)} = \dfrac{-\hat{e}_1 + \hat{e}_2}{\sqrt{2}}$, $\hat{n}^{(3)} = \dfrac{-\hat{e}_1 - \hat{e}_2 + 2\hat{e}_3}{\sqrt{6}}$

Problem 3.14

When referred to principal axes at **P**, the stress matrix in ksi units is

$$\left[t^*_{ij}\right] = \begin{bmatrix} 2 & 0 & 0 \\ 0 & 7 & 0 \\ 0 & 0 & 12 \end{bmatrix}$$

If the transformation matrix between the principal axes and axes $Px_1x_2x_3$ is

$$[a_{ij}] = \frac{1}{\sqrt{2}} \begin{bmatrix} -\frac{3}{5} & 1 & -\frac{4}{5} \\ a_{21} & a_{22} & a_{23} \\ -\frac{3}{5} & -1 & -\frac{4}{5} \end{bmatrix}$$

where a_{21}, a_{22}, and a_{23} are to be determined, calculate $[t_{ij}]$.

Answer

$$[t_{ij}] = \begin{bmatrix} 7 & 3 & 0 \\ 3 & 7 & 4 \\ 0 & 4 & 7 \end{bmatrix} \text{ ksi}$$

Problem 3.15

The stress matrix in MPa when referred to axes $Px_1x_2x_3$ is

$$[t_{ij}] = \begin{bmatrix} 3 & -10 & 0 \\ -10 & 0 & 30 \\ 0 & 30 & -27 \end{bmatrix}.$$

Determine

(a) the principal stresses, σ_I, σ_{II}, σ_{III},

(b) the principal stress directions.

Answer

(a) $\sigma_I = 23$ MPa, $\sigma_{II} = 0$ MPa, $\sigma_{III} = -47$ MPa

(b) $\hat{n}^{(1)} = -0.394\hat{e}_1 + 0.788\hat{e}_2 + 0.473\hat{e}_3$

$\hat{n}^{(2)} = 0.931\hat{e}_1 + 0.274\hat{e}_2 + 0.304\hat{e}_3$

$\hat{n}^{(3)} = 0.110\hat{e}_1 + 0.551\hat{e}_2 - 0.827\hat{e}_3$

Problem 3.16

At point **P**, the stress matrix relative to axes $Px_1x_2x_3$ is given in MPa by

$$[t_{ij}] = \begin{bmatrix} 5 & a & -a \\ a & 0 & b \\ -a & b & 0 \end{bmatrix}$$

where a and b are unspecified. At the same point relative to axes $Px_1^*x_2^*x_3^*$ the matrix is

$$[t_{ij}^*] = \begin{bmatrix} \sigma_I & 0 & 0 \\ 0 & 2 & 0 \\ 0 & 0 & \sigma_{III} \end{bmatrix}.$$

If the magnitude of the maximum shear stress at **P** is 5.5 MPa, determine σ_I and σ_{III}.

Answer

$\sigma_I = 7$ MPa, $\sigma_{III} = -4$ MPa

Problem 3.17

The state of stress at point **P** is given in ksi with respect to axes $Px_1x_2x_3$ by the matrix

$$[t_{ij}] = \begin{bmatrix} 1 & 0 & 2 \\ 0 & 1 & 0 \\ 2 & 0 & -2 \end{bmatrix}.$$

Determine

(a) the principal stress values and principal stress directions at **P**,

(b) the maximum shear stress value at **P**,

(c) the normal $\hat{n} = n_i \hat{e}_i$ to the plane at **P** on which the maximum shear stress acts.

Answer

(a) $\sigma_I = 2$ ksi, $\sigma_{II} = 1$ ksi, $\sigma_{III} = -3$ ksi
$\hat{n}^{(1)} = \dfrac{2\hat{e}_1 + \hat{e}_3}{\sqrt{5}}$, $\hat{n}^{(2)} = \hat{e}_2$, $\hat{n}^{(3)} = \dfrac{-\hat{e}_1 + 2\hat{e}_3}{\sqrt{5}}$

(b) $(\sigma_S)_{max} = \pm\ 2.5$ ksi

(c) $\hat{n} = \dfrac{\hat{e}_1 + 3\hat{e}_3}{\sqrt{10}}$

Problem 3.18

The stress tensor at **P** is given with respect to $Ox_1x_2x_3$ in matrix form with units of MPa by

$$[t_{ij}] = \begin{bmatrix} 4 & b & b \\ b & 7 & 2 \\ b & 2 & 4 \end{bmatrix}$$

where b is unspecified. If $\sigma_{III} = 3$ MPa and $\sigma_I = 2\sigma_{II}$, determine

(a) the principal stress values,

(b) the value of b,

(c) the principal stress direction of σ_{II} .

Answer

(a) $\sigma_I = 8$ MPa, $\sigma_{II} = 4$ MPa, $\sigma_{III} = 3$ MPa

(b) $b = 0$, (c) $\hat{n}^{(2)} = \hat{e}_1$

Problem 3.19

The state of stress at **P**, when referred to axes $Px_1x_2x_3$ is given in ksi units by the matrix

$$[t_{ij}] = \begin{bmatrix} 9 & 3 & 0 \\ 3 & 9 & 0 \\ 0 & 0 & 18 \end{bmatrix}$$

Determine

(a) the principal stress values at **P**,

(b) the unit normal $\hat{\mathbf{n}}^* = n_i \hat{\mathbf{e}}_i^*$ of the plane on which $\sigma_N = 12$ ksi and $\sigma_S = 3$ ksi.

Answer

(a) $\sigma_I = 18$ ksi, $\sigma_{II} = 12$ ksi, $\sigma_{III} = 6$ ksi

(b) $\hat{\mathbf{n}}^* = \dfrac{\hat{\mathbf{e}}_1^* + \sqrt{6}\hat{\mathbf{e}}_2^* + \hat{\mathbf{e}}_3^*}{2\sqrt{2}}$

Problem 3.20

Verify the result listed for Problem 3.19b above by use of Eq 3.63.

Problem 3.21

Sketch the Mohr's circles for the various stress states shown on the cube which is oriented along the coordinate axes.

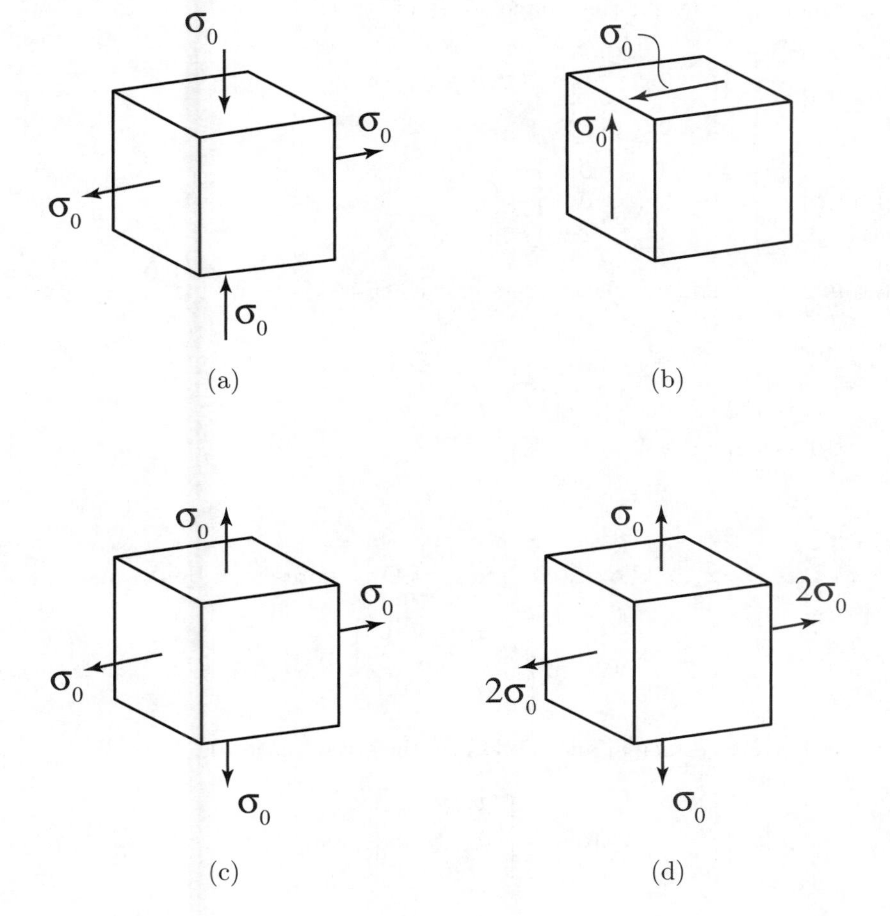

Problem 3.22
The state of stress referred to axes $Px_1x_2x_3$ is given in MPa by the matrix

$$[t_{ij}] = \begin{bmatrix} 9 & 12 & 0 \\ 12 & -9 & 0 \\ 0 & 0 & 5 \end{bmatrix} .$$

Determine

(a) the normal and shear components, σ_N and σ_S, respectively, on the plane at **P** whose unit normal is

$$\hat{n} = \frac{1}{5}(4\hat{e}_1 + 3\hat{e}_2) ,$$

(b) verify the result determined in (a) by a Mohr's circle construction similar to that shown in the figure from Example 3.5.

Answer

$\sigma_N = 14.04$ MPa, $\sigma_S = 5.28$ MPa

Problem 3.23
Sketch the Mohr's circles for the simple states of stress given by

$$\text{(a)}\ [t_{ij}] = \begin{bmatrix} \sigma_0 & 0 & \sigma_0 \\ 0 & \sigma_0 & 0 \\ \sigma_0 & 0 & \sigma_0 \end{bmatrix} ,$$

$$\text{(b)}\ [t_{ij}] = \begin{bmatrix} \sigma_0 & 0 & 0 \\ 0 & 2\sigma_0 & 0 \\ 0 & 0 & -\sigma_0 \end{bmatrix} ,$$

and determine the maximum shear stress in each case.

Answer

(a) $(\sigma_S)_{max} = \sigma_0$
(b) $(\sigma_S)_{max} = \frac{3}{2}\sigma_0$

Problem 3.24
Relative to axes $Ox_1x_2x_3$, the state of stress at O is represented by the matrix

$$[t_{ij}] = \begin{bmatrix} 6 & -3 & 0 \\ -3 & 6 & 0 \\ 0 & 0 & 0 \end{bmatrix} \text{[ksi]} .$$

Show that, relative to principal axes $Ox_1^*x_2^*x_3^*$, the stress matrix is

$$[t_{ij}] = \begin{bmatrix} 3 & 0 & 0 \\ 0 & 9 & 0 \\ 0 & 0 & 0 \end{bmatrix} \text{[ksi]} ,$$

and that these axes result from a rotation of 45° about the x_3 axis. Verify these results by Eq 3.75.

Problem 3.25

The stress matrix representation at **P** is given by

$$[t_{ij}] = \begin{bmatrix} 29 & 0 & 0 \\ 0 & -26 & 6 \\ 0 & 6 & 9 \end{bmatrix} \text{ [ksi]} .$$

Decompose this matrix into its spherical and deviator parts, and determine the principal deviator stress values.

Answer

$S_I = 25$ ksi, $S_{II} = 6$ ksi, $S_{III} = -31$ ksi

Problem 3.26

Let the second invariant of the stress deviator be expressed in terms of its principal values, that is, by

$$II_S = S_I S_{II} + S_{II} S_{III} + S_{III} S_I .$$

Show that this sum is the negative of two-thirds the sum of squares of the principal shear stresses, as given by Eq 3.60.

Problem 3.27

Verify the results presented in Eqs 3.89 and 3.90 for the octahedral shear stress.

Problem 3.28

At point **P** in a continuum body, the stress tensor components are given in MPa with respect to axes $Px_1x_2x_3$ by the matrix

$$[t_{ij}] = \begin{bmatrix} 1 & -3 & \sqrt{2} \\ -3 & 1 & -\sqrt{2} \\ \sqrt{2} & -\sqrt{2} & 4 \end{bmatrix} .$$

Determine

(a) the principal stress values σ_I, σ_{II}, and σ_{III}, together with the corresponding principal stress directions,

(b) the stress invariants I_T, II_T, and III_T,

(c) the maximum shear stress value and the normal to the plane on which it acts,

(d) the principal deviator stress values,

(e) the stress vector on the octahedral plane together with its normal and shear components,

(f) the stress matrix for axes rotated 60° counterclockwise with respect to the axis PQ, which makes equal angles relative to the coordinate axes $Px_1x_2x_3$.

Answer

(a) $\sigma_I = 6$ MPa, $\sigma_{II} = 2$ MPa, $\sigma_{III} = -2$ MPa

$$\hat{\mathbf{n}}^{(1)} = \frac{1}{2}\left(\hat{\mathbf{e}}_1 - \hat{\mathbf{e}}_2 + \sqrt{2}\hat{\mathbf{e}}_3\right)$$

$$\hat{\mathbf{n}}^{(2)} = \frac{1}{2}\left(\hat{\mathbf{e}}_1 - \hat{\mathbf{e}}_2 - \sqrt{2}\hat{\mathbf{e}}_3\right)$$

$$\hat{\mathbf{n}}^{(3)} = \frac{\hat{\mathbf{e}}_1 + \hat{\mathbf{e}}_2}{\sqrt{2}}$$

(b) $I_T = 6$ MPa, $II_T = -4$ MPa, $III_T = -24$ MPa

(c) $(\sigma_S)_{max} = 4$ MPa $\quad \hat{\mathbf{n}}_{max} = \dfrac{\left(1+\sqrt{2}\right)\hat{\mathbf{e}}_1 - \left(1-\sqrt{2}\right)\hat{\mathbf{e}}_2 + \sqrt{2}\hat{\mathbf{e}}_3}{2\sqrt{2}}$

(d) $S_I = 4$ MPa, $S_{II} = 0$ MPa, $S_{III} = -4$ MPa

(e) $\hat{\mathbf{t}}^{(\hat{\mathbf{n}})} = \dfrac{6\hat{\mathbf{e}}_1^* + 2\hat{\mathbf{e}}_2^* - 2\hat{\mathbf{e}}_3^*}{\sqrt{3}}$, $\sigma_N = 2$, $\sigma_{oct} = \sqrt{\frac{32}{3}}$

(f) $$\left[t'_{ij}\right] = \frac{1}{9}\begin{bmatrix} -12 & -12+3\sqrt{2} & -12-3\sqrt{2} \\ -12+3\sqrt{2} & 33-12\sqrt{2} & -3 \\ -12-3\sqrt{2} & -3 & 33+12\sqrt{2} \end{bmatrix} \text{ MPa}$$

Problem 3.29

In a continuum, the stress field relative to axes $Ox_1x_2x_3$ is given by

$$[t_{ij}] = \begin{bmatrix} x_1^2x_2 & x_1\left(1-x_2^2\right) & 0 \\ x_1\left(1-x_2^2\right) & \frac{1}{3}\left(x_2^3 - 3x_2\right) & 0 \\ 0 & 0 & 2x_3^2 \end{bmatrix}.$$

Determine

(a) the body force distribution if the equilibrium equations are to be satisfied throughout the field,

(b) the principal stresses at $\mathbf{P}(a, 0, 2\sqrt{a})$,

(c) the maximum shear stress at $\mathbf{P}$,

(d) the principal deviator stresses at $\mathbf{P}$.

Answer

(a) $b_1 = b_2 = 0$, $b_3 = -\dfrac{4x_3}{\rho}$

(b) $\sigma_I = 8a$, $\sigma_{II} = a$, $\sigma_{III} = -a$

(c) $(\sigma_S)_{max} = \pm 4.5a$

(d) $S_I = \dfrac{16}{3}a$, $S_{II} = -\dfrac{5}{3}a$, $S_{III} = -\dfrac{11}{3}a$

Problem 3.30

In describing the yield surface in plasticity the second invariant of the deviator stress, often denoted by J_2, plays an important role. Starting with the second invariant of the deviator stress

$$J_2 = II_S = -\left(S_{(1)}S_{(2)} + S_{(2)}S_{(3)} + S_{(1)}S_{(3)}\right)$$

derive the formula

$$J_2 = \frac{3}{2}\sigma_{oct}^2 \,.$$

Problem 3.31

Show that

$$J_2 = II_S = -(S_{(1)}S_{(2)} + S_{(2)}S_{(3)} + S_{(1)}S_{(3)}) = \frac{1}{2}\left(S_{(1)}^2 + S_{(2)}^2 + S_{(3)}^2\right)$$

where J_2 is the second invariant of the deviator stress and $S_{(1)}$, $S_{(2)}$, $S_{(3)}$ are its principal values.

Problem 3.32

Let the stress tensor components t_{ij} be derivable from the symmetric tensor field φ_{ij} by the equation $t_{ij} = \varepsilon_{iqk}\varepsilon_{jpm}\varphi_{km,qp}$. Show that, in the absence of body forces, the equilibrium equations are satisfied. Recall from Problem 2.15 that

$$\varepsilon_{iqk}\varepsilon_{jpm} = \begin{vmatrix} \delta_{ji} & \delta_{jq} & \delta_{jk} \\ \delta_{pi} & \delta_{pq} & \delta_{pk} \\ \delta_{mi} & \delta_{mq} & \delta_{mk} \end{vmatrix} .$$

Problem 3.33

Verify that $\partial t_{ij}/\partial t_{mn} = \delta_{im}\delta_{jn}$ and use this result (or otherwise) to show that

$$\frac{\partial II_S}{\partial t_{ij}} = -S_{ij} \,.$$

that is, the derivative of the second invariant of the deviatoric stress with respect to the stress components is equal to the negative of the corresponding component of the deviatoric stress.

4

Kinematics of Deformation and Motion

Kinematics is the study of motion. In Continuum Mechanics it has the same meaning as in dynamics or fluid mechanics. It is the basis for most all constitutive theories. Recall from Chapter 1 that a constitutive response, or material model, describes how stress develops according to strain or strain rate. Strain and strain rate are a measures of kinematics.

4.1 Particles, Configurations, Deformations and Motion

In continuum mechanics we consider *material bodies* in the form of solids, liquids and gases. Let us begin by describing the model we use to represent such bodies. For this purpose we define a material body $\mathcal{B}$ as the set of elements X, called *particles*, or *material points*, which can be put into a one-to-one correspondence with the points of a regular region of physical space. Note that whereas a particle of classical mechanics has an assigned mass, a continuum particle is essentially a material point for which a density is defined.

The specification of the position of all of the particles of $\mathcal{B}$ with respect to a fixed origin at some instant of time is said to define the *configuration* of the body at that instant. Mathematically, this is expressed by the mapping

$$\mathbf{x} = \boldsymbol{\kappa}(X) \tag{4.1}$$

in which the vector function $\boldsymbol{\kappa}$ assigns the position $\mathbf{x}$ relative to some origin of each particle X of the body. Assume that this mapping is uniquely invertible and differentiable as many times as required; in general two or three times will suffice. The inverse is written

$$X = \boldsymbol{\kappa}^{-1}(\mathbf{x}) \tag{4.2}$$

and identifies the particle X located at position $\mathbf{x}$.

A change in configuration is the result of a *displacement* of the body. For example, a *rigid-body displacement* is one consisting of a simultaneous translation and rotation which produces a new configuration but causes no changes in the size or shape of the body; only changes in its position and/or orientation. On the other hand, an arbitrary displacement will usually include both a rigid-body displacement and a *deformation* which results in a change in size, or shape, or possibly both.

A *motion* of body $\mathcal{B}$ is a continuous time sequence of displacements that carries the set of particles X into various configurations in a stationary space. Such a motion may be expressed by the equation

$$\mathbf{x} = \boldsymbol{\kappa}(X, t) \tag{4.3}$$

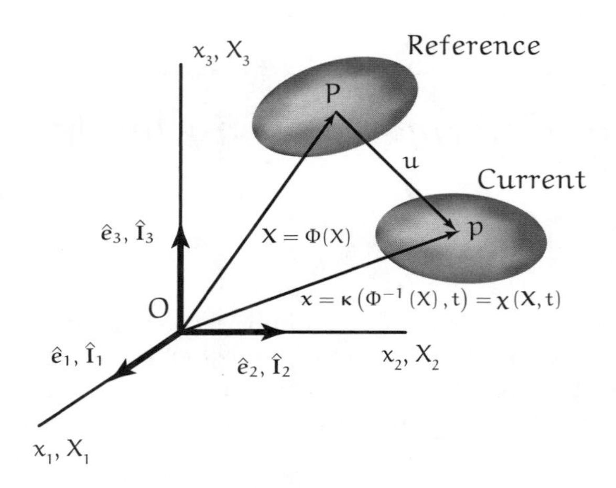

FIGURE 4.1
Position of typical particle in reference configuration X_A and current configuration x_i.

which gives the position $\mathbf{x}$ for each particle X for all times t, where t ranges from $-\infty$ to ∞. As with configuration mappings, we assume the motion function in Eq 4.3 is uniquely invertible and differentiable, so that we may write the inverse

$$X = \boldsymbol{\kappa}^{-1}(\mathbf{x}, t) \tag{4.4}$$

which identifies the particle X located at position $\mathbf{x}$ at time t.

We give special meaning to certain configurations of the body. In particular, we single out a *reference configuration* from which all displacements are reckoned. For the purpose it serves, the reference configuration needs not be one the body ever actually occupies. Often however, the *initial configuration*, that is, the one which the body occupies at time $t = 0$, is chosen as the reference configuration, and the ensuing deformations and motions related to it. The *current configuration* is that one which the body occupies at the current time t.

In developing the concepts of strain we confine attention to two specific configurations without any regard for the sequence by which the second configuration is reached from the first. It is customary to call the first (reference) state the *undeformed configuration*, and the second state the *deformed configuration*. Additionally, time is not a factor in deriving the various strain tensors so that both configurations are considered independent of time.

In fluid mechanics the idea of specific configurations has very little meaning since fluids do not possess a natural geometry, and because of this it is the *velocity field* of a fluid that assumes the fundamental kinematic role.

4.2 Material and Spatial Coordinates

Consider now the reference configuration prescribed by some mapping function $\boldsymbol{\Phi}$ such that the position vector $\mathbf{X}$ of particle X relative to the axes $OX_1X_2X_3$ of Fig. 4.1 is given by

$$\mathbf{X} = \mathbf{\Phi}(X) \tag{4.5}$$

In this case we may express $\mathbf{X}$ in terms of the base vectors $\hat{\mathbf{I}}_1, \hat{\mathbf{I}}_2, \hat{\mathbf{I}}_3$ shown in Fig. 4.1 by the equation

$$\mathbf{X} = X_A \hat{\mathbf{I}}_A \ , \tag{4.6}$$

and we call the components X_A the *material coordinates*, or sometimes the *referential coordinates*, of the particle X. Uppercase letters which are used as subscripts on material coordinates, or on any quantity expressed in terms of material coordinates, obey all the rules of indicial notation. It is customary to designate the material coordinates (that is, the position vector $\mathbf{X}$) of each particle as the *name*, or *label* of that particle, so that in all subsequent configurations every particle can be identified by the position $\mathbf{X}$ it occupied in the reference configuration. As usual, we assume an inverse mapping

$$X = \Phi^{-1}(\mathbf{X}) \ , \tag{4.7}$$

so that upon substitution of Eq 4.7 into Eq 4.3 we obtain

$$\mathbf{x} = \boldsymbol{\kappa}\left[\Phi^{-1}(\mathbf{X}), t\right] = \boldsymbol{\chi}(\mathbf{X}, t) \tag{4.8}$$

which defines the motion of the body in physical space relative to the reference configuration prescribed by the mapping function $\mathbf{\Phi}$.

Notice that Eq 4.8 maps the particle at $\mathbf{X}$ in the reference configuration onto the point $\mathbf{x}$ in the current configuration at time t as indicated in Fig. 4.1. With respect to the usual Cartesian axes $Ox_1x_2x_3$ the current position vector is

$$\mathbf{x} = x_i \hat{\mathbf{e}}_i \tag{4.9}$$

where the components x_i are called the *spatial coordinates* of the particle. Although it is not necessary to superpose the material and spatial coordinate axes as we have done in Fig. 4.1, it is convenient to do so, and there are no serious restrictions from this practice in the derivations which follow. We emphasize, however, that the material coordinates are used in conjunction with the reference configuration only, and the spatial coordinates serve for all other configurations. As already remarked the material coordinates are therefore time independent.

We may express Eq 4.8 in either a Cartesian component or a coordinate-free notation by the equivalent equations

$$x_i = \chi_i(X_A, t) \qquad \text{or} \qquad \mathbf{x} = \boldsymbol{\chi}(\mathbf{X}, t) \ . \tag{4.10}$$

It is common practice in continuum mechanics to write these equations in the alternative forms

$$x_i = x_i(X_A, t) \qquad \text{or} \qquad \mathbf{x} = \mathbf{x}(\mathbf{X}, t) \tag{4.11}$$

with the understanding that the symbol x_i (or $\mathbf{x}$) on the right hand side of the equation represents the *function* whose arguments are $\mathbf{X}$ and t, while the same symbol on the left-hand side represents the *value* of the function, that is, a point in space. We shall use this notation frequently in the text that follows.

Notice that as $\mathbf{X}$ ranges over its assigned values corresponding to the reference configuration, while t simultaneously varies over some designated interval of time, the vector function $\boldsymbol{\chi}$ gives the spatial position $\mathbf{x}$ occupied at any instant of time for every particle of the body. At a specific time, say at $t = t_1$, the function $\boldsymbol{\chi}$ defines the configuration

$$x_1 = \chi(X, t_1) \ . \tag{4.12}$$

In particular, at $t = 0$, Eq 4.10 defines the initial configuration which is often adopted as the reference configuration, and this results in the initial spatial coordinates being identical in value with the material coordinates, so that in this case

$$\mathbf{x} = \chi(\mathbf{X}, 0) = \mathbf{X} \tag{4.13}$$

at time $t = 0$.

If we focus attention on a specific particle X^P having the material position vector $\mathbf{X}^P$, Eq 4.10 takes the form

$$\mathbf{x}^P = \chi(\mathbf{X}^P, t) \tag{4.14}$$

and describes the *path*, or *trajectory* of that particle as a function of time. The *velocity* $\mathbf{v}^P$ of the particle along its path is defined as the time rate of change of position, or

$$\mathbf{v}^P = \frac{d\mathbf{x}^P}{dt} = \dot{\mathbf{x}}^P = \left(\frac{\partial \chi}{\partial t}\right)_{\mathbf{X}=\mathbf{X}^P} \tag{4.15}$$

where the notation in the last form indicates that the variable $\mathbf{X}$ is held constant in taking the partial derivative of χ. Also, as is standard practice, the super-positioned dot has been introduced to denote differentiation with respect to time. In an obvious generalization we may define the *velocity field* of the total body as the derivative

$$\mathbf{v} = \dot{\mathbf{x}} = \frac{d\mathbf{x}}{dt} = \frac{\partial \chi(\mathbf{X}, t)}{\partial t} = \frac{\partial \mathbf{x}(\mathbf{X}, t)}{\partial t} \ . \tag{4.16}$$

Similarly, the *acceleration field* is given by

$$\mathbf{a} = \dot{\mathbf{v}} = \ddot{\mathbf{x}} = \frac{d^2\mathbf{x}}{dt^2} = \frac{\partial^2 \chi(\mathbf{X}, t)}{\partial t^2} \ , \tag{4.17}$$

and the acceleration of any particular particle determined by substituting its material coordinates into Eq 4.17.

Of course, the individual particles of a body cannot execute arbitrary motions independent of one another. In particular, no two particles can occupy the same location in space at a given time (the axiom of impenetrability), and furthermore, in the smooth motions we consider here, any two particles arbitrarily close in the reference configuration remain arbitrarily close in all other configurations. For these reasons, the function χ in Eq 4.10 must be single-valued and continuous, and must possess continuous derivatives with respect to space and time to whatever order is required; usually to the second or third. Moreover, we require the inverse function χ^{-1} in the equation

$$\mathbf{X} = \chi^{-1}(\mathbf{x}, t) \tag{4.18}$$

to be endowed with the same properties as χ. Conceptually, Eq 4.18 allows us to "reverse" the motion and trace backwards to discover where the particle, now at $\mathbf{x}$, was located in the reference configuration. The mathematical condition that guarantees the existence of such an inverse function is the non-vanishing of the Jacobian determinant J. That is, for the equation

$$J = \left| \frac{\partial \chi_i}{\partial X_A} \right| \neq 0 \tag{4.19}$$

to be valid. This determinant may also be written as

$$J = \left| \frac{\partial x_i}{\partial X_A} \right| \ . \tag{4.20}$$

Example 4.1

Let the motion of a body be given by Eq 4.10 in component form as

$$x_1 = X_1 + t^2 X_2 \,,$$
$$x_2 = X_2 + t^2 X_1 \,,$$
$$x_3 = X_3 \,.$$

Determine

(a) the path of the particle originally at $\mathbf{X} = (1, 2, 1)$ and

(b) the velocity and acceleration components of the same particle when $t = 2\,\text{s}$.

Solution

(a) For the particle $\mathbf{X} = (1, 2, 1)$ the motion equations are

$$x_1 = 1 + 2t^2; \quad x_2 = 2 + t^2; \quad x_3 = 1$$

which upon elimination of the variable t gives $x_1 - 2x_2 = -3$ as well as $x_3 = 1$ so that the particle under consideration moves on a straight line path in the plane $x_3 = 1$.

(b) By Eqs 4.16 and 4.17 the velocity and acceleration fields are given in component form, respectively, by

$$\begin{aligned} v_1 &= 2tX_2 \\ v_2 &= 2tX_1 \\ v_3 &= 0 \end{aligned} \qquad \text{and} \qquad \begin{aligned} a_1 &= 2X_2 \\ a_2 &= 2X_1 \\ a_3 &= 0 \,, \end{aligned}$$

so that for the particle $\mathbf{X} = (1, 2, 1)$ at $t = 2$

$$\begin{aligned} v_1 &= 8 \\ v_2 &= 4 \\ v_3 &= 0 \end{aligned} \qquad \text{and} \qquad \begin{aligned} a_1 &= 4 \\ a_2 &= 2 \\ a_3 &= 0 \,. \end{aligned}$$

Example 4.2

Invert the motion equations of Example 4.1 to obtain $\mathbf{X} = \chi^{-1}(\mathbf{x}, t)$ and determine the velocity and acceleration components of the particle at $\mathbf{x} = (1, 0, 1)$ when $t = 2\text{s}$.

Solution

By inverting the motion equations directly we obtain

$$X_1 = \frac{x_1 - t^2 x_2}{1 - t^4}; \quad X_2 = \frac{x_2 - t^2 x_1}{1 - t^4}; \quad X_3 = x_3$$

which upon substitution into the velocity and acceleration expressions of Example 4.1 yields

$$v_1 = \frac{2t(x_2 - t^2 x_1)}{1 - t^4} \qquad a_1 = \frac{2(x_2 - t^2 x_1)}{1 - t^4}$$

$$v_2 = \frac{2t(x_1 - t^2 x_2)}{1 - t^4} \quad \text{and} \quad a_2 = \frac{2(x_1 - t^2 x_2)}{1 - t^4}$$

$$v_3 = 0 \qquad a_3 = 0 \,.$$

For the particle at $\mathbf{x} = (1, 0, 1)$ when $t = 2s$

$$v_1 = \frac{16}{15} \qquad a_1 = \frac{8}{15}$$

$$v_2 = -\frac{4}{15} \quad \text{and} \quad a_2 = -\frac{2}{15}$$

$$v_3 = 0 \qquad a_3 = 0 \,.$$

Example 4.3

Determine the Jacobian J for the motion of a continuum be given by

$$\begin{aligned} x_1 &= X_1 + f(t) \cos kX_3 \;, \\ x_2 &= X_2 + f(t) \sin kX_3 \;, \\ x_3 &= X_3 \;. \end{aligned}$$

Solution

Form and compute the determinant of Eq 4.20

$$\det |x_{i,A}| = \det \begin{vmatrix} \frac{\partial x_1}{\partial X_1} & \frac{\partial x_1}{\partial X_2} & \frac{\partial x_1}{\partial X_3} \\ \frac{\partial x_2}{\partial X_1} & \frac{\partial x_2}{\partial X_2} & \frac{\partial x_2}{\partial X_3} \\ \frac{\partial x_3}{\partial X_1} & \frac{\partial x_3}{\partial X_2} & \frac{\partial x_3}{\partial X_3} \end{vmatrix} = \det \begin{vmatrix} 1 & 0 & f(t)k \sin kX_3 \\ 0 & 1 & -f(t)k \cos kX_3 \\ 0 & 0 & 1 \end{vmatrix} = 1 \,.$$

4.3 Langrangian and Eulerian Descriptions

If a physical property of the body $\mathcal{B}$ such as its density ρ, or a kinematic property of its motion such as the velocity $\mathbf{v}$, is expressed in terms of the material coordinates $\mathbf{X}$, and the time t, we say that property is given by the *referential* or *material description*. When the referential configuration is taken as the actual configuration at time $t = 0$, this description is usually called the Lagrangian description. Thus, the equations

$$\rho = \rho(X_A, t) \qquad \text{or} \qquad \rho = \rho(\mathbf{X}, t)\ , \tag{4.21a}$$

and

$$v_i = v_i(X_A, t) \qquad \text{or} \qquad \mathbf{v} = \mathbf{v}(\mathbf{X}, t) \tag{4.21b}$$

chronicle a time history of these properties for each particle of the body. In contrast, if the properties ρ and $\mathbf{v}$ are given as functions of the spatial coordinates $\mathbf{x}$ and time t, we say that those properties are expressed by a *spatial description*, or as it is sometimes called, by the *Eulerian description*. In view of Eq 4.18 it is clear that Eq 4.21 may be converted to express the same properties in the spatial description. Accordingly, we write

$$\rho = \rho\,(\mathbf{X}, t) = \rho\left[\chi^{-1}\,(\mathbf{x}, t)\,, t\right] = \tilde{\rho}\,(\mathbf{x}, t) \tag{4.22a}$$

and

$$\mathbf{v} = \mathbf{v}\,(\mathbf{X}, t) = \mathbf{v}\left[\chi^{-1}\,(\mathbf{x}, t)\,, t\right] = \tilde{\mathbf{v}}\,(\mathbf{x}, t) \tag{4.22b}$$

where the tilde is added solely for the purpose of emphasizing that different functional forms result from the switch in variables. We note that in the material description attention is focused on what is happening to the individual particles during the motion, whereas in the spatial description the emphasis is directed to the events taking place at specific points in space.

Example 4.4

Let the motion equations be given in component form by the Lagrangian description

$$x_1 = X_1 e^t + X_3(e^t - 1)\ ,$$
$$x_2 = X_2 + X_3(e^t - e^{-t})\ ,$$
$$x_3 = X_3\ .$$

Determine the Eulerian description of this motion.

Solution

Notice first that for the given motion $x_1 = X_1$, $x_2 = X_2$ and $x_3 = X_3$ at $t = 0$, so that the initial configuration has been taken as the reference configuration. Because of the simplicity of these Lagrangian equations of the motion, we may substitute x_3 for X_3 into the first two equations and solve these directly to obtain the inverse equations

$$X_1 = x_1 e^{-t} + x_3(e^{-t} - 1)\ ,$$
$$X_2 = x_2 + x_3(e^{-t} - e^t)\ ,$$
$$X_3 = x_3\ .$$

Example 4.5

For the motion of Example 4.4 determine the velocity and acceleration fields, and express these in both Lagrangian and Eulerian forms.

Solution
From the given motion equations and the velocity definition Eq 4.16 we obtain the Lagrangian velocity components,

$$v_1 = X_1 e^t + X_3 e^t \ ,$$
$$v_2 = X_3(e^t + e^{-t}) \ ,$$
$$v_3 = 0 \ ,$$

and from Eq 4.17 the acceleration components

$$a_1 = (X_1 + X_3)e^t \ ,$$
$$a_2 = X_3(e^t - e^{-t}) \ ,$$
$$a_3 = 0 \ .$$

Therefore, by introducing the inverse mapping equations determined in Example 4.4 we obtain the velocity and acceleration equations in Eulerian form,

$$\begin{aligned} v_1 &= x_1 + x_3 \\ v_2 &= x_3(e^t + e^{-t}) \\ v_3 &= 0 \end{aligned} \qquad \text{and} \qquad \begin{aligned} a_1 &= x_1 + x_3 \\ a_2 &= x_3(e^t - e^{-t}) \\ a_3 &= 0 \ . \end{aligned}$$

4.4 The Displacement Field

As may be seen from Fig. 4.1, the typical particle of body $\mathcal{B}$ undergoes a *displacement*

$$\mathbf{u} = \mathbf{x} - \mathbf{X} \tag{4.23}$$

in the transition from the reference configuration to the current configuration. Because this relationship holds for all particles it is often useful to analyze deformation or motion in terms of the *displacement field* of the body. We may write the displacement vector $\mathbf{u}$ in component form by either of the equivalent expressions

$$\mathbf{u} = u_i \hat{\mathbf{e}}_i = u_A \hat{\mathbf{I}}_A \ . \tag{4.24}$$

Additionally with regard to the material and spatial descriptions we may interpret Eq 4.23 in either the material form

$$\mathbf{u}(\mathbf{X}, t) = \mathbf{x}(\mathbf{X}, t) - \mathbf{X} \ , \tag{4.25}$$

or the spatial form

$$\mathbf{u}(\mathbf{x}, t) = \mathbf{x} - \mathbf{X}(\mathbf{x}, t) \ . \tag{4.26}$$

In the first of this pair of equations we are describing the displacement that will occur to the particle that starts at $\mathbf{X}$, and in the second equation we present the displacement that the particle now at $\mathbf{x}$ has undergone. Recalling that since the material coordinates relate to positions in the reference configuration only, and hence are independent of time, we may take the time rate of change of displacement as an alternative definition for velocity. Thus

$$\frac{d\mathbf{u}}{dt} = \frac{d(\mathbf{x} - \mathbf{X})}{dt} = \frac{d\mathbf{x}}{dt} = \mathbf{v} \,. \tag{4.27}$$

Example 4.6

Obtain the displacement field for the motion of Example 4.4 in both material and spatial descriptions.

Solution

From the motion equations of Example 4.4, namely,

$$\begin{aligned} x_1 &= X_1 e^t + X_3(e^t - 1) \,, \\ x_2 &= X_2 + X_3(e^t - e^{-t}) \,, \\ x_3 &= X_3 \,, \end{aligned}$$

we may compute the displacement field in material form directly as

$$\begin{aligned} u_1 &= x_1 - X_1 = (X_1 + X_3)(e^t - 1) \,, \\ u_2 &= x_2 - X_2 = X_3(e^t - e^{-t}) \,, \\ u_3 &= x_3 - X_3 = 0 \,, \end{aligned}$$

and by using the inverse equations from Example 4.4, namely,

$$\begin{aligned} X_1 &= x_1 e^{-t} + x_3(e^{-t} - 1) \,, \\ X_2 &= x_2 + x_3(e^{-t} - e^t) \,, \\ X_3 &= x_3 \,, \end{aligned}$$

in Eq 4.26 we obtain the spatial description of the displacement field in component form

$$\begin{aligned} u_1 &= (x_1 + x_3)(1 - e^{-t}) \,, \\ u_2 &= x_3(e^t - e^{-t}) \,, \\ u_3 &= 0 \,. \end{aligned}$$

4.5 The Material Derivative

In this section let us consider any physical or kinematic property of a continuum body. It may be a scalar, vector or tensor property, and so we represent it by the general symbol $P_{ij\ldots}$ with the understanding that it may be expressed in either the material description

$$P_{ij\ldots} = P_{ij\ldots}(\mathbf{X}, t) \,, \tag{4.28a}$$

or in the spatial description

$$P_{ij\ldots} = P_{ij\ldots}(\mathbf{x}, t) \ , \tag{4.28b}$$

The *material derivative* of any such property is the time rate of change of that property for a specific collection of particles (one or more) of the continuum body. This derivative can be thought of as the rate at which $P_{ij\ldots}$ changes when measured by an observer attached to, and traveling with, the particle, or group of particles. We use the differential operator d/dt, or the superpositioned dot, to denote a material derivative, and note that velocity and acceleration as we have previously defined them are material derivatives.

When $P_{ij\ldots}$ is given in the material description of Eq 4.28a, the material derivative is simply the partial derivative with respect to time,

$$\frac{d}{dt}[P_{ij\ldots}(\mathbf{X}, t)] = \frac{\partial}{\partial t}[P_{ij\ldots}(\mathbf{X}, t)] \ , \tag{4.29}$$

since, as explained earlier, the material coordinates $\mathbf{X}$ are labels and do not change with time. If, however, $P_{ij\ldots}$ is given in the spatial form of Eq 4.28b we recognize that the specific collection of particles of interest will be changing position in space, and we must use the chain rule of differentiation of the calculus to obtain

$$\frac{d}{dt}[P_{ij\ldots}(\mathbf{x}, t)] = \frac{\partial}{\partial t}[P_{ij\ldots}(\mathbf{x}, t)] + \frac{\partial}{\partial x_k}[P_{ij\ldots}(\mathbf{x}, t)]\frac{dx_k}{dt} \ . \tag{4.30}$$

In this equation the first term on the right hand side gives the change occurring in the property at position $\mathbf{x}$, known as the *local rate of change*; the second term results from the particles changing position in space and is referred to as the *convective rate of change*. Since by Eq 4.16 the velocity is defined as $\mathbf{v} = d\mathbf{x}/dt$ (or $v_k = dx_k/dt$), Eq 4.30 may be written as

$$\frac{d}{dt}[P_{ij\ldots}(\mathbf{x}, t)] = \frac{\partial}{\partial t}[P_{ij\ldots}(\mathbf{x}, t)] + \frac{\partial}{\partial x_k}[P_{ij\ldots}(\mathbf{x}, t)]\, v_k \tag{4.31}$$

from which we deduce the *material derivative operator* for properties expressed in the *spatial description*

$$\frac{d}{dt} = \frac{\partial}{\partial t} + v_k\frac{\partial}{\partial x_k} \qquad \text{or} \qquad \frac{d}{dt} = \frac{\partial}{\partial t} + \mathbf{v}\cdot\boldsymbol{\nabla} \ . \tag{4.32}$$

The first form of Eq 4.32 is for rectangular Cartesian coordinates, while the second form is coordinate-free. The del operator ($\boldsymbol{\nabla}$) will always indicate partial derivatives with respect to the spatial variables unless specifically stated.

Example 4.7

Let a certain motion of a continuum be given by the component equations,

$$x_1 = X_1 e^{-t}, \quad x_2 = X_2 e^{t}, \quad x_3 = X_3 + X_2(e^{-t} - 1) \ ,$$

and let the temperature field of the body be given by the spatial description,

$$\theta = e^{-t}(x_1 - 2x_2 + 3x_3) \ ,$$

Determine the velocity field in spatial form, and using that, compute the material derivative $d\theta/dt$ of the temperature field.

Solution
Note again here that the initial configuration serves as the reference configuration so that Eq 4.13 is satisfied. When Eq 4.30 is used, the velocity components in material form are readily determined to be

$$v_1 = -X_1 e^{-t}, \quad v_2 = X_2 e^{t}, \quad v_3 = -X_2 e^{-t} \,.$$

Also, the motion equations can be inverted directly to give,

$$X_1 = x_1 e^{t}, \quad X_2 = x_2 e^{-t}, \quad X_3 = x_3 - x_2(e^{-2t} - e^{-t})$$

which upon substitution into the above velocity expressions yields the spatial components,

$$v_1 = -x_1, \quad v_2 = x_2, \quad v_3 = -x_2 e^{-2t} \,.$$

Therefore, we may now calculate $d\theta/dt$ in spatial form using Eq 4.31,

$$\frac{d\theta}{dt} = -e^{-t}(x_1 - 2x_2 + 3x_3) - x_1 e^{-t} - 2x_2 e^{-t} - 3x_3 e^{-t}$$

which may be converted to its material form using the original motion equations, resulting in

$$\frac{d\theta}{dt} = -2X_1 e^{-2t} - 3X_2(2e^{-2t} - e^{-t}) - 3X_3 e^{-t} \,.$$

An interesting and rather unique situation arises when we wish to determine the velocity field in spatial form by a direct application of Eq 4.31 to the displacement field in its spatial form. The following example illustrates the point.

Example 4.8

Verify the spatial velocity components determined in Example 4.7 by applying Eq 4.31 directly to the displacement components in spatial form for the motion in that example.

Solution
We may determine the displacement components in material form directly from the motion equations given in Example 4.7,

$$\begin{aligned} u_1 &= x_1 - X_1 = X_1(e^{-t} - 1) \,, \\ u_2 &= x_2 - X_2 = X_2(e^{t} - 1) \,, \\ u_3 &= x_3 - X_3 = X_2(e^{-t} - 1) \,, \end{aligned}$$

and, using the inverse equations $\mathbf{X} = \chi^{-1}(\mathbf{x}, t)$ computed in Example 4.7, we obtain the spatial displacements

$$\begin{aligned} u_1 &= x_1(1 - e^{t}) \,, \\ u_2 &= x_2(1 - e^{-t}) \,, \\ u_3 &= x_2(e^{-2t} - e^{-t}) \,. \end{aligned}$$

Therefore, substituting u_i for $P_{ij...}$ in Eq 4.31 yields

$$v_i = \frac{du_i}{dt} = \frac{\partial u_i}{\partial t} + v_k \frac{\partial u_i}{\partial x_k} ,$$

so that by differentiating the above displacement components

$$\begin{aligned} v_1 &= -x_1 e^t + v_1(1 - e^t) , \\ v_2 &= x_2 e^{-t} + v_2(1 - e^{-t}) , \\ v_3 &= -x_2(2e^{-2t} - e^{-t}) + v_2(e^{-2t} - e^{-t}) , \end{aligned}$$

which results in a set of equations having the desired velocity components on both sides of the equations. In general, this set of equations must be solved simultaneously. In this case the solution is quite easily obtained, yielding

$$v_1 = -x_1, \qquad v_2 = x_2, \qquad v_3 = -x_2 e^{-2t}$$

to confirm the results of Example 4.7.

Example 4.9

Cilia are motile cells that are responsible for locomotion of single-cell bodies, hearing, and moving fluid in the body. The cilia beat in a manner that propels the surrounding fluid. The motion of the cilia is complex. A model for the cilia motion is to enclose the tips by a flexible envelope that moves in a fashion similar to the tips of the cilia. This is reasonable given that in many biological processes fluid motions are slow. Two possible descriptions of the envelope motion are [1]

Case 1:

$$\begin{aligned} x_1 &= X_1^0 + \epsilon\alpha \cos(x_1 - ct) , \\ x_2 &= X_2^0 + \epsilon\beta \sin(x_1 - ct) , \end{aligned}$$

Case 2:

$$\begin{aligned} x_1 &= X_1^0 + \epsilon\alpha \cos\left(X_1^0 - ct\right) , \\ x_2 &= X_2^0 + \epsilon\beta \sin\left(X_1^0 - ct\right) . \end{aligned}$$

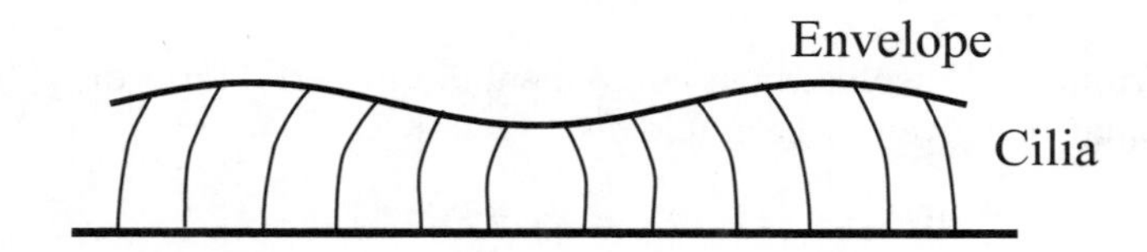

Envelope containing the tips of the beating cilia.

Here $\mathbf{x}$ is the position of the envelope and $\mathbf{X}^0$ is an initial position of the particles.

(a) Show that the both descriptions indicate that particles move in an elliptical path centered about $\left(X_1^0, X_2^0\right)$.

[1] see T. J. Lardner and W. J. Shack (1972) Cilia transport, *Bulletin of Mathematical Biophysics*, **34** (3) 325-335.

(b) Fluid mechanics has a no-slip condition that requires the velocity of the fluid be the same as the velocity of the envelope. Determine the velocity of the particles in the two descriptions.

Solution

(a) For description 1, the particle paths are

$$\frac{\left(x_1 - X_1^o\right)^2}{(\epsilon\beta)^2} + \frac{\left(x_2 - X_2^o\right)^2}{(\epsilon\alpha)^2} = \sin^2(x_1 - ct) + \cos^2(x_1 - ct) = 1 \; .$$

For description 2, the particle paths are

$$\frac{\left(x_1 - X_1^o\right)^2}{(\epsilon\beta)^2} + \frac{\left(x_2 - X_2^o\right)^2}{(\epsilon\alpha)^2} = \sin^2\left(X_1^o - ct\right) + \cos^2\left(X_1^o - ct\right) = 1 \; .$$

These are both descriptions of ellipses centered at $\left(X_1^o, X_2^o\right)$.

(b) For description 1, the velocity v_1 is

$$\begin{aligned} v_1 = \frac{dx_1}{dt} &= \left[\frac{\partial}{\partial t} + v_1\frac{\partial}{\partial x_1} + v_2\frac{\partial}{\partial x_2}\right]\left[X_1^o + \epsilon\alpha\cos(x_1 - ct)\right] \\ &= \frac{\partial}{\partial t}\left[X_1^o + \epsilon\alpha\cos(x_1 - ct)\right] + v_1\frac{\partial}{\partial x_1}\left[X_1^o + \epsilon\alpha\cos(x_1 - ct)\right] \\ &= c\epsilon\alpha\sin(x_1 - ct) - v_1\epsilon\alpha\sin(x_1 - ct) \; , \end{aligned}$$

and

$$\begin{aligned} v_1\left[1 + \epsilon\alpha\sin(x_1 - ct)\right] &= c\epsilon\alpha\sin(x_1 - ct) \; , \\ v_1 &= \frac{c\epsilon\alpha\sin(x_1 - ct)}{1 + \epsilon\alpha\sin(x_1 - ct)} \; . \end{aligned}$$

The velocity v_2 is

$$\begin{aligned} v_2 = \frac{dx_2}{dt} &= \left[\frac{\partial}{\partial t} + v_1\frac{\partial}{\partial x_1} + v_2\frac{\partial}{\partial x_2}\right]\left[X_2^o + \epsilon\beta\sin(x_1 - ct)\right] \\ &= \frac{\partial}{\partial t}\left[X_2^o + \epsilon\beta\sin(x_1 - ct)\right] + v_1\frac{\partial}{\partial x_1}\left[X_2^o + \epsilon\beta\sin(x_1 - ct)\right] \\ &= -c\epsilon\alpha\cos(x_1 - ct) + v_1\epsilon\beta\cos(x_1 - ct) \\ &= -c\epsilon\alpha\cos(x_1 - ct) + \epsilon\beta\cos(x_1 - ct)\frac{c\epsilon\alpha\sin(x_1 - ct)}{1 + \epsilon\alpha\sin(x_1 - ct)} \; , \end{aligned}$$

or

$$v_2 = \frac{-c\epsilon\beta\cos(x_1 - ct)}{1 + \epsilon\alpha\sin(x_1 - ct)} \; .$$

For description 2, the velocity v_1 is

$$\begin{aligned} v_1 = \frac{dx_1}{dt} &= \left[\frac{\partial}{\partial t} + v_1\frac{\partial}{\partial x_1} + v_2\frac{\partial}{\partial x_2}\right]\left[X_1^o + \epsilon\alpha\cos\left(X_1^o - ct\right)\right] \\ &= \frac{\partial}{\partial t}\left[X_1^o + \epsilon\alpha\cos\left(X_1^o - ct\right)\right] = c\epsilon\alpha\sin\left(X_1^o - ct\right) \; . \end{aligned}$$

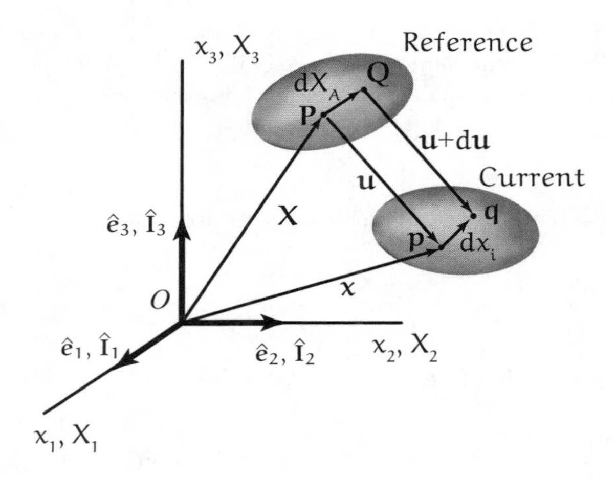

FIGURE 4.2
Vector dX_A, between points **P** and **Q** in reference configuration, becomes dx_i, between points **p** and **q**, in the current configuration. Displacement vector **u** is the vector between points **p** and **P**.

The velocity v_2 is

$$v_2 = \frac{dx_2}{dt} = \left[\frac{\partial}{\partial t} + v_1 \frac{\partial}{\partial x_1} + v_2 \frac{\partial}{\partial x_2}\right] \left[X_2^0 + \epsilon\beta \sin\left(X_1^0 - ct\right)\right]$$
$$= \frac{\partial}{\partial t}\left[X_2^0 + \epsilon\beta \sin\left(X_1^0 - ct\right)\right] = -c\epsilon\beta \cos\left(X_1^0 - ct\right) .$$

4.6 Deformation Gradients, Finite Strain Tensors

In deformation analysis we confine our attention to two stationary configurations and disregard any consideration for the particular sequence by which the final *deformed configuration* is reached from the initial *undeformed configuration*. Accordingly, the mapping function is not dependent upon time as a variable, so that Eq 4.10 takes the form

$$x_i = \chi_i(\mathbf{X}) \qquad \text{or} \qquad \mathbf{x} = \chi(\mathbf{X}) . \tag{4.33}$$

Consider, therefore, two neighboring particles of the body situated at the points **P** and **Q** in the undeformed configuration such that **Q** is located with respect to **P** by the relative differential position vector

$$d\mathbf{X} = dX_A \hat{\mathbf{I}}_A \tag{4.34}$$

as shown in Fig. 4.2. The magnitude squared of $d\mathbf{X}$ is

$$(dX)^2 = d\mathbf{X} \cdot d\mathbf{X} = dX_A\, dX_A . \tag{4.35}$$

Under the displacement field prescribed by the function χ of Eq 4.35 the particles originally at **P** and **Q** move to the positions **p** and **q**, respectively, in the deformed configuration such that their relative position vector is now

$$\mathbf{dx} = dx_i \hat{\mathbf{e}}_i \tag{4.36}$$

having a magnitude squared

$$(dx)^2 = \mathbf{dx} \cdot \mathbf{dx} = dx_i dx_i \ . \tag{4.37}$$

We assume the mapping function χ_i of Eq 4.33 is continuous so that

$$dx_i = \frac{\partial \chi_i}{\partial X_A} dX_A \ , \tag{4.38}$$

or as it is more often written,

$$dx_i = \frac{\partial \chi_i}{\partial X_A} dX_A = x_{i,A} dX_A \tag{4.39}$$

where

$$x_{i,A} \equiv F_{iA} \tag{4.40}$$

is called the *deformation gradient tensor* or simply the *deformation gradient*. The tensor **F** characterizes the local deformation at **X**, and may depend explicitly upon **X**, in which case the deformation is termed *inhomogeneous*. If **F** is independent of **X**, the deformation is called *homogeneous*. In symbolic notation Eq 4.39 appears in either of the forms

$$\mathbf{dx} = \mathbf{F} \cdot \mathbf{dX} \qquad \text{or} \qquad \mathbf{dx} = \mathbf{F}\mathbf{dX} \tag{4.41}$$

where, as indicated by the second equation, the dot is often omitted for convenience. In view of the smoothness conditions we have imposed on the mapping function χ we know that **F** is invertible so that the inverse $\mathbf{F}^{-1}$ exists such that

$$dX_A = X_{A,i} dx_i \qquad \text{or} \qquad \mathbf{dX} = \mathbf{F}^{-1} \cdot \mathbf{dx} \ . \tag{4.42}$$

In describing motions and deformations, several measures of deformation are commonly used. Let us consider first one based upon the change during the deformation in the magnitude squared of the distance between the particles originally at **P** and **Q**. Namely,

$$(dx)^2 - (dX)^2 = dx_i dx_i - dX_A dX_A$$

which from Eq 4.39 and the substitution property of the Kronecker delta, δ_{AB} may be developed as follows,

$$\begin{aligned}(dx)^2 - (dX)^2 &= (x_{i,A} dX_A)(x_{i,B} dX_B) - \delta_{AB} dX_A dX_B \\ &= (x_{i,A} x_{i,B} - \delta_{AB}) dX_A dX_B \\ &= (C_{AB} - \delta_{AB}) dX_A dX_B \end{aligned} \tag{4.43}$$

where the symmetric tensor

$$C_{AB} = x_{i,A} x_{i,B} \qquad \text{or} \qquad \mathbf{C} = \mathbf{F}^T \cdot \mathbf{F} \tag{4.44}$$

is called the *Green's deformation tensor*. From this we immediately define the *Lagrangian finite strain tensor* E_{AB} as

$$2E_{AB} = C_{AB} - \delta_{AB} \qquad \text{or} \qquad 2\mathbf{E} = \mathbf{C} - \mathbf{I} \tag{4.45}$$

where the factor of two is introduced for convenience in later calculations. Finally we can write,

$$(dx)^2 - (dX)^2 = 2E_{AB}dX_A dX_B = \mathbf{dX} \cdot 2\mathbf{E} \cdot \mathbf{dX} \,. \tag{4.46}$$

The difference $(dx)^2 - (dX)^2$ may also be developed in terms of the spatial variables in a similar way as

$$\begin{aligned}(dx)^2 - (dX)^2 &= \delta_{ij} dx_i dx_j - (X_{A,i} dx_i)(X_{A,j} dx_j) \\ &= (\delta_{ij} - X_{A,i} X_{A,j}) dx_i dx_j \\ &= (\delta_{ij} - c_{ij}) dx_i dx_j \end{aligned} \tag{4.47}$$

where the symmetric tensor

$$c_{ij} = X_{A,i} X_{A,j} \qquad \text{or} \qquad \mathbf{c} = (\mathbf{F}^{-1})^T \cdot (\mathbf{F}^{-1}) \tag{4.48}$$

is called the *Cauchy deformation tensor*. From Eq 4.48 we define the *Eulerian finite strain tensor* $\mathbf{e}$ as

$$2e_{ij} = (\delta_{ij} - c_{ij}) \qquad \text{or} \qquad 2\mathbf{e} = (\mathbf{I} - \mathbf{c}) \,, \tag{4.49}$$

so that now

$$(dx)^2 - (dX)^2 = 2e_{ij} dx_i dx_j = \mathbf{dx} \cdot 2\mathbf{e} \cdot \mathbf{dx} \,. \tag{4.50}$$

Both E_{AB} and e_{ij} are, of course, symmetric second-order tensors, as can be observed from their definitions.

For any two arbitrary differential vectors $\mathbf{dX}^{(1)}$ and $\mathbf{dX}^{(2)}$ which deform into $\mathbf{dx}^{(1)}$ and $\mathbf{dx}^{(2)}$, respectively, we have from Eq 4.41 together with Eqs 4.44 and 4.45,

$$\begin{aligned}\mathbf{dx}^{(1)} \cdot \mathbf{dx}^{(2)} &= \mathbf{F} \cdot \mathbf{dX}^{(1)} \cdot \mathbf{F} \cdot \mathbf{dX}^{(2)} = \mathbf{dX}^{(1)} \cdot \mathbf{F}^T \cdot \mathbf{F} \cdot \mathbf{dX}^{(2)} \\ &= \mathbf{dX}^{(1)} \cdot \mathbf{C} \cdot \mathbf{dX}^{(2)} = \mathbf{dX}^{(1)} \cdot (\mathbf{I} + 2\mathbf{E}) \cdot \mathbf{dX}^{(2)} \\ &= \mathbf{dX}^{(1)} \cdot \mathbf{dX}^{(2)} + \mathbf{dX}^{(1)} \cdot 2\mathbf{E} \cdot \mathbf{dX}^{(2)}. \end{aligned} \tag{4.51}$$

If $\mathbf{E}$ is identically zero (no strain), Eq 4.51 asserts that the lengths of all line elements are unchanged [we may choose $\mathbf{dX}^{(1)} = \mathbf{dX}^{(2)} = \mathbf{dX}$ so that $(dx)^2 = (dX)^2$], and in view of the definition $\mathbf{dx}^{(1)} \cdot \mathbf{dx}^{(2)} = dx^{(1)} dx^{(2)} \cos\theta$, the angle between any two elements will also be unchanged. Thus in the absence of strain, only a rigid body displacement can occur.

The Lagrangian and Eulerian finite strain tensors expressed by Eqs 4.45 and 4.49, respectively, are given in terms of the appropriate deformation gradients. These same tensors may also be developed in terms of displacement gradients. For this purpose we begin by writing Eq 4.25 in its time-independent form consistent with deformation analysis. In component notation, the material description is

$$u_i(X_A) = x_i(X_A) - X_i \,, \tag{4.52}$$

and the spatial description is

$$u_A(x_i) = x_A - X_A(x_i) \,. \tag{4.53}$$

From the first of these, Eq 4.45 becomes

$$2E_{AB} = x_{i,A} x_{i,B} - \delta_{AB} = (u_{i,A} + \delta_{iA})(u_{i,B} + \delta_{iB}) - \delta_{AB} \tag{4.54}$$

which reduces to

$$2E_{AB} = u_{A,B} + u_{B,A} + u_{i,A} u_{i,B} \,, \tag{4.55}$$

and from the second, Eq 4.49 becomes

$$2e_{ij} = \delta_{ij} - X_{A,i}X_{A,j} = \delta_{ij} - (\delta_{Ai} - u_{A,i})(\delta_{Aj} - u_{A,j}) \tag{4.56}$$

which reduces to

$$2e_{ij} = u_{i,j} + u_{j,i} - u_{A,i}u_{A,j} \,. \tag{4.57}$$

Example 4.10

Let the simple shear deformation $x_1 = X_1$; $x_2 = X_2 + kX_3$; $x_3 = X_3 + kX_2$, where k is a constant, be applied to the small cube of edge dimensions dL shown in the figure below. Draw the deformed shape of face $ABGH$ of the cube and determine the difference $(dx)^2 - (dX)^2$ for the diagonals AG, BH and OG of the cube.

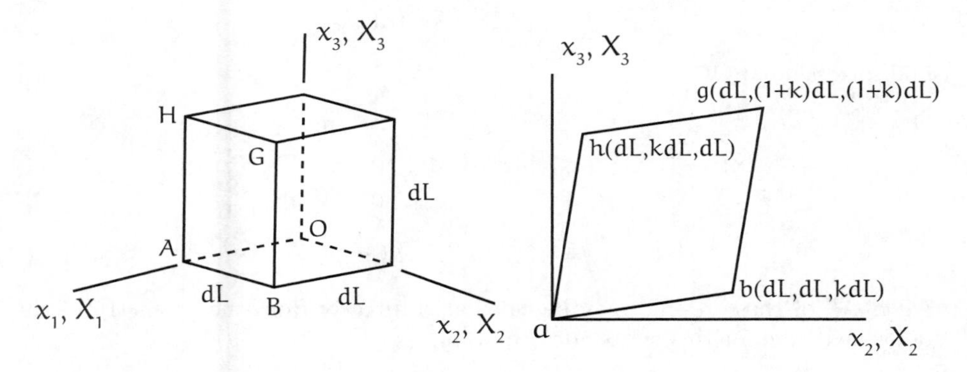

Reference position of cube undergoing simple shear.

Small cube geometry.

Solution

From the mapping equations directly, the origin O is seen to remain in place, and the particles originally at points A, B, G and H are displaced to the points $a(dL, 0, 0)$, $b(dL, dL, kdL)$, $g(dL, (1+k)dL, (1+k)dL)$ and $h(dL, kdL, dL)$, respectively, so that particles in planes parallel to the X_2X_3 remain in those planes, and the square face $ABGH$ becomes the diamond shaped parallelogram $abgh$ shown above. Also from the mapping equations and Eq 4.40, we see that the deformation gradient $\mathbf{F}$ has the matrix form

$$[F_{iA}] = \begin{bmatrix} 1 & 0 & 0 \\ 0 & 1 & k \\ 0 & k & 1 \end{bmatrix} ,$$

and since $\mathbf{C} = \mathbf{F}^T \cdot \mathbf{F}$

$$[C_{AB}] = \begin{bmatrix} 1 & 0 & 0 \\ 0 & 1+k^2 & 2k \\ 0 & 2k & 1+k^2 \end{bmatrix}$$

from which we determine $2\mathbf{E} = \mathbf{C} - \mathbf{I}$,

$$[2E_{AB}] = \begin{bmatrix} 0 & 0 & 0 \\ 0 & k^2 & 2k \\ 0 & 2k & k^2 \end{bmatrix} .$$

In general, $(dx)^2 - (dX)^2 = \mathbf{dX} \cdot 2\mathbf{E} \cdot \mathbf{dX}$ so that for diagonal AG,

$$(dx)^2 - (dX)^2 = [0, dL, dL] \begin{bmatrix} 0 & 0 & 0 \\ 0 & k^2 & 2k \\ 0 & 2k & k^2 \end{bmatrix} \begin{bmatrix} 0 \\ dL \\ dL \end{bmatrix} = 2(2k + k^2)(dL)^2 \ .$$

For diagonal BH,

$$(dx)^2 - (dX)^2 = [0, -dL, dL] \begin{bmatrix} 0 & 0 & 0 \\ 0 & k^2 & 2k \\ 0 & 2k & k^2 \end{bmatrix} \begin{bmatrix} 0 \\ -dL \\ dL \end{bmatrix} = 2(-2k + k^2)(dL)^2 \ ,$$

and for diagonal OG,

$$(dx)^2 - (dX)^2 = [dL, dL, dL] \begin{bmatrix} 0 & 0 & 0 \\ 0 & k^2 & 2k \\ 0 & 2k & k^2 \end{bmatrix} \begin{bmatrix} dL \\ dL \\ dL \end{bmatrix} = 2(2k + k^2)(dL)^2 \ ,$$

Note: All of these results may be calculated directly from the geometry of the deformed cube for this simple deformation.

4.7 Infinitesimal Deformation Theory

If the numerical values of all the components of the displacement and the displacement gradient tensors are very small we may neglect the squares and products of these quantities in comparison to the gradients themselves so that Eqs 4.55 and 4.57 reduce to

$$2E_{AB} = u_{A,B} + u_{B,A} \ , \tag{4.58}$$

and

$$2e_{ij} = u_{i,j} + u_{j,i} \ . \tag{4.59}$$

These expressions are known as the *linearized* Lagrangian and Eulerian strain tensors, respectively. Furthermore, to the same order of approximation,

$$\frac{\partial u_i}{\partial X_A} = \frac{\partial u_i}{\partial x_k}\frac{\partial x_k}{\partial X_A} = \frac{\partial u_i}{\partial x_k}\left(\frac{\partial u_k}{\partial X_A} + \delta_{kA}\right) \approx \frac{\partial u_i}{\partial x_k}\delta_{kA} \tag{4.60}$$

where we have used the relationship

$$\frac{\partial x_k}{\partial X_A} = \frac{\partial u_k}{\partial X_A} + \delta_{kA}$$

obtained by differentiating Eq 4.52. Therefore, to the first order of approximation for the case of small displacement gradients, it is unimportant whether we differentiate the displacement components with respect to the material or spatial coordinates. In view of this, we may display the equivalent relative displacement gradients for small deformation theory as either $u_{i,A}$ or $u_{i,j}$. Similarly, it can be shown that in the linear theory $u_{A,B}$ and $u_{A,j}$ are equivalent. It follows that to the same order of approximation, from Eqs 4.58 and 4.59,

$$E_{AB} \approx e_{ij}\delta_{iA}\delta_{jB} \ , \tag{4.61}$$

and it is customary to define a single infinitesimal strain tensor for which we introduce the symbol ϵ_{ij} as

$$2\epsilon_{ij} = \frac{\partial u_i}{\partial X_A}\delta_{Aj} + \frac{\partial u_j}{\partial X_B}\delta_{Bi} = \frac{\partial u_i}{\partial x_j} + \frac{\partial u_j}{\partial x_i} = u_{i,j} + u_{j,i} \ . \tag{4.62}$$

Because the strain tensors E_{AB}, e_{ij}, and ϵ_{ij} are all symmetric, second-order tensors, the entire development for principal strains, strain invariants and principal strain directions may be carried out exactly as was done for the stress tensor in Chapter 3. Thus, taking, ϵ_{ij} as the typical tensor of the group, we summarize these results by displaying its matrix relative to principal axes in the alternative forms,

$$[\epsilon^*_{ij}] = \begin{bmatrix} \epsilon_1 & 0 & 0 \\ 0 & \epsilon_2 & 0 \\ 0 & 0 & \epsilon_3 \end{bmatrix} = \begin{bmatrix} \epsilon_I & 0 & 0 \\ 0 & \epsilon_{II} & 0 \\ 0 & 0 & \epsilon_{III} \end{bmatrix} \tag{4.63}$$

together with the strain invariants

$$I_\epsilon = \epsilon_{ii} = \text{tr}\,\boldsymbol{\epsilon} = \epsilon_I + \epsilon_{II} + \epsilon_{III} \ , \tag{4.64a}$$

$$II_\epsilon = \frac{1}{2}\left(\epsilon_{ii}\epsilon_{jj} - \epsilon_{ij}\epsilon_{ij}\right) = \epsilon_I\epsilon_{II} + \epsilon_{II}\epsilon_{III} + \epsilon_{III}\epsilon_I \ , \tag{4.64b}$$

$$III_\epsilon = \varepsilon_{ijk}\epsilon_{1i}\epsilon_{2j}\epsilon_{3k} = \epsilon_I\epsilon_{II}\epsilon_{III} \ . \tag{4.64c}$$

The components of $\boldsymbol{\epsilon}$ have specific physical interpretations which we now consider. Within the context of small deformation theory we express Eq 4.46 in the modified form

$$(dx)^2 - (dX)^2 = 2\epsilon_{ij}\,dX_i\,dX_j = d\mathbf{X}\cdot 2\boldsymbol{\epsilon}\cdot d\mathbf{X} \tag{4.65}$$

which, upon factoring the left hand side and dividing by $(dX)^2$, becomes

$$\frac{dx - dX}{dX}\cdot\frac{dx + dX}{dX} = 2\epsilon_{ij}\frac{dX_i}{dX}\frac{dX_j}{dX} \ .$$

But $dX_i/dX = N_i$, a unit vector in the direction of $d\mathbf{X}$, and for small deformations we may assume $(dx + dX)/dX \approx 2$, so that

$$\frac{dx - dX}{dX} = \epsilon_{ij}N_iN_j = \hat{\mathbf{N}}\cdot\boldsymbol{\epsilon}\cdot\hat{\mathbf{N}} \ . \tag{4.66}$$

The scalar ratio on the left-hand side of this equation is clearly the change in length per unit original length for the element in the direction of $\hat{\mathbf{N}}$. It is known as the *longitudinal strain*, or the *normal strain* and we denote it by $e_{(\hat{\mathbf{N}})}$. If, for example, $\hat{\mathbf{N}}$ is taken in the X_1 direction so that $\hat{\mathbf{N}} = \hat{\mathbf{I}}_1$, then

$$e_{(\hat{\mathbf{I}}_1)} = \hat{\mathbf{I}}_1\cdot\boldsymbol{\epsilon}\cdot\hat{\mathbf{I}}_1 = \epsilon_{11} \ .$$

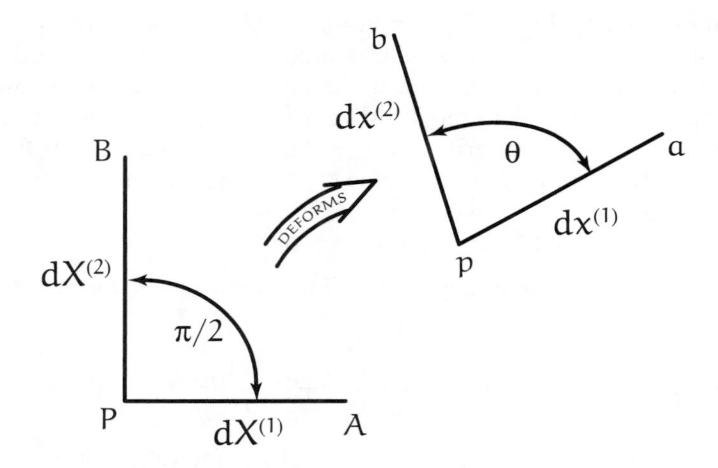

FIGURE 4.3
The right angle between line segments AP and BP in the reference configuration becomes θ, the angle between segments ap and bp, in the deformed configuration.

Likewise, for $\hat{\mathbf{N}} = \hat{\mathbf{I}}_2$, or $\hat{\mathbf{N}} = \hat{\mathbf{I}}_3$ the normal strains are found to be ϵ_{22} and ϵ_{33}, respectively. Thus the diagonal elements of the small (infinitesimal) strain tensor represent normal strains in the coordinate directions.

To gain an insight into the physical meaning of the off-diagonal elements of the infinitesimal strain tensor we consider differential vectors $d\mathbf{X}^{(1)}$ and $d\mathbf{X}^{(2)}$ at position $\mathbf{P}$ which are deformed into vectors $d\mathbf{x}^{(1)}$ and $d\mathbf{x}^{(2)}$, respectively. In this case, Eq 4.51 may be written,

$$d\mathbf{x}^{(1)} \cdot d\mathbf{x}^{(2)} = d\mathbf{X}^{(1)} \cdot d\mathbf{X}^{(2)} + d\mathbf{X}^{(1)} \cdot 2\boldsymbol{\epsilon} \cdot d\mathbf{X}^{(2)} \tag{4.67}$$

which, if we choose $d\mathbf{X}^{(1)}$ and $d\mathbf{X}^{(2)}$ perpendicular to one another, reduces to

$$d\mathbf{x}^{(1)} \cdot d\mathbf{x}^{(2)} = dx^{(1)} dx^{(2)} \cos\theta = d\mathbf{X}^{(1)} \cdot 2\boldsymbol{\epsilon} \cdot d\mathbf{X}^{(2)} \tag{4.68}$$

where θ is the angle between the deformed vectors as shown in Fig. 4.3. If we let, $\theta = \frac{\pi}{2} - \gamma$, the angle γ measures the small change in the original right angle between $d\mathbf{X}^{(1)}$ and $d\mathbf{X}^{(2)}$. Also

$$\cos\theta = \cos\left(\frac{\pi}{2} - \gamma\right) = \sin\gamma \approx \gamma$$

since γ is very small for infinitesimal deformations. Therefore, assuming as before that $dx^{(1)} \approx dX^{(1)}$ and $dx^{(2)} \approx dX^{(2)}$ because of small deformations

$$\gamma \approx \cos\theta = \frac{d\mathbf{X}^{(1)}}{dx^{(1)}} \cdot 2\boldsymbol{\epsilon} \cdot \frac{d\mathbf{X}^{(2)}}{dx^{(2)}} \approx \hat{\mathbf{N}}_{(1)} \cdot 2\boldsymbol{\epsilon} \cdot \hat{\mathbf{N}}_{(2)} \,. \tag{4.69}$$

Here, if we take $\hat{\mathbf{N}}_{(1)} = \hat{\mathbf{I}}_1$ and $\hat{\mathbf{N}}_{(2)} = \hat{\mathbf{I}}_2$ and designate the angle γ as γ_{12} , we obtain

$$\gamma_{12} = 2\,[1,0,0] \begin{bmatrix} \epsilon_{11} & \epsilon_{12} & \epsilon_{13} \\ \epsilon_{12} & \epsilon_{22} & \epsilon_{23} \\ \epsilon_{13} & \epsilon_{23} & \epsilon_{33} \end{bmatrix} \begin{bmatrix} 0 \\ 1 \\ 0 \end{bmatrix} = 2\epsilon_{12} \,, \tag{4.70}$$

so that by choosing the undeformed vector pairs in Eq 4.68 in coordinate directions we may generalize Eq 4.70 to obtain

$$\gamma_{ij} = 2\epsilon_{ij} \qquad (i \neq j)\ . \tag{4.71}$$

This establishes the relationship between the off-diagonal components of ϵ_{ij} and the so-called *engineering shear strain* components γ_{ij}, which represent the changes in the original right angles between the coordinate axes in the undeformed configuration. Note that since ϵ can be defined with respect to any set of Cartesian axes at $\mathbf{P}$, this result holds for any pair of perpendicular vectors at that point. In engineering texts, the infinitesimal strain tensor is frequently written in matrix form as

$$[\epsilon_{ij}] = \begin{bmatrix} \epsilon_{11} & \frac{1}{2}\gamma_{12} & \frac{1}{2}\gamma_{13} \\ \frac{1}{2}\gamma_{12} & \epsilon_{22} & \frac{1}{2}\gamma_{23} \\ \frac{1}{2}\gamma_{13} & \frac{1}{2}\gamma_{23} & \epsilon_{33} \end{bmatrix}\ . \tag{4.72}$$

If $\hat{\mathbf{N}}_{(1)}$ and $\hat{\mathbf{N}}_{(2)}$ are chosen in principal strain directions, Eq 4.69 becomes

$$\gamma = \hat{\mathbf{N}}_{(1)} \cdot 2\epsilon^* \cdot \hat{\mathbf{N}}_{(2)} = 0 \tag{4.73}$$

from which we may generalize to conclude that principal strain directions remain orthogonal under infinitesimal deformation. Therefore, a small rectangular parallelpiped of undeformed edge dimensions $dX^{(1)}$, $dX^{(2)}$ and $dX^{(3)}$ taken in the principal strain directions will be deformed into another rectangular parallelpiped having edge lengths

$$dx^{(i)} = (1 + \epsilon_{(i)})dX^{(i)}, \quad (i = 1, 2, 3) \tag{4.74}$$

as shown in Fig. 4.4, where $\epsilon_{(i)}$ are the normal strains in principal directions. The change in volume per unit original volume of the parallelpiped is

$$\begin{aligned} \frac{\Delta \mathcal{V}}{\mathcal{V}} &= \frac{[1+\epsilon_{(1)}]dX^{(1)}[1+\epsilon_{(2)}]dX^{(2)}[1+\epsilon_{(3)}]dX^{(3)} - dX^{(1)}dX^{(2)}dX^{(3)}}{dX^{(1)}dX^{(2)}dX^{(3)}} \\ &\approx \epsilon_{(1)} + \epsilon_{(2)} + \epsilon_{(3)} \end{aligned} \tag{4.75}$$

where terms involving products of the principal strains have been neglected. The ratio $\Delta\mathcal{V}/\mathcal{V}$, being the first invariant of ϵ, is called the *cubical dilatation*. We shall denote it by the symbol e, and write

$$e = \Delta V/V = \epsilon_{ii} = I_{\epsilon}\ . \tag{4.76}$$

Because ϵ is a symmetrical second-order tensor the development of Mohr's circles for small strain, as well as the decomposition of ϵ into its *spherical* and *deviator* component tensors follows in much the same way as the analogous concepts for stress in Chapter 3. One distinct difference is that for the Mohr's circles, the shear strain axis (ordinate) has units of $\frac{1}{2}\gamma$ as shown by the typical diagram of Fig. 4.5. The infinitesimal *spherical strain tensor* is represented by a diagonal matrix having equal elements denoted by $\epsilon_M = \frac{1}{3}\epsilon_{ii} = \frac{1}{3}e$, known as the *mean normal strain*. The *infinitesimal deviator strain tensor*, η is defined by

$$\eta_{ij} = \epsilon_{ij} - \frac{1}{3}\delta_{ij}\epsilon_{kk} = \epsilon_{ij} - \delta_{ij}\epsilon_M\ , \tag{4.77}$$

and in matrix form

$$\begin{bmatrix} \eta_{11} & \eta_{12} & \eta_{13} \\ \eta_{21} & \eta_{22} & \eta_{23} \\ \eta_{31} & \eta_{32} & \eta_{33} \end{bmatrix} = \begin{bmatrix} \epsilon_{11} - \epsilon_M & \epsilon_{12} & \epsilon_{13} \\ \epsilon_{12} & \epsilon_{22} - \epsilon_M & \epsilon_{23} \\ \epsilon_{13} & \epsilon_{23} & \epsilon_{33} - \epsilon_{33} \end{bmatrix}\ . \tag{4.78}$$

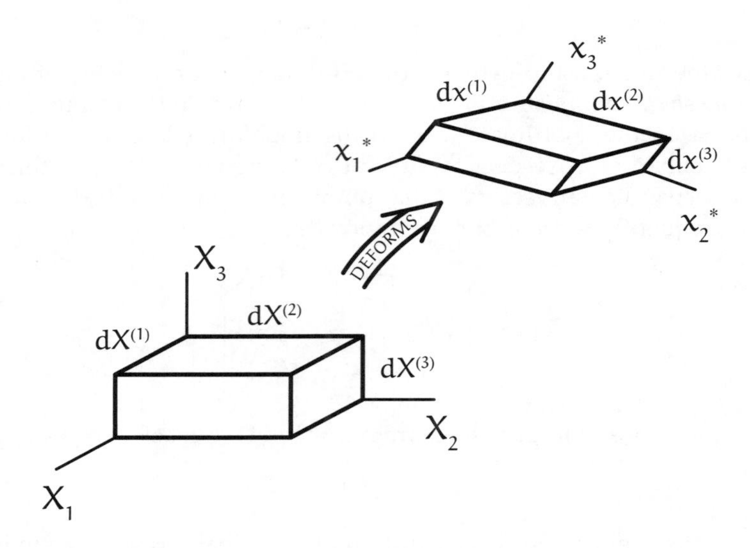

FIGURE 4.4
A rectangular parallelpiped with edge lengths $dX^{(1)}$, $dX^{(2)}$ and $dX^{(3)}$ in the reference configuration becomes a skewed parallelpiped with edge lengths $dx^{(1)}$, $dx^{(2)}$ and $dx^{(3)}$ in the deformed configuration.

Note that as with its stress counterpart, the first invariant of the deviator strain is zero, or

$$\eta_{ii} = 0 \,, \tag{4.79}$$

and the principal deviator strains are given by

$$\eta_{(q)} = \epsilon_{(q)} - \epsilon_M, \qquad (q = 1, 2, 3) \tag{4.80}$$

where $\epsilon_{(q)}$ is a principal value of the infinitesimal strain tensor.

A state of *plane strain* parallel to the X_1X_2 plane exists at **P** if

$$\epsilon_{33} = \gamma_{13} = \gamma_{31} = \gamma_{23} = \gamma_{32} = 0 \tag{4.81}$$

at that point. Also, plane strain relative to the X_1X_2 plane in the continuum body as a whole exists if Eq 4.81 is satisfied everywhere in the body, and if in addition, the remaining non-zero components are independent of X_3. With respect to axes $OX_1'X_2'X_3'$ rotated about X_3 by the angle θ relative to $OX_1X_2X_3$ as shown by Fig. 4.6, the transformation equations for plane strain (analogous to Eq 3.75 for plane stress) follow the tensor transformation formula, Eq 2.64. In conjunction with the table of direction cosines of Table 4.1 the results are

$$\epsilon'_{11} = \frac{\epsilon_{11} + \epsilon_{22}}{2} + \frac{\epsilon_{11} - \epsilon_{22}}{2} \cos 2\theta + \frac{\gamma_{12}}{2} \sin 2\theta \,, \tag{4.82a}$$

$$\epsilon'_{22} = \frac{\epsilon_{11} + \epsilon_{22}}{2} - \frac{\epsilon_{11} - \epsilon_{22}}{2} \cos 2\theta - \frac{\gamma_{12}}{2} \sin 2\theta \,, \tag{4.82b}$$

$$\gamma'_{12} = -(\epsilon_{11} - \epsilon_{22}) \sin 2\theta + \gamma_{12} \cos 2\theta \,. \tag{4.82c}$$

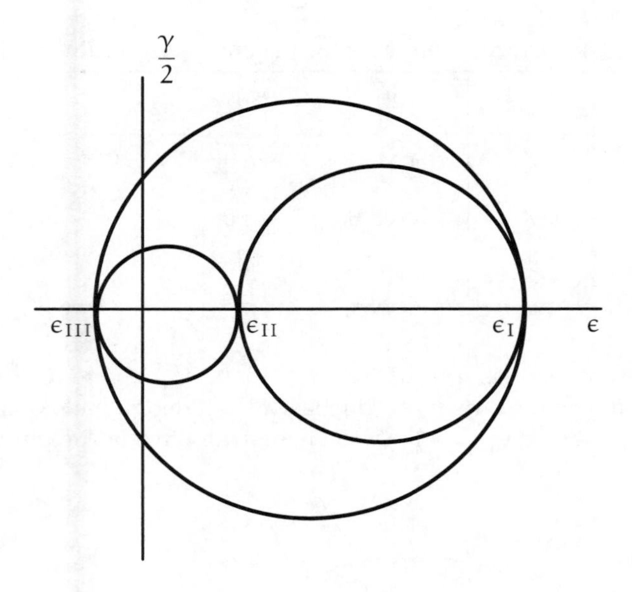

FIGURE 4.5
Typical Mohr's circle for strain.

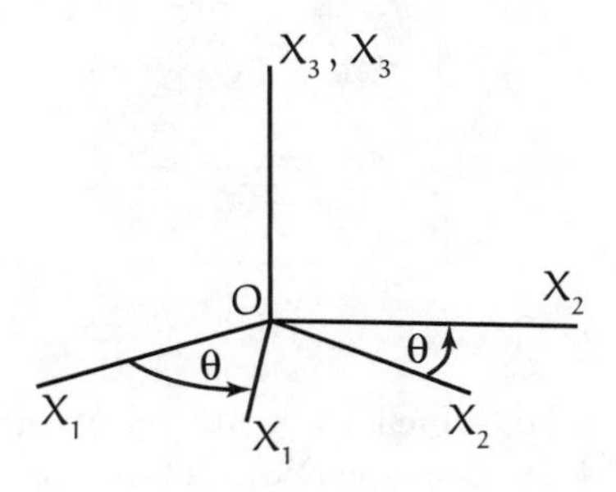

FIGURE 4.6
Rotation of axes for plane strain.

Also, the non-zero principal strain values for plane strain are given by

$$\epsilon_{(1)}, \epsilon_{(2)} = \frac{\epsilon_{11} + \epsilon_{22}}{2} \pm \sqrt{\left(\frac{\epsilon_{11} - \epsilon_{22}}{2}\right)^2 + \left(\frac{\gamma_{12}}{2}\right)^2}\,. \tag{4.83}$$

Because shear strains are very difficult to measure experimentally, the state of strain at a point is usually determined by recording three separate longitudinal strains at the point (using a strain gauge rosette) and substituting these values into Eqs 4.82 to calculate γ_{12}.

Example 4.11

A delta rosette has the shape of an equilateral triangle, and records longitudinal strains in the directions x_1, x_1' and x_1'' shown in the sketch. If the measured

TABLE 4.1
Transformation table for general plane strain.

	x_1	x_2	x_3
x_1'	$\cos\theta$	$\sin\theta$	0
x_2'	$-\sin\theta$	$\cos\theta$	0
x_3'	0	0	1

strains in these directions are $\epsilon_{11} = -3 \times 10^{-4}$, $\epsilon'_{11} = 4 \times 10^{-4}$, and $\epsilon''_{11} = 2 \times 10^{-4}$ where the units are m/m (dimensionless), determine ϵ_{22}, γ_{12} and ϵ'_{22}. Show that $\epsilon_{11} + \epsilon_{22} = \epsilon'_{11} + \epsilon'_{22}$ as the first strain invariant requires.

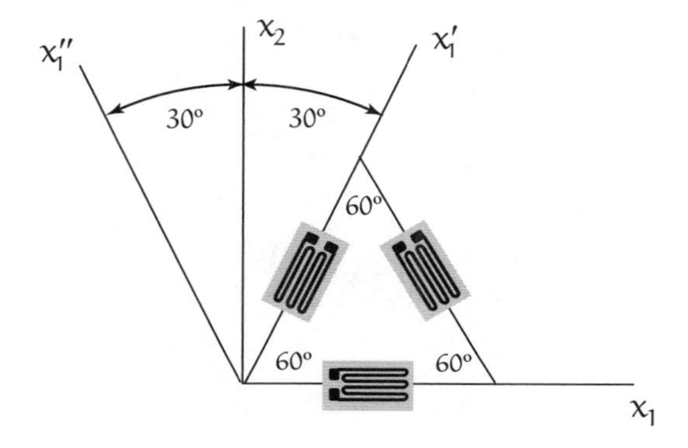

Solution
We need only Eq 4.82 here, which we write for x_1' and x_1'' in turn (omitting the common factor 10^{-4} for convenience). Thus, for $\theta = 60°$, and $\theta = 120°$, respectively, we have

$$4 = \frac{-3 + \epsilon_{22}}{2} + \frac{-3 - \epsilon_{22}}{2}\left(-\frac{1}{2}\right) + \frac{\gamma_{12}}{2}\frac{\sqrt{3}}{2} ,$$

$$2 = \frac{-3 + \epsilon_{22}}{2} + \frac{-3 - \epsilon_{22}}{2}\left(-\frac{1}{2}\right) - \frac{\gamma_{12}}{2}\frac{\sqrt{3}}{2} .$$

Adding these two equations to eliminate γ_{12} we determine $\epsilon_{22} = 5$; subtracting the second from the first to eliminate ϵ_{22} we determine $\gamma_{12} = 4/\sqrt{3}$. Next, using $\theta = 150°$ we determine ϵ'_{22} from Eq 4.82

$$\epsilon'_{22} = \frac{-3 + 5}{2} + \frac{-3 - 5}{2} + \frac{2}{\sqrt{3}}\left(-\frac{\sqrt{3}}{2}\right) = -2 ,$$

and by the first invariant of the small strain tensor we check that

$$\epsilon_{11} + \epsilon_{22} = -3 + 5 = \epsilon'_{11} + \epsilon'_{22} = 4 - 2 = 2 .$$

Consider once more the two neighboring particles which were at positions **P** and **Q** in the undeformed configuration, and are now at positions **p** and **q**, respectively, in the deformed configuration (see Fig. 4.2). In general, an arbitrary displacement will include both deformation (strain) and rigid body displacements. To separate these we consider the differential displacement vector $d\mathbf{u}$. Assuming conditions on the displacement field that guarantee the existence of a derivative, the displacement differential du_i is written

$$du_i = \left(\frac{\partial u_i}{\partial X_j}\right)_{\mathbf{P}} dX_j \tag{4.84}$$

where the derivative is evaluated at **P** as indicated by the notation. From this we may define the *unit relative displacement* of the particle at **Q** with respect to the one at **P** by the equation

$$\frac{du_i}{dX} = \frac{du_i}{dX_j}\frac{dX_j}{dX} = \frac{du_i}{dX_j}N_j \tag{4.85}$$

where N_j is the unit vector in the direction from **P** toward **Q**. By decomposing the displacement gradient in Eq 4.84 into its symmetric and skew-symmetric parts we obtain

$$\begin{aligned} du_i &= \left[\frac{1}{2}\left(\frac{\partial u_i}{\partial X_j} + \frac{\partial u_j}{\partial X_i}\right) + \frac{1}{2}\left(\frac{\partial u_i}{\partial X_j} - \frac{\partial u_j}{\partial X_i}\right)\right] dX_j \\ &= (\epsilon_{ij} + \omega_{ij})dX_j \end{aligned} \tag{4.86}$$

in which ϵ_{ij} is recognized as the *infinitesimal strain tensor*, and ω_{ij} is called the *infinitesimal rotation tensor*.

If ϵ_{ij} happens to be identically zero, there is no strain, and the displacement is a rigid body displacement. For this case we define the rotation vector

$$\omega_i = \frac{1}{2}\varepsilon_{ijk}\omega_{kj} \tag{4.87}$$

which may be readily inverted since $\omega_{kj} = -\omega_{jk}$ to yield

$$\omega_{ij} = \varepsilon_{kji}\omega_k \, . \tag{4.88}$$

Therefore, Eq 4.86 with $\epsilon_{ij} \equiv 0$ becomes

$$du_i = \varepsilon_{kjl}\omega_k dX_j = \varepsilon_{ikj}\omega_k dX_j \qquad \text{or} \qquad d\mathbf{u} = \boldsymbol{\omega} \times d\mathbf{X} \, , \tag{4.89}$$

so that the relative differential displacement is seen to be the result of a rigid body rotation about the axis of the rotation vector ω. On the other hand, if $\omega_{ij} \equiv 0$, the relative displacement will be the result of *pure strain*.

A comment is appropriate about the notation ω_{ij} for the skew–symmetric part of the displacement gradient in the linear theory. It might have been convenient to denote this quantity as $\boldsymbol{W} = w_{ij}\hat{\mathbf{e}}_i \otimes \hat{\mathbf{e}}_j$; however, this notation has been reserved to represent the spin tensor in Section 4.11. The spin tensor is the skew–symmetric part of the velocity gradient. It is a bit awkward to have the tensor ω_{ij} and its axial vector ω_j denoted with $\boldsymbol{\omega}$, but this is not as bad as having the same symbol representing the infinitesimal rotation and spin tensors.

4.8 Compatibility Equations

The compatibility equations merit a section because of the important role played in linear elasticity problems. Very often, solving a linear elasticity problem (see Chapter 6) involves solving the compatibility equations discussed below.

If we consider the six independent strain-displacement relations, Eq 4.62

$$\frac{\partial u_i}{\partial x_j} + \frac{\partial u_j}{\partial x_i} = 2\epsilon_{ij}$$

as a system of partial differential equations for determining the three displacement components u_i (assuming the ϵ_{ij} are known as functions of x_i), the system is over-determined, and we cannot in general find three single-valued functions $u_i = u_i(x_j)$ satisfying the six partial differential equations. Therefore, some restrictive conditions must be imposed upon the strain components (actually upon derivatives of the strain components) if the equations above are to be satisfied by a single-valued displacement field. Such conditions are expressed by the strain *compatibility equations*

$$\epsilon_{ij,km} + \epsilon_{km,ij} - \epsilon_{ik,jm} - \epsilon_{jm,ik} = 0 \; . \tag{4.90}$$

There are $3^4 = 81$ equations in all (four free indices) in Eq 4.90 but only six of these are distinct

$$\begin{aligned}
&\epsilon_{11,23} + \epsilon_{23,11} - \epsilon_{12,13} - \epsilon_{13,12} = 0 \; , \\
&\epsilon_{22,31} + \epsilon_{31,22} - \epsilon_{23,21} - \epsilon_{21,23} = 0 \; , \\
&\epsilon_{33,12} + \epsilon_{12,33} - \epsilon_{31,32} - \epsilon_{32,31} = 0 \; , \\
&2\epsilon_{12,12} - \epsilon_{11,22} - \epsilon_{22,11} = 0 \; , \\
&2\epsilon_{23,23} - \epsilon_{22,33} - \epsilon_{33,22} = 0 \; , \\
&2\epsilon_{31,31} - \epsilon_{33,11} - \epsilon_{11,33} = 0 \; .
\end{aligned} \tag{4.91}$$

For plane strain in the x_1-x_2 plane, the six unique equations in Eq 4.90 reduce to a single equation,

$$\epsilon_{11,22} + \epsilon_{22,11} = 2\epsilon_{12,12} \; , \tag{4.92}$$

which may be easily verified as a necessary condition by a simple differentiation of Eq 4.62 for a range of two on the indices i and j.

It may be shown that the compatibility equations, either Eq 4.90 or Eq 4.91, are both necessary and sufficient for a single-valued displacement field of a body occupying a simply connected domain. The compatibility equations seem daunting, but it is not too difficult to demonstrate the necessity and sufficiency.

For the necessity, given the strain $\epsilon_{ij} = \frac{1}{2}(u_{i,j} + u_{j,i})$ with continuously differentiable displacements we write that

$$\operatorname{curl}\operatorname{curl}\boldsymbol{\epsilon} = 0 \; . \tag{4.93}$$

This equation is a more compact way of writing Eqs 4.90 and 4.91. By the definition of linear strain and the continuity of the displacements, we see

$$\varepsilon_{ijk}\varepsilon_{lmn}\epsilon_{jm,kn} = \varepsilon_{ijk}\varepsilon_{lmn}\left(u_{j,mkn} + u_{m,jkn}\right) = 0$$

where the first displacement term, symmetric in mn, cancels with skew–symmetric ε_{lmn}, and the second displacement term, symmetric in jk, cancels with skew–symmetric ε_{ijk}.

As with many proofs, sufficiency is more difficult to demonstrate. Here it must be shown that starting from curl curl $\boldsymbol{\epsilon} = 0$ that the linear strain $\epsilon_{ij} = \frac{1}{2}(u_{i,j} + u_{j,i})$ can be constructed. To start, let

$$\mathbf{A} = \text{curl}\,\boldsymbol{\epsilon} \tag{4.94}$$

from which compatibility in the form of Eq 4.93 gives

$$\text{curl}\,\mathbf{A} = 0\,. \tag{4.95}$$

Since $\boldsymbol{\epsilon}$ is symmetric,

$$\text{tr}\,\mathbf{A} = 0 \tag{4.96}$$

based on the identity Eq 2.91 from Example 2.17. From Eqs 4.95 and 4.96 there exists a skew–symmetric tensor $\boldsymbol{\omega}$ such that [2]

$$\mathbf{A} = -\,\text{curl}\,\boldsymbol{\omega}\,. \tag{4.97}$$

Equating Eqs 4.97 and 4.94 on $\mathbf{A}$ gives

$$\text{curl}\,(\boldsymbol{\epsilon} + \boldsymbol{\omega}) = 0 \tag{4.98}$$

from which [3]

$$\boldsymbol{\epsilon} + \boldsymbol{\omega} = \boldsymbol{\nabla}\mathbf{u}\,. \tag{4.99}$$

Taking the symmetric part of this equation gives us the strain

$$\boldsymbol{\epsilon} = \frac{1}{2}\left(\boldsymbol{\nabla}\mathbf{u} + \boldsymbol{\nabla}\mathbf{u}^T\right)\,, \tag{4.100a}$$

or

$$\epsilon_{ij} = \frac{1}{2}\left(u_{i,j} + u_{j,i}\right)\,. \tag{4.100b}$$

The compatibility equations give a convenient way to compute the displacements from the strain components. A line integral form of the compatibility equations gives

$$\begin{aligned} u_i(\xi) &= \int_{\xi^0}^{\xi} \left\{\epsilon_{ij}(\hat{\xi}) + \left(\xi_k - \hat{\xi}_k\right)\left[\epsilon_{ij,k}(\hat{\xi}) - \epsilon_{jk,i}(\hat{\xi})\right]\right\} d\hat{\xi}_j \\ &\quad + u_j(\xi^0) + w_{ij}(\xi^0)\left(\xi_j - \xi_j^0\right)\,. \end{aligned} \tag{4.101}$$

The last two terms represent an infinitesimal rigid motion and are often omitted.

Example 4.12

For a given state of strain, ϵ_{ij}, the compatibility equations are necessary and sufficient conditions for unique displacements.

(a) The compatibility condition may be written as $\varepsilon_{kmn}\varepsilon_{pqr}\epsilon_{mq,nr} = 0$ where ε_{pqr} is the alternating tensor. Show that the compatibility equation in this form is symmetric in k and p. This means that there are only six independent conditions.

[2] This is not a difficult step, however, it is best to refer to ME Gurtin, "The Linear Theory of Elasticty", in *Mechanics of Solids*, Volume II, Editor C Truesdell, Springer-Verlag, c. 1973, pp17-18 and p 40.

[3] *ibid*, p 17

(b) For displacements to be path independent the integrand of Eq 4.101 must be exact. If

$$U_{ij} = \epsilon_{ij}(\hat{\xi}) + \left(\xi_k - \hat{\xi}_k\right)\left[\epsilon_{ij,k}(\hat{\xi}) - \epsilon_{jk,i}(\hat{\xi})\right] ,$$

a necessary and sufficient condition for the integral to be exact is

$$U_{ij,m} + U_{im,j} = 0 .$$

Show this condition leads to

$$\epsilon_{ij,km} + \epsilon_{km,ij} - \epsilon_{jk,im} - \epsilon_{im,jk} = 0 .$$

(c) If the displacement field has three continuous derivatives, show that $\varepsilon_{kmn}\varepsilon_{pqr}\epsilon_{mq,nr} = 0$.

Solution

(a) This can be shown through index manipulation. Noting strain is symmetric and twice continuously differentiable and m, n, q and r are dummy indices:

$$\varepsilon_{kmn}\varepsilon_{pqr}\epsilon_{mq,nr} = \varepsilon_{kmn}\varepsilon_{pqr}\epsilon_{qm,rn} = \varepsilon_{kqr}\varepsilon_{pmn}\epsilon_{mq,nr} .$$

This expression is a symmetric, second-order tensor having six independent terms.

(b) Differentiation and substitution leads to

$$\begin{aligned} \epsilon_{ij,m} &- \delta_{km}\left(\epsilon_{ij,k} - \epsilon_{jk,i}\right) - \epsilon_{im,j} + \delta_{kj}\left(\epsilon_{im,k} - \epsilon_{mk,i}\right) \\ &+ \left(\xi_k - \hat{\xi}_k\right)\left[\epsilon_{ij,km} - \epsilon_{jk,im} - \epsilon_{im,kj} + \epsilon_{mk,ij}\right] = 0 . \end{aligned}$$

Since strain is symmetric and ξ and $\hat{\xi}_k$ can be chosen arbitrarily

$$\epsilon_{ij,km} + \epsilon_{km,ij} - \epsilon_{jk,im} - \epsilon_{im,jk} = 0 .$$

(c) Substitute the strain-displacement relations into the compatibility equation

$$\varepsilon_{kmn}\varepsilon_{pqr}\epsilon_{mq,nr} = \tfrac{1}{2}\varepsilon_{kmn}\varepsilon_{pqr}\left(u_{m,qnr} + u_{q,mnr}\right) .$$

The alternating tensors are skew-symmetric in mn and qr. Additionally, the displacement is assumed to be continuously differentiable and so order of differentiation is immaterial. Thus, the parenthetical term is symmetric in mn and qr. The product of skew-symmetric and symmetric tensors is zero. So,

$$\varepsilon_{kmn}\varepsilon_{pqr}\epsilon_{mq,nr} = \tfrac{1}{2}\varepsilon_{kmn}\varepsilon_{pqr}\left(u_{m,qnr} + u_{q,mnr}\right) = 0 .$$

4.9 Stretch Ratios

Referring again to Fig. 4.2, define the ratio of the magnitudes of $\mathbf{dx}$ and $\mathbf{dX}$ to be the stretch ratio, Λ (or simply the *stretch*). In particular, for the differential element in the direction of the unit vector $\hat{\mathbf{N}}$ at $\mathbf{P}$, we write

$$\Lambda_{(\hat{\mathbf{N}})} = \frac{dx}{dX} \tag{4.102}$$

where dx is the deformed magnitude of $\mathbf{dX} = dX\hat{\mathbf{N}}$. As a matter of convenience we often prefer to work with stretch-squared values,

$$\Lambda^2_{(\hat{\mathbf{N}})} = \left(\frac{dx}{dX}\right)^2 . \tag{4.103}$$

Thus, from Eqs 4.41 and 4.44,

$$(dx)^2 = \mathbf{dx} \cdot \mathbf{dx} = \mathbf{F} \cdot \mathbf{dX} \cdot \mathbf{F} \cdot \mathbf{dX} = \mathbf{dX} \cdot \mathbf{C} \cdot \mathbf{dX}\,, \tag{4.104}$$

so that after dividing by $(dX)^2$,

$$\Lambda^2_{(\hat{\mathbf{N}})} = \frac{\mathbf{dX}}{dX} \cdot \mathbf{C} \cdot \frac{\mathbf{dX}}{dX} = \hat{\mathbf{N}} \cdot \mathbf{C} \cdot \hat{\mathbf{N}} \tag{4.105}$$

for the element originally in the direction of $\hat{\mathbf{N}}$.

In an analogous way, we define the stretch ratio, $\lambda_{(\hat{\mathbf{n}})}$ in the direction of $\hat{\mathbf{n}} = \mathbf{dx}/dx$ at $\mathbf{p}$ by the equation,

$$\frac{1}{\lambda_{(\hat{\mathbf{n}})}} = \frac{dX}{dx} . \tag{4.106}$$

Here, recalling from Eq 4.42 that $\mathbf{dX} = \mathbf{F}^{-1} \cdot \mathbf{dx}$ and by using Eq 4.48, we obtain

$$(dX)^2 = \mathbf{dX} \cdot \mathbf{dX} = \mathbf{F}^{-1} \cdot \mathbf{dx} \cdot \mathbf{F}^{-1} \cdot \mathbf{dx} = \mathbf{dx} \cdot \mathbf{c} \cdot \mathbf{dx} \tag{4.107}$$

which upon dividing by $(dx)^2$ becomes

$$\frac{1}{\lambda^2_{(\hat{\mathbf{n}})}} = \frac{\mathbf{dx}}{dx} \cdot \mathbf{c} \cdot \frac{\mathbf{dx}}{dx} = \hat{\mathbf{n}} \cdot \mathbf{c} \cdot \hat{\mathbf{n}} . \tag{4.108}$$

In general, $\Lambda_{(\hat{\mathbf{N}})} \neq \lambda_{(\hat{\mathbf{n}})}$. However if $\hat{\mathbf{n}}$ is a unit vector in the direction that $\hat{\mathbf{N}}$ assumes in the deformed configuration, the two stretches are the same. For $\hat{\mathbf{N}} = \hat{\mathbf{I}}_1$,

$$\Lambda^2_{(\hat{\mathbf{I}}_1)} = \hat{\mathbf{I}}_1 \cdot \mathbf{C} \cdot \hat{\mathbf{I}}_1 = C_{11} = 1 + 2E_{11}\,, \tag{4.109}$$

and for $\hat{\mathbf{n}} = \hat{\mathbf{e}}_1$,

$$\frac{1}{\lambda^2_{(\hat{\mathbf{e}}_1)}} = \hat{\mathbf{e}}_1 \cdot \mathbf{c} \cdot \hat{\mathbf{e}}_1 = c_{11} = 1 - 2e_{11} \tag{4.110}$$

with analogous expressions for $\hat{\mathbf{N}}$ and $\hat{\mathbf{n}}$ in the other coordinate directions.

Consider next the unit extension (longitudinal strain) in any direction $\hat{\mathbf{N}}$ at $\mathbf{P}$. This may be expressed in terms of the stretch as

$$e_{(\hat{\mathbf{N}})} = \frac{dx - dX}{dX} = \Lambda_{(\hat{\mathbf{N}})} - 1 = \sqrt{\hat{\mathbf{N}} \cdot \mathbf{C} \cdot \hat{\mathbf{N}}} - 1 . \tag{4.111}$$

Notice that the unit extension is zero when the stretch is unity, as occurs with a rigid body displacement. If $\hat{\mathbf{N}} = \hat{\mathbf{I}}_1$,

$$e_{(\hat{\mathbf{I}}_1)} = \sqrt{\hat{\mathbf{I}}_1 \cdot \mathbf{C} \cdot \hat{\mathbf{I}}_1} - 1 = \sqrt{C_{11}} - 1 = \sqrt{1 + 2E_{11}} - 1 , \tag{4.112}$$

or, solving for E_{11},

$$E_{11} = e_{(\hat{\mathbf{I}}_1)} + \frac{1}{2} e^2_{(\hat{\mathbf{I}}_1)} . \tag{4.113}$$

For small deformation theory where $E_{11} \to \epsilon_{11}$, and for which $e^2_{(\hat{\mathbf{N}})}$ may be neglected in comparison to $e_{(\hat{\mathbf{N}})}$, the above equation asserts that $E_{11} = \epsilon_{11} = e_{(\hat{\mathbf{I}}_1)}$.

The change in angle between any two line elements may also be given in terms of stretch. Let $d\mathbf{X}^{(1)}$ and $d\mathbf{X}^{(2)}$ be arbitrary vectors which become $d\mathbf{x}^{(1)}$ and $d\mathbf{x}^{(2)}$, respectively, during a deformation. By the dot product, $d\mathbf{x}^{(1)} \cdot d\mathbf{x}^{(2)} = dx^{(1)}dx^{(2)} \cos\theta$, we may compute the angle θ between $d\mathbf{x}^{(1)}$ and $d\mathbf{x}^{(2)}$ from its cosine, which, with the help of Eq 4.104, takes the form,

$$\cos\theta = \frac{d\mathbf{x}^{(1)}}{dx^{(1)}} \cdot \frac{d\mathbf{x}^{(2)}}{dx^{(2)}} = \frac{\mathbf{F} \cdot d\mathbf{X}^{(1)}}{\sqrt{d\mathbf{X}^{(1)} \cdot \mathbf{C} \cdot d\mathbf{X}^{(1)}}} \cdot \frac{\mathbf{F} \cdot d\mathbf{X}^{(2)}}{\sqrt{d\mathbf{X}^{(2)} \cdot \mathbf{C} \cdot d\mathbf{X}^{(2)}}} ,$$

or upon dividing the numerator and denominator by the scalar product $dX^{(1)}dX^{(2)}$ and making use of Eqs 4.105 and 4.44 we obtain

$$\cos\theta = \frac{\hat{\mathbf{N}}_1 \cdot \mathbf{C} \cdot \hat{\mathbf{N}}_2}{\Lambda_{(\hat{\mathbf{N}}_1)}\Lambda_{(\hat{\mathbf{N}}_2)}} . \tag{4.114}$$

Thus, for elements originally in the $\hat{\mathbf{I}}_1$ and $\hat{\mathbf{I}}_2$ directions, the angle between them in the deformed configuration may be determined from

$$\cos\theta_{12} = \frac{C_{12}}{\Lambda_{(\hat{\mathbf{I}}_1)}\Lambda_{(\hat{\mathbf{I}}_2)}} = \frac{C_{12}}{\sqrt{C_{11}C_{22}}} . \tag{4.115}$$

In a similar fashion, from $d\mathbf{X}^{(1)} \cdot d\mathbf{X}^{(2)} = dX^{(1)}dX^{(2)} \cos\Theta$, where Θ is the angle between $d\mathbf{X}^{(1)}$ and $d\mathbf{X}^{(2)}$, we obtain from Eqs 4.107 and 4.108,

$$\begin{aligned}\cos\Theta &= \frac{d\mathbf{X}^{(1)}}{dX^{(1)}} \cdot \frac{d\mathbf{X}^{(2)}}{dX^{(2)}} = \frac{\hat{\mathbf{n}}_1 \cdot \mathbf{c} \cdot \hat{\mathbf{n}}_2}{\sqrt{\hat{\mathbf{n}}_1 \cdot \mathbf{c} \cdot \hat{\mathbf{n}}_1}\sqrt{\hat{\mathbf{n}}_2 \cdot \mathbf{c} \cdot \hat{\mathbf{n}}_2}} \\ &= \lambda_{(\hat{\mathbf{n}}_1)}\lambda_{(\hat{\mathbf{n}}_2)}(\hat{\mathbf{n}}_1 \cdot \mathbf{c} \cdot \hat{\mathbf{n}}_2)\end{aligned}$$

which gives the original angle between elements in the directions $\hat{\mathbf{n}}_1$ and $\hat{\mathbf{n}}_2$ of the current configuration.

Example 4.13

A homogeneous deformation is given by the mapping equations, $x_1 = X_1 - X_2 + X_3$, $x_2 = X_2 - X_3 + X_1$ and $x_3 = X_3 - X_1 + X_2$. Determine

(a) the stretch ratio in the direction of $\hat{\mathbf{N}}_1 = (\hat{\mathbf{I}}_1 + \hat{\mathbf{I}}_2)/\sqrt{2}$, and
(b) the angle θ_{12} in the deformed configuration between elements that were originally in the directions of $\hat{\mathbf{N}}_1$, and $\hat{\mathbf{N}}_2 = \hat{\mathbf{I}}_2$.

Solution
For the given deformation (as the student should verify),

$$[F_{iA}] = \begin{bmatrix} 1 & -1 & 1 \\ 1 & 1 & -1 \\ -1 & 1 & 1 \end{bmatrix} \quad \text{and} \quad [C_{AB}] = \begin{bmatrix} 3 & -1 & -1 \\ -1 & 3 & -1 \\ -1 & -1 & 3 \end{bmatrix} .$$

(a) Therefore, from Eq 4.105,

$$\Lambda^2_{(\hat{\mathbf{N}}_1)} = \begin{bmatrix} \frac{1}{\sqrt{2}} & \frac{1}{\sqrt{2}} & 0 \end{bmatrix} \begin{bmatrix} 3 & -1 & -1 \\ -1 & 3 & -1 \\ -1 & -1 & 3 \end{bmatrix} \begin{bmatrix} \frac{1}{\sqrt{2}} \\ \frac{1}{\sqrt{2}} \\ 0 \end{bmatrix} = 2 ,$$

and

$$\Lambda_{(\hat{\mathbf{N}}_1)} = \sqrt{2} .$$

(b) For $\hat{\mathbf{N}}_2 = \hat{\mathbf{I}}_2$, $\Lambda^2_{(\hat{\mathbf{I}}_2)} = \hat{\mathbf{I}}_2 \cdot \mathbf{C} \cdot \hat{\mathbf{I}}_2 = C_{22} = 3$ so that from Eq 4.114, using the result in part (a),

$$\cos\theta_{12} = \frac{(\hat{\mathbf{I}}_1 + \hat{\mathbf{I}}_2)/\sqrt{2} \cdot \mathbf{C} \cdot \hat{\mathbf{I}}_2}{\sqrt{2}\sqrt{3}} = \frac{2/\sqrt{2}}{\sqrt{6}} ,$$

and $\theta_{12} = 54.7°$. Thus, the original 45° angle is enlarged by 9.7°.

It is evident from Eq 4.115 that if the coordinate axes are chosen in the principal directions of $\mathbf{C}$ the deformed angle θ_{12} is a right angle ($C_{12} = 0$ in this case) and there has been no change in the angle between elements in the X_1 and X_2 directions. By the same argument, any three mutually perpendicular *principal* axes of $\mathbf{C}$ at $\mathbf{P}$ are deformed into three mutually perpendicular axes at $\mathbf{p}$. Consider, therefore, the volume element of a rectangular parallelepiped whose edges are in the principal directions of $\mathbf{C}$ (and thus also of $\mathbf{E}$). Since there is no shear strain between any two of these edges the new volume is still a rectangular parallelopiped, and in the edge directions $\hat{\mathbf{N}}_i$ $(i = 1, 2, 3)$ the unit strains are

$$e_{(\hat{\mathbf{N}}_i)} = \Lambda_{(\hat{\mathbf{N}}_i)} - 1 \qquad (i = 1, 2, 3) , \tag{4.116}$$

so that

$$dx^{(i)} = dX^{(i)} + dX^{(i)}[\Lambda_{(\hat{\mathbf{N}}_i)} - 1] = dX^{(i)}\Lambda_{(\hat{\mathbf{N}}_i)} , \quad (i = 1, 2, 3) \tag{4.117}$$

and the ratio of the deformed volume to the original becomes

$$\frac{dV}{dV^o} = \frac{dx^{(1)}dx^{(2)}dx^{(3)}}{dX^{(1)}dX^{(2)}dX^{(3)}} = \Lambda_{(\hat{\mathbf{N}}_1)}\Lambda_{(\hat{\mathbf{N}}_2)}\Lambda_{(\hat{\mathbf{N}}_3)} \tag{4.118a}$$

which, when Eq 4.105 is used, becomes,

$$\frac{d\mathcal{V}}{d\mathcal{V}^o} = \sqrt{C_{(1)}C_{(2)}C_{(3)}} = \sqrt{\text{III}_C}\,. \tag{4.118b}$$

The importance of the second form of Eq 4.118b is that it is an invariant expression and can be calculated without reference to the principal axes of **C**.

Example 4.14

Determine the volume ratio $d\mathcal{V}/d\mathcal{V}^o$ for the deformation of Example 4.13 using Eq 4.118a, and verify using Eq 4.118b.

Solution

As the student should show, a set of principal axes for the **C** tensor of Example 4.13 are $\hat{\mathbf{N}}_1 = (\hat{\mathbf{I}}_1 + \hat{\mathbf{I}}_2 + \hat{\mathbf{I}}_3)/\sqrt{3}$, $\hat{\mathbf{N}}_2 = (\hat{\mathbf{I}}_1 - \hat{\mathbf{I}}_2)/\sqrt{2}$, and $\hat{\mathbf{N}}_3 = (\hat{\mathbf{I}}_1 + \hat{\mathbf{I}}_2 - 2\hat{\mathbf{I}}_3)/\sqrt{6}$. Thus from Eq 4.105 the principal stretches are $\Lambda_{(\hat{\mathbf{N}}_1)} = 1$, $\Lambda_{(\hat{\mathbf{N}}_2)} = 2$, and $\Lambda_{(\hat{\mathbf{N}}_3)} = 2$, respectively. Using these results, Eq 4.118a gives $d\mathcal{V}/d\mathcal{V}^o = 4$. By Eq 4.118b,

$$\text{III}_C = \det \mathbf{C} = \begin{vmatrix} 3 & -1 & -1 \\ -1 & 3 & -1 \\ -1 & -1 & 3 \end{vmatrix} = 16\,.$$

and $d\mathcal{V}/d\mathcal{V}^o = \sqrt{16} = 4$.

4.10 Rotation Tensor, Stretch Tensors

In Chapter 2 we noted that an arbitrary second-order tensor may be resolved by an additive decomposition into its symmetric and skew-symmetric parts. Here, we introduce a multiplicative decomposition known as the *polar decomposition* by which any non-singular tensor can be decomposed into a product of two component tensors. Recall that the deformation gradient **F** is a non-singular (invertible) tensor. Because of this nonsingularity, the deformation gradient can be decomposed into either of the two products

$$\mathbf{F} = \mathbf{R} \cdot \mathbf{U} = \mathbf{V} \cdot \mathbf{R} \tag{4.119}$$

where **R** is the orthogonal *rotation tensor*, and **U** and **V** are symmetric, positive-definite tensors called the *right* and *left stretch tensors*, respectively. Moreover, **U** and **V** have the same eigenvalues (see Problem 4.32).

The deformation gradient can be thought of as a mapping of the infinitesimal vector $d\mathbf{X}$ of the reference configuration into the infinitesimal vector $d\mathbf{x}$ of the current configuration. Note that the first decomposition in Eq 4.119 replaces the linear transformation $d\mathbf{x} = \mathbf{F} \cdot d\mathbf{X}$ of Eq 4.41 by two sequential transformations,

$$d\mathbf{x}' = \mathbf{U} \cdot d\mathbf{X} \tag{4.120a}$$

followed by

$$d\mathbf{x} = \mathbf{R} \cdot d\mathbf{x}' \, . \tag{4.120b}$$

The tensor $\mathbf{U}$ has three positive eigenvalues, $U_{(1)}$, $U_{(2)}$ and $U_{(3)}$ called the principal stretches, and associated with each is a principal stretch direction, $\hat{\mathbf{N}}_1$, $\hat{\mathbf{N}}_2$ and $\hat{\mathbf{N}}_3$, respectively. These unit vectors form an orthogonal triad known as the *right principal directions of stretch*. By the transformation Eq 4.120a, line elements along the directions $\hat{\mathbf{N}}_i$, $(i = 1, 2, 3)$ are stretched by an amount $U_{(i)}$, $(i = 1, 2, 3)$, respectively, with no change in direction. This is followed by a rigid body rotation given by Eq 4.120b. The second decomposition of Eq 4.120b reverses the sequence; first a rotation by $\mathbf{R}$, then the stretching by $\mathbf{V}$. In a general deformation, a rigid body translation may also be involved as well as the rotation and stretching described here.

As a preliminary to determining the rotation and stretch tensors, we note that an arbitrary tensor $\mathbf{T}$ is positive definite if $\mathbf{v} \cdot \mathbf{T} \cdot \mathbf{v} > 0$ for all vectors $\mathbf{v} \neq 0$. A necessary and sufficient condition for $\mathbf{T}$ to be positive definite is for all its eigenvalues to be positive. In this regard, consider the tensor $\mathbf{C} = \mathbf{F}^T \cdot \mathbf{F}$. Inasmuch as $\mathbf{F}$ is non-singular ($\det \mathbf{F} \neq 0$) and $\mathbf{F} \cdot \mathbf{v} \neq 0$ if $\mathbf{v} \neq 0$, we see that $(\mathbf{F} \cdot \mathbf{v}) \cdot (\mathbf{F} \cdot \mathbf{v})$ is a sum of squares and hence greater than zero. Thus

$$(\mathbf{F} \cdot \mathbf{v}) \cdot (\mathbf{F} \cdot \mathbf{v}) = \mathbf{v} \cdot \mathbf{F}^T \cdot \mathbf{F} \cdot \mathbf{v} = \mathbf{v} \cdot \mathbf{C} \cdot \mathbf{v} > 0 \, , \tag{4.121}$$

and $\mathbf{C}$ is positive definite. Furthermore,

$$(\mathbf{F}^T \cdot \mathbf{F})^T = \mathbf{F}^T \cdot (\mathbf{F}^T)^T = \mathbf{F}^T \cdot \mathbf{F} \tag{4.122}$$

which proves that $\mathbf{C}$ is also symmetric. By the same arguments we may show that $\mathbf{c} = (\mathbf{F}^{-1})^T \cdot (\mathbf{F}^{-1})$ is also symmetric and positive definite.

Now let $\mathbf{C}$ be given in principal axes form by the matrix

$$[C^*_{AB}] = \begin{bmatrix} C_{(1)} & 0 & 0 \\ 0 & C_{(2)} & 0 \\ 0 & 0 & C_{(3)} \end{bmatrix} \tag{4.123}$$

and let $[a_{MN}]$ be the orthogonal transformation that relates the components of $\mathbf{C}^*$ to the components of C in any other set of axes through the equation expressed here in both indicial and matrix form

$$C^*_{AB} = a_{AQ} a_{BP} C_{QP} \qquad \text{or} \qquad \mathbf{C}^* = \mathcal{A} \mathbf{C} \mathcal{A}^T \, . \tag{4.124}$$

We define $\mathbf{U}$ as the square root of $\mathbf{C}$, that is, $\mathbf{U} = \sqrt{\mathbf{C}}$ or $\mathbf{U} \cdot \mathbf{U} = \mathbf{C}$, and since the principal values $C_{(i)}$, $(i = 1, 2, 3)$ are all positive we may write

$$\left[\sqrt{C^*_{AB}}\right] = \begin{bmatrix} \sqrt{C_{(1)}} & 0 & 0 \\ 0 & \sqrt{C_{(2)}} & 0 \\ 0 & 0 & \sqrt{C_{(3)}} \end{bmatrix} = [U^*_{AB}] \, , \tag{4.125}$$

and, as is obvious, the inverse $(\mathbf{U}^*)^{-1}$ by

$$[(U^*_{AB})^{-1}] = \begin{bmatrix} 1/\sqrt{C_{(1)}} & 0 & 0 \\ 0 & 1/\sqrt{C_{(2)}} & 0 \\ 0 & 0 & 1/\sqrt{C_{(3)}} \end{bmatrix} \, . \tag{4.126}$$

Note that both $\mathbf{U}$ and $\mathbf{U}^{-1}$ are symmetric positive definite tensors given by

$$U_{AB} = a_{QA} a_{PB} U^*_{QP} \quad \text{or} \quad \mathbf{U} = \mathcal{A}^T \mathbf{U}^* \mathcal{A} , \tag{4.127}$$

and

$$U^{-1}_{AB} = a_{QA} a_{PB} (U^*_{QP})^{-1} \quad \text{or} \quad \mathbf{U}^{-1} = \mathcal{A}^T (\mathbf{U}^*)^{-1} \mathcal{A} \tag{4.128}$$

respectively.

Therefore, from the first decomposition in Eq 4.119,

$$\mathbf{R} = \mathbf{F} \cdot \mathbf{U}^{-1} , \tag{4.129}$$

so that

$$\mathbf{R}^T \cdot \mathbf{R} = (\mathbf{F} \cdot \mathbf{U}^{-1})^T \cdot (\mathbf{F} \cdot \mathbf{U}^{-1}) = (\mathbf{U}^{-1})^T \cdot \mathbf{F}^T \cdot \mathbf{F} \cdot \mathbf{U}^{-1} \tag{4.130}$$

$$= \mathbf{U}^{-1} \cdot \mathbf{C} \cdot \mathbf{U}^{-1} = \mathbf{U}^{-1} \cdot \mathbf{U} \cdot \mathbf{U} \cdot \mathbf{U}^{-1} = \mathbf{I} \tag{4.131}$$

which shows that $\mathbf{R}$ is proper orthogonal.

The second decomposition in Eq 4.119 may be confirmed by a similar development using $\mathbf{C}^{-1} = \mathbf{F} \cdot \mathbf{F}^T = \mathbf{V}^2$.

Example 4.15

A homogeneous deformation is given by the equations; $x_1 = 2X_1 - 2X_2$, $x_2 = X_1 + X_2$ and $x_3 = X_3$. Determine the polar decomposition $\mathbf{F} = \mathbf{R} \cdot \mathbf{U}$ for this deformation.

Solution

The matrix form of the tensor $F_{iA} = x_{i,A}$ is easily determined to be,

$$[F_{iA}] = \begin{bmatrix} 2 & -2 & 0 \\ 1 & 1 & 0 \\ 0 & 0 & 1 \end{bmatrix}$$

from which we calculate $\mathbf{C} = \mathbf{F}^T \cdot \mathbf{F}$,

$$[C_{AB}] = \begin{bmatrix} 5 & -3 & 0 \\ -3 & 5 & 0 \\ 0 & 0 & 1 \end{bmatrix} .$$

In principal axes form this matrix becomes

$$[C^*_{AB}] = \begin{bmatrix} 8 & 0 & 0 \\ 0 & 2 & 0 \\ 0 & 0 & 1 \end{bmatrix}$$

with an orthogonal transformation matrix found to be

$$[a_{MN}] = \begin{bmatrix} \frac{1}{\sqrt{2}} & -\frac{1}{\sqrt{2}} & 0 \\ \frac{1}{\sqrt{2}} & \frac{1}{\sqrt{2}} & 0 \\ 0 & 0 & 1 \end{bmatrix} .$$

This is found by calculating the eigenvectors of $\mathbf{C}$. Therefore, from Eqs 4.125 and 4.126

$$[U^*_{AB}] = \begin{bmatrix} 2\sqrt{2} & 0 & 0 \\ 0 & \sqrt{2} & 0 \\ 0 & 0 & 1 \end{bmatrix} \quad \text{and} \quad [(U^*_{AB})^{-1}] = \frac{1}{2\sqrt{2}} \begin{bmatrix} 1 & 0 & 0 \\ 0 & 2 & 0 \\ 0 & 0 & 2\sqrt{2} \end{bmatrix},$$

and by use of the transformation equations Eqs 4.127 and 4.128 (as the student should verify), we determine

$$[U_{AB}] = \begin{bmatrix} 3/\sqrt{2} & -\sqrt{2} & 0 \\ -\sqrt{2} & 3\sqrt{2} & 0 \\ 0 & 0 & 1 \end{bmatrix},$$

and

$$[U^{-1}_{AB}] = \frac{1}{4\sqrt{2}} \begin{bmatrix} 3 & 1 & 0 \\ 1 & 3 & 0 \\ 0 & 0 & 4\sqrt{2} \end{bmatrix}.$$

Finally, from Eq 4.129,

$$[R_{AB}] = \begin{bmatrix} 2 & -2 & 0 \\ 1 & 1 & 0 \\ 0 & 0 & 1 \end{bmatrix} \begin{bmatrix} 3 & 1 & 0 \\ 1 & 3 & 0 \\ 0 & 0 & 4\sqrt{2} \end{bmatrix} \frac{1}{4\sqrt{2}}$$

$$= \begin{bmatrix} 1/\sqrt{2} & -1/\sqrt{2} & 0 \\ 1/\sqrt{2} & 1/\sqrt{2} & 0 \\ 0 & 0 & 1 \end{bmatrix}$$

It is readily confirmed using these results that $\mathbf{F} = \mathbf{RU}$ and that $\mathbf{R}^T \cdot \mathbf{R} = \mathbf{I}$.

4.11 Velocity Gradient, Rate of Deformation, Vorticity

Let the velocity field of a continuum be given in some region of space by $v_i = v_i(\mathbf{x}, t)$. The *spatial velocity gradient* is defined by

$$L_{ij} = \frac{\partial v_i}{\partial x_j}. \tag{4.132}$$

An additive decomposition of this tensor into its symmetric and skew-symmetric parts is written as

$$L_{ij} = d_{ij} + w_{ij} \tag{4.133}$$

where the symmetric portion

$$d_{ij} = \frac{1}{2}\left(\frac{\partial v_i}{\partial x_j} + \frac{\partial v_j}{\partial x_i}\right) \tag{4.134}$$

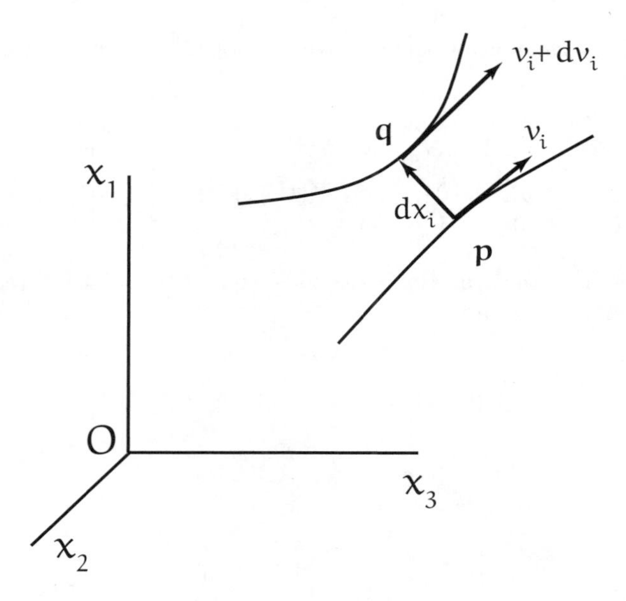

FIGURE 4.7
Differential velocity field at point $\mathbf{p}$.

is the *rate of deformation tensor*, and the skew-symmetric portion

$$w_{ij} = \frac{1}{2}\left(\frac{\partial v_i}{\partial x_j} - \frac{\partial v_j}{\partial x_i}\right) \tag{4.135}$$

is the *vorticity*, or *spin tensor*. This decomposition makes no assumption on the velocity gradient components being small, and is valid for finite components $\partial v_i/\partial x_j$.

Consider the velocity components at two neighboring points $\mathbf{p}$ and $\mathbf{q}$. Let the particle currently at $\mathbf{p}$ have a velocity v_i, and the particle at $\mathbf{q}$ a velocity $v_i + dv_i$ as shown in Fig. 4.7. Thus the particle at $\mathbf{q}$ has a velocity relative to the particle at $\mathbf{p}$ of

$$dv_i = \frac{\partial v_i}{\partial x_j} dx_j \quad \text{or} \quad d\mathbf{v} = \mathbf{L} \cdot d\mathbf{x}\,. \tag{4.136}$$

Note that

$$\frac{\partial v_i}{\partial x_j} = \frac{\partial v_i}{\partial X_A}\frac{\partial X_A}{\partial x_j} = \frac{d}{dt}\left(\frac{\partial x_i}{\partial X_A}\right)\frac{\partial X_A}{\partial x_j}\,, \tag{4.137}$$

or in symbolic notation

$$\mathbf{L} = \dot{\mathbf{F}} \cdot \mathbf{F}^{-1} \tag{4.138}$$

where we have used the fact that material time derivatives and material gradients commute. Therefore,

$$\dot{\mathbf{F}} = \mathbf{L} \cdot \mathbf{F}\,. \tag{4.139}$$

Consider next the stretch ratio $\Lambda = dx/dX$ where Λ is as defined in Eq 4.105, that is, the stretch of the line element dX initially along $\hat{\mathbf{N}}$ and currently along $\hat{\mathbf{n}}$. By the definition of the deformation gradient, $dx_i = x_{i,A}\, dX_A$, along with the unit vectors $n_i = dx_i/dx$ and $N_A = dX_A/dX$ we may write

$$dx\, n_i = x_{i,A}\, dX\, N_A$$

which becomes (after dividing both sides by the scalar dX)

$$n_i \Lambda = x_{i,A} N_A \qquad \text{or} \qquad \hat{\mathbf{n}}\Lambda = \mathbf{F} \cdot \hat{\mathbf{N}} \,. \tag{4.140}$$

If we take the material derivative of this equation (using the symbolic notation for convenience),

$$\dot{\hat{\mathbf{n}}}\Lambda + \hat{\mathbf{n}}\dot{\Lambda} = \dot{\mathbf{F}} \cdot \hat{\mathbf{N}} = \mathbf{L} \cdot \mathbf{F} \cdot \hat{\mathbf{N}} = \mathbf{L} \cdot \hat{\mathbf{n}}\Lambda \,,$$

so that

$$\dot{\hat{\mathbf{n}}} + \hat{\mathbf{n}}\dot{\Lambda}/\Lambda = \mathbf{L} \cdot \hat{\mathbf{n}} \,. \tag{4.141}$$

By forming the inner product of this equation with $\hat{\mathbf{n}}$ we obtain

$$\hat{\mathbf{n}} \cdot \dot{\hat{\mathbf{n}}} + \hat{\mathbf{n}} \cdot \hat{\mathbf{n}}\dot{\Lambda}/\Lambda = \hat{\mathbf{n}} \cdot \mathbf{L} \cdot \hat{\mathbf{n}} \,.$$

But $\hat{\mathbf{n}} \cdot \hat{\mathbf{n}} = 1$ and so $\dot{\hat{\mathbf{n}}} \cdot \hat{\mathbf{n}} = 0$, resulting in

$$\dot{\Lambda}/\Lambda = \hat{\mathbf{n}} \cdot \mathrm{L} \cdot \hat{\mathbf{n}} \qquad \text{or} \qquad \dot{\Lambda}/\Lambda = v_{i,j} n_i n_j \tag{4.142}$$

which represents the *rate of stretching per unit stretch* of the element that originated in the direction of $\hat{\mathbf{N}}$, and is in the direction of $\hat{\mathbf{n}}$ of the current configuration. Note further that Eq 4.142 may be simplified since $\boldsymbol{W}$ is skew-symmetric which means that

$$L_{ij} n_i n_j = (d_{ij} + w_{ij}) n_i n_j = d_{ij} n_i n_j \,,$$

and so

$$\dot{\Lambda}/\Lambda = \hat{\mathbf{n}} \cdot \mathrm{D} \cdot \hat{\mathbf{n}} \qquad \text{or} \qquad \dot{\Lambda}/\Lambda = d_{ij} n_i n_j \,. \tag{4.143}$$

For example, for the element in the x_1 direction, $\hat{\mathbf{n}} = \hat{\mathbf{e}}_1$ and

$$\dot{\Lambda}/\Lambda = \begin{bmatrix} 1 & 0 & 0 \end{bmatrix} \begin{bmatrix} d_{11} & d_{12} & d_{13} \\ d_{12} & d_{22} & d_{23} \\ d_{13} & d_{23} & d_{33} \end{bmatrix} \begin{bmatrix} 1 \\ 0 \\ 0 \end{bmatrix} = d_{11} \,.$$

Likewise, for $\hat{\mathbf{n}} = \hat{\mathbf{e}}_2$, $\dot{\Lambda}/\Lambda = d_{22}$ and for $\hat{\mathbf{n}} = \hat{\mathbf{e}}_3$, $\dot{\Lambda}/\Lambda = d_{33}$. Thus the diagonal elements of the rate of deformation tensor represent *rates of extension*, or *rates of stretching* in the coordinate (spatial) directions.

In order to interpret the off-diagonal elements of the rate of deformation tensor we consider two arbitrary differential vectors $dx_i^{(1)}$ and $dx_i^{(2)}$ at $\mathbf{p}$. The material derivative of the inner product of these two vectors is (using the superposed dot to indicate differentiation with respect to time,

$$\begin{aligned}
\overline{dx_i^{(1)} dx_i^{(2)}}^{\,\cdot} &= \overline{dx_i^{(1)}}^{\,\cdot} dx_i^{(2)} + dx_i^{(1)} \overline{dx_i^{(2)}}^{\,\cdot} \\
&= dv_i^{(1)} dx_i^{(2)} + dx_i^{(1)} dv_i^{(2)} \\
&= v_{i,j} dx_j^{(1)} dx_i^{(2)} + dx_i^{(1)} v_{i,j} dx_j^{(2)} \\
&= (v_{i,j} + v_{j,i})\, dx_i^{(1)} dx_j^{(2)} \\
&= 2 d_{ij} dx_i^{(1)} dx_j^{(2)} \,.
\end{aligned} \tag{4.144}$$

But $dx_i^{(1)} dx_i^{(2)} = dx^{(1)} dx^{(2)} \cos\theta$, and

$$\begin{aligned}
\overline{dx^{(1)} dx^{(2)} \cos\theta}^{\,\cdot} &= \overline{dx_i^{(1)}}^{\,\cdot} dx_i^{(2)} \cos\theta + dx^{(1)} \overline{dx^{(2)}}^{\,\cdot} \cos\theta - dx^{(1)} dx^{(2)} \dot{\theta} \sin\theta \\
&= \left(\left\{ \frac{\overline{dx^{(1)}}^{\,\cdot}}{dx^{(1)}} + \frac{\overline{dx^{(2)}}^{\,\cdot}}{dx^{(2)}} \right\} \cos\theta - \dot{\theta} \sin\theta \right) dx^{(1)} dx^{(2)} \,.
\end{aligned} \tag{4.145}$$

Equating Eqs 4.144 and 4.145 gives

$$2d_{ij}dx_i^{(1)}dx_i^{(2)} = \left(\left\{\frac{\overline{dx^{(1)}}^{\cdot}}{dx^{(1)}} + \frac{\overline{dx^{(2)}}^{\cdot}}{dx^{(2)}}\right\}\cos\theta - \dot{\theta}\sin\theta\right)dx^{(1)}dx^{(2)}\,. \tag{4.146}$$

If $dx_i^{(1)} = dx_i^{(2)} = dx_i$, then $\theta = 0$, and $\cos\theta = 1$, $\sin\theta = 0$ and $dx^{(1)} = dx^{(2)} = dx$ so that Eq 4.146 reduces to

$$d_{ij}\frac{dx_i}{dx}\frac{dx_j}{dx} = d_{ij}n_in_j = \frac{\overline{dx}^{\cdot}}{dx} \tag{4.147}$$

which is seen to be the *rate of extension per unit length* of the element currently in the direction of n_i (compare with Eq 4.143). If, however, $dx_i^{(1)}$ is perpendicular to $dx_i^{(2)}$, so that $\theta = \frac{\pi}{2}$, $\cos\theta = 0$, $\sin\theta = 1$, then Eq 4.146 becomes

$$2d_{ij}n_i^{(1)}n_i^{(2)} = \hat{\mathbf{n}}_1\cdot 2\mathbf{D}\cdot\hat{\mathbf{n}}_2 = -\dot{\theta}\,. \tag{4.148}$$

This rate of decrease in the angle θ is a measure of the *shear rate* between the elements in the directions of $\hat{\mathbf{n}}_1$ and $\hat{\mathbf{n}}_2$. In the engineering literature it is customary to define the *rate of shear* as *half* the change (increase or decrease) between two material line elements instantaneously at right angles to one another. Thus for $\hat{\mathbf{n}}_1 = \hat{\mathbf{e}}_1$ and $\hat{\mathbf{n}}_2 = \hat{\mathbf{e}}_2$,

$$-\frac{1}{2}\dot{\theta}_{12} = \hat{\mathbf{e}}_1\cdot \mathrm{D}\cdot\hat{\mathbf{e}}_2 = d_{12}\,,$$

and, in general, the off-diagonal elements of the rate of deformation tensor are seen to represent shear rates for the three pairs of coordinate axes.

Because $\mathbf{D}$ is a symmetric, second-order tensor, the derivation of principal values, principal directions, a Mohr's circles representation, a rate of deformation deviator tensor, etc., may be carried out as with all such tensors. Also, it is useful to develop the relationship between $\mathbf{D}$ and the material derivative of the strain tensor $\mathbf{E}$. Recall that

$$2\mathbf{E} = \mathbf{C} - \mathbf{I} = \mathbf{F}^T\cdot\mathbf{F} - \mathbf{I}\,,$$

so that, using Eq 4.139

$$\begin{aligned}2\dot{\mathbf{E}} &= \dot{\mathbf{F}}^T\cdot\mathbf{F} + \mathbf{F}^T\cdot\dot{\mathbf{F}} = (\mathbf{L}\cdot\mathbf{F})^T\cdot\mathbf{F} + \mathbf{F}^T\cdot(\mathbf{L}\cdot\mathbf{F})\\ &= \mathbf{F}^T\cdot\mathbf{L}^T\cdot\mathbf{F} + \mathbf{F}^T\cdot\mathbf{L}\cdot\mathbf{F} = \mathbf{F}^T\cdot(\mathbf{L}^T + \mathbf{L})\cdot\mathbf{F} = \mathbf{F}^T\cdot(2\mathbf{D})\cdot\mathbf{F}\,,\end{aligned}$$

or

$$\dot{\mathbf{E}} = \mathbf{F}^T\cdot\mathbf{D}\cdot\mathbf{F}\,. \tag{4.149}$$

Note also that from $u_i + X_i = x_i$ we have $u_{i,A} + \delta_{i,A} = x_{i,A}$ and if the displacement gradients $u_{i,A}$ are very small, $u_{i,A} \ll 1$ and may be neglected, then $\delta_{i,A} \approx x_{i,A}$ $(\mathbf{I} \approx \mathbf{F})$, and of course, $\mathbf{F}^T = \mathbf{I}^T = \mathbf{I}$. At the same time for $u_{i,A}$ very small in magnitude, by Eq 4.61, $\mathbf{E} \approx \boldsymbol{\epsilon}$ and Eq 4.74 reduces to

$$\dot{\boldsymbol{\epsilon}} = \mathbf{I}\cdot\mathbf{D}\cdot\mathbf{I} = \mathbf{D} \tag{4.150}$$

for the infinitesimal theory. Finally, taking the material derivative of the difference $(dx)^2 - (dX)^2 = d\mathbf{X}\cdot 2\mathbf{E}\cdot d\mathbf{X}$, and noting that $\overline{(dx)^2 - (dX)^2}^{\cdot} = \overline{(dx)^2}^{\cdot}$ since $\overline{(dX)^2}^{\cdot} = 0$, we obtain

$$\overline{(dx)^2}^{\cdot} = d\mathbf{X}\cdot 2\dot{\mathbf{E}}\cdot d\mathbf{X} = d\mathbf{X}\cdot\mathbf{F}^T\cdot 2\mathbf{D}\cdot\mathbf{F}\cdot d\mathbf{X} = d\mathbf{x}\cdot 2\mathbf{D}\cdot d\mathbf{x} \tag{4.151}$$

which shows that the local motion at some point $\mathbf{x}$ is a rigid body motion if and only if $\mathbf{D} = 0$ at $\mathbf{x}$.

Solving Eq 4.141 for $\dot{n}_i$ and using Eq 4.143 we may write

$$\dot{n}_i = v_{i,j}n_j - n_i\dot{\Lambda}/\Lambda = (d_{ij} + w_{ij})n_j - d_{qk}n_q n_k n_i \; .$$

If now, n_i is chosen along a principal direction of $\mathbf{D}$ so that $d_{ij}n_j^{(p)} = D_{(p)}n_i^{(p)}$ $(p = 1,2,3)$ where $D_{(p)}$ represents a principal value of $\mathbf{D}$, then

$$\dot{n}_i = D_{(p)}\dot{n}_i^{(p)} + w_{ij}\dot{n}_j^{(p)} - D_{(p)}n_q^{(p)}n_q^{(p)}n_i^{(p)} = w_{ij}n_j^{(p)} \tag{4.152}$$

since $n_q^{(p)}n_q^{(p)} = 1$. Because a unit vector can change only in direction, Eq 4.152 indicates that w_{ij} gives the *rate of change in direction* of the principal axes of $\mathbf{D}$. Hence the names, *vorticity*, or *spin* given to $\mathbf{W}$. Additionally, we associate with $\mathbf{W}$ the vector

$$w_i = \frac{1}{2}\varepsilon_{ijk}v_{k,j} \qquad \text{or} \qquad \mathbf{w} = \frac{1}{2}\operatorname{curl}\mathbf{v} \tag{4.153}$$

called the *vorticity vector*, by the following calculation,

$$\begin{aligned} \varepsilon_{pqi}w_i &= \frac{1}{2}\varepsilon_{pqi}\varepsilon_{ijk}v_{k,j} = \frac{1}{2}(\delta_{pj}\delta_{qk} - \delta_{pk}\delta_{qj})v_{k,j} \\ &= \frac{1}{2}(v_{q,p} - v_{p,q}) = w_{qp} \; . \end{aligned} \tag{4.154}$$

Thus if $\mathbf{D} \equiv 0$ so that $L_{ij} = w_{ij}$, it follows that $dv_i = L_{ij}dx_j = w_{ij}dx_j = \varepsilon_{jik}w_k dx_j$ and since $\varepsilon_{jik} = -\varepsilon_{ijk} = \varepsilon_{ikj}$,

$$dv_i = \varepsilon_{ijk}w_j dx_k \qquad \text{or} \qquad d\mathbf{v} = \mathbf{w} \times d\mathbf{x} \tag{4.155}$$

according to which the relative velocity in the vicinity of $\mathbf{p}$ corresponds to a rigid body rotation about an axis through $\mathbf{p}$. The vector $\mathbf{w}$ indicates the angular velocity, the direction and the sense of this rotation.

To summarize the physical interpretation of the velocity gradient $\mathbf{L}$, we note that it effects a separation of the *local instantaneous* motion into two parts:

1. The so-called logarithmic rates of stretching, $D_{(p)}$, $(p = 1,2,3)$, that is the eigenvalues of $\mathbf{D}$ along the mutually orthogonal principal axes of $\mathbf{D}$,

$$\dot{\Lambda}/\Lambda = \frac{d(\ln\Lambda)}{dt} = d_{ij}n_i^{(p)}n_j^{(p)} = n_i^{(p)}D_{(p)}n_i^{(p)} = D_{(p)} \; ,$$

   and

2. A rigid body rotation of the principal axes of $\mathbf{D}$ with angular velocity $\mathbf{w}$.

Example 4.16

Consider a continuum having a shearing motion defined by the mapping

$$\begin{aligned} x_1 &= X_1 + \kappa X_2 \; , \\ x_2 &= X_2 \; , \\ x_3 &= X_3 \; . \end{aligned}$$

(a) Invert the mapping to find X_A in terms of x_i

(b) Determine $x_{j,B}$ and $X_{A,j}$

(c) Compute $X_{A,j}x_{j,B}$ and $x_{j,B}X_{B,i}$

(d) Use the result of (c) to determine the indicial expression for $\overline{\dot{X_{A,i}}}$ for all general motions (not just this pure shear case).

Solution

(a) The mapping can be inverted, find X_A in terms of x_i, by inspection to get

$$\begin{aligned} X_1 &= x_1 - \kappa x_2 \ , \\ X_2 &= x_2 \ , \\ X_3 &= x_3 \ . \end{aligned}$$

(b) Partial differentiation of the two forms of the motion give us $\mathbf{F} = x_{j,B}$ and $\mathbf{F}^{-1} = X_{A,j}$

$$x_{j,B} = \begin{bmatrix} 1 & \kappa & 0 \\ 0 & 1 & 0 \\ 0 & 0 & 1 \end{bmatrix} \quad \text{and} \quad X_{A,j} = \begin{bmatrix} 1 & -\kappa & 0 \\ 0 & 1 & 0 \\ 0 & 0 & 1 \end{bmatrix} .$$

(c) Multiplying these two expressions will solidify that $X_{A,j}x_{j,B}$ is equivalent to $\mathbf{F}^{-1}\mathbf{F}$ and $x_{j,B}X_{B,i}$ is equivalent to $\mathbf{F}\mathbf{F}^{-1}$. In indicial notation this fact is sometimes not obvious when so many new symbols have been introduced.

$$X_{A,j}x_{j,B} = \begin{bmatrix} 1 & -\kappa & 0 \\ 0 & 1 & 0 \\ 0 & 0 & 1 \end{bmatrix} \begin{bmatrix} 1 & \kappa & 0 \\ 0 & 1 & 0 \\ 0 & 0 & 1 \end{bmatrix} = \begin{bmatrix} 1 & \kappa - \kappa & 0 \\ 0 & 1 & 0 \\ 0 & 0 & 1 \end{bmatrix} = \delta_{AB}$$

$$x_{j,B}X_{B,i} = \begin{bmatrix} 1 & \kappa & 0 \\ 0 & 1 & 0 \\ 0 & 0 & 1 \end{bmatrix} \begin{bmatrix} 1 & -\kappa & 0 \\ 0 & 1 & 0 \\ 0 & 0 & 1 \end{bmatrix} = \begin{bmatrix} 1 & -\kappa + \kappa & 0 \\ 0 & 1 & 0 \\ 0 & 0 & 1 \end{bmatrix} = \delta_{ij}$$

(d) Take the material derivative of the first form noting that δ_{AB} is a constant, and use Eq 4.138

$$\overline{\dot{X_{A,j}}}x_{j,B} = -X_{A,j}\overline{\dot{x_{j,B}}} = -X_{A,j}L_{jm}x_{m,B} \ .$$

Multiply both sides by $X_{B,i}$ which is the inverse of $x_{j,B}$ to get $\overline{\dot{X_{A,j}}}$:

$$\begin{aligned} \overline{\dot{X_{A,j}}}x_{j,B}X_{B,i} = \overline{\dot{X_{A,j}}}\delta_{ji} = \overline{\dot{X_{A,i}}} &= -X_{A,j}L_{jm}x_{m,B}X_{B,i} \\ &= -X_{A,j}L_{jm}\delta_{mi} = -X_{A,j}L_{ji} \ . \end{aligned}$$

Thus,

$$\overline{\dot{X_{A,i}}} = -X_{A,j}L_{ji} = -X_{A,j}v_{j,i} \ .$$

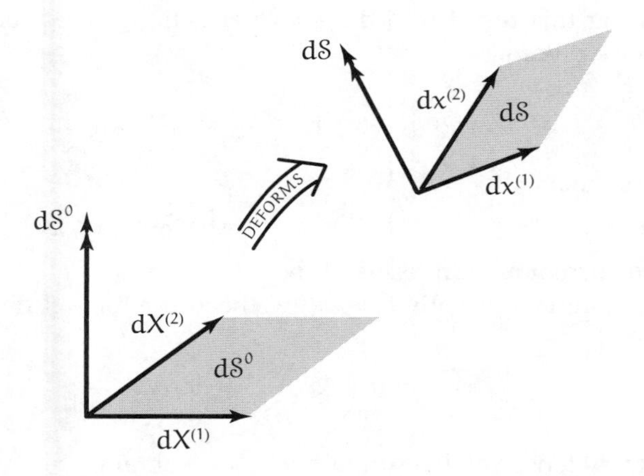

FIGURE 4.8
Area $d\mathcal{S}^0$ between vectors $d\mathbf{X}^{(1)}$ and $d\mathbf{X}^{(2)}$ in the reference configuration becomes $d\mathcal{S}$ between $d\mathbf{x}^{(1)}$ and $d\mathbf{x}^{(2)}$ in the deformed configuration.

4.12 Material Derivative of Line Elements, Areas, Volumes

Consider first the material derivative of the differential line element $d\mathbf{x} = \mathbf{F} \cdot d\mathbf{X}$. Clearly, $\dot{\overline{d\mathbf{x}}} = \dot{\mathbf{F}} \cdot d\mathbf{X}$ and by Eq 4.139,

$$\dot{\overline{d\mathbf{x}}} = \dot{\mathbf{F}} \cdot d\mathbf{X} = \mathbf{L} \cdot \mathbf{F} \cdot d\mathbf{X} = \mathbf{L} \cdot d\mathbf{x} \quad \text{or} \quad \dot{\overline{dx_i}} = v_{i,j}\, dx_j \ . \tag{4.156}$$

Note further, that from Eq 4.156 the material derivative of the dot product $d\mathbf{x} \cdot d\mathbf{x}$ is

$$\dot{\overline{d\mathbf{x} \cdot d\mathbf{x}}} = 2d\mathbf{x} \cdot \dot{\overline{d\mathbf{x}}} = d\mathbf{x} \cdot 2\mathbf{L} \cdot d\mathbf{x} = d\mathbf{x} \cdot 2(\mathbf{D} + \mathbf{W}) \cdot d\mathbf{x} = d\mathbf{x} \cdot 2\mathbf{D} \cdot d\mathbf{x}$$

in agreement with Eq 4.151.

It remains to develop expressions for the material derivatives of area and volume elements. Consider the plane area defined in the reference configuration by the differential line elements $dX_A^{(1)}$ and $dX_A^{(2)}$ as shown in Fig. 4.8. The parallelogram area $d\mathcal{S}^0$ may be represented by the vector

$$d\mathcal{S}_A^0 = \varepsilon_{ABC}\, dX_B^{(1)}\, dX_C^{(2)} \ . \tag{4.157}$$

As a result of the motion $\mathbf{x} = \mathbf{x}(\mathbf{X}, t)$ this area is carried into the current area $d\mathcal{S}_i$ shown in Fig. 4.8 and is given by

$$d\mathcal{S}_i = \varepsilon_{ijk}\, dx_j^{(1)}\, dx_k^{(2)} = \varepsilon_{ijk} x_{j,B}\, dX_B^{(1)} x_{k,C}\, dX_C^{(2)} \tag{4.158}$$

which upon multiplication by $x_{i,A}$ results in

$$x_{i,A}\, d\mathcal{S}_i = \varepsilon_{ijk} x_{i,A} x_{j,B} x_{k,C}\, dX_B^{(1)}\, dX_C^{(2)} = \varepsilon_{ijk} F_{iA} F_{jB} F_{kC}\, dX_B^{(1)}\, dX_C^{(2)} \ .$$

Recall that $\det\mathbf{F} = J$ (the Jacobian) and from Eq 2.43

$$\varepsilon_{ijk} F_{iA} F_{jB} F_{kC} = \varepsilon_{ABC} \det \mathbf{F} = \varepsilon_{ABC} J \ .$$

Therefore, by inserting this result into the above equation for $x_{i,A}d\mathcal{S}_i$ and multiplying both sides by $X_{A,q}$, we obtain,

$$x_{i,A}X_{A,q}d\mathcal{S}_i = \varepsilon_{ABC}JdX_B^{(1)}dX_C^{(2)}X_{A,q}$$

But $x_{i,A}X_{A,q} = \delta_{iq}$ so that

$$\delta_{iq}d\mathcal{S}_i = d\mathcal{S}_q = X_{A,q}Jd\mathcal{S}_A^0 \tag{4.159}$$

which expresses the current area in terms of the original area.

To determine the material derivative of $d\mathcal{S}_i$, we need the following identity

$$\dot{\overline{\det \mathbf{A}}} = \operatorname{tr}\left(\dot{\mathbf{A}} \cdot \mathbf{A}^{-1}\right) \det\mathbf{A} \tag{4.160}$$

where $\mathbf{A}$ is an arbitrary tensor. Substituting $\mathbf{F}$ for $\mathbf{A}$ we obtain

$$\dot{\overline{\det \mathbf{F}}} = \dot{J} = (\det \mathbf{F})\ \operatorname{tr}(\dot{\mathbf{F}} \cdot \mathbf{F}^{-1}) = J\ \operatorname{tr}(\mathbf{L})\ ,$$

or

$$\dot{J} = Jv_{i,i} = J\operatorname{div}\mathbf{v}\ . \tag{4.161}$$

Noting that Eq 4.159 may be written

$$d\mathcal{S}_q = JX_{A,q}d\mathcal{S}_A^0$$

and using symbolic notation to take advantage of Eq 4.160 we obtain

$$d\boldsymbol{\mathcal{S}} = J(\mathbf{F}^{-1})^T \cdot d\boldsymbol{\mathcal{S}}^0 = Jd\boldsymbol{\mathcal{S}}^0 \cdot \mathbf{F}^{-1}\ ,$$

and so

$$d\boldsymbol{\mathcal{S}} \cdot \mathbf{F} = Jd\boldsymbol{\mathcal{S}}^0$$

which upon differentiating becomes

$$d\dot{\boldsymbol{\mathcal{S}}} \cdot \mathbf{F} + d\boldsymbol{\mathcal{S}} \cdot \dot{\mathbf{F}} = \dot{J}d\boldsymbol{\mathcal{S}}^0 = J(\operatorname{tr}\mathbf{L})d\boldsymbol{\mathcal{S}}^0\ ,$$

$$d\dot{\boldsymbol{\mathcal{S}}} + d\boldsymbol{\mathcal{S}} \cdot \dot{\mathbf{F}} \cdot \mathbf{F}^{-1} = J(\operatorname{tr}\mathbf{L})d\boldsymbol{\mathcal{S}}^0 \cdot \mathbf{F}^{-1} = (\operatorname{tr}\mathbf{L})d\boldsymbol{\mathcal{S}}\ ,$$

and finally

$$d\dot{\boldsymbol{\mathcal{S}}} = (\operatorname{tr}\mathbf{L})\, d\boldsymbol{\mathcal{S}} - d\boldsymbol{\mathcal{S}} \cdot \mathbf{L} \quad \text{or} \quad d\dot{\mathcal{S}}_i = v_{k,k}d\mathcal{S}_i - d\mathcal{S}_j v_{j,i} \tag{4.162}$$

which gives the rate of change of the current element of area in terms of the current area, the trace of the velocity gradient, and of the components of $\mathbf{L}$.

Consider next the volume element defined in the referential configuration by the box product,

$$d\mathcal{V}^0 = d\mathbf{X}^{(1)} \cdot d\mathbf{X}^{(2)} \times d\mathbf{X}^{(3)} = \varepsilon_{ABC}dX_A^{(1)}dX_B^{(2)}dX_C^{(3)} = \left[d\mathbf{X}^{(1)}, d\mathbf{X}^{(2)}, d\mathbf{X}^{(3)}\right]$$

and pictured in Fig. 4.9, and let the deformed volume element, also shown in Fig. 4.9 be given by

$$d\mathcal{V} = d\mathbf{x}^{(1)} \cdot d\mathbf{x}^{(2)} \times d\mathbf{x}^{(3)} = \varepsilon_{ijk}dx_i^{(1)}dx_j^{(2)}dx_k^{(3)} = \left[d\mathbf{x}^{(1)}, d\mathbf{x}^{(2)}, d\mathbf{x}^{(3)}\right]\ .$$

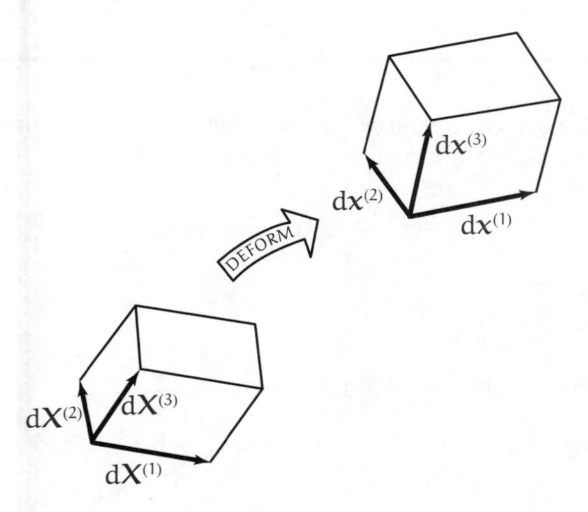

FIGURE 4.9
Volume of parallelpiped defined by vectors $\mathbf{dX}^{(1)}$, $\mathbf{dX}^{(2)}$ and $\mathbf{dX}^{(3)}$ in the reference configuration deforms into volume defined by parallelpiped defined by vectors $\mathbf{dx}^{(1)}$, $\mathbf{dx}^{(2)}$ and $\mathbf{dx}^{(3)}$ in the deformed configuration.

For the motion $\mathbf{x} = \mathbf{x}(\mathbf{X}, t)$, $\mathbf{dx} = \mathbf{F} \cdot \mathbf{dX}$ so the current volume is the box product

$$\begin{aligned} d\mathcal{V} &= \left[\mathbf{F} \cdot \mathbf{dX}^{(1)}, \mathbf{F} \cdot \mathbf{dX}^{(2)}, \mathbf{F} \cdot \mathbf{dX}^{(3)}\right] = \varepsilon_{ijk} x_{i,A} x_{j,B} x_{k,C} dX_A^{(1)} dX_B^{(2)} dX_C^{(3)} \\ &= \det(\mathbf{F}) \left[\mathbf{dX}^{(1)}, \mathbf{dX}^{(2)}, \mathbf{dX}^{(3)}\right] = J d\mathcal{V}^0 \end{aligned} \tag{4.163}$$

which gives the current volume element in terms of its original size. Since $J \neq 0$ ($\mathbf{F}$ is invertible), we have either $J < 0$ or $J > 0$. Mathematically, $J < 0$ is possible, but physically it corresponds to a negative volume, so we reject it. Henceforth, we assume $J > 0$. If $J = 1$, then $d\mathcal{V} = d\mathcal{V}^0$ and the volume magnitude is preserved. If J is equal to unity for all $\mathbf{X}$ we say the motion is isochoric.

To determine the time rate of change of $d\mathcal{V}$ we take the material derivative as follows,

$$\dot{\overline{d\mathcal{V}}} = \dot{J} d\mathcal{V}^0 = J \operatorname{tr}(\mathbf{L}) d\mathcal{V}^0 = J v_{i,i} d\mathcal{V}^0 = v_{i,i} d\mathcal{V} . \tag{4.164}$$

Thus a necessary and sufficient condition for a motion to be isochoric is that

$$v_{i,i} = \operatorname{div} \mathbf{v} = 0 . \tag{4.165}$$

In summary, we observe that the deformation gradient $\mathbf{F}$ governs the stretch of a line element, the change of an area element, and the change of a volume element. But it is the velocity gradient $\mathbf{L}$ that determines the rate at which these changes occur.

Example 4.17

To derive the material derivative of the Jacobian, $\dot{J}$, the identity Eq 4.160 was used in the text. As an alternative derivation, use

$$J = \frac{1}{6}\varepsilon_{ABC}\varepsilon_{ijk}F_{iA}F_{jB}F_{kC}$$

to derive Eq 4.161.

Solution

Differentiate and use Eq 4.139, Eq 2.43 and Eq 2.7b

$$\begin{aligned}
\dot{J} &= \tfrac{1}{6}\varepsilon_{ABC}\varepsilon_{ijk}\left(\dot{F}_{iA}F_{jB}F_{kC} + F_{iA}\dot{F}_{jB}F_{kC} + F_{iA}F_{jB}\dot{F}_{kC}\right) \\
&= \tfrac{1}{6}\varepsilon_{ABC}\varepsilon_{ijk}\left(L_{im}F_{mA}F_{jB}F_{kC} + F_{iA}L_{jm}F_{mB}F_{kC} + F_{iA}F_{jB}L_{km}F_{mC}\right) \\
&= \tfrac{1}{6}\varepsilon_{ijk}\left(L_{im}\varepsilon_{mjk}J + L_{jm}\varepsilon_{imk}J + L_{km}\varepsilon_{ijm}J\right) \\
&= \tfrac{1}{6}\left(2\delta_{im}L_{im} + 2\delta_{mj}L_{jm} + 2\delta_{km}L_{km}\right)J \\
&= JL_{ii} = Jv_{i,i}\ .
\end{aligned}$$

Problems

Problem 4.1

The motion of a continuous medium is specified by the component equations

$$\begin{aligned} x_1 &= \tfrac{1}{2}(X_1 + X_2)e^{t} + \tfrac{1}{2}(X_1 - X_2)e^{-t} \ , \\ x_2 &= \tfrac{1}{2}(X_1 + X_2)e^{t} - \tfrac{1}{2}(X_1 - X_2)e^{-t} \ , \\ x_3 &= X_3 \ . \end{aligned}$$

(a) Show that the Jacobian determinant J does not vanish, and solve for the inverse equations $\mathbf{X} = \mathbf{X}(\mathbf{x}, t)$.

(b) Calculate the velocity and acceleration components in terms of the material coordinates.

(c) Using the inverse equations developed in part (a), express the velocity and acceleration components in terms of spatial coordinates.

Answer

(a) $J = \cosh^2 t - \sinh^2 t = 1$
$X_1 = \frac{1}{2}(x_1 + x_2)e^{-t} + \frac{1}{2}(x_1 - x_2)e^{t}$
$X_2 = \frac{1}{2}(x_1 + x_2)e^{-t} - \frac{1}{2}(x_1 - x_2)e^{t}$
$X_3 = x_3$

(b) $v_1 = \frac{1}{2}(X_1 + X_2)e^{t} - \frac{1}{2}(X_1 - X_2)e^{-t}$
$v_2 = \frac{1}{2}(X_1 + X_2)e^{t} + \frac{1}{2}(X_1 - X_2)e^{-t}$
$v_3 = 0$
$a_1 = \frac{1}{2}(X_1 + X_2)e^{t} + \frac{1}{2}(X_1 - X_2)e^{-t}$
$a_2 = \frac{1}{2}(X_1 + X_2)e^{t} - \frac{1}{2}(X_1 - X_2)e^{-t}$
$a_3 = 0$

(c) $v_1 = x_2,\ v_2 = x_1,\ v_3 = 0$
$a_1 = x_1,\ a_2 = x_2,\ a_3 = 0$

Problem 4.2

Let the motion of a continuum be given in component form by the equations

$$\begin{aligned} x_1 &= X_1 + X_2 t + X_3 t^2 \ , \\ x_2 &= X_2 + X_3 t + X_1 t^2 \ , \\ x_3 &= X_3 + X_1 t + X_2 t^2 \ . \end{aligned}$$

(a) Show that $J \neq 0$, and solve for the inverse equations.

(b) Determine the velocity and acceleration

(1) at time $t = 1$ s for the particle which was at point $(2.75, 3.75, 4.00)$ when $t = 0.5$ s.

(2) at time $t = 2$ s for the particle which was at point $(1, 2, -1)$ when $t = 0$.

Answer

(a) $J = (1 - t^3)^2$
$X_1 = (x_1 - x_2 t)/(1 - t^3)$
$X_2 = (x_2 - x_3 t)/(1 - t^3)$
$X_3 = (x_3 - x_1 t)/(1 - t^3)$

(b) (1) $\mathbf{v} = 8\hat{\mathbf{e}}_1 + 5\hat{\mathbf{e}}_2 + 5\hat{\mathbf{e}}_3,\ \mathbf{a} = 6\hat{\mathbf{e}}_1 + 2\hat{\mathbf{e}}_2 + 4\hat{\mathbf{e}}_3$
(2) $\mathbf{v} = -2\hat{\mathbf{e}}_1 + 3\hat{\mathbf{e}}_2 + 9\hat{\mathbf{e}}_3,\ \mathbf{a} = -2\hat{\mathbf{e}}_1 + 2\hat{\mathbf{e}}_2 + 4\hat{\mathbf{e}}_3$

Problem 4.3

A continuum body has a motion defined by the equations

$$x_1 = X_1 + 2X_2 t^2 \ ,$$
$$x_2 = X_2 + 2X_1 t^2 \ ,$$
$$x_3 = X_3 \ .$$

(a) Determine the velocity components at $t = 1.5$ s of the particle which occupied the point $(2, 3, 4)$ when $t = 1.0$ s.

(b) Determine the equation of the path along which the particle designated in part (a) moves.

(c) Calculate the acceleration components of the same particle at time $t = 2$ s.

Answer

(a) $v_1 = 2,\ v_2 = 8,\ v_3 = 0$
(b) $4x_1 - x_2 = 5$ in the plane $x_3 = 4$
(c) $a_1 = 4/3,\ a_2 = 16/3,\ a_3 = 0.$

Problem 4.4

If the motion $\mathbf{x} = \mathbf{x}(\mathbf{X}, t)$ is given in component form by the equations

$$x_1 = X_1(1 + t), \quad x_2 = X_2(1 + t)^2, \quad x_3 = X_3(1 + t^2) \ ,$$

determine expressions for the velocity and acceleration components in terms of both Lagrangian and Eulerian coordinates.

Answer

$v_1 = X_1 = x_1/(1 + t)$

$v_2 = 2X_2(1 + t) = 2x_2/(1 + t)$

$v_3 = 2X_3 t = 2x_3 t/(1 + t^2)$

$a_1 = 0$

$a_2 = 2X_2 = 2x_2/(1 + t)^2$

$a_3 = 2X_3 = 2x_3/(1 + t^2)$

Problem 4.5

The Lagrangian description of a continuum motion is given by

$$\begin{aligned} x_1 &= X_1 e^{-t} + X_3(e^{-t} - 1) \ , \\ x_2 &= X_2 e^{t} - X_3(1 - e^{-t}) \ , \\ x_3 &= X_3 e^{t} \ . \end{aligned}$$

Show that these equations are invertible and determine the Eulerian description of the motion.

Answer

$$\begin{aligned} X_1 &= x_1 e^{t} - x_3(e^{t} - 1) \\ X_2 &= x_2 e^{-t} + x_3(e^{-2t} - e^{-3t}) \\ X_3 &= x_3 e^{-t} \end{aligned}$$

Problem 4.6

A velocity field is given in Lagrangian form by

$$v_1 = 2t + X_1, \quad v_2 = X_2 e^{t}, \quad v_3 = X_3 - t \ .$$

Integrate these equations to obtain $\mathbf{x} = \mathbf{x}(\mathbf{X}, t)$ with $\mathbf{x} = \mathbf{X}$ at $t = 0$, and using that result compute the velocity and acceleration components in the Eulerian (spatial) form.

Answer

$$\begin{aligned} v_1 &= (x_1 + 2t + t^2)/(1 + t) \\ v_2 &= x_2 \\ v_3 &= (2x_3 - 2t - t^2)/2(1 + t) \\ a_1 &= 2, \ a_2 = x_2, \ a_3 = -1 \end{aligned}$$

Problem 4.7

If the motion of a continuous medium is given by

$$\begin{aligned} x_1 &= X_1 e^{t} - X_3(e^{t} - 1) \ , \\ x_2 &= X_2 e^{-t} + X_3(1 - e^{-t}) \ , \\ x_3 &= X_3 \ , \end{aligned}$$

determine the displacement field in both material and spatial descriptions.

Answer

$$\begin{aligned} u_1 &= (X_1 - X_3)(e^{t} - 1) = (x_1 - x_3)(1 - e^{-t}) \\ u_2 &= (X_2 - X_3)(e^{-t} - 1) = (x_2 - x_3)(1 - e^{t}) \\ u_3 &= 0 \end{aligned}$$

Problem 4.8
The temperature field in a continuum is given by the expression

$$\theta = e^{-3t}/x^2 \quad \text{where} \quad x^2 = x_1^2 + x_2^2 + x_3^2 \;.$$

The velocity field of the medium has components

$$v_1 = x_2 + 2x_3, \quad v_2 = x_3 - x_1, \quad v_3 = x_1 + 3x_2 \;.$$

Determine the material derivative $d\theta/dt$ of the temperature field.

Answer

$$d\theta/dt = -e^{-3t}(3x^2 + 6x_1x_3 + 8x_2x_3)/x^4$$

Problem 4.9
In a certain region of a fluid the flow velocity has components

$$v_1 = A(x_1^3 + x_1x_2^2)e^{-kt}, \quad v_2 = A(x_1^2x_2 + x_2^3)e^{-kt}, \quad v_3 = 0$$

where A and k are constants. Use the (spatial) material derivative operator to determine the acceleration components at the point $(1, 1, 0)$ when $t = 0$.

Answer

$$a_1 = -2A(k - 5A),\; a_2 = -A(k - 5A),\; a_3 = 0$$

Problem 4.10
A displacement field is given in terms of the spatial variables and time by the equations

$$u_1 = x_2t^2, \quad u_2 = x_3t, \quad u_3 = x_1t \;.$$

Using the (spatial) material derivative operator, determine the velocity components.

Answer

$$v_1 = (2x_2t + x_3t^2 + x_1t^3)/(1 - t^4)$$
$$v_2 = (x_3 + x_1t + 2x_2t^3)/(1 - t^4)$$
$$v_3 = (x_1 + 2x_2t^2 + x_3t^3)/(1 - t^4)$$

Problem 4.11
For the motion given by the equations

$$x_1 = X_1 \cos\omega t + X_2 \sin\omega t \;,$$
$$x_2 = -X_1 \sin\omega t + X_2 \cos\omega t \;,$$
$$x_3 = (1 + kt)X_3 \;,$$

where ω and k are constants, determine the displacement field in Eulerian form.

Answer

$$u_1 = x_1(1 - \cos\omega t) + x_2 \sin\omega t$$
$$u_2 = -x_1 \sin\omega t + x_2(1 - \cos\omega t)$$
$$u_3 = x_3 kt/(1 + kt)$$

Problem 4.12

Show that the displacement field for the motion analyzed in Problem 4.1 has the Eulerian form

$$u_1 = x_1 - (x_1 + x_2)e^{-t}/2 - (x_1 - x_2)e^{t}/2 ,$$
$$u_2 = -x_2 - (x_1 + x_2)e^{-t}/2 + (x_1 - x_2)e^{t}/2 ,$$

and by using the material derivative operator $(du_i/dt = \partial u_i/\partial t + v_j \partial u_i/\partial x_j)$, verify the velocity and acceleration components calculated in Problem 4.1.

Problem 4.13

The Lagrangian description of a deformation is given by

$$x_1 = X_1 + X_3(e^2 - e^{-2}) ,$$
$$x_2 = X_2 - X_3(e^2 - 1) ,$$
$$x_3 = X_3 e^2 .$$

Determine the components of the deformation matrix F_{iA} and from it show that the Jacobian J does not vanish. Invert the mapping equations to obtain the Eulerian description of the deformation.

Answer

$$J = e^2$$
$$X_1 = x_1 - x_3(1 - e^{-4})$$
$$X_2 = x_2 + x_3(1 - e^{-2})$$
$$X_3 = x_3 e^{-2}$$

Problem 4.14

A *homogeneous* deformation has been described as one for which all of the deformation and strain tensors are independent of the coordinates, and may therefore be expressed in general by the displacement field $u_i = A_{ij}X_j$ where the A_{ij} are constants (or in the case of a motion, functions of time). Show that for a homogeneous deformation with the A_{ij} constant:

(a) plane material surfaces remain plane,

(b) straight line particle elements remain straight,

(c) material surfaces which are spherical in the reference configuration become ellipsoidal surfaces in the deformed configuration.

Problem 4.15

An *infinitesimal* homogeneous deformation $u_i = A_{ij}X_j$ is one for which the constants A_{ij} are so small that their products may be neglected. Show that for two sequential infinitesimal deformations the total displacement is the sum of the individual displacements regardless of the order in which the deformations are applied.

Problem 4.16

For the homogeneous deformation defined by

$$\begin{aligned} x_1 &= \alpha X_1 + \beta X_2 \ , \\ x_2 &= -\beta X_1 + \alpha X_2 \ , \\ x_3 &= \mu X_3 \ , \end{aligned}$$

where α, β and μ are constants, calculate the Lagrangian finite strain tensor $\mathbf{E}$. Show that if $\alpha = \cos\theta$, $\beta = \sin\theta$ and $\mu = 1$ the strain is zero and the mapping corresponds to a rigid body rotation of magnitude θ about the X_3 axis.

Answer

$$E_{AB} = \tfrac{1}{2}\begin{bmatrix} \alpha^2+\beta^2-1 & 0 & 0 \\ 0 & \alpha^2+\beta^2-1 & 0 \\ 0 & 0 & \mu^2-1 \end{bmatrix}$$

Problem 4.17

Given the deformation defined by

$$x_1 = X_1, \quad x_2 = X_2 + \frac{1}{2}X_3^2, \quad x_3 = X_3$$

(a) Sketch the deformed shape of the unit square OABC in the plane $X_1 = 0$.

(b) Determine the differential vectors $\mathbf{dx}^{(2)}$ and $\mathbf{dx}^{(3)}$ which are the deformed vectors resulting from $\mathbf{dX}^{(2)} = dX^{(2)}\hat{\mathbf{I}}_2$ and $\mathbf{dX}^{(3)} = dX^{(3)}\hat{\mathbf{I}}_3$, respectively, that were originally at corner C.

(c) Calculate the dot product $\mathbf{dx}^{(2)} \cdot \mathbf{dx}^{(3)}$, and from it determine the change in the original right angle between $\mathbf{dX}^{(2)}$ and $\mathbf{dX}^{(3)}$ at C due to the deformation.

(d) Compute the stretch Λ at B in the direction of the unit normal

$$\hat{N} = \left(\hat{\mathbf{I}}_2 + \hat{\mathbf{I}}_3\right)/\sqrt{2}\ .$$

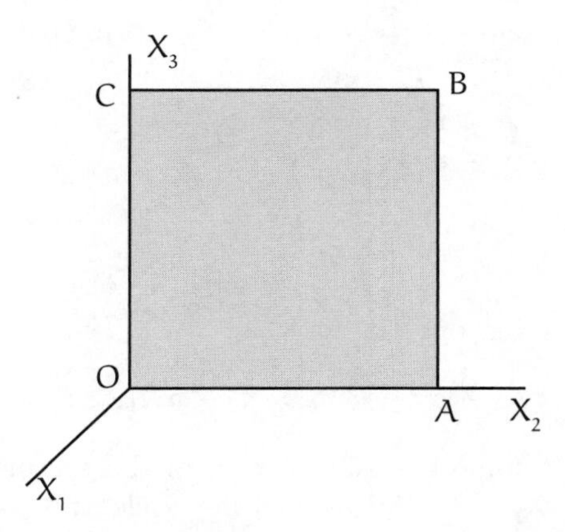

Unit square OABC in the reference configuration.

Answer

(b) $d\mathbf{x}^{(2)} = dX^{(2)}\hat{\mathbf{e}}_2$, $d\mathbf{x}^{(3)} = dX^{(3)}\left(\hat{\mathbf{e}}_2 + \hat{\mathbf{e}}_3\right)$
(c) $\Delta\theta = -45^{\circ}$
(d) $\Lambda_{(\hat{\mathbf{N}})} = \sqrt{2.5}$

Problem 4.18

Given the deformation expressed by

$$x_1 = X_1 + AX_2^2, \quad x_2 = X_2, \quad x_3 = X_3 - AX_2^2$$

where A is a constant (not necessarily small), determine the finite strain tensors $\mathbf{E}$ and $\mathbf{e}$, and show that if the displacements are small so that $\mathbf{x} \approx \mathbf{X}$, and if squares of A may be neglected, both tensors reduce to the infinitesimal strain tensor $\boldsymbol{\epsilon}$.

Answer

$$[\epsilon_{ij}] = \begin{bmatrix} 0 & Ax_2 & 0 \\ Ax_2 & 0 & -Ax_2 \\ 0 & -Ax_2 & 0 \end{bmatrix}$$

Problem 4.19

For the infinitesimal homogeneous deformation $x_i = X_i + A_{ij}X_j$ where the constants A_{ij} are very small, determine the small strain tensor $\boldsymbol{\epsilon}$, and from it the longitudinal (normal) strain in the direction of the unit vector $\hat{\mathbf{N}} = \left(\hat{\mathbf{I}}_1 - \hat{\mathbf{I}}_3\right)/\sqrt{2}$.

Answer

$$2e_{(\hat{\mathbf{N}})} = A_{11} - A_{13} - A_{31} + A_{33}$$

Problem 4.20

A deformation is defined by

$$x_1 = X_1/\left(X_1^2 + X_2^2\right), \quad x_2 = X_2/\left(X_1^2 + X_2^2\right), \quad x_3 = X_3 \ .$$

Determine the deformation tensor $\mathbf{C}$ together with its principal values.

Answer

$$C_{(1)} = C_{(2)} = \left(X_1^2 + X_2^2\right)^{-2}, \quad C_{(3)} = 1$$

Problem 4.21

For the deformation field given by

$$x_1 = X_1 + \alpha X_2, \quad x_2 = X_2 - \alpha X_1, \quad x_3 = X_3$$

where α is a constant, determine the matrix form of the tensors $\mathbf{E}$ and $\mathbf{e}$, and show that the circle of particles $X_1^2 + X_2^2 = 1$ deforms into the circle $x_1^2 + x_2^2 = 1 + \alpha^2$.

Answer

$$[E_{ij}] = \frac{1}{2}\begin{bmatrix} \alpha^2 & 0 & 0 \\ 0 & \alpha^2 & 0 \\ 0 & 0 & 0 \end{bmatrix}$$

$$[e_{ij}] = \frac{1}{2\left(1+\alpha^2\right)^2}\begin{bmatrix} -\alpha^2 & 0 & 0 \\ 0 & -\alpha^2 & 0 \\ 0 & 0 & 0 \end{bmatrix}$$

Problem 4.22

Let the deformation of a continuum be given by the equations

$$x_1 = X_1 + kX_2^2, \quad x_2 = X_2 - kX_1^2, \quad x_3 = X_3$$

where k is a constant. Determine the Lagrangian finite strain tensor $\mathbf{E}$, and from it, assuming k is very small, deduce the infinitesimal strain tensor $\boldsymbol{\epsilon}$. Verify this by calculating the displacement field and using the definition $2\epsilon_{ij} = u_{i,j} + u_{j,i}$ for the infinitesimal theory.

Problem 4.23

Given the displacement field

$$u_1 = AX_2X_3, \quad u_2 = AX_3^2, \quad u_3 = AX_1^2$$

where A is a very small constant, determine

(a) the components of the infinitesimal strain tensor $\boldsymbol{\epsilon}$, and the infinitesimal rotation tensor $\boldsymbol{\omega}$.

(b) the principal values of $\boldsymbol{\epsilon}$, at the point $(1, 1, 0)$.

Answer

(a) $\epsilon_{11} = \epsilon_{22} = \epsilon_{33} = 0$, $\epsilon_{12} = AX_3$, $\epsilon_{13} = \frac{1}{2}A(X_2 + 2X_1)$ and $\epsilon_{23} = AX_3$
$\omega_{11} = \omega_{22} = \omega_{33} = 0$, $\omega_{12} = -\omega_{21} = \frac{1}{2}AX_3$,
$\omega_{13} = -\omega_{31} = \frac{1}{2}AX_1 - AX_1$ and $\omega_{23} = -\omega_{32} = AX_3$

(b) $\epsilon_{(I)} = \frac{3}{2}A$, $\epsilon_{(II)} = 0$, $\epsilon_{(III)} = -\frac{3}{2}A$

Problem 4.24

A 45° strain rosette measures longitudinal strains along the X_1, X_2, and X_1' axes shown below. At point O the strains recorded are

$$\epsilon_{11} = 6 \times 10^{-4}, \quad \epsilon_{22} = 4 \times 10^{-4} \quad \text{and} \quad \epsilon_{11}' = 8 \times 10^{-4}$$

Determine the shear strain γ_{12} at O, together with ε_{22}', and verify that $\varepsilon_{11}+\varepsilon_{22} = \varepsilon_{11}'+\varepsilon_{22}'$. See Eq 4.82.

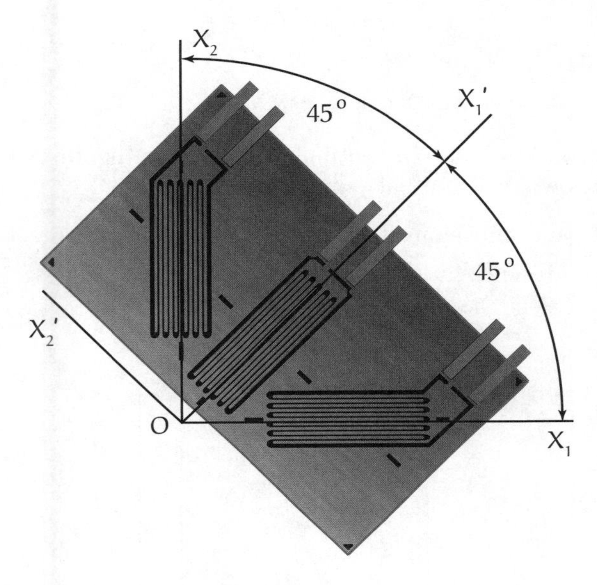

Problem 4.25

By a direct expansion of Eq 4.87, $2\omega_i = \varepsilon_{ijk}\omega_{kj}$, show that $\omega_1 = \omega_{32} = -\omega_{23}$, etc. Also, show that only if A is a very small constant does the mapping

$$x_1 = X_1 - AX_2 + AX_3 \ ,$$
$$x_2 = X_2 - AX_3 + AX_1 \ ,$$
$$x_3 = X_3 - AX_1 + AX_2 \ ,$$

represent a rigid body rotation ($\mathbf{E} \equiv 0$). Additionally, determine the infinitesimal rotation tensor ω_{ij} in this case; from it, using the result proven above, deduce the rotation vector ω_i.

Answer

$$\boldsymbol{\omega} = A(\hat{\mathbf{e}}_1 + \hat{\mathbf{e}}_2 + \hat{\mathbf{e}}_3)$$

Problem 4.26

For the displacement field

$$u_1 = kX_1X_2, \quad u_2 = kX_1X_2, \quad u_3 = 2k(X_1 + X_2)X_3$$

where k is a very small constant, determine the rotation tensor $\boldsymbol{\omega}$, and show that it has only one real principal value at the point $(0, 0, 1)$.

Answer

$$\omega_{(1)} = 0,\ \omega_{(2)} = -\omega_{(3)} = ik\sqrt{2}, \text{ where } i = \sqrt{-1}$$

Problem 4.27

Let the deformation

$$x_1 = X_1 + AX_2X_3\ ,$$
$$x_2 = X_2 + AX_3^2\ ,$$
$$x_3 = X_3 + AX_1^2\ ,$$

where A is a constant be applied to a continuum body. For the unit square of material line elements OBCD as shown, calculate at point C

(a) the stretch and unit elongation for the element in the direction of diagonal OC,

(b) the change in the right angle at C if $A = 1$; if $A = 0.1$.

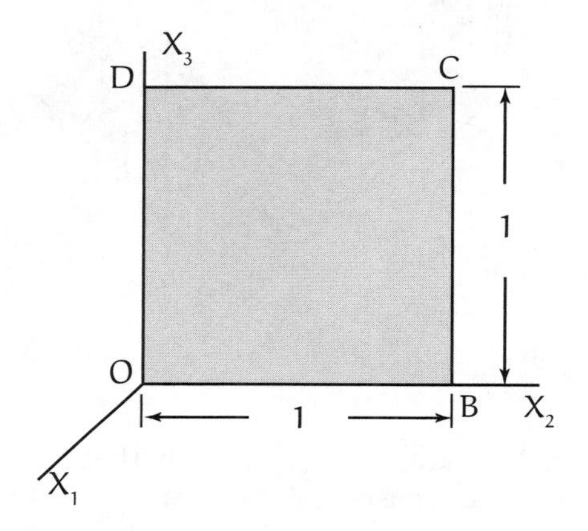

Unit square OBCD in the reference configuration.

Answer

(a) $\Lambda^2_{(OC)} = 1 + 2A + 4A^2$, $e_{(OC)} = \sqrt{1 + 2A + 4A^2} - 1$

(b) $\Delta\theta_{(A=1)} = 60°$, $\Delta\theta_{(A=0.1)} = 11.77°$

Problem 4.28

For the homogeneous deformation expressed by the equations

$$x_1 = \sqrt{2}X_1 + \frac{3\sqrt{2}}{4}X_2\ ,$$
$$x_2 = -X_1 + \frac{3}{4}X_2 + \frac{\sqrt{2}}{4}X_3\ ,$$
$$x_3 = X_1 - \frac{3}{4}X_2 + \frac{\sqrt{2}}{4}X_3\ ,$$

determine

(a) the unit normal $\hat{\mathbf{n}}$ for the line element originally in the direction of $\hat{\mathbf{N}} = \left(\hat{\mathbf{I}}_1 - \hat{\mathbf{I}}_2 + \hat{\mathbf{I}}_3\right)/\sqrt{3}$.

(b) the stretch $\Lambda_{(\hat{\mathbf{N}})}$ of this element.

(c) the maximum and minimum stretches at the point $X_1 = 1$, $X_2 = 0$, $X_3 = -2$ in the reference configuration.

Answer

(a) $$\hat{\mathbf{n}} = \frac{\sqrt{2}\hat{\mathbf{e}}_1 + \left(\sqrt{2}-7\right)\hat{\mathbf{e}}_2 + \left(\sqrt{2}+7\right)\hat{\mathbf{e}}_3}{\sqrt{104}}$$

(b) $\Lambda_{(\hat{\mathbf{N}})} = 1.472$

(c) $\Lambda_{(max)} = 2$; $\Lambda_{(min)} = 0.5$

Problem 4.29

Let the deformation

$$x_1 = a_1(X_1 + 2X_2), \quad x_2 = a_2 X_2, \quad x_3 = a_3 X_3$$

where a_1, a_2, and a_3 are constants be applied to the unit cube of material shown in the sketch. Determine

(a) the deformed length l of diagonal OC,

(b) the angle between edges OA and OG after deformation,

(c) the conditions which the constants must satisfy for the deformation to be possible if

 (1) the material is incompressible,

 (2) the angle between elements OC and OB is to remain unchanged.

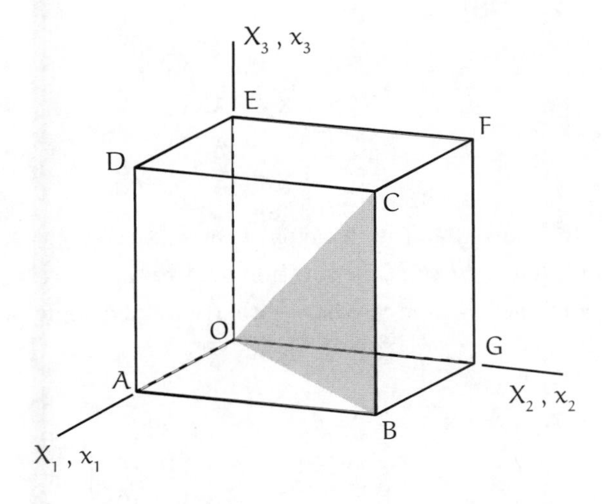

Unit cube having diagonal OC.

Answer

$$\text{(a) } l^2 = 9a_1^2 + a_2^2 + a_3^2$$

$$\text{(b) } \cos\theta = \frac{2a_1}{\sqrt{4a_1^2 + a_2^2}}$$

$$\text{(c) (1) } a_1 a_2 a_3 = 1,$$
$$\text{(2) } 9a_1^2 + a_2^2 = 2a_3^2$$

Problem 4.30

A homogeneous deformation is defined by

$$x_1 = \alpha X_1 + \beta X_2, \quad x_2 = -\alpha X_1 + \beta X_2, \quad x_3 = \mu X_3$$

where α, β and μ are constants. Determine

(a) the magnitudes and directions of the principal stretches,

(b) the matrix representation of the rotation tensor **R**,

(c) the direction of the axis of the rotation vector, and the magnitude of the angle of rotation.

Answer

$$\text{(a) } \Lambda_{(1)}^2 = \Lambda_{(\hat{\mathbf{e}}_1)}^2 = 2\alpha^2, \quad \Lambda_{(2)}^2 = \Lambda_{(\hat{\mathbf{e}}_2)}^2 = 2\beta^2, \quad \Lambda_{(3)}^2 = \Lambda_{(\hat{\mathbf{e}}_3)}^2 = \mu^2$$

$$\text{(b) } [R_{ij}] = \begin{bmatrix} \frac{1}{\sqrt{2}} & \frac{1}{\sqrt{2}} & 0 \\ -\frac{1}{\sqrt{2}} & \frac{1}{\sqrt{2}} & 0 \\ 0 & 0 & 1 \end{bmatrix}$$

$$\text{(c) } \hat{\mathbf{n}} = \hat{\mathbf{I}}_3; \ \Phi = 45°$$

Problem 4.31

Consider the deformation field

$$x_1 = X_1 - AX_2 + AX_3\ ,$$
$$x_2 = X_2 - AX_3 + AX_1\ ,$$
$$x_3 = X_3 - AX_1 + AX_2\ ,$$

where A is a constant. Show that the principal values of the right stretch tensor have a multiplicity of two, and that the axis of the rotation tensor is along $\hat{\mathbf{N}} = \left(\hat{\mathbf{I}}_1 + \hat{\mathbf{I}}_2 + \hat{\mathbf{I}}_3\right)/\sqrt{3}$. Determine the matrix of the rotation vector together with the angle of rotation ϕ.

Answer

$$\Lambda_{(1)} = 1, \ \Lambda_{(2)} = \Lambda_{(3)} = \sqrt{1 + 3A^2} = \beta$$

$$[R_{ij}] = \frac{1}{3\beta}\begin{bmatrix} \beta + 2 & \beta - 1 - 3A & \beta - 1 + 3A \\ \beta - 1 + 3A & \beta + 2 & \beta - 1 - 3A \\ \beta - 1 - 3A & \beta - 1 + 3A & \beta + 2 \end{bmatrix}$$

$$\phi = \cos^{-1}(1/\beta)$$

Problem 4.32

For the deformation field

$$x_1 = \sqrt{3}X_1 + X_2\ ,$$
$$x_2 = 2X_2\ ,$$
$$x_3 = X_3\ .$$

determine

(a) the matrix representation of the rotation tensor $\mathbf{R}$,

(b) the right stretch tensor $\mathbf{U}$ and the left stretch tensor $\mathbf{V}$, then show that the principal values of $\mathbf{U}$ and $\mathbf{V}$ are equal,

(c) the direction of the axis of rotation and the magnitude of the angle of rotation.

Answer

(a) $[R_{ij}] = \dfrac{1}{2\sqrt{2}}\begin{bmatrix} \sqrt{3}+1 & \sqrt{3}-1 & 0 \\ -\sqrt{3}+1 & \sqrt{3}+1 & 0 \\ 0 & 0 & 2\sqrt{2} \end{bmatrix}$

(b) $\Lambda_{(1)} = \sqrt{6}$, $\Lambda_{(2)} = \sqrt{2}$, $\Lambda_{(3)} = 1$

(c) $\hat{\mathbf{N}} = \hat{\mathbf{I}}_3$; $\phi = 15°$

Problem 4.33

Let a displacement field be given by

$$u_1 = \frac{1}{4}(X_3 - X_2),\quad u_2 = \frac{1}{4}(X_1 - X_3),\quad u_3 = \frac{1}{4}(X_2 - X_1)\ .$$

Determine

(a) the volume ratio dV/dV^0,

(b) the change in the right angle between line elements originally along the unit vectors $\hat{\mathbf{N}}_1 = \left(3\hat{\mathbf{I}}_1 - 2\hat{\mathbf{I}}_2 - \hat{\mathbf{I}}_3\right)/\sqrt{14}$ and $\hat{\mathbf{N}}_2 = \left(\hat{\mathbf{I}}_1 + 4\hat{\mathbf{I}}_2 - 5\hat{\mathbf{I}}_3\right)/\sqrt{42}$. Explain your answer.

Answer

(a) $dV/dV^0 = 1.1875$

(b) $\Delta\theta = 0°$

Problem 4.34

Consider again the deformation given in Example 4.15, namely

$$x_1 = 2(X_1 - X_2),\ x_2 = X_1 + X_2,\ x_3 = X_3\ .$$

Determine

(a) the left stretch tensor $\mathbf{V}$,

(b) the direction normals of the principal stretches of $\mathbf{V}$.

Answer

(a) $$[V_{AB}] = \begin{bmatrix} 2\sqrt{2} & 0 & 0 \\ 0 & \sqrt{2} & 0 \\ 0 & 0 & 1 \end{bmatrix}$$

(b) $\hat{\mathbf{N}}_1 = \hat{\mathbf{I}}_1$, $\hat{\mathbf{N}}_2 = \hat{\mathbf{I}}_2$, $\hat{\mathbf{N}}_3 = \hat{\mathbf{I}}_3$

Problem 4.35

A deformation field is expressed by

$$x_1 = \mu(X_1 \cos \beta X_3 + X_2 \sin \beta X_3) \; ,$$
$$x_2 = \mu(-X_1 \sin \beta X_3 + X_2 \cos \beta X_3) \; ,$$
$$x_3 = \nu X_3 \; ,$$

where μ, β, and ν are constants.

(a) Determine the relationship between these constants if the deformation is to be a possible one for an incompressible medium.

(b) If the above deformation is applied to the circular cylinder shown by the sketch, determine

(1) the deformed length l in terms of L, the dimension a, and the constants μ, β, and ν of an element of the lateral surface which has unit length and is parallel to the cylinder axis in the reference configuration, and

(2) the initial length L of a line element on the lateral surface which has unit length and is parallel to the cylinder axis after deformation.

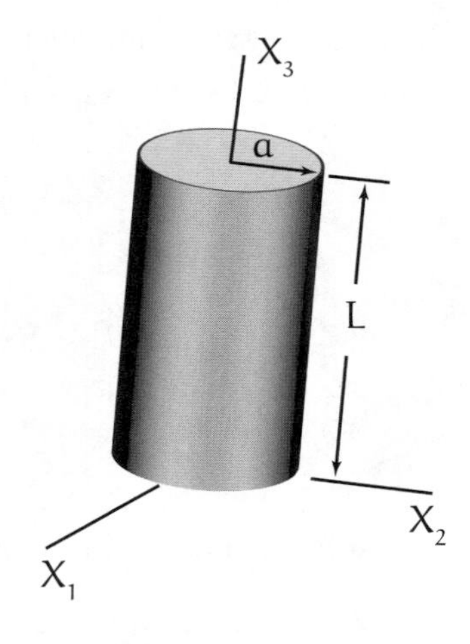

Answer

(a) $\mu^2\nu = 1$

(b) (1) $l = \sqrt{\mu^2\beta^2 a^2 + \nu^2}$

(2) $L = \frac{1}{\nu}\sqrt{\beta^2 a^2 + 1}$

Problem 4.36

A velocity field is defined in terms of the spatial coordinates and time by the equations,

$$v_1 = 2tx_1 \sin x_3, \quad v_2 = 2tx_2 \cos x_3, \quad v_3 = 0 .$$

At the point $(1, -1, 0)$ at time $t = 1$, determine

(a) the rate of deformation tensor and the vorticity tensor,

(b) the stretch rate per unit length in the direction of the normal $\hat{n} = (\hat{e}_1 + \hat{e}_2 + \hat{e})/\sqrt{3}$,

(c) the maximum stretch rate per unit length and the direction in which it occurs,

(d) the maximum shear strain rate.

Answer

(a) $[d_{ij}] = \begin{bmatrix} 0 & 0 & 1 \\ 0 & 2 & 0 \\ 1 & 0 & 0 \end{bmatrix}, \quad [w_{ij}] = \begin{bmatrix} 0 & 0 & 1 \\ 0 & 0 & 0 \\ -1 & 0 & 0 \end{bmatrix}$

(b) $\dot{\Lambda}/\Lambda = 4/3$

(c) $\left(\dot{\Lambda}/\Lambda\right)_{max} = 2$, $\hat{n} = \hat{e}_2$

(d) $\dot{\gamma}_{max} = 1.5$

Problem 4.37

Let N_A and n_i denote direction cosines of a material line element in the reference and current configurations, respectively. Beginning with Eq 4.140, $n_i\Lambda = x_{i,A}N_A$, and using the indicial notation throughout, show that

(a) $\dot{\Lambda}/\Lambda = D_{ij}n_i n_j$,

(b) $\ddot{\Lambda}/\Lambda = Q_{ij}n_i n_j + \dot{n}_i n_j$ where $Q_{ij} = \frac{1}{2}(a_{i,j} + a_{j,i})$ with a_i being the components of acceleration.

Problem 4.38

In a certain region of flow the velocity components are

$$v_1 = \left(x_1^3 + x_1 x_2^2\right) e^{-kt}, \quad v_2 = \left(x_2^3 - x_1^2 x_2\right) e^{-kt}, \quad v_3 = 0$$

where k is a constant, and t is time in s. Determine at the point $(1, 1, 1)$ when $t = 0$,

(a) the components of acceleration,

(b) the principal values of the rate of deformation tensor,

(c) the maximum shear rate of deformation.

Answer

(a) $a_1 = 2(4 - k)$, $a_2 = -4$, $a_3 = 0$
(b) $D_{(1)} = 4$, $D_{(2)} = 2$, $D_{(3)} = 0$
(c) $\dot{\gamma}_{max} = \pm 2$

Problem 4.39
For the motion

$$x_1 = X_1 \ ,$$
$$x_2 = X_2 e^t + X_1(e^t - 1) \ ,$$
$$x_3 = X_1(e^t - e^{-t}) + X_3 \ ,$$

determine the velocity field $v_i = v_i(\mathbf{x})$, and show that for this motion

(a) $\mathbf{L} = \dot{\mathbf{F}} \cdot \mathbf{F}^{-1}$,
(b) $\mathbf{D} = \dot{\boldsymbol{\epsilon}}$ at $t = 0$.

Problem 4.40
Determine an expression for the material derivative $d(\ln dx)/dt$ in terms of the rate of deformation tensor $\mathbf{D}$ and the unit normal $\hat{\mathbf{n}} = d\mathbf{x}/dx$.

Answer

$$d(\ln dx)/dt = \hat{\mathbf{n}} \cdot \mathbf{D} \cdot \hat{\mathbf{n}}$$

Problem 4.41
A velocity field is given in spatial form by

$$v_1 = x_1 x_3, \quad v_2 = x_2^2 t, \quad v_3 = x_2 x_3 t \ .$$

(a) Determine the vorticity tensor $\mathbf{W}$ and the vorticity vector $\boldsymbol{w}$.
(b) Verify the equation $\varepsilon_{pqi} w_i = w_{qp}$ for the results of part (a).
(c) Show that at the point $(1, 0, 1)$ when $t = 1$, the vorticity tensor has only one real root.

Answer

(a) $w_1 = \frac{1}{2} x_3 t$, $w_2 = \frac{1}{2} x_1$, $w_3 = 0$
(c) $W_{(1)} = 0$, $W_{(2)} = -W_{(3)} = i/\sqrt{2}$, where $i = \sqrt{-1}$

Problem 4.42
Consider the velocity field

$$v_1 = e^{x_3 - ct} \cos \omega t, \quad v_2 = e^{x_3 - kt} \sin \omega t, \quad v_3 = c$$

where c and ω are constants.

(a) Show that the speed of every particle is constant.
(b) Determine the acceleration components a_i.
(c) Calculate the logarithmic stretching, $\dot{\Lambda}/\Lambda = d(\ln \Lambda)/dt$, for the element in the current configuration in the direction of $\hat{\mathbf{n}} = (\hat{\mathbf{e}}_1 + \hat{\mathbf{e}}_2)/\sqrt{2}$ at $\mathbf{x} = 0$.

Answer

(b) $a_1 = -\omega v_2$, $a_2 = \omega v_1, a_3 = 0$
(c) $\frac{1}{2}e^{-ct}\cos\omega t$

Problem 4.43
Show that the velocity field

$$v_1 = 1.5x_3 - 3x_2, \quad v_2 = 3x_1 - x_3, \quad v_3 = x_2 - 1.5x_1$$

corresponds to a rigid body rotation, and determine the axis of spin (the vorticity vector).

Answer

$$\mathbf{w} = \hat{\mathbf{e}}_1 + 1.5\hat{\mathbf{e}}_2 + 3\hat{\mathbf{e}}_3$$

Problem 4.44
For the steady velocity field

$$v_1 = x_1^2 x_2, \quad v_2 = 2x_2^2 x_3, \quad v_3 = 3x_1 x_2 x_3$$

determine the rate of extension at $(2, 0, 1)$ in the direction of the unit vector $(4\hat{\mathbf{e}}_1 - 3\hat{\mathbf{e}}_3)/5$.

Answer

$$\dot{\Lambda}/\Lambda = -\frac{48}{25}$$

Problem 4.45
Prove that $d(\ln J)/dt = \operatorname{div}\mathbf{v}$ and, in particular, verify that this relationship is satisfied for the motion

$$x_1 = X_1 + ktX_3, \quad x_2 = X_2 + ktX_3, \quad x_3 = X_3 - kt(X_1 + X_2)$$

where k is a constant.

Answer

$$\dot{J}/J = \operatorname{div}\mathbf{v} = 4k^2 t/(1 + 2k^2 t^2)$$

Problem 4.46
Prove the identity

$$\left(\frac{x_{i,A}}{J}\right)_{,i} = 0\,, \qquad J = \det(x_{i,A})$$

Problem 4.47

Equation 4.151 gives the material derivative of dx^2 in terms of d_{ij} . Using that equation as the starting point, show that $d^2(dx^2)/dt^2$ is given in terms of d_{ij} and its time derivative by

$$d^2(dx^2)/dt^2 = 2(\dot{d}_{ij} + v_{k,i}d_{kj} + v_{k,j}d_{ik})dx_i dx_j \ .$$

Problem 4.48

A continuum body in the form of the unit cube shown by the sketch undergoes the homogeneous deformation

$$x_1 = \lambda_1 X_1, \quad x_2 = \lambda_2 X_2, \quad x_3 = \lambda_3 X_3$$

where λ_1, λ_2, and λ_3 are constants.

Determine the relationships among λ_1, λ_2, and λ_3 if

(a) the length of diagonal OC remains unchanged,

(b) the rectangular area ABFE remains unchanged,

(c) the triangular area ACE remains unchanged.

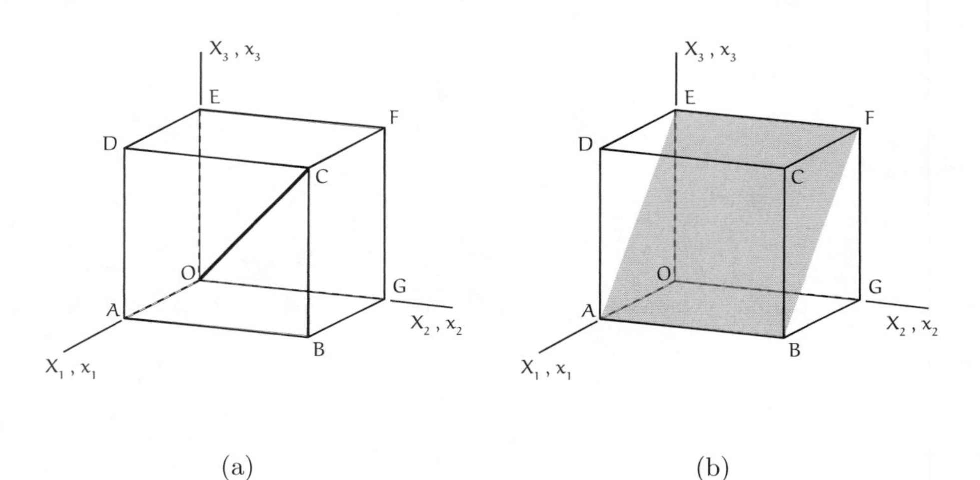

(a) (b)

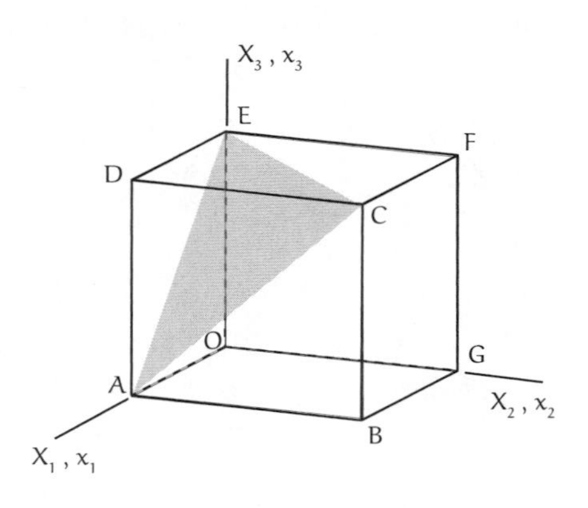

(c)

Answer

(a) $\lambda_1^2 + \lambda_2^2 + \lambda_3^2 = 3$
(b) $\lambda_2^2 \left(\lambda_1^2 + \lambda_3^2\right) = 2$
(c) $\lambda_1^2\lambda_2^2 + \lambda_2^2\lambda_3^2 + \lambda_3^2\lambda_1^2 = 3$

Problem 4.49

Let the unit cube shown in Problem 4.47 be given the motion

$$x_1 = X_1 + \frac{1}{2}t^2 X_2, \quad x_2 = X_2 + \frac{1}{2}t^2 X_1, \quad x_3 = X_3$$

Determine, at time t,

(a) the rate-of-change of area ABFE,
(b) the volume of the body.

Answer

(a) $\dot{\overline{dS}} = t^3\hat{e}_1/(1 - \frac{1}{4}t^4) - t\hat{e}_2 - t^3\hat{e}_3$
(b) $\mathcal{V} = (1 - \frac{1}{4}t^4)$

Problem 4.50

For the homogeneous deformation

$$x_1 = X_1 + \alpha X_2 + \alpha\beta X_3 \ ,$$
$$x_2 = \alpha\beta X_1 + X_2 + \beta^2 X_3 \ ,$$
$$x_3 = X_1 + X_2 + X_3 \ ,$$

where α and β are constants, determine the relationship between these constants if the deformation is isochoric.

Answer

$\beta = (\alpha^2 + \alpha)/(\alpha^2 + \alpha + 1)$

Problem 4.51

Show that for any velocity field $\mathbf{v}$ derived from a vector potential $\boldsymbol{\psi}$ by $\mathbf{v} = \mathrm{curl}\,\boldsymbol{\psi}$, the flow is isochoric. Also, for the velocity field

$$v_1 = a x_1 x_3 - 2x_3, \quad v_2 = -b x_2 x_3, \quad v_3 = 2x_1 x_2$$

determine the relationship between the constants a and b if the flow is isochoric.

Answer

$a = b$

5

Fundamental Laws and Equations

A number of the fundamental laws of continuum mechanics are expressions of the conservation of some physical quantity. These *balance laws*, as they are often called, are applicable to all material continua and result in equations that must always be satisfied. In this introductory text, we consider only the conservation laws dealing with mass, linear and angular momentum, and energy. With respect to energy, we shall first develop a purely mechanical energy balance and follow that by an energy balance that includes both mechanical and thermal energies, that is, a statement of the first law of thermodynamics. In addition to that, the Clausius-Duhem form of the second law of thermodynamics is covered.

The balance laws are usually formulated in the context of global (integral) relationships derived by a consideration of the conservation of some property of the body as a whole. As explained in Chapter 1, the global equations may then be used to develop associated *field equations* which are valid at all points within the body and on its boundary. For example, we shall derive the local equations of motion from a global statement of the conservation of linear momentum.

Constitutive equations, which reflect the internal constitution of a material, define specific types of material behavior. They are fundamental in the sense that they serve as the starting point for studies in the disciplines of elasticity, plasticity and various idealized fluids. Constitutive equations establish the relationship between kinematics and stresses for a material. These equations are the topic of the final section of this chapter.

Before we begin a discussion of the global conservation laws, it is useful to develop expressions for the material derivatives of certain integrals.

5.1 Material Derivatives of Line, Surface and Volume Integrals

Let any scalar, vector or tensor property of the collection of particles occupying the current volume $\mathcal{V}$ be represented by the integral

$$P_{ij\ldots}(t) = \int_{\mathcal{V}} P^*_{ij\ldots}(\mathbf{x}, t)\, d\mathcal{V} \tag{5.1}$$

where $P^*_{ij\ldots}$ represents the distribution of the property per unit volume and has continuous derivatives as necessary. The ellipsis in the subscript indicate that $P^*_{ij\ldots}$ could be any order tensor. The material derivative of this property is given in both spatial and material form, using Eq 4.163, by

$$\dot{P}_{ij\ldots}(t) = \frac{d}{dt}\int_{\mathcal{V}} P^*_{ij\ldots}(\mathbf{x}, t)\, d\mathcal{V} = \frac{d}{dt}\int_{\mathcal{V}^0} P^*_{ij\ldots}[\mathbf{x}(\mathbf{X}, t), t] J\, d\mathcal{V}^0 \,.$$

Since $\mathcal{V}^0$ is a fixed volume in the referential configuration, the differentiation and integration commute and the differentiation can be performed inside the integral sign. Thus, from Eq 4.161, using the notation $\dot{\overline{[\cdot]}}$ to indicate differentiation with respect to time,

$$\int_{\mathcal{V}^0} \dot{\overline{\left[P^*_{ij\ldots}(\mathbf{X},t)J\right]}}\, d\mathcal{V}^0 = \int_{\mathcal{V}^0} \left(\dot{P}^*_{ij\ldots} J + P^*_{ij\ldots}\dot{J}\right)\, d\mathcal{V}^0$$

$$= \int_{\mathcal{V}^0} \left(\dot{P}^*_{ij\ldots} + v_{k,k} P^*_{ij\ldots}\right) J\, d\mathcal{V}^0 \,,$$

and converting back to the spatial formulation

$$\dot{P}_{ij\ldots}(t) = \int_{\mathcal{V}} \left[\dot{P}^*_{ij\ldots}(\mathbf{x},t) + v_{k,k} P^*_{ij\ldots}(\mathbf{x},t)\right]\, d\mathcal{V} \,. \tag{5.2}$$

With the help of the material derivative operator given in Eq 4.32, this equation may be written (we omit listing the independent variables $\mathbf{x}$ and t for convenience),

$$\dot{P}_{ij\ldots}(t) = \int_{\mathcal{V}} \left[\frac{\partial P^*_{ij\ldots}}{\partial t} + v_k \frac{\partial P^*_{ij\ldots}}{\partial x_k} + v_{k,k} P^*_{ij\ldots}\right]\, d\mathcal{V}$$

$$= \int_{\mathcal{V}} \left[\frac{\partial P^*_{ij\ldots}}{\partial t} + \left(v_k P^*_{ij\ldots k}\right)_{,k}\right]\, d\mathcal{V}$$

which upon application of the divergence theorem becomes

$$\dot{P}_{ij\ldots}(t) = \int_{\mathcal{V}} \frac{\partial P^*_{ij\ldots}}{\partial t}\, d\mathcal{V} + \int_{\mathcal{S}} v_k P^*_{ij\ldots} n_k\, d\mathcal{S} \,. \tag{5.3}$$

This equation gives the time rate of change of the property $P_{ij\ldots}$ as the sum of the amount created in the volume $\mathcal{V}$, plus the amount entering through the bounding surface $\mathcal{S}$, and is often spoken of as the *transport theorem*.

Time derivatives of integrals over material surfaces and material curves may also be derived in an analogous fashion. First, we consider a tensorial property $Q_{ij\ldots}$ of the particles which make up the current surface $\mathcal{S}$, as given by

$$Q_{ij\ldots}(t) = \int_{\mathcal{S}} Q^*_{ij\ldots}(\mathbf{x},t)\, d\mathcal{S}_p = \int_{\mathcal{S}} Q^*_{ij\ldots}(\mathbf{x},t) n_p\, d\mathcal{S} \tag{5.4}$$

where $Q^*_{ij\ldots}(\mathbf{x},t)$ is the distribution of the property over the surface. From Eq 4.162, we have in Eulerian form (again omitting the variables $\mathbf{x}$ and t),

$$\dot{Q}_{ij\ldots}(t) = \int_{\mathcal{S}} \left(\dot{Q}^*_{ij\ldots} + v_{k,k} Q^*_{ij\ldots}\right)\, d\mathcal{S}_p - \int_{\mathcal{S}} Q^*_{ij\ldots} v_{q,p}\, d\mathcal{S}_q$$

$$= \int_{\mathcal{S}} \left[\left(\dot{Q}^*_{ij\ldots} + v_{k,k} Q^*_{ij\ldots}\right)\delta_{pq} - Q^*_{ij\ldots} v_{q,p}\right]\, d\mathcal{S}_q \,. \tag{5.5}$$

Similarly, for properties of particles lying on the spatial curve $\mathcal{C}$ and expressed by the line integral

$$R_{ij\ldots}(t) = \int_{\mathcal{C}} R^*_{ij\ldots}(\mathbf{x},t)\, dx_p \tag{5.6}$$

we have, using Eq 4.156,

$$\dot{R}_{ij\ldots}(t) = \int_{\mathcal{C}} \dot{R}^*_{ij\ldots}\, dx_p + \int_{\mathcal{C}} v_{p,q} R^*_{ij\ldots}\, dx_q$$

$$= \int_{\mathcal{C}} \left(\dot{R}^*_{ij\ldots}\delta_{pq} + v_{p,q} R^*_{ij\ldots}\right) dx_q \,. \tag{5.7}$$

5.2 Conservation of Mass, Continuity Equation

Every material body, as well as every portion of such a body, is endowed with a non-negative, scalar measure, called the *mass* of the body, or of the portion under consideration. Physically, the mass is associated with the inertia property of the body, that is, its tendency to resist a change in motion. The measure of mass may be a function of the space variables and time. If Δm is the mass of a small volume ΔV in the current configuration, and if we assume that Δm is absolutely continuous, the limit

$$\rho = \lim_{\Delta V \to 0} \frac{\Delta m}{\Delta V} \tag{5.8}$$

defines the scalar field $\rho = \rho(\mathbf{x}, t)$ called the *mass density* of the body for that configuration at time t. Therefore, the mass m of the entire body is given by

$$m = \int_V \rho(\mathbf{x}, t)\, dV \,. \tag{5.9}$$

In the same way, we define the mass of the body in the referential (initial) configuration in terms of the density field $\rho_0 = \rho_0(\mathbf{X}, t)$ by the integral

$$m = \int_{V^0} \rho_0(\mathbf{X}, t)\, dV^0 \,. \tag{5.10}$$

The law of *conservation of mass* asserts that the mass of a body, or of any portion of the body, is invariant under motion, that is, remains constant in every configuration. Thus, the material derivative of Eq 5.9 is zero,

$$\dot{m} = \frac{d}{dt}\int_V \rho(\mathbf{x}, t)\, dV = 0 \tag{5.11}$$

which upon application of Eq 5.2 with $P^*_{ij\ldots} \equiv \rho$ becomes,

$$\dot{m} = \int_V (\dot{\rho} + \rho v_{i,i})\, dV = 0 \,, \tag{5.12}$$

and since V is an arbitrary part of the continuum, the integrand here must vanish, resulting in

$$\dot{\rho} + \rho v_{i,i} = 0 \tag{5.13}$$

which is known as the *continuity equation* in Eulerian form. But the material derivative of ρ can be written as

$$\dot{\rho} = \frac{\partial \rho}{\partial t} + v_i \frac{\partial \rho}{\partial x_i} \,,$$

so that Eq 5.13 may be rewritten in the alternative forms,

$$\frac{\partial \rho}{\partial t} + v_i \frac{\partial \rho}{\partial x_i} + \rho v_{i,i} = 0 \,, \qquad (5.14a)$$

or

$$\frac{\partial \rho}{\partial t} + (\rho v_i)_{,i} = 0 \,. \qquad (5.14b)$$

If the density of the individual particles is constant so that $\dot{\rho} = 0$, the material is said to be incompressible, and thus it follows from Eq 5.13 that

$$v_{i,i} = 0 \qquad \text{or} \qquad \operatorname{div} \mathbf{v} = 0 \qquad (5.15)$$

for incompressible media.

Since the law of conservation of mass requires the mass to be the same in all configurations, we may derive the continuity equation from a comparison of the expressions for m in the referential and current configurations. Therefore, if we equate Eqs 5.9 and 5.10,

$$m = \int_{\mathcal{V}} \tilde{\rho}(\mathbf{x}, t)\, d\mathcal{V} = \int_{\mathcal{V}^o} \rho_0(\mathbf{X}, t)\, d\mathcal{V}^o \,, \qquad (5.16)$$

and, noting that for the motion $\mathbf{x} = \mathbf{x}(\mathbf{X}, t)$, we have

$$\int_{\mathcal{V}} \rho\,[\mathbf{x}\,(\mathbf{X}, t)\,, t]\; d\mathcal{V} = \int_{\mathcal{V}^o} \rho\,(\mathbf{X}, t)\, J\, d\mathcal{V}^o \,.$$

The tilde in Eq 5.16 is there because the functional form may be different when written as a function of $\mathbf{x}$ and t than the functional form written in terms of $\mathbf{X}$ and t. Now if we substitute the right-hand side of this equation for the left-hand side of Eq 5.16 and collect terms,

$$\int_{\mathcal{V}^o} [\rho(\mathbf{X}, t) J - \rho_0(\mathbf{X}, t)]\; d\mathcal{V}^o = 0 \,.$$

But $\mathcal{V}^o$ is arbitrary, and so in the material description

$$\rho J = \rho_0 \,, \qquad (5.17a)$$

and, furthermore, $\dot{\rho}_0 = 0$, from which we conclude that

$$\dot{\overline{(\rho J)}} = 0 \,. \qquad (5.17b)$$

Equations 5.17 are called the Lagrangian, or material, form of the continuity equation.

Example 5.1

Show that the spatial form of the continuity equation follows from the material form.

Solution

Carrying out the indicated differentiation in Eq 5.17b,

$$\dot{\overline{(\rho J)}} = \dot{\rho} J + \rho \dot{J} = 0$$

and by Eq 4.161,

$$\dot{J} = v_{i,i} J$$

so now

$$\overline{(\rho J)}^{\boldsymbol{\cdot}} = J\,(\dot{\rho} + \rho v_{i,i}) = 0\ .$$

But $J = \det \mathbf{F} \neq 0$ (an invertible tensor), which requires $\dot{\rho} + \rho v_{i,i} = 0$ the spatial continuity equation.

As a consequence of the continuity equation, we are able to derive a useful result for the material derivative of the integral in Eq 5.1 when $P^*_{ij\ldots}$ is equal to the product $\rho A^*_{ij\ldots}$, where $A^*_{ij\ldots}$ is the distribution of any property *per unit mass*. Accordingly, let

$$\dot{P}_{ij\ldots}(t) = \frac{d}{dt}\int_{\mathcal{V}} A^*_{ij\ldots}(\mathbf{x},t)\rho\, d\mathcal{V} = \frac{d}{dt}\int_{\mathcal{V}^0} A^*_{ij\ldots}(\mathbf{X},t)\rho J\, d\mathcal{V}^0$$

$$= \int_{\mathcal{V}^0}\left[\dot{A}^*_{ij\ldots}\,(\rho J) + A^*_{ij\ldots}\overline{(\rho J)}^{\boldsymbol{\cdot}}\right] d\mathcal{V}^0$$

which, because of Eq 5.17b, reduces to

$$\dot{P}^*_{ij\ldots}(t) = \int_{\mathcal{V}^0} \dot{A}^*_{ij\ldots}\rho\, d\mathcal{V}^0 = \int_{\mathcal{V}} \dot{A}^*_{ij\ldots}\rho\, d\mathcal{V}\ ,$$

and so

$$\frac{d}{dt}\int_{\mathcal{V}} A^*_{ij\ldots}(\mathbf{x},t)\rho\, d\mathcal{V} = \int_{\mathcal{V}} \dot{A}^*_{ij\ldots}\rho\, d\mathcal{V}\ . \tag{5.18}$$

We shall have numerous occasions to make use of this very important equation.

5.3 Linear Momentum Principle, Equations of Motion

Let a material continuum body having a current volume $\mathcal{V}$ and bounding surface $\mathcal{S}$ be subjected to surface traction $t_i^{(\hat{\mathbf{n}})}$ and distributed body forces ρb_i as shown in Fig. 5.1. In addition, let the body be in motion under the velocity field $v_i = v_i(\mathbf{x},t)$. The linear momentum of the body is defined by the vector

$$P_i(t) = \int_{\mathcal{V}} \rho v_i\, d\mathcal{V}\ , \tag{5.19}$$

and the *principle of linear momentum* states that the time rate of change of the linear momentum is equal to the resultant force acting on the body. Therefore, in global form, with reference to Fig. 5.1,

$$\frac{d}{dt}\int_{\mathcal{V}} \rho v_i\, d\mathcal{V} = \int_{\mathcal{S}} t_i^{(\hat{\mathbf{n}})}\, d\mathcal{S} + \int_{\mathcal{V}} \rho b_i\, d\mathcal{V}\ , \tag{5.20}$$

and because $t_i^{(\hat{\mathbf{n}})} = t_{ji}n_j$, we can convert the surface integral to a volume integral having the integrand $t_{ji,j}$. By the use of Eq 5.18 on the left-hand side of Eq 5.20 we have, after collecting terms,

$$\int_{\mathcal{V}} (\rho\dot{v}_i - t_{ji,j} - \rho b_i)\, d\mathcal{V} = 0 \tag{5.21}$$

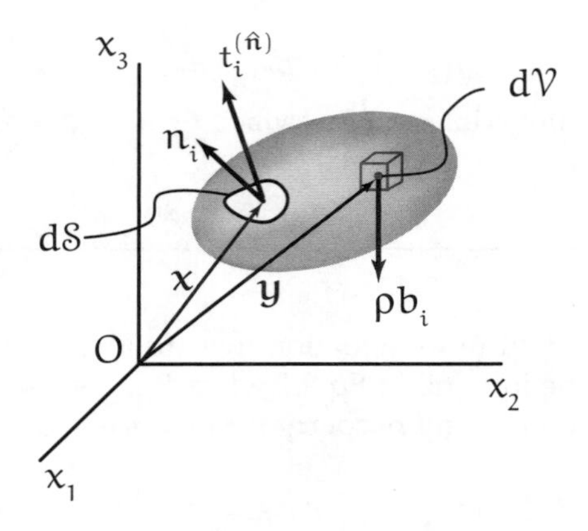

FIGURE 5.1
Material body in motion subjected to body and surface forces.

where $\dot{v}_i$ is the *acceleration field* of the body. Again, $\mathcal{V}$ is arbitrary and so the integrand must vanish, and we obtain,

$$t_{ji,j} + \rho b_i = \rho \dot{v}_i \tag{5.22}$$

which are known as the *local equations of motion* in Eulerian form.

When the velocity field is zero, or constant so that $\dot{v}_i = 0$, the equations of motion reduce to the equilibrium equations,

$$t_{ji,j} + \rho b_i = 0 \tag{5.23}$$

which are important in solid mechanics, especially elastostatics.

5.4 Piola-Kirchhoff Stress Tensors, Lagrangian Equations of Motion

As mentioned in the previous section, the equations of motion Eq 5.22 are in Eulerian form. These equations may also be cast in the referential form based upon the Piola-Kirchhoff tensor, which we now introduce.

Recall that in Section 3.3 we defined the stress components t_{ij} of the Cauchy stress tensor $\mathbf{T}$ as the ith component of the stress vector $t_i^{(\hat{e}_j)}$ acting on the material surface having the unit normal $\hat{n} = \hat{e}_j$. Notice that this unit normal is defined in the current configuration. It is also possible to define a stress vector that is referred to a material surface in the reference configuration and from it construct a stress tensor that is associated with that configuration. In doing this, we parallel the development in Section 3.3 for the Cauchy stress tensor associated with the current configuration.

Let the vector $\mathbf{p}^{0(\hat{\mathbf{N}})}$be defined as the stress vector referred to the area element $\Delta \mathcal{S}^0$ in the plane perpendicular to the unit normal $\hat{\mathbf{N}} = N_A \hat{\mathbf{I}}_A$. Just as we defined the Cauchy stress vector in Eq 3.5, we write

$$\lim_{\Delta \mathcal{S}^0 \to 0} \frac{\Delta f}{\Delta \mathcal{S}^0} = \frac{df}{d\mathcal{S}^0} = \mathbf{p}^{0(\hat{\mathbf{N}})} \tag{5.24}$$

where $\Delta \mathbf{f}$ is the resultant force acting on the material surface which in the reference configuration was $\Delta \mathcal{S}^0$.

The principle of linear momentum can also be written in terms of quantities which are referred to in the referential configuration as

$$\int_{\mathcal{S}^0} \mathbf{p}^{0(\hat{\mathbf{N}})}(\mathbf{X},t)\, d\mathcal{S}^0 + \int_{\mathcal{V}^0} \rho_0 \mathbf{b}^0(\mathbf{X},t)\, d\mathcal{V}^0 = \int_{\mathcal{V}^0} \rho_0 \mathbf{a}^0(\mathbf{X},t)\, d\mathcal{V}^0 \tag{5.25}$$

where $\mathcal{S}^0$, $\mathcal{V}^0$ and ρ_0 are the material surface, volume and density, respectively, referred to the reference configuration. The superscript zero after the variable is used to emphasize the fact that the function is written in terms of the reference configuration. For example,

$$a_i(\mathbf{x},t) = a_i\left[\boldsymbol{\chi}(\mathbf{X},t),t)\right] = a_i^0(\mathbf{X},t)\ .$$

Notice that, since all quantities are in terms of material coordinates, we have moved the differential operator d/dt of Eq 3.8 inside the integral to give rise to the acceleration $\mathbf{a}^0$. In a similar procedure to that carried out in Section 3.2, we apply Eq 5.25 to portions I and II of the body (as defined in Fig. 3.2(a)) and to the body as a whole to arrive at the equation

$$\int_{\mathcal{S}^0} \left[\mathbf{p}^{0(\hat{\mathbf{N}})} + \mathbf{p}^{0(-\hat{\mathbf{N}})}\right] d\mathcal{S}^0 = 0\ . \tag{5.26}$$

This equation must hold for arbitrary portions of the body surface, and so

$$\mathbf{p}^{0(\hat{\mathbf{N}})} = -\mathbf{p}^{0(-\hat{\mathbf{N}})} \tag{5.27}$$

which is the analog of Eq 3.10.

The stress vector $\mathbf{p}^{0(\hat{\mathbf{N}})}$ can be written out in components associated with the referential coordinate planes as

$$\mathbf{p}^{0(\hat{\mathbf{I}}_A)} = p_i^{0(\hat{\mathbf{I}}_A)} \hat{\mathbf{e}}_i, \qquad (A = 1,2,3)\ . \tag{5.28}$$

This describes the components of the stress vector $\mathbf{p}^{0(\hat{\mathbf{N}})}$ with respect to the referential coordinate planes, and to determine its components with respect to an arbitrary plane defined by the unit vector $\hat{\mathbf{N}}$, we apply a force balance to an infinitesimal tetrahedron of the body. As we let the tetrahedron shrink to the point, we have

$$p_i^{0(\hat{\mathbf{N}})} = p_i^{0(\hat{\mathbf{I}}_A)} N_A\ , \tag{5.29}$$

and defining

$$P_{Ai} \equiv p_i^{0(\hat{\mathbf{I}}_A)} \tag{5.30}$$

we obtain

$$p_i^{0(\hat{\mathbf{N}})} = P_{Ai} N_A \tag{5.31}$$

where P_{Ai} are the components of the *first Piola-Kirchhoff* stress tensor. These represent the x_i components of the force per unit area of a surface whose referential normal is $\hat{\mathbf{N}}$.

Using the first Piola-Kirchhoff stress tensor, we can derive the equations of motion, and hence the equilibrium equations in the referential formulation. Starting with Eq 5.25, we introduce Eq 5.31 to obtain

$$\int_{\mathcal{S}^0} P_{Ai} N_A\, d\mathcal{S}^0 + \int_{\mathcal{V}^0} \rho_0 b_i^0\, d\mathcal{V}^0 = \int_{\mathcal{V}^0} \rho_0 a_i^0\, d\mathcal{V}^0\ , \tag{5.32}$$

and using the divergence theorem on the surface integral, we consolidate Eq 5.32 as

$$\int_{\mathcal{V}^0} (P_{Ai,A} + \rho_0 b_i^0 - \rho_0 a_i^0)\, d\mathcal{V}^0 = 0\,.$$

This equation must hold for arbitrary portions of the body so that the integrand is equal to zero, or

$$P_{Ai,A} + \rho_0 b_i^0 = \rho_0 a_i^0 \tag{5.33a}$$

which are the *equations of motion* in referential form. If the acceleration field is zero, these equations reduce to the *equilibrium equations* in referential form

$$P_{Ai,A} + \rho_0 b_i^0 = 0\,. \tag{5.33b}$$

We note that the partial derivatives of the Piola-Kirchhoff stress components are with respect to the material coordinates because this stress tensor is referred to a surface in the reference configuration.

Equilibrium also requires a balance of moments about every point. Summing moments about the origin (Fig. 5.1 may be useful in visualizing this operation) gives us

$$\int_{\mathcal{S}^0} \varepsilon_{ijk} x_j p_k^{0(\hat{\mathbf{N}})}\, d\mathcal{S}^0 + \int_{\mathcal{V}^0} \varepsilon_{ijk} x_j \rho_0 b_k^0\, d\mathcal{V}^0 = 0 \tag{5.34}$$

which reduces to

$$\int_{\mathcal{V}^0} \varepsilon_{ijk} \left[(x_j P_{Ak})_{,A} + x_j \rho_0 b_k^0\right] d\mathcal{V}^0 = 0$$

where we have used Eq 5.31 and the divergence theorem. Carrying out the indicated partial differentiation we obtain

$$\int_{\mathcal{V}^0} \varepsilon_{ijk} \left[x_{j,A} P_{Ak} + x_j \left(P_{Ak,A} + \rho_0 b_k^0\right)\right] dV^0 = 0$$

and by Eq 5.33b this reduces to

$$\int_{\mathcal{V}^0} (\varepsilon_{ijk} x_{j,A} P_{Ak})\, d\mathcal{V}^0 = 0 \tag{5.35}$$

since the term in parentheses is zero on account of the balance of momentum. Again, this equation must hold for all portions $\mathcal{V}^0$ of the body, so the integrand must vanish, giving

$$\varepsilon_{ijk} x_{j,A} P_{Ak} = 0 \tag{5.36}$$

Following a similar argument to that presented in Section 3.4, we conclude that Eq 5.36 implies

$$x_{j,A} P_{Ak} = x_{k,A} P_{Aj}\,. \tag{5.37}$$

If we now introduce the definition for s_{AB}

$$P_{Ai} = x_{i,B} s_{BA}\,, \tag{5.38}$$

and substitute into Eq 5.37 we observe that

$$s_{AB} = s_{BA} \tag{5.39}$$

which is called the *second Piola-Kirchhoff* stress tensor, or sometimes the *symmetric Piola-Kirchhoff* stress tensor.

The Piola-Kirchhoff stresses can be related to the Cauchy stress by considering the differential force exerted on an element of deformed surface $d\mathcal{S}$ as

$$df_i = t_{ji} n_j \, d\mathcal{S} \,. \tag{5.40}$$

This force can also be written in terms of the first Piola-Kirchhoff stress tensor as

$$df_i = P_{Ai} N_A \, d\mathcal{S}^0 \,. \tag{5.41}$$

Recall from Eq 4.159 that the surface element in the deformed configuration is related to the surface element in the reference configuration by

$$n_q \, dS = X_{A,q} J N_A \, d\mathcal{S}^0 \,.$$

Using this, along with Eqs 5.40 and 5.41, we obtain

$$df_i = t_{ji} n_j \, dS = t_{ji} X_{A,j} J N_A \, dS^0 = P_{Ai} N_A \, d\mathcal{S}^0 \tag{5.42}$$

which can be rewritten as

$$(t_{ji} X_{A,j} J - P_{Ai}) N_A \, d\mathcal{S}^0 = 0 \,. \tag{5.43}$$

From this we see that the Cauchy stress and the first Piola-Kirchhoff stress are related through

$$J t_{ji} = P_{Ai} x_{j,A} \,. \tag{5.44}$$

Also, from Eq 5.38 we can write

$$J t_{ji} = x_{j,A} x_{i,B} s_{AB} \tag{5.45}$$

which relates the Cauchy stress to the second Piola-Kirchhoff stress.

In Chapter 4 we showed that the difference between Eulerian and Lagrangian strains disappears when linear deformations are considered. Here, we will show that in linear theories the distinction between Cauchy and Piola-Kirchhoff stress measures is not necessary.

To show the equivalence of Cauchy and Piola-Kirchhoff stresses in linear theories, we have to recall some kinematic results from Section 4.7 and also derive a few more. Introducing a positive number ε that is a measure of smallness such that the displacement gradients $u_{i,A}$ are of the same order of magnitude as ε, we may write

$$u_{i,A} = O(\varepsilon) \quad \text{as} \quad \varepsilon \to 0 \,. \tag{5.46}$$

As we discovered in Section 4.7, the Eulerian and Lagrangian strains are equivalent as $\varepsilon \to 0$, so from Eqs 4.58 and 4.59 we have

$$E_{AB} \delta_{iA} \delta_{jB} = e_{ij} = O(\varepsilon) \,.$$

Examination of Eqs 5.44 and 5.45 relates stress measures t_{ji}, P_{Ai}, and s_{AB}. To discuss this relationship in the linear case, we must find an expression for the Jacobian as we let $\varepsilon \to 0$. Starting with the definition of J in the form

$$J = \frac{1}{6} \varepsilon_{ijk} \varepsilon_{ABC} F_{iA} F_{jB} F_{kC}$$

we substitute $F_{iA} = u_{i,A} + \delta_{iA}$, etc., to get

$$J = \frac{1}{6} \varepsilon_{ijk} \varepsilon_{ABC} (u_{i,A} + \delta_{i,A})(u_{j,B} + \delta_{jB})(u_{k,C} + \delta_{kC}) \,.$$

Carrying out the algebra and after some manipulation of the indices

$$J = \frac{1}{6}\varepsilon_{ijk}\varepsilon_{ABC}\left[\delta_{iA}\delta_{jB}\delta_{kC} + 3u_{k,C}\delta_{iA}\delta_{jB} + O(\varepsilon^2)\right]$$

where terms on the order of ε^2 and higher have not been written out explicitly. Since $\varepsilon_{ijk}\varepsilon_{ijk} = 6$ and $\varepsilon_{ijk}\varepsilon_{ijC} = 2\delta_{kC}$,

$$J = 1 + u_{k,k} + O(\varepsilon^2) \,. \tag{5.47}$$

Now we can evaluate Eqs 5.44 and 5.45 as $\varepsilon \to 0$, that is, for the case of a linear theory. Since $u_{k,k}$ is $O(\varepsilon)$.

$$t_{ji} + O(\epsilon) = P_{Ai}\delta_{Aj} + O(\varepsilon) \,. \tag{5.48}$$

With a similar argument for Eq 5.45, we find

$$t_{ji} = S_{AB}\delta_{Ai}\delta_{Bj} \quad \text{as} \quad \varepsilon \to 0 \,. \tag{5.49}$$

Equations 5.48 and 5.49 demonstrate that, in linear theory, Cauchy, Piola-Kirchhoff and symmetric Piola-Kirchhoff stress measures are all equivalent.

5.5 Moment of Momentum (Angular Momentum) Principle

Moment of momentum is the phrase used to designate the moment of the linear momentum with respect to some point. This vector quantity is also frequently called the *angular momentum* of the body. The principle of angular momentum states that the time rate of change of the moment of momentum of a body with respect to a given point is equal to the moment of the surface and body forces with respect to that point. For the body shown in Fig 5.1, if we take the origin as the point of reference, the angular momentum principle has the mathematical form

$$\frac{d}{dt}\int_{\mathcal{V}} \varepsilon_{ijk}x_j\rho v_k \, d\mathcal{V} = \int_{\mathcal{S}} \varepsilon_{ijk}x_j t_k^{(\hat{n})} \, d\mathcal{S} + \int_{V} \varepsilon_{ijk}x_j\rho b_k \, d\mathcal{V} \,. \tag{5.50}$$

Making use of Eq 5.18 in taking the derivative on the left-hand side of the equation and applying the divergence theorem to the surface integral, after introducing the identity $t_k^{(\hat{n})} = t_{qk}n_q$, results in

$$\int_{\mathcal{V}} \varepsilon_{ijk}[x_j(\rho\dot{v}_k - t_{qk,q} - \rho b_k) - t_{jk}] \, d\mathcal{V} = 0$$

which reduces to

$$\int_{\mathcal{V}} \varepsilon_{ijk}t_{kj} \, d\mathcal{V} = 0 \tag{5.51}$$

because of Eq 5.22 (the equations of motion) and the sign-change property of the permutation symbol. Again, with $\mathcal{V}$ arbitrary, the integrand must vanish so that

$$\varepsilon_{ijk}t_{kj} = 0 \tag{5.52}$$

which by direct expansion demonstrates that $t_{kj} = t_{jk}$, and the stress tensor is symmetric. Note that in formulating the angular momentum principle by Eq 5.50 we have assumed that no body or surface couples act on the body. If any such concentrated moments do act, the material is said to be a *polar* material, and the symmetry property of $\mathbf{T}$ no longer holds. But as mentioned in Chapter 3, this is a rather specialized situation and we shall not consider it here.

5.6 Law of Conservation of Energy, The Energy Equation

The statement we adopt for the law of conservation of energy is the following: the material time derivative of the kinetic plus internal energies is equal to the sum of the rate of work of the surface and body forces, plus all other energies that enter or leave the body per unit time. Other energies may include, for example, thermal, electrical, magnetic or chemical energies. In this text we consider only mechanical and thermal energies, and also, we require the continuum material to be non-polar (free of body or traction couples).

If only mechanical energy is considered, the energy balance can be derived from the equations of motion, (Eq 5.22). Here we take a different approach and proceed as follows. By definition, the *kinetic energy* of the material occupying an arbitrary volume $\mathcal{V}$ of the body in Fig. 5.1 is

$$K(t) = \frac{1}{2}\int_{\mathcal{V}} \rho \mathbf{v}\cdot\mathbf{v}\, d\mathcal{V} = \frac{1}{2}\int_{\mathcal{V}} \rho v_i v_i \, d\mathcal{V} \, . \tag{5.53}$$

Also, the *mechanical power*, or *rate of work* of the body and surface forces shown in the figure is defined by the scalar

$$P(t) = \int_{\mathcal{S}} t_i^{(\hat{\mathbf{n}})} v_i \, d\mathcal{S} + \int_{\mathcal{V}} \rho b_i v_i \, d\mathcal{V} \, . \tag{5.54}$$

Consider now, the material derivative of the kinetic energy integral

$$\dot{K} = \frac{d}{dt}\int_{\mathcal{V}} \frac{1}{2}\rho v_i v_i \, d\mathcal{V} = \frac{1}{2}\int_{\mathcal{V}} \rho \overline{(v_i v_i)}^{\boldsymbol{\cdot}} \, d\mathcal{V}$$

$$= \int_{\mathcal{V}} \rho (v_i \dot{v}_i) \, d\mathcal{V} = \int_{\mathcal{V}} v_i \left(t_{ji,j} + \rho b_i\right) d\mathcal{V}$$

where Eq 5.22 has been used to obtain the final form of the integrand. But $v_i t_{ji,j} = (v_i t_{ji})_{,j} - v_{i,j} t_{ji}$ and so

$$\dot{K} = \int_{\mathcal{V}} \left[\rho b_i v_i + (v_i t_{ij})_{,j} - v_{i,j} t_{ij}\right] d\mathcal{V}$$

which, if we convert the middle term by the divergence theorem and make use of the decomposition $v_{i,j} = d_{ij} + w_{ij}$, may be written

$$\dot{K} = \int_{\mathcal{V}} \rho b_i v_i \, d\mathcal{V} + \int_{\mathcal{S}} t_i^{(\hat{\mathbf{n}})} v_i \, d\mathcal{S} - \int_{\mathcal{V}} t_{ij} d_{ij} \, d\mathcal{V} \, . \tag{5.55}$$

By the definition Eq 5.54 this may be expressed as

$$\dot{K} + S = P \tag{5.56}$$

where the integral

$$S = \int_{\mathcal{V}} t_{ij} d_{ij} \, d\mathcal{V} = \int_{\mathcal{V}} (\mathbf{T} : \mathbf{D}) \, d\mathcal{V} \tag{5.57}$$

is known as the stress work, and its integrand $t_{ij} d_{ij}$ as the *stress power*. The balance of mechanical energy given by Eq 5.56 shows that, of the total work done by the external forces, a portion goes toward increasing the kinetic energy, and the remainder appears as work done by the internal stresses.

In general, S cannot be expressed as the material derivative of a volume integral, that is,

$$S \neq \frac{d}{dt} \int_{\mathcal{V}} (\cdots)\, d\mathcal{V} \tag{5.58}$$

because there is no known function we could insert as the integrand of this equation. However, in the special situation when

$$S = \dot{U} = \frac{d}{dt} \int_{\mathcal{V}} \rho u \, d\mathcal{V} = \int_{\mathcal{V}} \rho \dot{u} \, d\mathcal{V} \tag{5.59}$$

where U is called the *internal energy* and u the *specific internal energy*, or *energy density* (per unit mass), Eq 5.56 becomes

$$\frac{d}{dt} \int_{\mathcal{V}} \rho \left(\frac{1}{2} v_i v_i + u \right) d\mathcal{V} = \int_{\mathcal{V}} \rho b_i v_i \, d\mathcal{V} = \int_{\mathcal{S}} t_i^{(\hat{n})} v_i \, d\mathcal{S} , \tag{5.60a}$$

or, briefly,

$$\dot{K} + \dot{U} = P . \tag{5.60b}$$

Note that Eq 5.60 indicates that part of the external work P causes an increase in kinetic energy, and the remainder is stored as internal energy. As we shall see in Chapter 6, ideal elastic materials respond to forces in this fashion. [1]

For a *thermomechanical continuum*, we represent the rate at which thermal energy is added to a body by

$$Q = \int_{\mathcal{V}} \rho r \, d\mathcal{V} - \int_{\mathcal{S}} q_i n_i \, d\mathcal{S} . \tag{5.61}$$

The scalar field r specifies the rate at which heat per unit mass is produced by internal sources and is known as the *heat supply*. The vector q_i, called the *heat flux* vector, is a measure of the rate at which heat is conducted into the body per unit area per unit time across the element of surface $d\mathcal{S}$ whose outward normal is n_i (hence the minus sign in Eq 5.61). The heat flux q_i is often assumed to obey *Fourier's law of heat conduction;*

$$q_i = -\kappa \theta_{,i} \qquad \text{or} \qquad \mathbf{q} = -\kappa \boldsymbol{\nabla} \theta \tag{5.62}$$

where κ is the thermal conductivity and $\theta_{,i}$ is the temperature gradient. But, since not all materials obey this conduction "law," it is not universally valid.

With the addition of the thermal energy consideration, the complete energy balance requires modification of Eq 5.60 which now takes the form

$$\dot{K} + \dot{U} = P + Q , \tag{5.63a}$$

or, when written out in detail,

$$\frac{d}{dt} \int_{\mathcal{V}} \rho \left(\frac{1}{2} v_i v_i + u \right) d\mathcal{V} = \int_{\mathcal{V}} \rho \left(b_i v_i + r \right) d\mathcal{V} + \int_{\mathcal{S}} \left[t_i^{(\hat{n})} v_i - q_i n_i \right] d\mathcal{S} . \tag{5.63b}$$

[1]The symbol u is used for specific internal energy because of its widespread acceptance in the literature. There appears to be very little chance that it might be misinterpreted in this context as the magnitude of the displacement vector $\mathbf{u}$.

If we convert the surface integral to a volume integral and make use of the equations of motion (Eq 5.22), the reduced form of Eq 5.63b is readily seen to be

$$\int_{\mathcal{V}} (\rho\dot{u} - t_{ij}d_{ij} - \rho r + q_{i,i})\, d\mathcal{V} = 0 \quad \text{or} \quad \int_{\mathcal{V}} (\rho\dot{u} - \mathbf{T} : \mathbf{D} - \rho r + \boldsymbol{\nabla} \cdot \mathbf{q})\, d\mathcal{V}\,, \tag{5.64a}$$

or, briefly,

$$S = \dot{U} - Q \tag{5.64b}$$

which is sometimes referred to as the *thermal energy balance*, in analogy with Eq 5.56 that relates to the *mechanical energy balance*. Thus, we observe that the rate of work of the internal forces equals the rate at which internal energy is increasing minus the rate at which heat enters the body. As usual for an arbitrary volume $\mathcal{V}$, by the argument which is standard now, upon setting the integrand of Eq 5.64a equal to zero, we obtain the field equation,

$$\rho\dot{u} - t_{ij}d_{ij} - \rho r + q_{i,i} = 0 \quad \text{or} \quad \rho\dot{u} - \mathbf{T} : \mathbf{D} - \rho r + \operatorname{div} \mathbf{q} = 0 \tag{5.65}$$

which is called the *energy equation*.

In summary, then, the mechanical energy balance Eq 5.55 is derivable directly from the equations of motion (linear momentum principle) and is but one part of the complete energy picture. When thermal energy is included, the global balance Eq 5.63 is a statement of the *first law of thermodynamics*.

5.7 Entropy and the Clausius-Duhem Equation

The conservation of energy as formulated in Section 5.6 is a statement of the interconvertibility of heat and work. However, there is not total interconvertibility for irreversible processes. For instance, the case of mechanical work being converted to heat via friction is understood, but the converse does not hold. That is, heat cannot be utilized to directly generate work. This, of course, is the motivation for the *second law of thermodynamics*.

Continuum mechanics uses the second law in a different way than classical thermodynamics. In that discipline, the second law is used to draw restrictions on the direction of the flow of heat and energy. In the Kelvin-Plank statement, a device cannot be constructed to operate in a cycle and produce no other effect besides mechanical work through the exchange of heat with a single reservoir. Alternatively, in the Clausius statement, it is impossible to construct a device operating in a cycle and producing no effect other than the transfer of heat from a cooler body to a hotter body (van Wylen and Sonntag, 1965). In continuum mechanics, a statement of the Second Law is made to place restrictions on continua. However, in the case of continuum mechanics the restrictions are placed on the material response functions called *constitutive responses*.

In this section, a thermodynamic parameter called *entropy* is introduced as a way to link mechanical and thermal responses. Using this parameter the second law of thermodynamics is stated in the form of the *Clausius-Duhem* equation. This equation is used in later sections to place functional restrictions on postulated constitutive responses for various materials.

At any given state for the continuum there are various quantities that effect the internal energy. These might be the volume of an ideal gas or the components of the deformation gradient of a solid. In the case of the deformation gradient, the nine components represent a deformation in the body that is storing energy. The collection of these parameters is called the *thermodynamic substate* and will be denoted by $v_1, v_2, \cdots, v_n$.[2]

While the thermodynamic substate influences the internal energy of the body it does not completely define it. Assume that the substate plus an additional independent scalar parameter, η, is sufficient to define the internal energy. This definition may be made in the form of

$$u = f(\eta, v_1, v_2, \cdots, v_n) \tag{5.66}$$

which is often referred to as the *caloric equation of state*. Parameter η is called the *specific entropy*. Since the internal energy is unambiguously defined once entropy is adjoined to the substate, the combination η plus $v_1, v_2, \cdots, v_n$ constitutes the *thermodynamic state*.

Temperature is the result of the change in internal energy with respect to entropy

$$\theta = \frac{\partial u}{\partial \eta} . \tag{5.67}$$

Furthermore, partial differentiation of the internal energy with respect to the thermodynamic substate variables results in thermodynamic tensions

$$\tau_a = \frac{\partial u}{\partial v_a} . \tag{5.68}$$

The preceding equations can be used to write a differential form of the internal energy as follows:

$$du = \theta d\eta + \sum_a \tau_a dv_a . \tag{5.69}$$

From Eqs 5.66 and 5.67 we see that both temperature and thermodynamic tensions are functions of entropy and the substate parameters.

Assuming that all the functions defined in this section are continuously differentiable as many times as necessary, it is possible to solve for entropy in terms of temperature

$$\eta = \eta\,(\theta, v_a) . \tag{5.70}$$

This result may be substituted into the caloric equation of state to yield internal energy as a function of temperature and substate parameters

$$u = u\,(\theta, v_a) . \tag{5.71}$$

Using this result in Eq 5.68 allows the definition of the thermal equations of state

$$\tau_a = \tau_a\,(\theta, v_a) \tag{5.72}$$

which inverts to give the substate parameters

$$v_a = v_a\,(\theta, \tau_a) . \tag{5.73}$$

[2]Notation warning: Do not confuse these thermodynamic substate parameters with particle velocity

The principles of thermodynamics are often posed in terms of thermodynamic potentials which may be defined as follows:

internal energy	$u\,,$	(5.74a)
free energy	$\psi = u - \eta\theta\,,$	(5.74b)
enthalpy	$\chi = u - \sum_a \tau_a v_a\,,$	(5.74c)
free enthalpy	$\zeta = \chi - \eta\theta = u - \eta\theta - \sum_a \tau_a v_a\,.$	(5.74d)

These potentials are related through the relationship

$$u - \psi + \zeta - \chi = 0\,. \tag{5.75}$$

All of the energy potentials may be written in terms of any one of the following independent variable sets

$$\eta, v_a; \quad \theta, v_a; \quad \eta, \tau_a; \quad \theta, \tau_a\,. \tag{5.76}$$

In order to describe the motion of a purely mechanical continuum the function $x_i = x_i(X_A, t)$ is needed. Adding the thermodynamic response requires the addition of temperature, θ, or, equivalently, entropy, η, both being a function of position and time

$$\theta = \theta(X_A, t) \qquad \text{or} \qquad \eta = \eta(X_A, t) \tag{5.77}$$

When considered for a portion $\mathcal{P}$ of the body, the total entropy is given as

$$H = \int_{\mathcal{P}} \rho\eta \, d\mathcal{V}\,, \tag{5.78}$$

and the entropy production in the portion $\mathcal{P}$ is given by

$$\Gamma = \int_{\mathcal{P}} \rho\gamma \, d\mathcal{V} \tag{5.79}$$

where the scalar γ is the specific *entropy production.* The second law can be stated as follows: the time rate-of-change in the entropy equals the change in entropy due to heat supply, heat flux entering the portion, plus the internal entropy production. For a portion $\mathcal{P}$ of the body, this is written as

$$\frac{d}{dt}\int_{\mathcal{P}} \rho\eta \, dV = \int_{\mathcal{P}} \frac{\rho r}{\theta} \, d\mathcal{V} - \int_{\partial\mathcal{P}} \frac{q_i n_i}{\theta} \, d\mathcal{S} + \int_{\mathcal{P}} \rho\gamma \, d\mathcal{V}\,. \tag{5.80}$$

The entropy production is always positive which leads to a statement of the second law in the form of the *Clausius-Duhem inequality*

$$\frac{d}{dt}\int_{\mathcal{P}} \rho\eta \, d\mathcal{V} \geqslant \int_{\mathcal{P}} \frac{\rho r}{\theta} \, d\mathcal{V} - \int_{\partial\mathcal{P}} \frac{q_i n_i}{\theta} \, d\mathcal{S}\,. \tag{5.81}$$

This global form can easily be posed locally by the now-familiar procedures. Applying the divergence theorem to the heat flux term yields

$$\int_{\partial\mathcal{P}} \frac{q_i n_i}{\theta} \, d\mathcal{S} = \int_{\mathcal{P}} \left(\frac{q_i}{\theta}\right)_{,i} d\mathcal{V}\,.$$

Furthermore, the differentiation of the entropy term is simplified by the fact that it is a specific quantity (see Section 5.2, Eq 5.18). Thus, we write

$$\int_{\mathcal{P}} \left[\rho\dot{\eta} - \rho\frac{r}{\theta} + \left(\frac{q_i}{\theta}\right)_{,i} \right] dV \geqslant 0 \,, \tag{5.82}$$

and since this must hold for all arbitrary portions of the body, and the integrand is continuous, then

$$\rho\dot{\eta} - \rho\frac{r}{\theta} + \left(\frac{q_i}{\theta}\right)_{,i} = \rho\dot{\eta} - \rho\frac{r}{\theta} + \frac{q_{i,i}}{\theta} - \frac{1}{\theta^2} q_i \theta_{,i} \geqslant 0 \,.$$

Thus, the local form of the Clausius-Duhem equation is

$$\rho\theta\dot{\eta} - \rho r + q_{i,i} - \frac{1}{\theta} q_i \theta_{,i} \geqslant 0 \,. \tag{5.83a}$$

Often, the gradient of the temperature is written as $g_i = \theta_{,i}$ in which case Eq 5.83a becomes

$$\rho\theta\dot{\eta} - \rho r + q_{i,i} - \frac{1}{\theta} q_i g_i \geqslant 0 \,. \tag{5.83b}$$

Combining this result with Eq 5.65 brings the stress power and internal energy into the expression giving a reduced form of the Clausius-Duhem

$$\rho\theta\dot{\eta} - \rho\dot{u} + d_{ij} t_{ij} - \frac{1}{\theta} q_i g_i \geqslant 0 \,. \tag{5.84}$$

One final form of the Clausius-Duhem equation is obtained by using Eq 5.74b to obtain the local dissipation inequality

$$-\rho\left(\dot{\psi} + \eta\dot{\theta}\right) + d_{ij} t_{ij} - \frac{1}{\theta} q_i g_i \geqslant 0 \,. \tag{5.85}$$

Finally, the fundamental laws are summarized in a Table 5.1. This table shows the global and local forms of conservation of mass, linear momentum, angular momentum, energy and the Clausius-Duhem equation. Details on these fundamental equations discussed in preceding sections provide the foundation from which engineering problems are solved. All that is needed in addition to these fundamental equations is a constitutive response.

5.8 The General Balance Law

The above discussion leads to a general expression for the balance law.[3] The balance law is an integral expression that applies to the body as a whole or some part of the body. The expression for the balance law is

$$\frac{d}{dt}\int_{\mathcal{P}} \Psi(\mathbf{x}, t)\, dV = \int_{\partial\mathcal{P}} \Gamma(\mathbf{x}, t)\, dS + \int_{\mathcal{P}} \Sigma(\mathbf{x}, t)\, dV \,. \tag{5.86}$$

[3]In the literature, some authors use master balance law rather than general balance law.

TABLE 5.1
Fundamental equations in global and local forms.

	Global Form	Local Form
Conservation of Mass	$\dot{m} = \int_{\mathcal{V}} \rho(\mathbf{x}, t) d\mathcal{V} = 0$	$\dot{\rho} + v_{i,i}\rho = 0$ $\frac{\partial \rho}{\partial t} + (\rho v_i)_{,i} = 0$ $\rho J = \rho_0$
Linear Momentum	$\frac{d}{dt}\int_{\mathcal{V}} \rho v_i d\mathcal{V} = \int_{\mathcal{S}} t_i^{(\hat{n})} d\mathcal{S} + \int_{\mathcal{V}} \rho b_i d\mathcal{V}$	$\rho \dot{v}_i = t_{ji,j} + \rho b_i$
Angular Momentum	$\frac{d}{dt}\int_{\mathcal{V}} \varepsilon_{ijk} x_j (\rho v_k) d\mathcal{V} = \int_{\mathcal{S}} \varepsilon_{ijk} x_j t_k^{(\hat{n})} d\mathcal{S} + \int_{\mathcal{V}} \varepsilon_{ijk} x_j (\rho b_k) d\mathcal{V}$	$\varepsilon_{ijk} t_{ij} = 0$ or $t_{ij} = t_{ji}$
Conservation of Energy	$\frac{d}{dt}\int_{\mathcal{V}} \left(\frac{1}{2} v_i v_i + u\right) d\mathcal{V} =$ $\int_{\mathcal{V}} (\rho b_i v_i + r)\, d\mathcal{V} + \int_{\mathcal{S}} \left(t_i^{(\hat{n})} v_i - q_i n_i\right) d\mathcal{S}$	$\rho \dot{u} - t_{ij} d_{ij} - \rho r + q_{i,i} = 0$
Clausius-Duhem Equation	$\frac{d}{dt}\int_{\mathcal{V}} \rho \eta d\mathcal{V} \geqslant \int_{\mathcal{V}} \rho \frac{r}{\theta} d\mathcal{V} - \int_{\mathcal{S}} \frac{q_i}{\theta} n_i d\mathcal{S}$	$\rho \dot{\eta} - \rho \frac{r}{\theta} + \left(\frac{q_i}{\theta}\right)_{,i} \geqslant 0$ $\rho \theta \dot{\eta} - \rho \dot{u} + t_{ij} d_{ij} - \frac{1}{\theta} q_i \theta_{,i} \geqslant 0$ $\dot{\psi} + \eta \dot{\theta} - t_{ij} d_{ij} - \frac{1}{\theta} q_i \theta_{,i} \geqslant 0$

TABLE 5.2
Identification of quantities in the balance laws.

Balance Law	Property	Flux	Source
Mass	ρ	0	0
Momentum	$\rho\mathbf{v}$	$\mathbf{t}^{(\hat{\mathbf{n}})}(\mathbf{x}, t,)$	$\rho\mathbf{b}$
Angular Momentum	$\mathbf{x} \times \rho\mathbf{v}$	$\mathbf{x} \times \mathbf{t}^{(\hat{\mathbf{n}})}(\mathbf{x}, t)$	$\mathbf{x} \times \rho\mathbf{b}$
Energy	$\frac{1}{2}\rho\mathbf{v} \cdot \mathbf{v} + u$	$\mathbf{t}^{(\hat{\mathbf{n}})}(\mathbf{x}, t) \cdot \mathbf{v} - \mathbf{q} \cdot \mathbf{n}$	$\rho\mathbf{b} + r$
Entropy	$\rho\eta$	$-\dfrac{\mathbf{q} \cdot \mathbf{n}}{\theta}$	$\dfrac{\rho r}{\theta}$

Physically this expression states that the time rate of change of the field Ψ is equal to the flux of Γ through the boundary $\partial\mathcal{P}$ plus the rate of increase or generation, Σ, of Ψ within the volume $\mathcal{V}$. Σ is the source of Ψ within the volume. Similarly, by changing the equality in Eq 5.86, the balance law can be recast as an inequality with a similar interpretation.

$$\frac{d}{dt}\int_{\mathcal{P}} \Psi(\mathbf{x}, t)\, d\mathcal{V} \geqslant \int_{\partial\mathcal{P}} \Gamma(\mathbf{x}, t)\, d\mathcal{S} + \int_{\mathcal{P}} \Sigma(\mathbf{x}, t)\, d\mathcal{V}\,. \tag{5.87}$$

These expressions are in the current or deformed configuration.

If the flux depends on the normal to the boundary, $\hat{\mathbf{n}}$, the flux, $\Gamma(\mathbf{x}, t, \hat{\mathbf{n}})$, will be a tensor one order higher than the field quantity Ψ or the source term Σ. The surface integral in Eq 5.86 can be converted to a volume integral by using the divergence theorem Eq 2.93. This leads through the localization principle to a partial differential equation that applies pointwise to determine the field $\Psi(\mathbf{x}, t)$.

$$\frac{d\Psi}{dt} + \Psi\,\text{div}\,\mathbf{v} = \text{div}\,\Gamma + \Sigma\,. \tag{5.88}$$

Table 5.2 shows the relationship of the quantities in the various balance laws to the master balance law.

The discussion to this point focused on a continuum body with a fixed set of material points. This gives rise to the balance of mass that has no mass flux or source terms. One very important class of materials that is gaining increasing importance is biological materials. In biological systems, mass increases by the growth of additional material. This requires a recasting of the continuum balance equations to accommodate the addition of mass through the growth of new material.

Growth can occur in biological systems through either volumetric growth as in soft tissue or by surface growth as in bone. The focus in the following will be on materials that exhibit volumetric growth. The mass balance equation requires an added source term to model the volumetric growth

$$\frac{d}{dt}\int_{\mathcal{P}} \rho\, d\mathcal{V} = \int_{\mathcal{P}} \rho c\, d\mathcal{V} \tag{5.89}$$

where c is the mass growth per unit density. This results in a local balance of mass equation

$$\frac{d\rho}{dt} + \rho v_{k,k} = \rho c \qquad \text{or} \qquad \frac{d\rho}{dt} + \rho\,\text{div}\,\mathbf{v} = \rho c\,. \tag{5.90}$$

The balance of linear momentum requires an additional source term to accommodate the momentum that the added mass brings to the system

$$\frac{d}{dt}\int_{\mathcal{P}} \rho v_i \, d\mathcal{V} = \int_{\partial\mathcal{P}} t_i^{(\hat{n})} \, d\mathcal{S} + \int_{\mathcal{P}} \rho b_i \, d\mathcal{V} + \int_{\mathcal{P}} \rho c v_i \, d\mathcal{V} \,, \tag{5.91}$$

or

$$\frac{d}{dt}\int_{\mathcal{P}} \rho \mathbf{v} \, d\mathcal{V} = \int_{\partial\mathcal{P}} \mathbf{t}^{(\hat{n})} \, d\mathcal{S} + \int_{\mathcal{P}} \rho \mathbf{b} \, d\mathcal{V} + \int_{\mathcal{P}} \rho c \mathbf{v} \, d\mathcal{V} \,.$$

To determine the local form of the balance of linear momentum, it is necessary to determine the derivative of an integral quantity per unit mass

$$\frac{d}{dt}\int_{\mathcal{P}} \rho f \, d\mathcal{V} = \int_{\mathcal{P}} \left(\rho \dot{f} + \rho c f\right) d\mathcal{V} \,. \tag{5.92}$$

The last result follows from Eq 5.90 and is equivalent to Eq 5.18 when c is zero. This result together with $t_i^{(\hat{n})} = t_{ji} n_j$ yields the local balance of linear momentum

$$\rho \frac{dv_i}{dt} = t_{ji,j} + \rho b_i \,. \tag{5.93}$$

The local form for the balance of linear momentum is unchanged by the addition of the mass. Balance of angular momentum also requires an additional source term for the added system mass. That is

$$\frac{d}{dt}\int_{\mathcal{P}} \varepsilon_{ijk} x_j \rho v_k \, d\mathcal{V} = \int_{\partial\mathcal{P}} \varepsilon_{ijk} x_j t_k^{(\hat{n})} \, d\mathcal{S} + \int_{\mathcal{P}} \varepsilon_{ijk} x_j \rho \left(b_k + c v_k\right) d\mathcal{V} \,, \tag{5.94}$$

or

$$\frac{d}{dt}\int_{\mathcal{P}} (\mathbf{x} - \mathbf{0}) \times \rho \mathbf{v} \, d\mathcal{V} = \int_{\partial\mathcal{P}} (\mathbf{x} - \mathbf{0}) \times \mathbf{t}^{(\hat{n})} \, d\mathcal{S} + \int_{\mathcal{P}} (\mathbf{x} - \mathbf{0}) \times \rho \left(\mathbf{b} + c\mathbf{v}\right) d\mathcal{V} \,.$$

This results in a symmetric stress tensor upon using the balance of mass and linear momentum. This is the case since the volumetric growth is not coupled to other mechanisms like species diffusion or micromolecular processes.

The energy and entropy equations must be modified for the body as growing mass is added. The added mass brings with it an associated kinetic energy and internal energy. The assumption that the added material is mechanically equivalent to the existing material at the point gives

$$\begin{aligned} \frac{d}{dt}\int_{\mathcal{V}} \rho \left(\frac{1}{2} v_i v_i + u\right) d\mathcal{V} &= \int_{\mathcal{P}} \rho \left(b_i v_i + r\right) d\mathcal{V} + \int_{\partial\mathcal{P}} \left(t_i^{(\hat{n})} v_i - q_i n_i\right) d\mathcal{S} \\ &+ \int_{\mathcal{P}} \rho c \left(\frac{1}{2} v_i v_i + u\right) d\mathcal{V} + \int_{\mathcal{P}} \rho c \bar{h} \, d\mathcal{V} \,. \end{aligned} \tag{5.95}$$

The final term represents a rate of growth energy needed to account for any additional energy required to create material with the same internal and kinetic energy as the original material. The local form of the energy equation is

$$\rho \dot{u} - t_{ij} d_{ij} - \rho r + q_{i,i} - \rho c \bar{h} = 0 \quad \text{or} \quad \rho \dot{u} - \mathbf{T} : \mathbf{D} - \rho r + \operatorname{div} \mathbf{q} - \rho c \bar{h} = 0 \,, \tag{5.96}$$

where balance of mass, linear momentum and symmetry of the stress tensor were used in arriving at this result.

The entropy inequality is

$$\frac{d}{dt}\int_{\mathcal{P}} \rho\eta \, d\mathcal{V} \geqslant \int_{\mathcal{P}} \frac{\rho r}{\theta} \, d\mathcal{V} - \int_{\partial\mathcal{P}} \frac{q_i n_i}{\theta} \, d\mathcal{S} + \int_{\mathcal{P}} \rho c \left(\eta + \frac{\bar{\bar{h}}}{\theta}\right) d\mathcal{V} \, .$$

The added mass brings additional entropy and an additional entropy production, $\bar{\bar{h}}$, like that in the energy equation. Using the standard procedure, the local form of the entropy inequality is

$$\rho\theta\dot{\eta} - \rho r + q_{i,i} - \frac{1}{\theta} q_i \theta_{,i} - \rho c \frac{\bar{\bar{h}}}{\theta} \geqslant 0 \, . \tag{5.97}$$

The specific free energy, $\psi = u - \eta\theta$, allows the entropy production rate to be eliminated from 5.97 with Eq 5.96 giving the energy rate $\dot{u}$. This yields an entropy production inequality of

$$-\left(\rho\dot{\psi} - \eta\dot{\theta}\right) - t_{ij} d_{ij} - \frac{1}{\theta} q_i \theta_{,i} + \rho c h \geqslant 0 \, .$$

The value of the variable h is $h = \bar{h} - \bar{\bar{h}}$. The reduced entropy equation can be used to restrict constitutive equations for this special class of biologically growing material.

5.9 Restrictions on Elastic Materials by the Second Law of Thermodynamics

In general, the thermomechanical continuum body must be specified by response functions that involve mechanical and thermodynamic quantities. To completely specify the continuum, a thermodynamic process must be defined. For a continuum body $\mathcal{B}$ having material points X a thermodynamic process is described by eight functions of the material point and time. These functions would be as follows:

1. Spatial position $x_i = \chi_i(X, t)$,
2. Stress tensor $t_{ij} = t_{ij}(X, t)$,
3. Body force per unit mass $b_i = b_i(X, t)$,
4. Specific internal energy $u = u(X, t)$,
5. Heat flux vector $q_i = q_i(X, t)$,
6. Heat supply per unit mass $r = r(X, t)$,
7. Specific entropy $\eta = \eta(X, t)$,
8. Temperature (always positive) $\theta = \theta(\mathbf{X}, t)$.

A set of these eight functions which are compatible with the balance of linear momentum and the conservation of energy makes up a thermodynamic process. These two balance laws are given in their local form in Eqs 5.22 and 5.65 and are repeated below in a slightly different form:

$$\begin{aligned} t_{ji,j} - \rho\dot{v}_i &= -\rho b_i \, , \\ \rho\dot{u} - t_{ij} d_{ij} + q_{i,i} &= \rho r \, . \end{aligned} \tag{5.98}$$

In writing the balance laws this way, the external influences on the body, heat supply and body force, have been placed on the right-hand side of the equal signs. From this it is noted that it is sufficient to specify $x_i, t_{ij}, u, q_i, \eta$,and θ and the remaining two process functions r and b_i are determined from Eqs 5.98.

One of the uses for the Clausius-Duhem form of the second law is to infer restrictions on the constitutive responses. Taking Eq 5.85 as the form of the Clausius-Duhem equation we see that functions for stress, free energy, entropy, and heat flux must be specified. The starting point for a constitutive response for a particular material is the principle of equipresence (Coleman and Mizel, 1963):

> An independent variable present in one constitutive equation of a material should be so present in all, unless its presence is in direct contradiction with the assumed symmetry of the material, the principle of material objectivity, or the laws of thermodynamics.

For an elastic material, it is assumed that the response functions will depend on the deformation gradient, the temperature, and the temperature gradient. Thus, we assume

$$t_{ij} = \tilde{t}_{ij}\left(F_{iA}, \theta, g_i\right), \qquad \psi = \tilde{\psi}\left(F_{iA}, \theta, g_i\right),$$
$$\eta = \tilde{\eta}\left(F_{iA}, \theta, g_i\right), \qquad q = \tilde{q}\left(F_{iA}, \theta, g_i\right). \tag{5.99}$$

These response functions are written to distinguish between the functions and their value. A superposed tilde is used to designate the response function rather than the response value. If an independent variable of one of the response functions is shown to contradict material symmetry, material frame indifference, or the Clausius-Duhem inequality it is removed from that function's list.

In using Eq 5.85, the derivative of ψ must be formed in terms of its independent variables

$$\dot{\psi} = \frac{\partial \tilde{\psi}}{\partial F_{iA}} \dot{F}_{iA} + \frac{\partial \tilde{\psi}}{\partial \theta} \dot{\theta} + \frac{\partial \tilde{\psi}}{\partial g_i} \dot{g}_i \,. \tag{5.100}$$

This equation is simplified by using Eq 4.139 to replace the time derivative of the deformation gradient in terms of the velocity gradient and deformation gradient

$$\dot{\psi} = \frac{\partial \tilde{\psi}}{\partial F_{iA}} L_{ij} F_{jA} + \frac{\partial \tilde{\psi}}{\partial \theta} \dot{\theta} + \frac{\partial \tilde{\psi}}{\partial g_i} \dot{g}_i \,. \tag{5.101}$$

Substitution of Eq 5.100 into 5.85 and factoring common terms results in

$$-\rho \left(\frac{\partial \tilde{\psi}}{\partial \theta} + \tilde{\eta} \right) \dot{\theta} + \left(t_{ij} - \rho \frac{\partial \tilde{\psi}}{\partial F_{iA}} F_{jA} \right) L_{ij} - \rho \frac{\partial \tilde{\psi}}{\partial g_i} \dot{g}_i - \frac{1}{\theta} \tilde{q}_i g_i \geqslant 0 \,. \tag{5.102}$$

Note that in writing Eq 5.102 the stress power has been written as $t_{ij} L_{ij}$ rather than $t_{ij} d_{ij}$. This can be done because stress is symmetric and adding the skew–symmetric part of L_{ij} is essentially adding zero to the inequality. The velocity gradient is used because the partial derivative of the free energy with respect to the deformation gradient times the transposed deformation gradient is not, in general, symmetric.

The second law must hold for every thermodynamic process, which means a special case may be chosen which might result in further restrictions placed on the response functions. That this is the case may be demonstrated by constructing displacement and temperature fields as such a special case. Define the deformation and temperature field as follows:

$$x_i = \chi\left(X_A, t\right) = Y_A + A_{iA}\left(t\right)\left[X_A - Y_A\right] \ ,$$

$$\theta = \theta\left(X_A, t\right) = \alpha\left(t\right) + \left[A_{Ai}\left(t\right) a_i\left(t\right)\right]\left[X_A - Y_A\right] . \tag{5.103}$$

Here, X_A and Y_A are the positions in the reference configuration of material points **X** and **Y**, and function $A_{iA}(t)$ is an invertible tensor, $a_i(t)$ is a time dependent vector, and $\alpha(t)$ is a scalar function of time. At the spatial position Y_A, the following is readily computed

$$\theta\left(Y_A, t\right) = \alpha\left(t\right) , \tag{5.104}$$

$$F_{iA}\left(Y_A, t\right) = A_{iA}\left(t\right) . \tag{5.105}$$

Note that Eq 5.103 may be written in terms of the current configuration as

$$\theta\left(y_i, t\right) = \alpha\left(t\right) + a_i\left(t\right)\left[x_i - y_i\right] . \tag{5.106}$$

Thus, the gradient of the temperature at material point **Y** is written as

$$\theta_{,i} = g_{,i} = a_i\left(t\right) . \tag{5.107}$$

From Eqs 5.104, 5.105, and 5.106 it is clear that quantities θ, g_i, and F_{iA} can be independently chosen. Furthermore, the time derivatives of these quantities may also be arbitrarily chosen. Because of the assumed continuity on the response functions, it is possible to arbitrarily specify functions u, g_i, and F_{iA} and their time derivatives.

Returning to an elastic material and Eq 5.101, for a given material point in the continuum, consider the case where the velocity gradient, L_{ij}, is identically zero and the temperature is constant. This means $L_{ij} = 0$ and $\dot{\theta} = 0$. Furthermore, assume the temperature gradient to be some arbitrary constant $g_i = g_i^0$. Eq 5.102 becomes

$$-\rho\frac{\partial\tilde{\psi}\left(F_{iA}, \theta_0, g_i^0\right)}{\partial g_i}\dot{g}_i - \frac{1}{\theta}\tilde{q}_i\left(F_{iA}, \theta_0, g_i^0\right) g_i \geqslant 0 .$$

Again, take advantage of the fact that the second law must hold for all processes. Since $\dot{g}_i$ is arbitrary it may be chosen to violate the inequality. Thus, the temperature gradient time derivative coefficient must be zero

$$\frac{\partial\tilde{\psi}\left(F_{iA}, \theta_0, g_i^0\right)}{\partial g_i} = 0 . \tag{5.108}$$

Since g_i^0 was taken to be an arbitrary temperature gradient, Eq 5.108 implies that the free energy is not a function of the temperature gradient. That is,

$$\frac{\partial\tilde{\psi}\left(F_{iA}, \theta_0, g_i\right)}{\partial g_i} = 0$$

which immediately leads to $\psi = \tilde{\psi}\left(F_{iA}, \theta\right)$. This fact eliminates the third term of Eq 5.102.

Further information about the constitutive assumptions can be deduced by applying additional special cases to the now reduced Eq 5.102. For the next special process, consider an arbitrary material point at an arbitrary time in which $L_{ij} = 0$ and $\dot{g} = 0$, but the temperature gradient is an arbitrary constant $g_i = g_i^0$. For this case, the Clausius-Duhem inequality is written as

$$-\rho\left(\frac{\partial\tilde{\psi}\left(F_{kB}, \theta\right)}{\partial\theta} + \tilde{\eta}\left(F_{kB}, \theta, g_k^0\right)\right)\dot{\theta} - \frac{1}{\theta}\tilde{q}_i\left(F_{kB}, \theta, g_k^0\right) g_i^0 \geqslant 0 \tag{5.109}$$

which must hold for all temperature rates, $\dot{\theta}$. Thus, the entropy response function may be solved in terms of the free energy

$$\tilde{\eta}\left(F_{kB}, \theta, g_k^0\right) = -\frac{\partial\tilde{\psi}\left(F_{kB}, \theta\right)}{\partial\theta} ,$$

and since the free energy is only a function of the deformation gradient and temperature the entropy must be a function of only those two as well. That is,

$$\eta = \tilde{\eta}\left(F_{kB}, \theta\right) = -\frac{\partial \tilde{\psi}}{\partial \theta} . \tag{5.110}$$

One more application of the Clausius-Duhem inequality for a special process will lead to an expression for the Cauchy stress in terms of the free energy. For this process, select the temperature gradient to be an arbitrary constant and the time rate-of-change of the temperature gradient to be identically zero. Eq 5.102 becomes

$$\left(\tilde{t}_{ij}\left(F_{kB}, \theta, g_k^0\right) - \rho \frac{\partial \tilde{\psi}}{\partial F_{iA}} F_{jA}\right) L_{ij} - \frac{1}{\theta} \tilde{q}_i\left(F_{kB}, \theta, g_k^0\right) g_i^0 \geqslant 0 \tag{5.111}$$

which must hold for all velocity gradients L_{ij}. Picking a L_{ij} that would violate the inequality unless the coefficient of L_{ij} vanishes implies

$$t_{ij} = \tilde{t}_{ij}\left(F_{kB}, \theta\right) = \rho \frac{\partial \tilde{\psi}}{\partial F_{iA}} F_{jA} . \tag{5.112}$$

In arriving at Eq 5.111 it was noted that the free energy had already been shown to be independent of the temperature gradient by virtue of Eq 5.108. Thus, g_k was dropped from the independent variable list in Eq 5.111.

Finally, substituting the results of Eqs 5.112, 5.110, and 5.108 into 5.102 a last restriction for an elastic material is found to be

$$\tilde{q}_i g_i \leqslant 0 . \tag{5.113}$$

5.10 Invariance

The concept of *invariance* has been discussed in Section 2.5 with respect to tensors. A tensor quantity is one which remains invariant under admissible coordinate transformations. For example, all the different stress components represented by Mohr's circle refer to a single stress state. This invariance is crucial for consolidating different stress components into a yield criterion such as the maximum shearing stress (or Tresca criterion).

Invariance plays another important role in continuum mechanics. Requiring a continuum to be invariant with regards to reference frame, or have unchanged response when a superposed rigid body motion is applied to all material points, has significant results associated with them. The most important of these consequences might be restrictions placed on constitutive models as discussed in the next section.

There are two basic methods for examining invariance of constitutive response functions: *material frame indifference* and *superposed rigid body motion*. In the first, a continuum body's response to applied forces or prescribed motion must be the same as observed from two different reference frames. The body and the applied forces remain the same; only the observers reference frame changes. In superposing a rigid body motion to the body the observer maintains the same reference frame. Here, each material point has a superposed motion added to it. The forces applied to the body are rotated with the superposed motion.

Both of these methods produce the same restrictions on constitutive responses within the context of this book. In this text, the method of superposed rigid body motion will be presented as the means for enforcing invariance.

The superposition of a rigid motion along with a time shift can be applied to the basic definition of the motion

$$x_i = \chi_i(X_A, t) \ . \tag{5.114}$$

From this state a superposed rigid body motion is applied maintaining all relative distances between material points. Also, since the motion is a function of time, a time shift is imposed on the motion. After the application of the superposed motion, the position of the material point becomes x_i^+ at time $t^+ = t + a$ where a is a constant. A superscript "+" is used for quantities having the superposed motion. Some literature uses a superscript "$*$" to denote this, but since we use this to represent principal stress quantities the "+" is used. The motion in terms of the superposed motion is written as

$$x_i^+ = \chi_i^+(X_A, t) \ . \tag{5.115}$$

Assuming sufficient continuity, the motion can be written in terms of the current configuration since $X_A = \chi^{-1}(x_i, t)$. That is to say

$$x_i^+ = \chi_i^+(X_A, t) = \tilde{\chi}_i^+(x_i, t) \tag{5.116}$$

where $\tilde{\chi}_i^+$ is written because the substitution of $X_A = \chi^{-1}(x_i, t)$ results in a different function than χ_i^+.

To represent the relative distance between two particles, a second material point is selected. In an analogous manner, it is straight-forward to determine

$$y_i^+ = \chi_i^+(Y_A, t) = \tilde{\chi}_i^+(y_i, t) \ . \tag{5.117}$$

Relative distance between material particles X_A and Y_A is written as

$$\left(x_i^+ - y_i^+\right)\left(x_i^+ - y_i^+\right) = \left[\tilde{\chi}_i^+(x_i, t) - \tilde{\chi}_i^+(y_i, t)\right]\left[\tilde{\chi}_i^+(x_i, t) - \tilde{\chi}_i^+(y_i, t)\right] \tag{5.118}$$

which is the dot product of the vector between positions x_i and y_i. Since the material points X_A and Y_A were arbitrarily chosen, quantities x_i and y_i are independent. Subsequent differentiation of Eq 5.118 with respect to x_i then y_i results in

$$\frac{\partial \tilde{\chi}_i^+(x_j, t)}{\partial x_p}\frac{\partial \tilde{\chi}_i^+(y_j, t)}{\partial y_q} = \delta_{pq} \ . \tag{5.119}$$

Since this must hold for all pairs of material points, it is possible to set

$$\frac{\partial \tilde{\chi}_i^+(x_j, t)}{\partial x_p} = \frac{\partial \tilde{\chi}_i^+(y_j, t)}{\partial y_p} = Q_{ip}(t) \ . \tag{5.120}$$

Use of Eq 5.120 in Eq 5.119 shows that the matrix $Q_{ip}(t)$ is orthogonal. Furthermore, since the special case of the superposed motion as a null motion, that is $\tilde{\chi}_i^+(x_i, t) = x_i$, then matrix $Q_{ip}(t)$ must be proper orthogonal having $Q_{ip}(t)Q_{iq}(t) = \delta_{pq}$ and $\det(Q_{ip}) = +1$.

To come up with a particular form for the superposed motion, Eq 5.120 may be spatially integrated to obtain

$$\tilde{\chi}_i^+(x_i, t) = a(t) + Q_{im}(t)\, x_m \ , \tag{5.121a}$$

or

$$x_i^+ = a_i + Q_{im} x_m \ . \tag{5.121b}$$

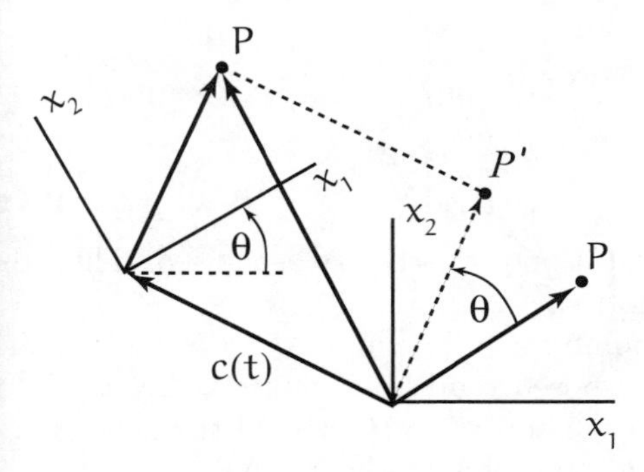

FIGURE 5.2
Reference frames $Ox_1x_2x_3$ and $O^+x_1^+x_2^+x_3^+$ differing by a superposed rigid body motion.

Vector a_i may be written in the alternative form

$$a_i(t) = c_i^+(t^+) - Q_{im}(t)\,c_m(t) \tag{5.122}$$

yielding

$$x_i^+ = c_i^+ + Q_{im}[x_m - c_m] \tag{5.123}$$

where

$$Q_{im}Q_{in} = Q_{mi}Q_{ni} = \delta_{mn} \quad \text{and} \quad \det(Q_{ij}) = 1\,. \tag{5.124}$$

A similar development of the superposed motion can be obtained by assuming two Cartesian reference frames $Ox_1x_2x_3$ and $O^+x_1^+x_2^+x_3^+$ which are separated by vector $c_i(t)$ and rotated by an admissible coordinate transformation defined by Q_{im} (Malvern, 1969). Rather than integrating differential Eq 5.120, the superposed motion can be written as

$$p_i^+ = c_i(t) + Q_{im}(t)\,p_m \tag{5.125}$$

where vectors p_i^+ and p_m are defined as shown in Fig. 5.2. Here, Q_{im} is simply the matrix of the direction cosines between $Ox_1x_2x_3$ and $O^+x_1^+x_2^+x_3^+$.

Example 5.2

Show that the superposed rigid body motion defined by Eq 5.123 is distance and angle preserving.

Solution
Consider the distance squared between material points X_A and Y_A in terms of the superposed motion

$$\left(x_i^+ - y_i^+\right)\left(x_i^+ - y_i^+\right) = Q_{im}(x_m - y_m)\,Q_{in}(x_n - y_n)$$

$$= Q_{im}Q_{in}(x_m - y_m)(x_n - y_n)\;.$$

Since Q_{im} is orthogonal

$$\left(x_i^+ - y_i^+\right)\left(x_i^+ - y_i^+\right) = \delta_{nm}\left(x_m - y_m\right)\left(x_n - y_n\right)$$

$$= \left(x_m - y_m\right)\left(x_m - y_m\right)$$

where the delta substitution property has been used. Thus, distance is preserved in the superposed motion.

Three material points, X_A, Y_A, and Z_A, are used to show that angles are preserved in the superposed rigid body motion. Let θ^+ be the angle included between vectors $x_i^+ - y_i^+$ and $x_i^+ - z_i^+$. Use of the definition of the dot product gives a convenient way to represent the angle θ^+

$$\cos\theta^+ = \frac{x_i^+ - y_i^+ \; x_i^+ - z_i^+}{|x_n^+ - y_n^+|\,|x_n^+ - z_n^+|}\ .$$

Note the "n" indices in the denominator do not participate in the summation since they are inside the vector magnitude operator. Direct substitution for Eq 5.121 followed by utilizing the orthogonality of Q_{im} and use of results from the first part of this example gives

$$\cos\theta^+ = \frac{Q_{im}\left(x_m - y_m\right) Q_{in}\left(x_n - z_n\right)}{|x_n - y_n||x_n - z_n|}$$

$$= \frac{Q_{im}Q_{in}\left(x_m - y_m\right)\left(x_n - z_n\right)}{|x_n - y_n||x_n - z_n|}$$

$$= \frac{\delta_{mn}\left(x_m - y_m\right)\left(x_n - z_n\right)}{|x_n - y_n||x_n - z_n|}$$

$$= \frac{\left(x_m - y_m\right)\left(x_m - z_m\right)}{|x_n - y_n||x_n - z_n|}$$

$$= \cos\theta\ .$$

Next, consider how superposed rigid body motion affects the continuum's velocity. Define the velocity in the superposed configuration as the time derivative with respect to t^+

$$v_i^+ = \dot{x}_i^+ = \frac{dx_i^+}{dt^+}\ . \tag{5.126}$$

By the result of Eq 5.121 along with the chain rule, the velocity is given by

$$v_i^+ = \frac{d}{dt^+}\left[a_i\left(t\right) + Q_{im}\left(t\right)x_m\right]$$

$$= \frac{d}{dt}\left[a_i\left(t\right) + Q_{im}\left(t\right)x_m\right]\frac{dt}{dt^+}\ .$$

Recall the definition $t^+ = t + a$ from which it is obvious that $dt/dt^+ = 1$. As a result of this, an expression for the velocity under a superposed rigid body motion is obtained by taking the time derivatives of the bracketed term of the preceding equation:

$$v_i^+ = \dot{a}_i(t) + \dot{Q}_{im}(t)\,x_m + Q_{im}\dot{x}_m$$

$$= \dot{a}_i(t) + \dot{Q}_{im}(t)\,x_m + Q_{im}v_m\,. \tag{5.127}$$

Define

$$\Omega_{ij}(t) = \dot{Q}_{im}(t)\,Q_{jm}(t)\,, \tag{5.128}$$

or, written an alternative way by post-multiplying by Q_{jk}

$$\Omega_{ij}(t)\,Q_{jk}(t) = \dot{Q}_{im}(t)\,Q_{jm}(t)\,Q_{jk}(t) = \dot{Q}_{ik}(t)\,. \tag{5.129}$$

Substituting Eq 5.129 into the second of Eq 5.127 yields an expression for the velocity field of a continuum undergoing a superposed rigid body rotation

$$v_i^+ = \dot{a}_i + \Omega_{ij}Q_{jk}x_k + Q_{im}v_m\,. \tag{5.130}$$

Note that Ω_{ij} is skew–symmetric by taking the time derivative of the orthogonality condition for Q_{ij}. That is,

$$\frac{d}{dt}[Q_{im}Q_{in}] = \frac{d}{dt}[1] = 0\,.$$

When the derivatives are taken

$$\dot{Q}_{im}Q_{in} + Q_{im}\dot{Q}_{in} = 0\,,$$

and the definition of $\dot{Q}_{im}$ used from Eq 5.129 substituted into this expression yields

$$Q_{im}[\Omega_{ji} + \Omega_{ij}]Q_{jm} = 0\,.$$

This is true only if the bracketed term is zero, thus

$$\Omega_{ij} = -\Omega_{ji}\,. \tag{5.131}$$

The fact that Ω_{ij} is skew–symmetric means that it has an axial vector defined by

$$\omega_k = -\frac{1}{2}\varepsilon_{ijk}\Omega_{ij} \tag{5.132}$$

which may be inverted to give

$$\Omega_{ij} = -\varepsilon_{ijk}\omega_k\,. \tag{5.133}$$

In the case of rigid body dynamics, the axial vector ω_k can be shown to be the angular velocity of the body (see Problem 5.37).

For later use in constitutive modeling, various kinematic quantities' properties under superposed rigid body motions will be needed. Here, a derivation of the superposed rigid body motion's effect on vorticity and rate-of-deformation tensors will be demonstrated.

Start with the velocity given in Eq 5.130 and substitute for x_i using Eq 5.121b to write

$$v_i^+ = \dot{a}_i + \Omega_{ij}[x_j^+ - a_j] + Q_{ij}v_j \tag{5.134}$$

$$= Q_{ij} + \dot{Q}_{im}x_m + \dot{a}_i - \Omega_{ij}a_j \tag{5.135}$$

$$= Q_{ij}v_j + \dot{Q}_{im}x_m + c_i \tag{5.136}$$

where Eq 5.128 has been used in going from Eq 5.134 to 5.135, and $c_i = \dot{a} - \Omega_{ij}a_j$ has been used in going from Eq 5.135 to 5.136. Writing the velocity in this form allows for a more convenient computation of the velocity gradient in the superposed reference frame

$$\frac{\partial v_i^+}{\partial x_m^+} = \Omega_{ij}\frac{\partial x_j^+}{\partial x_m^+} + Q_{ij}\frac{\partial v_j}{\partial x_m^+} \; . \tag{5.137}$$

Note that

$$\frac{\partial x_j^+}{\partial x_m^+} = \delta_{jm} \; ,$$

and

$$\frac{\partial v_j}{\partial x_m^+} = \frac{\partial v_j}{\partial x_n}\frac{\partial x_n}{\partial x_m^+} = \frac{\partial v_j}{\partial x_n}\frac{\partial}{\partial x_m^+}[Q_{kn}\left(x_k^+ - a_k\right)]$$

where the last substitution comes from solving for x_k^+ in Eq 5.121b. Use of these in Eq 5.137 followed by the delta substitution property yields

$$\frac{\partial v_i^+}{x_m^+} = \Omega_{im} + Q_{ij}Q_{mn}v_{j,n} = \Omega_{im} + Q_{ij}Q_{mn}[d_{jn} + w_{jn}] \tag{5.138}$$

where d_{jn} and w_{jn} are the rate-of-deformation and vorticity, respectively. In the superposed motion the velocity gradient may be decomposed into symmetric and skew–symmetric parts

$$\frac{\partial v_i^+}{\partial x_m^+} = d_{im}^+ + w_{im}^+$$

which are the rate-of-deformation and vorticity in the superposed rigid body frame. From Eqs 5.138 and 5.131 it is clear that

$$d_{ij}^+ = Q_{im}Q_{jn}d_{mn} \; , \tag{5.139}$$

and

$$w_{ij}^+ = \Omega_{ij} + Q_{im}Q_{jn}w_{mn} \; . \tag{5.140}$$

Recalling the general transformation equations introduced in Chapter 2 it would be expected that vectors and second-order tensors transform according to the generic formulae

$$u_i^+ = Q_{im}u_m, \qquad U_{ij}^+ = Q_{im}Q_{jn}U_{mn} \; .$$

But preceding results show us that this is not the case. Velocity and vorticity transform in more complex ways. All that is left to ready ourselves for the study of constitutive equation theory is to determine how stress transforms under superposed rigid body motion.

The transformation of the stress vector and the stress components is not as clear cut as the kinematic quantities demonstrated above. This is a result of starting with the stress vector which is a force. It is assumed that forces transform as a generic vector in the form of

$$t_i^+ = Q_{ij}t_j \; .$$

Since stress is a measure of force per unit area, a regression back to kinematic results is necessary to determine how a differential area transforms under superposed rigid body motion. A differential element of area in the current configuration may be written as

$$da_k = J\frac{\partial X_K}{\partial x_k}dA_K \; .$$

Under the superposed rigid body motion the differential area element is written as

$$da_k^+ = da^+ n_k^+ J^+ \frac{\partial X_K}{\partial x_k^+} dA_K \tag{5.141}$$

where reference quantities do not change, da^+ is the infinitesimal area, and n_k^+ is the unit normal vector to the area all in the superposed rigid body motion state. This expression will be further reduced in several steps.

First, the Jacobian can be shown to transform according to $J^+ = J$ (see Problem 5.33). Next, the quantity $\partial X_K / \partial x_k^+$ is the inverse of the deformation gradient as represented in the superposed rigid body rotated frame. Application of the chain rule yields

$$\frac{\partial X_K}{\partial x_k^+} = \frac{\partial X_K}{\partial x_j} \frac{\partial x_j}{\partial x_k^+} \ . \tag{5.142}$$

The last term of this equation is evaluated by referring to Eq 5.121b and solving for x_k by pre-multiplying by Q_{ik}. Differentiation of the result shows that the last term of Eq 5.142 is Q_{ik}. Thus

$$\frac{\partial X_K}{\partial x_k^+} = \frac{\partial X_K}{\partial x_j} Q_{kj} \ ,$$

or

$$(F_{kK})^{-1} = F_{jK}^{-1} Q_{kj} \ . \tag{5.143}$$

Substitution of the results from the preceding paragraph into Eq 5.141 results in

$$da_k^+ = da^+ n_k^+ = Q_{kj} da_j = Q_{kj} (da) n_j \ . \tag{5.144}$$

Squaring the second and fourth terms of the above equation and equating them leads to $(da^+)^2 = (da)^2$, and since area is always a positive number

$$da^+ = da \ . \tag{5.145}$$

With the use of Eqs 5.144 and 5.145, it is evident that

$$n_k^+ = Q_{kj} n_j \ . \tag{5.146}$$

All that remains is to determine how the stress components transform under superposed rigid body motion. All the results are now in place to find this transformation. In the superposed rigid body motion frame, the stress vector t_i^+ can be written as

$$t_i^+ = t_{ij}^+ n_j^+ = t_{ij}^+ Q_{jk} n_k \tag{5.147}$$

where Eq 5.146 has been used. The assumed transformation for the stress vector yields

$$t_i^+ = Q_{ij} t_j = Q_{ij} t_{jk} n_k \tag{5.148}$$

where, as in Eq 5.147, Cauchy's stress formula has been used. Equating the stress vector in the superposed rigid body reference frame leads to

$$\left(t_{ij}^+ Q_{jk} - Q_{ij} t_{jk} \right) n_k = 0$$

which holds for all n_k. Thus, the terms in parenthesis must equal zero. Multiplying the remaining terms by Q_{mk} results in the following expression for stress component transformation under superposed rigid body motion:

$$t_{ij}^+ = Q_{im} Q_{jn} t_{mn} \ . \tag{5.149}$$

In plasticity, as well as explicit finite element formulations, the stress constitutive response is usually formulated in an incremental form. This means that the stress rate is used. The stress rate must be objective meaning

$$\dot{t}^{+}_{ij} = \dot{Q}_{im}Q_{jn}t_{mn} + Q_{im}\dot{Q}_{jn}t_{mn} + Q_{im}Q_{jn}\dot{t}_{mn} \,. \tag{5.150}$$

It is clear from this equation that the stress rate is not objective even though the stress is objective. This result is serious since the stress rate as shown in Eq 5.150 could not be used as a response function or in the independent variable list of a response function. Luckily, there are several ways to express a form of the stress rate in an invariant manner.

One way to obtain an objective stress rate is found from using the spin tensor w_{ij}. Using Eqs 5.140 and 5.128 to solve for $\dot{Q}_{ij}$ in terms of the spin

$$\dot{Q}_{ip} = w^{+}_{ij}Q_{jp} - Q_{im}w_{mp} \tag{5.151}$$

which is then substituted into Eq 5.150 to give

$$\begin{aligned}\dot{t}^{+}_{ij} &= w^{+}_{iq}Q_{qm}t_{mn}Q_{jn} + Q_{im}t_{mn}Q_{qn}w^{+}_{jq} \\ &\quad + Q_{im}[\dot{t}_{mn} - w_{nq}t_{mq} - w_{mq}t_{qn}]Q_{jn} \,.\end{aligned} \tag{5.152}$$

Placing all quantities referred to the superposed rigid body motion to the left-hand side of the equal sign and using Eq 5.149 results in

$$\dot{t}^{+}_{ij} - w^{+}_{iq}t^{+}_{qj} - t^{+}_{iq}w_{jq} = Q_{im}[\dot{t}_{mn} - w_{mq}t_{qn} - t_{mq}w_{nq}]Q_{jn} \,. \tag{5.153}$$

It is clear that the quantity $\dot{t}_{mn} - w_{nq}t_{mq} - w_{mq}t_{qn}$ is objective which leads to the definition of the so called *Jaumann stress rate*

$$\overset{\nabla}{t}_{ij} = \dot{t}_{ij} - W_{iq}t_{qj} - t_{iq}W_{jq} \,. \tag{5.154}$$

There are several other stress rate definitions satisfying objectivity. For example, the *Green-Naghdi stress rate* is given as

$$\overset{\Delta}{t}_{ij} = \dot{t}_{ij} + \Omega_{iq}t_{qj} + t_{iq}\Omega_{jq} \tag{5.155}$$

where Ω_{iq} is defined in Eq 5.128.

5.11 Restrictions on Constitutive Equations from Invariance

It was seen in Section 5.9 that the second law of thermodynamics places restrictions on the form of constitutive response functions. Material frame indifference, or superposed rigid body motion, may also place restrictions on the independent variables of the response function as was stated in the principle of equipresence. In Section 5.10, the behavior of many of the quantities used in continuum mechanics undergoing a superposed rigid body motion was presented. This section will examine the response functions of bodies undergoing a superposed rigid body motion. In particular, what restrictions are placed on the constitutive independent variables given in Eq 5.99.

Under a superposed rigid body motion, scalars are unaffected allowing the following to be written

$$u^+ = u, \qquad \eta^+ = \eta, \qquad \theta^+ = \theta \; . \tag{5.156}$$

This being the case, it is clear from Eq 5.74a through 5.74d that ψ, χ, and ζ would be unaffected by the superposed rigid body motion. The remaining quantities of the response functions of Eq 5.99 and their independent variables are affected in different ways from the superposed rigid body motion. Under a superposed rigid body motion, these functions transform as follows (some of these are repeated from Section 5.10 or repeated in slightly different form):

$$F^+_{iA} = Q_{ij} F_{jA} \; , \tag{5.157}$$

$$\dot{F}^+_{iA} = Q_{ij} \dot{F}_{jA} + \Omega_{ij} Q_{jk} F_{kA} \; , \tag{5.158}$$

$$L^+_{im} = Q_{ij} L_{jn} Q_{nm} + \Omega_{im} \; , \tag{5.159}$$

$$t^+_{ij} = Q_{ik} t_{kl} Q_{lj} \; , \tag{5.160}$$

$$q^+_i = Q_{ij} q_j \; , \tag{5.161}$$

$$g^+_i = Q_{ij} g_j \; , \tag{5.162}$$

where Ω_{ij} is a skew–symmetric tensor defined by Eq 5.128.

Constitutive equations are objective if and only if they transform under a superposed rigid body motion as follows:

$$u^+ \left(\eta^+, F^+_{iA}\right) = u\left(\eta, F_{iA}\right) \; , \tag{5.163a}$$

$$\theta^+ \left(\eta^+, F^+_{iA}\right) = \theta\left(\eta, F_{iA}\right) \; , \tag{5.163b}$$

$$t^+_{ij} \left(\eta^+, F^+_{mA}\right) = Q_{ik} t_{kl}\left(\eta, F_{mA}\right) Q_{lk} \; , \tag{5.163c}$$

$$q^+_i \left(\eta^+, F^+_{iA}, g^+_i, L^+_{kl}\right) = Q_{im} q_m\left(\eta, F_{iA}, g_i, L_{kl}\right) \; . \tag{5.163d}$$

In writing these equations it is noted that the roles of η and θ can be interchanged because of assumed continuity. Also, restricting the independent variable list of Eqs 5.163a through 5.163c to the entropy (or temperature) and the deformation gradient has been shown to be a general case (Coleman and Mizel, 1964). Finally, the set of constitutive functions may be in terms of different response functions. That is, the elastic material considered in Section 5.9 had response functions ψ, t_{ij}, η, and θ_i postulated. However, it could have just as easily been postulated as a function of u, θ, t_{ij}, and q_i.

As an example of how Eqs 5.163 could restrict the independent variables consider a fluid whose stress response function is assumed to be a function of density, ρ, velocity, v_i, and velocity gradient, L_{ij}. With these assumptions, the restrictions of Eq 5.163c would be

$$t^+_{ij} \left(\rho^+, v^+_k, d^+_{mn}, w^+_{mn}\right) = Q_{ip} t_{pl}\left(\rho, v_k, d_{mn}, w_{mn}\right) Q_{lj} \tag{5.164}$$

where the velocity gradient has been decomposed into its symmetric and skew–symmetric parts, d_{mn} and w_{mn}, respectively. Using the results of Problem 5.34, Eqs 5.136, 5.139, and

5.140 results in

$$\tilde{t}_{ij}\left(\rho, v_k + c_k, Q_{mp} d_{pq} Q_{qn}, Q_{mp} w_{pq} Q_{qn} + \Omega_{mn}\right) = Q_{ip} \tilde{t}_{pl}\left(\rho, v_k, d_{mn}, w_{mn}\right) Q_{lj} \,. \tag{5.165}$$

Since Eq 5.165 must hold for all motions, a specific rigid body rotation may be chosen to reduce the constitutive assumption of Eq 5.164. For this purpose, suppose that $Q_{ij} = \delta_{ij}$ and thus $\dot{Q}_{ij} = 0$. Using this motion and Eq 5.165 implies

$$\tilde{t}_{ij}\left(\rho, v_k + c_k, d_{mn}, w_{mn}\right) = \tilde{t}_{ij}\left(\rho, v_k, d_{mn}, w_{mn}\right) \tag{5.166}$$

where it is noted that Eq 5.128 has been used. In this case, vector c_k is simply equal to $\dot{a}_k$ the time derivative of the superposed rigid body motion integration factor (see Eq 5.121). This arbitrary nature would allow for Eq 5.166 to be violated if the stress function has velocity as an independent variable. Thus, velocity must be removed from the independent variable list for stress leaving

$$t_{ij} = \tilde{t}_{ij}\left(\rho, d_{mn}, w_{mn}\right) \,. \tag{5.167}$$

Again, with the modified response function Eq 5.167, the invariance condition under superposed rigid body motion may be written as

$$\tilde{t}_{ij}\left(\rho, Q_{mp} d_{pq} Q_{qn}, Q_{mp} w_{pq} Q_{qn} + \Omega_{mn}\right) = Q_{ip} \tilde{t}_{pl}\left(\rho, d_{mn}, w_{mn}\right) Q_{lj} \tag{5.168}$$

which must hold for all motions. Select a motion such that $Q_{ij} = \delta_{ij}$ as before, but now require $\Omega_{ij} \neq 0$. Substitution of this into Eq 5.168 leaves

$$\tilde{t}_{ij}\left(\rho, d_{mn}, w_{mn} + \Omega_{mn}\right) = \tilde{t}_{ij}\left(\rho, d_{mn}, w_{mn}\right) \tag{5.169}$$

as the invariance requirement on the stress. For this to be true for all motions, the skew–symmetric part of the velocity gradient must not be an independent variable.

Applying the superposed rigid body motion to the twice reduced response function will yield further information. In this case,

$$\tilde{t}_{ij}\left(\rho, Q_{mp} d_{pq} Q_{qn}\right) = Q_{ip} \tilde{t}_{pl}\left(\rho, d_{mn}\right) Q_{lj} \,. \tag{5.170}$$

Since Q_{ij} is a proper orthogonal tensor, and this equation would hold if Q_{ij} were replaced by its negative, so the stress must be an isotropic function of d_{mn}. The most general form of a second-order, isotropic tensor function of d_{mn} may now represent the stress response

$$t_{ij} = -p\left(\rho\right) \delta_{ij} + \lambda\left(\rho\right) d_{kk} \delta_{ij} + 2\mu\left(\rho\right) d_{ij} \tag{5.171}$$

where p, λ, and μ are functions of density and would represent the viscosity coefficients.

5.12 Constitutive Equations

The global balance laws and resulting field equations developed earlier in this chapter are applicable to all continuous media, but say nothing about the response of specific materials to force or temperature loadings. To fill this need, we introduce *constitutive equations* which specify the mechanical and thermal properties of particular materials based upon their internal constitution. Mathematically, the usefulness of these constitutive equations is to describe the relationships among the kinematic, mechanical, and thermal field equations and to permit the formulations of well-posed problems in continuum mechanics. Physically, the constitutive equations define various idealized materials which serve as models for the behavior of real materials. However, it is not possible to write down one equation which is capable of representing a given material over its entire range of application, since many materials behave quite differently under changing levels of loading, such as, elastic-plastic response due to increasing stress. And so in this sense it is perhaps better to think of constitutive equations as being representative of a particular *behavior* rather than of a particular *material.*

In previous sections of this Chapter, the foundations of constitutive assumptions have been addressed. On that fundamental level, it is not possible to discuss in detail the fundamental derivation of all the constitutive models that an engineer would want to be familiar with. Instead, the theoretical background and definitions were given along with a few specific cases examined. This allows the inductive student to grasp the concepts presented.

Since only a few constitutive models were considered from a fundamental basis, this section is devoted to a brief survey of constitutive equations. This acts as an introduction for subsequent chapters when various constitutive models are discussed as applications of continuum mechanics.

A brief listing of some well-known constitutive equations is as follows:

(a) the stress-strain equations for a *linear elastic solid* assuming infinitesimal strains,

$$t_{ij} = C_{ijkm}\epsilon_{km} \tag{5.172}$$

where the C_{ijkm} are the elastic constants representing the properties of the body. For *isotropic* behavior, Eq 5.172 takes the special form

$$t_{ij} = \lambda\delta_{ij}\epsilon_{kk} + 2\mu\epsilon_{ij} \tag{5.173}$$

in which λ and μ are coefficients that express the elastic properties of the material.

(b) the *linear viscous fluid,*

$$\tau_{ij} = K_{ijmn}d_{mn} \tag{5.174}$$

where τ_{ij} is the shearing stress in the fluid and the constants K_{ijmn} represents its viscous properties. For a *Newtonian fluid,*

$$\tau_{ij} = \lambda^*\delta_{ij}d_{kk} + 2\mu^* d_{ij} \tag{5.175}$$

where λ^* and μ^* are viscosity coefficients.

(c) *plastic* stress-strain equation,

$$d\epsilon_{ij}^{P} = S_{ij}d\lambda \tag{5.176}$$

where $d\epsilon_{ij}^{P}$ is the *plastic strain increment,* S_{ij} the deviator stress, and $d\lambda$ a proportionality constant.

(d) *linear viscoelastic differential-operator* equations

$$\{P\}S_{ij} = 2\{Q\}\eta_{ij} \tag{5.177a}$$

$$t_{ii} = 3K\epsilon_{ii} \tag{5.177b}$$

where $\{P\}$ and $\{Q\}$ are differential time operators of the form

$$\{P\} = \sum_{i=0}^{N} p_i \frac{\partial^i}{\partial t^i} \tag{5.178a}$$

$$\{Q\} = \sum_{i=0}^{M} q_i \frac{\partial^i}{\partial t^i} \tag{5.178b}$$

and in which the coefficients p_i and q_i (not necessarily constants) represent the viscoelastic properties. Also, K is the *bulk modulus*. Note further that this pair of equations specifies separately the deviatoric and volumetric responses.

(e) *linear viscoelastic integral* equations

$$\eta_{ij} = \int_0^t \psi_s(t - t')\frac{\partial S_{ij}}{\partial t'}dt' \tag{5.179a}$$

$$\epsilon_{ii} = \int_0^t \psi_v(t - t')\frac{\partial t_{ij}}{\partial t'}dt' \tag{5.179b}$$

where the properties are represented by ψ_s and ψ_v, the *shear* and *volumetric* creep functions, respectively.

In formulating a well-posed problem in continuum mechanics, we need the field equations together with whatever equations of state are necessary, plus the appropriate constitutive equations and boundary conditions. As a point of reference we list again, as a group, the important field equations in both indicial and symbolic notation:

(a) the continuity equation (Eq 5.14)

$$\frac{\partial \rho}{\partial t} + (\rho v_k)_{,k} = 0 \quad \text{or} \quad \frac{\partial \rho}{\partial t} + \boldsymbol{\nabla}\cdot(\rho \mathbf{v}) = 0 \tag{5.180}$$

(b) the equations of motion (Eq 5.22)

$$t_{ji,j} + \rho b_i = \rho \dot{v}_i \quad \text{or} \quad \boldsymbol{\nabla}\cdot \mathbf{T} + \rho \mathbf{b} = \rho \dot{\mathbf{v}} \tag{5.181}$$

(c) the energy equation (Eq 5.65)

$$\rho\dot{u} - t_{ij}d_{ij} - \rho r + q_{i,i} = 0 \quad \text{or} \quad \rho\dot{u} - \mathbf{T}:\mathbf{D} - \rho r + \boldsymbol{\nabla}\cdot \mathbf{q} = 0 \tag{5.182}$$

If we assume the body forces b_i and distributed heat sources r are prescribed, the above collection consists of *five* independent equations involving *fourteen* unknowns, namely, ρ, v_i, t_{ij}, q_i and u. In addition, in a non-isothermal situation, the entropy η and temperature field $\theta = \theta(\mathbf{x}, t)$ have to be taken into consideration. For the isothermal theory, eleven equations are needed in conjunction with the five field equations listed above. Of these, six are constitutive equations, three are temperature-heat conduction equations (Fourier's law), and two are equations of state.

If the mechanical and thermal fields are uncoupled and isothermal conditions prevail, the continuity equation along with the equations of motion and the six constitutive equations provide a determinate set of ten kinematic-mechanical equations for the ten unknowns ρ, v_i and t_{ij}. It is with such rather simple continuum problems that we shall concern ourselves in subsequent chapters.

References

[1] Carlson, D. E. (1984), "Linear thermoelasticity," in S. Flugge's *Handbuch der Physik*, Vol. VIa/2 (edited by C. Truesdell), Springer-Verlag, pp. 297–345

[2] Coleman, B. D. and Noll, W. (1963), "The thermodynamics of elastic materials with heat conduction and viscosity," *Arch. Rational Mech. Anal.*, Vol. 13, pp. 167–178 (4)

[3] Coleman, B. D. and Mizel, V. J. (1964), "Existence of caloric equations of state in thermodynamics," *J. Chem. Phys.*, Vol. 40, pp. 1116–1125 (4)

[4] Dienes, J. K. (1979), "On the Analysis of Rotation and Stress Rate in Deforming Bodies," *Acta Mech.*, Vol. 32, pp. 217-232

[5] Flanagan, D. P. and Taylor, L. M., (1987), "An Accurate Numerical Algorithm for Stress Integration with Finite Rotations," *Computer Methods in Applied Mechanics and Engineering*, Vol. 62, pp. 305-320

[6] Green, A. E. and Naghdi, P. M., (1979), "A note on invariance under superposed rigid body motions," *J. of Elasticity*, Vol. 9, pp. 1–8

[7] Johnson, G. C. and Bammann, D. C., (1984), "A discussion of stress rates in finite deformation problems," *Int J Solids and Structures*, Vol. 20, pp 725-737

[8] Malvern, L. E. (1969), *Introduction to the Mechanics of a Continuous Medium*, Prentice-Hall, Inc., Englewood Cliffs, NJ

[9] Naghdi, P. M. (1984), "The theory of shells and plates," in S. Flugge's *Handbuch der Physik*, Vol. II (edited by C. Truesdell), Springer-Verlag, pp. 425–640

[10] Van Wylen, G. J. and Sonntag, R. E. (1965), *Fundamentals of Classic Thermodymanics*, Wiley, Inc., New York

Problems

Problem 5.1

Determine the material derivative of the flux of any vector property Q_i^* through the spatial area $\mathcal{S}$. Specifically, show that

$$\frac{d}{dt}\int_{\mathcal{S}} Q_i^* n_i \, d\mathcal{S} = \int_{\mathcal{S}} \left(\dot{Q}_i^* + Q_i^* v_{k,k} - Q_k^* v_{i,k}\right) n_i \, d\mathcal{S}$$

in agreement with Eq 5.5.

Problem 5.2

Let the property $P_{ij\ldots}^*$ in Eq 5.1 be the scalar 1 so that the integral in that equation represents the instantaneous volume $\mathcal{V}$. Show that in this case

$$\dot{P}_{ij\ldots} = \frac{d}{dt}\int_{\mathcal{V}} d\mathcal{V} = \int_{\mathcal{V}} v_{i,i} \, d\mathcal{V} \, .$$

Problem 5.3

Verify the identity

$$\varepsilon_{ijk} a_{k,j} = 2\left(\dot{w}_i + w_i w_{j,j} - w_j v_{i,j}\right) \, ,$$

and by using this identity as well as the result of Problem 5.1, prove that the material derivative of the vorticity flux equals one half the flux of the curl of the acceleration; that is, show that

$$\frac{d}{dt}\int_{\mathcal{S}} w_i n_i \, d\mathcal{S} = \frac{1}{2}\int_{\mathcal{S}} \varepsilon_{ijk} a_{k,j} n_i \, d\mathcal{S} \, .$$

Problem 5.4

Making use of the divergence theorem of Gauss together with the identity

$$\frac{\partial w_i}{\partial t} = \frac{1}{2}\varepsilon_{ijk} a_{k,j} - \varepsilon_{ijk}\varepsilon_{kmq}\left(w_m v_q\right)_{,j}$$

show that

$$\frac{\partial}{\partial t}\int_{\mathcal{V}} w_i \, d\mathcal{V} = \int_{\mathcal{S}} \left(\frac{1}{2}\varepsilon_{ijk} a_k + w_j v_i - w_i v_j\right) n_j \, d\mathcal{S} \, .$$

Problem 5.5

Show that the material derivative of the vorticity of the material contained in a volume $\mathcal{V}$ is given by

$$\frac{d}{dt}\int_{\mathcal{V}} w_i \, d\mathcal{V} = \int_{\mathcal{S}} \left(\frac{1}{2}\varepsilon_{ijk} a_k + w_j v_i\right) n_j \, d\mathcal{S} \, .$$

Problem 5.6

Given the velocity field

$$v_1 = a x_1 - b x_2, \quad v_2 = b x_1 + a x_2, \quad v_3 = c\sqrt{x_1^2 + x_2^2}$$

where a, b, and c are constants, determine

(a) whether or not the continuity equation is satisfied,
(b) whether the motion is isochoric.

Answer

(a) only when $\rho = \rho_0 e^{-2at}$ (b) only if $a = 0$

Problem 5.7
For a certain continuum at rest, the stress is given by

$$t_{ij} = -p_0 \delta_{ij}$$

where p_0 is a constant. Use the continuity equation to show that for this case the stress power may be expressed as

$$t_{ij} d_{ij} = \frac{p_0 \dot{\rho}}{\rho} .$$

Problem 5.8
Consider the motion $x_i = (1 + t/k) X_i$ where k is a constant. From the conservation of mass and the initial condition $\rho = \rho_0$ at $t = 0$, determine ρ as a function of ρ_0, t, and k.

Answer

$$\rho = \frac{\rho_0 k^3}{(k+t)^3}$$

Problem 5.9
By combining Eqs 5.17b and 5.13, verify the result presented in Eq 4.161.

Problem 5.10
Using the identity

$$\varepsilon_{ijk} a_{k,j} = 2 \left(\dot{w}_i + w_i v_{j,j} - w_j v_{i,j} \right)$$

as well as the continuity equation, show that

$$\frac{d}{dt}\left(\frac{w_i}{\rho}\right) = \frac{\varepsilon_{ijk} a_{k,j} + 2 w_j v_{i,j}}{2\rho} .$$

Problem 5.11
State the equations of motion and from them show by the use of the material derivative

$$\dot{v}_i = \frac{\partial v_i}{\partial t} + v_j v_{i,j} ,$$

and the continuity equation that

$$\frac{\partial (\rho v_i)}{\partial t} = \left(t_{ij} - \rho v_i v_j \right)_{,j} + \rho b_i .$$

Problem 5.12

Determine the form which the equations of motion take if the stress components are given by $t_{ij} = -p\delta_{ij}$ where $p = p(\mathbf{x}, t)$.

Answer

$$\rho a_i = -p_{,i} + \rho b_i$$

Problem 5.13

Let a material continuum have the constitutive equation

$$t_{ij} = \alpha\delta_{ij} d_{kk} + 2\beta d_{ij}$$

where α and β are constants. Determine the form which the equations of motion take in terms of the velocity gradients for this material.

Answer

$$\rho\dot{v}_i = \rho b_i + (\alpha + \beta)v_{j,ij} + \beta v_{i,jj}$$

Problem 5.14

Assume that distributed body moments m_i act throughout a continuum in motion. Show that the equations of motion are still valid in the form of Eq 5.22, but that the angular momentum principle now requires

$$\varepsilon_{ijk} t_{jk} + m_i = 0$$

implying that the stress tensor can no longer be taken as symmetric.

Problem 5.15

For a rigid body rotation about the origin, the velocity field may be expressed by $v_i = \varepsilon_{ijk}\omega_j x_k$ where ω_j is the angular velocity vector. Show that for this situation the angular momentum principle is given by

$$M_i = \dot{\overline{(\omega_j I_{ij})}}$$

where M_i is the total moment about the origin of all surface and body forces, and I_{ij} is the moment of inertia of the body defined by the tensor

$$I_{ij} = \int_V \rho\,(\delta_{ij} x_k x_k - x_i x_j)\; dV\,.$$

Problem 5.16

Determine expressions for the stress power $t_{ij} d_{ij}$ in terms of

(a) the first Piola-Kirchoff stress tensor,

(b) the second Piola-Kirchoff stress tensor.

Answer

(a) $t_{ij}d_{ij} = \rho\dot{F}_{iA}P_{iA}/\rho_0$ (b) $t_{ij}d_{ij} = \rho s_{AB}\dot{C}_{AB}/2\rho_0$

Problem 5.17

Show that, for a rigid body rotation about the origin, the kinetic energy integral Eq 5.53 reduces to the form given in rigid body dynamics, that is,

$$K = \frac{1}{2}\omega_i\omega_j I_{ij}$$

where I_{ij} is the inertia tensor defined in Problem 5.15.

Problem 5.18

Show that one way to express the rate of change of kinetic energy of the material currently occupying the volume $\mathcal{V}$ is by the equation

$$\dot{K} = \int_{\mathcal{V}} \rho b_i v_i \, d\mathcal{V} - \int_{\mathcal{V}} t_{ij}v_{i,j} \, d\mathcal{V} + \int_{\mathcal{S}} v_i t_i^{(\hat{n})} \, d\mathcal{S}$$

and give an interpretation of each of the above integrals.

Problem 5.19

Consider a continuum for which the stress is $t_{ij} = -p_0\delta_{ij}$ and which obeys the heat conduction law $q_i = -\kappa\theta_{,i}$. Show that for this medium the energy equation takes the form

$$\rho\dot{u} = -p_0 v_{i,i} + \rho r + \kappa\theta_{,ii} \; .$$

Problem 5.20

If mechanical energy only is considered, the energy balance can be derived from the equations of motion. Thus, by forming the scalar product of each term of Eq 5.22 with the velocity v_i and integrating the resulting equation term-by-term over the volume $\mathcal{V}$, we obtain the energy equation. Verify that one form of the result is

$$\frac{1}{2}\rho\overline{(\mathbf{v}\cdot\mathbf{v})}^{\cdot} + \mathrm{tr}\,(\mathbf{T}\cdot\mathbf{D}) - \rho\mathbf{b}\cdot\mathbf{v} - \mathrm{div}\,(\mathbf{T}\cdot\mathbf{v}) = 0 \; .$$

Problem 5.21

Show that for a continuum body experiencing volumetric growth

$$\frac{d}{dt}\int_P \rho f \, d\mathcal{V} = \int_P \left(\rho\dot{f} + \rho c f\right) \, d\mathcal{V} \; .$$

Problem 5.22

Using the local form of the balance of mass, Eq 5.90, show that the balance of linear momentum for a material with volumetric growth is Eq 5.93

$$\rho\frac{dv_i}{dt} = t_{ji,j} + \rho b_i \; .$$

Problem 5.23
Using the local form of the balance of mass, Eq 5.90, and linear momentum, Eq 5.93, show that the stress in a volumetrically growing body is symmetric.

Problem 5.24
If a continuum has the constitutive equation

$$t_{ij} = -p\delta_{ij} + \alpha d_{ij} + \beta d_{ik} d_{kj}$$

where p, α and β are constants, and if the material is incompressible ($d_{ii} = 0$), show that

$$t_{ii} = -3p - 2\beta II_D$$

where II_D is the second invariant of the rate of deformation tensor.

Problem 5.25
Starting with Eq 5.173 for isotropic elastic behavior, show that

$$t_{ii} = (3\lambda + 2\mu)\epsilon_{ii} ,$$

and using this result, deduce that

$$\epsilon_{ij} = \frac{1}{2\mu}\left(t_{ij} - \frac{\lambda}{3\lambda + 2\mu}\delta_{ij} t_{kk}\right) .$$

Problem 5.26
For a Newtonian fluid, the constitutive equation is given by

$$\begin{aligned} t_{ij} &= -p\delta_{ij} + \tau_{ij} \\ &= -p\delta_{ij} + \lambda^* \delta_{ij} d_{kk} + 2\mu^* d_{ij} \quad \text{(see Eq 5.175)} . \end{aligned}$$

By substituting this constitutive equation into the equations of motion, derive the equation

$$\rho \dot{v}_i = \rho b_i - p_{,i} + (\lambda^* + \mu^*) v_{j,ji} + \mu^* v_{i,jj} .$$

Problem 5.27
Combine Eqs 5.177a and Eq 5.177b into a single viscoelastic constitutive equation having the form

$$t_{ij} = \delta_{ij}\{R\}\epsilon_{kk} + \{S\}\epsilon_{ij}$$

where the linear time operators $\{R\}$ and $\{S\}$ are given, respectively,

$$\{R\} = \frac{3K\{P\} - 2\{Q\}}{3\{P\}} \quad \text{and} \quad \{S\} = \frac{2\{Q\}}{\{P\}} .$$

Problem 5.28
Assume a viscoelastic medium is governed by the constitutive equations Eq 5.177. Let a slender bar of such material be subjected to the axial tension stress $t_{11} = \sigma_0 f(t)$ where σ_0 is a constant and $f(t)$ some function of time. Assuming that $t_{22} = t_{33} = t_{13} = t_{23} = t_{12} = 0$, determine ϵ_{11}, ϵ_{22}, and ϵ_{33} as functions of $\{P\}$, $\{Q\}$, and $f(t)$.

Answer

$$\epsilon_{11} = \frac{3K\{P\} + \{Q\}}{9K\{Q\}} \sigma_0 f(t)$$

$$\epsilon_{22} = \epsilon_{33} = \frac{2\{Q\} - 3K\{P\}}{18K\{Q\}} \sigma_0 f(t)$$

Problem 5.29

Use the definition of the free energy along with the reduced form of the Clausius-Duhem equation to derive the local dissipation inequality.

Problem 5.30

The constitutive model for a compressible, viscous, and heat-conducting material is defined by

$$\psi = \tilde{\psi}\left(\theta, g_k, F_{iA}, \dot{F}_{iA}\right) ,$$
$$\eta = \tilde{\eta}\left(\theta, g_k, F_{iA}, \dot{F}_{iA}\right) ,$$
$$t_{ij} = \tilde{t}_{ij}\left(\theta, g_k, F_{iA}, \dot{F}_{iA}\right) ,$$
$$q_i = \tilde{q}_i\left(\theta, g_k, F_{iA}, \dot{F}_{iA}\right) .$$

Deduce the following restrictions on these constitutive response functions:

(a) $\dfrac{\partial \tilde{\psi}\left(\theta, g_k, F_{kB}, \dot{F}_{kB}\right)}{\partial g_i} = 0,$

(b) $\dfrac{\partial \tilde{\psi}\left(\theta, g_k, F_{kB}, \dot{F}_{kB}\right)}{\partial \dot{F}_{iA}} = 0,$

(c) $\tilde{\eta}(\theta, F_{iA}) = -\dfrac{\partial \tilde{\psi}(\theta, F_{kB})}{\partial \theta},$

(d) $\tilde{t}_{ij} = \rho F_{jA} \dfrac{\partial \tilde{\psi}(\theta, F_{kB})}{\partial F_{iA}},$

(e) $-\dfrac{1}{\theta}\tilde{q}_i(\theta, g_k, F_{iA}, 0)\, g_i \geqslant 0.$

Problem 5.31

Assume the constitutive relationships

$$u = \tilde{u}(C_{AB}, \eta) ,$$
$$\theta = \tilde{\theta}(C_{AB}, \eta) ,$$
$$t_{ij} = \tilde{t}_{ij}(C_{AB}, \eta) ,$$
$$q_i = \tilde{q}_i(C_{AB}, \eta, g_k) ,$$

for an elastic material. Use the Clausius-Duhem inequality to show

$$\begin{aligned}
\theta &= \frac{\partial \tilde{u}}{\partial \eta} \, , \\
u &= \tilde{u}\left(C_{AB}, \eta\right) \, , \\
t_{ij} &= \frac{1}{2}\rho F_{iA}\left(\frac{\partial \tilde{u}}{\partial C_{AB}} + \frac{\partial \tilde{u}}{\partial C_{BA}}\right) F_{jB} \, , \\
\tilde{q}_i g_i &\geqslant 0 \, .
\end{aligned}$$

Problem 5.32

Use the basic kinematic result of superposed rigid body motion given in Eq 5.121b to show the following:

(a) $\dfrac{\partial x_i^+}{\partial x_k} = Q_{ik}$,

(b) $F_{iA}^+ = Q_{ij} F_{jA}$.

Problem 5.33

Show that the Jacobian transforms as follows under a superposed rigid body motion:

$$J^+ = J \, .$$

Problem 5.34

Utilize the result of Problem 5.33 along with the law of conservation of mass to show that $\rho^+ = \rho$.

Problem 5.35

Show that the gradient of the stress components transforms under superposed rigid body motion as follows:

$$\frac{\partial t_{ij}^+}{\partial x_j^+} = Q_{im} \frac{\partial t_{mn}}{\partial x_n} \, .$$

Problem 5.36

Use the superposed rigid body motion definitions to show the following relationships:

(a) $C_{AB}^+ = C_{AB}$,

(b) $U_{AB}^+ = U_{AB}$,

(c) $R_{iA}^+ = Q_{ij} R_{jA}$,

(d) $B_{ij}^+ = Q_{im} B_{mk} Q_{kj}$.

Problem 5.37

In the context of rigid body dynamics, consider the motion defined by

$$\mathbf{x}\left(\mathbf{X}, t\right) = \mathbf{X}$$

along with

$$\mathbf{x} = \mathbf{Q}^{\mathsf{T}}(\mathrm{t})\left[\mathbf{x}^{+} - \mathbf{a}(\mathrm{t})\right]$$

show that

$$\mathbf{v}^{+} = \dot{\mathbf{a}}(\mathrm{t}) + \boldsymbol{\omega}(\mathrm{t}) \times \left[\mathbf{x}^{+} - \mathbf{a}(\mathrm{t})\right]$$

where $\boldsymbol{\omega}$ is the angular velocity of the body.

6

Linear Elasticity

Elastic behavior is characterized by the following two conditions: (1) where the stress in a material is a unique function of the strain, and (2) where the material has the property for complete recovery to a "natural" shape upon removal of the applied forces. If the behavior of a material is not elastic, we say that it is *inelastic*. Also, we acknowledge that elastic behavior may be *linear* or *non-linear*.

6.1 Elasticity, Hooke's Law, Strain Energy

Figure 6.1 shows geometrically these behavior patterns by simple stress-strain curves, with the relevant loading and unloading paths indicated. For many engineering applications, especially those involving structural materials such as metals and concrete, the conditions for elastic behavior are realized, and for these cases the theory of elasticity offers a very useful and reliable model for design. Symbolically, we write the constitutive equation for elastic behavior in its most general form as

$$\mathsf{T} = \mathbf{G}(\boldsymbol{\epsilon}) \tag{6.1}$$

where $\mathbf{G}$ is a symmetric tensor-valued function and $\boldsymbol{\epsilon}$ is any one of the various strain tensors we introduced earlier. However, for the response function $\mathbf{G}$ in this text we consider only that case of Eq 6.1 for which the stress is a *linear* function of strain. Also, we assume that, in the deformed material, the displacement gradients are everywhere small compared with unity. Thus, the distinction between the Lagrangian and Eulerian descriptions is negligible, and following the argument of Eq 4.60 we make use of the infinitesimal

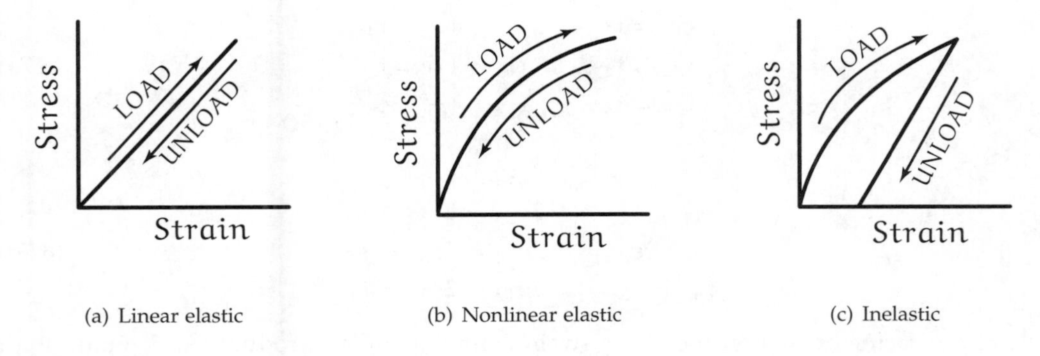

(a) Linear elastic (b) Nonlinear elastic (c) Inelastic

FIGURE 6.1
Uniaxial loading-unloading stress-strain curves for various material behaviors.

strain tensor defined in Eq 4.62, which we repeat here:

$$\epsilon_{ij} = \frac{1}{2}\left(\frac{\partial u_i}{\partial x_j} + \frac{\partial u_j}{\partial x_i}\right) = \frac{1}{2}\left(u_{i,j} + u_{j,i}\right) . \tag{6.2}$$

Within the context of the above assumptions, we write the constitutive equation for linear elastic behavior as

$$t_{ij} = C_{ijkm}\epsilon_{km} \quad \text{or} \quad \mathbf{T} = \mathbf{C}\boldsymbol{\epsilon} \tag{6.3}$$

where the tensor of elastic coefficients C_{ijkm} has $3^4 = 81$ components. However, due to the symmetry of both the stress and strain tensors, it is clear that

$$C_{ijkm} = C_{jikm} = C_{ijmk} \tag{6.4}$$

which reduces the 81 possibilities to 36 distinct coefficients at most.

We may demonstrate the tensor character of **C** by a consideration of the elastic constitutive equation when expressed in a rotated (primed) coordinate system in which it has the form

$$t'_{ij} = C'_{ijkm}\epsilon'_{km} . \tag{6.5}$$

But by the transformation laws for second-order tensors, along with Eq 6.3,

$$\begin{aligned} t'_{ij} &= a_{iq}a_{js}t_{qs} = a_{iq}a_{js}C_{qskm}\epsilon_{km} \\ &= a_{iq}a_{js}C_{qskm}a_{pk}a_{nm}\epsilon'_{pn} \end{aligned}$$

which by a direct comparison with Eq 6.5 provides the result

$$C'_{ijkm} = a_{iq}a_{js}a_{pk}a_{nm}C_{qskm} , \tag{6.6}$$

that is, the transformation rule for a fourth-order Cartesian tensor.

In general, the C_{ijkm} coefficients may depend upon temperature, but here we assume *adiabatic* (no heat gain or loss) and *isothermal* (constant temperature) conditions. We also ignore strain-rate effects and consider the components C_{ijkm} to be at most a function of position. If the elastic coefficients are constants, the material is said to be *homogeneous*. These constants are those describing the elastic properties of the material. The constitutive law given by Eq 6.3 is known as the generalized Hooke's law.

For certain purposes it is convenient to write Hooke's law using a single subscript on the stress and strain components and double subscripts on the elastic constants. To this end, we define

$$\begin{aligned} t_{11} &= t_1 , & t_{23} &= t_{32} = t_4 , \\ t_{22} &= t_2 , & t_{13} &= t_{31} = t_5 , \\ t_{33} &= t_3 , & t_{12} &= t_{21} = t_6 , \end{aligned} \tag{6.7a}$$

and

$$\begin{aligned} \epsilon_{11} &= \epsilon_1 , & 2\epsilon_{23} &= 2\epsilon_{32} = \epsilon_4 , \\ \epsilon_{22} &= \epsilon_2 , & 2\epsilon_{13} &= 2\epsilon_{31} = \epsilon_5 , \\ \epsilon_{33} &= \epsilon_3 , & 2\epsilon_{12} &= 2\epsilon_{21} = \epsilon_6 , \end{aligned} \tag{6.7b}$$

where the factor of two on the shear strain components is introduced in keeping with Eq 4.71. From these definitions, Hooke's law is now written

$$t_\alpha = \mathcal{C}_{\alpha\beta}\epsilon_\beta \quad \text{or} \quad \mathbf{T} = \mathcal{C}\boldsymbol{\epsilon} \tag{6.8}$$

with Greek subscripts having a range of six. Note the font change in $\mathcal{C}_{\alpha\beta}$ to that of a matrix. This is done to account for the fact that the reduced representation stiffness matrix does not transform as a tensor. In matrix form Eq 6.8 appears as

$$\begin{bmatrix} t_1 \\ t_2 \\ t_3 \\ t_4 \\ t_5 \\ t_6 \end{bmatrix} = \begin{bmatrix} \mathcal{C}_{11} & \mathcal{C}_{12} & \mathcal{C}_{13} & \mathcal{C}_{14} & \mathcal{C}_{15} & \mathcal{C}_{16} \\ \mathcal{C}_{21} & \mathcal{C}_{22} & \mathcal{C}_{23} & \mathcal{C}_{24} & \mathcal{C}_{25} & \mathcal{C}_{26} \\ \mathcal{C}_{31} & \mathcal{C}_{32} & \mathcal{C}_{33} & \mathcal{C}_{34} & \mathcal{C}_{35} & \mathcal{C}_{36} \\ \mathcal{C}_{41} & \mathcal{C}_{42} & \mathcal{C}_{43} & \mathcal{C}_{44} & \mathcal{C}_{45} & \mathcal{C}_{46} \\ \mathcal{C}_{51} & \mathcal{C}_{52} & \mathcal{C}_{53} & \mathcal{C}_{54} & \mathcal{C}_{55} & \mathcal{C}_{56} \\ \mathcal{C}_{61} & \mathcal{C}_{62} & \mathcal{C}_{63} & \mathcal{C}_{64} & \mathcal{C}_{65} & \mathcal{C}_{66} \end{bmatrix} \begin{bmatrix} \epsilon_1 \\ \epsilon_2 \\ \epsilon_3 \\ \epsilon_4 \\ \epsilon_5 \\ \epsilon_6 \end{bmatrix} . \tag{6.9}$$

We repeat that the array of the 36 constants $\mathcal{C}_{\alpha\beta}$ does not constitute a tensor.

In view of our assumption to neglect thermal effects at this point, the energy balance Eq 5.65 is reduced to the form

$$\dot{u} = \frac{1}{\rho} t_{ij} d_{ij} \tag{6.10a}$$

which for small-deformation theory, by Eq 4.150, becomes

$$\dot{u} = \frac{1}{\rho} t_{ij} \dot{\epsilon}_{ij} \tag{6.10b}$$

The internal energy u in these equations is purely mechanical and is called the *strain energy* (per unit mass). Recall now that, by the continuity equation in Lagrangian form, $\rho_0 = \rho J$ and also that to the first order of approximation

$$J = \det F = \det\left(\delta_{iA} + \frac{\partial u_i}{\partial X_A}\right) = 1 + \frac{\partial u_i}{\partial X_A} . \tag{6.11}$$

Therefore, from our assumption of small displacement gradients, namely $\partial u_i/\partial X_A \ll 1$, we may take $J \approx 1$ in the continuity equation to give $\rho = \rho_0$ a constant in Eqs 6.10a and 6.10b.

For elastic behavior under the assumptions we have imposed, the strain energy is a function of the strain components only, and we write

$$u = u(\epsilon_{ij}) \tag{6.12}$$

so that

$$\dot{u} = \frac{\partial u}{\partial \epsilon_{ij}} \dot{\epsilon}_{ij} , \tag{6.13}$$

and by a direct comparison with Eq 6.10b we obtain

$$\frac{1}{\rho} t_{ij} = \frac{\partial u}{\partial \epsilon_{ij}} . \tag{6.14}$$

The *strain energy density*, W (strain energy per unit volume) is defined by

$$W = \rho_0 u , \tag{6.15}$$

and since $\rho = \rho_0$, a constant, under the assumptions we have made, it follows from Eq 6.14 that

$$t_{ij} = \rho \frac{\partial u}{\partial \epsilon_{ij}} = \frac{\partial W}{\partial \epsilon_{ij}} . \tag{6.16}$$

It is worthwhile noting at this point that elastic behavior is sometimes defined on the basis of the existence of a strain energy function from which the stresses may be determined by the differentiation in Eq 6.16. A material defined in this way is called a *hyperelastic* material. The stress is still a unique function of strain so that this energy approach is compatible with our earlier definition of elastic behavior. Thus, in keeping with our basic restriction to infinitesimal deformations, we shall develop the linearized form of Eq 6.16. Expanding W about the origin, we have

$$W(\epsilon_{ij}) = W(0) + \frac{\partial W(0)}{\partial \epsilon_{ij}} \epsilon_{ij} + \frac{1}{2} \frac{\partial^2 W(0)}{\partial \epsilon_{ij} \partial \epsilon_{km}} \epsilon_{ij} \epsilon_{km} + \cdots , \tag{6.17}$$

and, from Eq 6.16,

$$t_{ij} = \frac{\partial W}{\partial \epsilon_{ij}} = \frac{\partial W(0)}{\partial \epsilon_{ij}} + \frac{\partial^2 W(0)}{\partial \epsilon_{ij} \partial \epsilon_{km}} \epsilon_{km} + \cdots . \tag{6.18}$$

It is customary to assume that there are no residual stresses in the unstrained state of the material so that $t_{ij} = 0$ when $\epsilon_{ij} = 0$. Thus, by retaining only the linear term of the above expansion, we may express the linear elastic constitutive equation as

$$t_{ij} = \frac{\partial^2 W(0)}{\partial \epsilon_{ij} \partial \epsilon_{km}} \epsilon_{km} = C_{ijkm} \epsilon_{km} \tag{6.19}$$

based on the strain energy function. This equation appears to be identical to Eq 6.3, but there is one very important difference between the two – not only do we have the symmetries expressed by Eq 6.4, but now we also have

$$C_{ijkm} = C_{kmij} \tag{6.20}$$

due to the fact that

$$\frac{\partial^2 W(0)}{\partial \epsilon_{ij} \partial \epsilon_{km}} = \frac{\partial^2 W(0)}{\partial \epsilon_{km} \partial \epsilon_{ij}} .$$

Thus, the existence of a strain energy function reduces the number of distinct components of C_{ijkm} from 36 to 21. Further reductions for special types of elastic behavior are obtained from material symmetry properties in the next section. Note that by substituting Eq 6.19 into Eq 6.17 and assuming a linear stress-strain relation, we may now write

$$W(\epsilon_{ij}) = \frac{1}{2} C_{ijkm} \epsilon_{ij} \epsilon_{km} = \frac{1}{2} t_{ij} \epsilon_{ij} \tag{6.21a}$$

which in the notation of Eq 6.8 becomes

$$W(\epsilon_{\alpha}) = \frac{1}{2} \mathcal{C}_{\alpha\beta} \epsilon_{\alpha} \epsilon_{\beta} = \frac{1}{2} t_{\alpha} \epsilon_{\alpha} , \tag{6.21b}$$

and by the symmetry condition we have only 21 distinct constants out of the 36 possible.

6.2 Hooke's Law for Isotropic Media, Elastic Constants

If the behavior of a material is elastic under a given set of circumstances, it is customarily spoken of as an elastic material when discussing that situation even though under a different set of circumstances its behavior may not be elastic. Furthermore, if a body's elastic properties as described by the coefficients C_{ijkm} are the same in every set of reference axes at any point for a given situation, we call it an *isotropic elastic material*. For such materials, the constitutive equation has only two elastic constants. A material that is not isotropic is called *anisotropic*; we shall define some of these based upon the degree of elastic symmetry each possesses.

In general, an isotropic tensor is defined as one whose components are unchanged by any orthogonal transformation from one set of Cartesian axes to another. Zero-order tensors of any order – and all zeroth-order tensors (scalars) – are isotropic, but there are no first-order isotropic tensors (vectors). The unit tensor **I**, having Kronecker deltas as components, and any scalar multiple of **I** are the only second-order isotropic tensors (see Problem 6.5). The only nontrivial third-order isotropic tensor is the permutation symbol. The most general fourth-order isotropic tensor may be shown to have a form in terms of Kronecker deltas which we now introduce as the prototype for **C**, namely,

$$C_{ijkm} = \lambda\delta_{ij}\delta_{km} + \mu\left(\delta_{ik}\delta_{jm} + \delta_{im}\delta_{jk}\right) + \beta\left(\delta_{ik}\delta_{jm} - \delta_{im}\delta_{jk}\right) \tag{6.22}$$

where λ, μ, and β are scalars. But by Eq 6.4, $C_{ijkm} = C_{jikm} = C_{ijmk}$. This implies that β must be zero for the stated symmetries since by interchanging i and j in the expression

$$\beta(\delta_{ik}\delta_{jm} - \delta_{im}\delta_{jk}) = \beta(\delta_{jk}\delta_{im} - \delta_{jm}\delta_{ik})$$

we see that $\beta = -\beta$ and, consequently, $\beta = 0$. Therefore, inserting the reduced Eq 6.22 into Eq 6.3, we have

$$t_{ij} = (\lambda\delta_{ij}\delta_{km} + \mu\delta_{ik}\delta_{jm} + \mu\delta_{im}\delta_{jk})\,\epsilon_{km}\ .$$

But by the substitution property of δ_{ij}, this reduces to

$$t_{ij} = \lambda\delta_{ij}\epsilon_{kk} + 2\mu\epsilon_{ij} \tag{6.23}$$

which is *Hooke's law* for *isotropic elastic behavior*. As mentioned earlier, we see that for isotropic elastic behavior the 21 constants of the generalized law have been reduced to two, λ and μ, known as the *Lamé constants*. Note that for an isotropic elastic material $C_{ijkm} = C_{kmij}$; that is, an isotropic elastic material is necessarily hyperelastic.

Example 6.1

Show that for an isotropic linear elastic solid the principal axes of the stress and strain tensors coincide, and develop an expression for the relationship among their principal values.

Solution

Let $\hat{n}^{(q)}$ $(q = 1,2,3)$ be unit normals in the principal directions of ϵ_{ij}, and associated with these normals the corresponding principal values are $\epsilon_{(q)}$ where $(q = 1,2,3)$. From Eq 6.23 we form the dot products

$$\begin{aligned} t_{ij}n_j^{(q)} &= (\lambda\delta_{ij}\epsilon_{kk} + 2\mu\epsilon_{ij})\,n_j^{(q)} \\ &= \lambda n_i^{(q)}\epsilon_{kk} + 2\mu\epsilon_{ij}n_j^{(q)}\ . \end{aligned}$$

But $n_j^{(q)}$ and $\epsilon_{(q)}$ satisfy the fundamental equation for the eigenvalue problem, namely, $\epsilon_{ij}n_j^{(q)} = \epsilon_{(q)}\delta_{ij}n_j^{(q)}$, so that now

$$\begin{aligned} t_{ij}n_j^{(q)} &= \lambda\epsilon_{kk}n_i^{(q)} + 2\mu\epsilon_{(q)}n_i^{(q)} \\ &= \left[\lambda\epsilon_{kk} + 2\mu\epsilon_{(q)}\right] n_i^{(q)} , \end{aligned}$$

and because $\epsilon_{kk} = \epsilon_{(1)} + \epsilon_{(2)} + \epsilon_{(3)}$ is the first invariant of strain, it is constant for all $n_i^{(q)}$ so that

$$t_{ij}n_j^{(q)} = \left\{\left(\lambda\left[\epsilon_{(1)} + \epsilon_{(2)} + \epsilon_{(3)}\right] + 2\mu\epsilon_{(q)}\right)\right\} n_j^{(q)} .$$

This indicates that $n_i^{(q)}$ $(q = 1, 2, 3)$ are principal directions of stress also, with principal stress values

$$t_{(q)} = \lambda\left[\epsilon_{(1)} + \epsilon_{(2)} + \epsilon_{(3)}\right] + 2\mu\epsilon_{(q)} \quad (q = 1, 2, 3) .$$

We may easily invert Eq 6.23 to express the strain components in terms of the stresses. To this end, we first determine ϵ_{ii} in terms of t_{ii} from Eq 6.23 by setting $i = j$ to yield

$$t_{ii} = 3\lambda\epsilon_{kk} + 2\mu\epsilon_{ii} = (3\lambda + 2\mu)\epsilon_{ii} . \tag{6.24}$$

Now, by solving Eq 6.23 for ϵ_{ij} and substituting from Eq 6.24, we obtain the inverse form of the isotropic constitutive equation,

$$\epsilon_{ij} = \frac{1}{2\mu}\left(t_{ij} + \frac{\lambda}{3\lambda + 2\mu}\delta_{ij}t_{kk}\right) . \tag{6.25}$$

By a formal – although admittedly not obvious – rearrangement of this equation, we may write

$$\epsilon_{ij} = \frac{\lambda + \mu}{\mu\left(3\lambda + 2\mu\right)}\left\{\left[1 + \frac{\lambda}{2\left(\lambda + \mu\right)}\right] t_{ij} - \frac{\lambda}{2\left(\lambda + \mu\right)}\delta_{ij}t_{kk}\right\} \tag{6.26}$$

from which, if we define,

$$E = \frac{\mu\left(3\lambda + 2\mu\right)}{\lambda + \mu} \tag{6.27a}$$

and

$$\nu = \frac{\lambda}{2\left(\lambda + \mu\right)} , \tag{6.27b}$$

we obtain the following form of Hooke's law for isotropic behavior in terms of the engineering constants E and ν,

$$\epsilon_{ij} = \frac{1}{E}\left[(1 + \nu)\, T_{ij} - \nu\delta_{ij}T_{kk}\right] . \tag{6.28}$$

Here E is called *Young's modulus*, or simply the *modulus of elasticity*, and ν is known as *Poisson's ratio*. By suitable combinations of these two constants, we may define two additional constants of importance in engineering elasticity. First, the *shear modulus*, or *modulus of rigidity*, is defined as

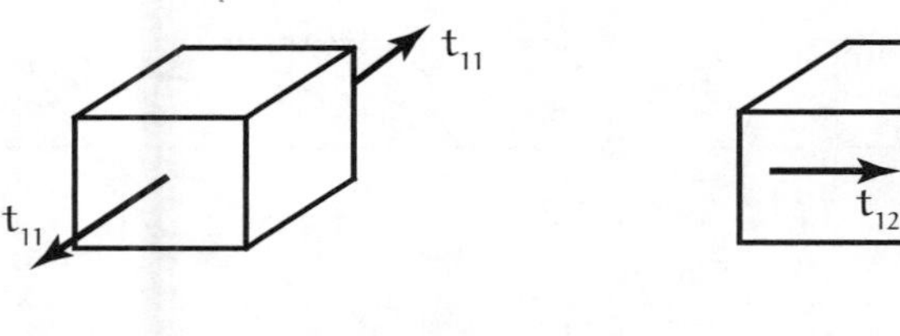

(a) Uniaxial tension (b) Simple shear

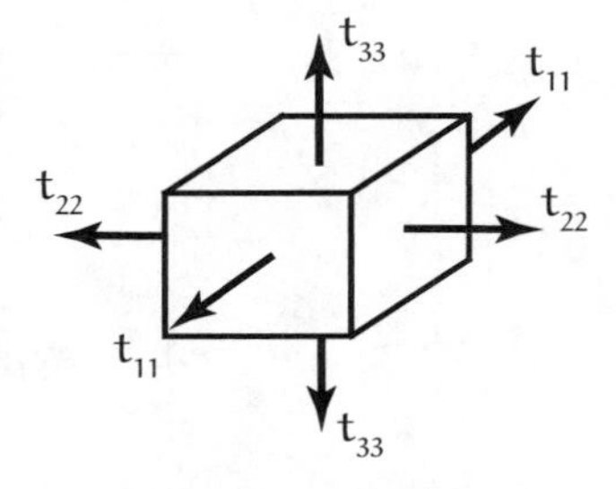

(c) Uniform triaxial tension

FIGURE 6.2
Simple stress states.

$$G = \frac{E}{2(1+\nu)} = \mu \tag{6.29a}$$

which, as noted, is identical to the Lamé constant μ. Second, the bulk modulus is defined as

$$K = \frac{E}{3(1-2\nu)} \; . \tag{6.29b}$$

For isotropic elastic materials, any two elastic constants completely define the material's response. In addition to that, any elastic constant can be determined in terms of any two other constants. A listing of all elastic constants in terms of other pairs of constants is given in Table 6.1.

The physical interpretations of the constants E, ν, G, and K introduced above can be determined from a consideration of the special states of stress displayed in Fig. 6.2. In the case of a uniaxial state of stress (tension or compression), say in the x_1 direction with $t_{11} = \pm\sigma_0$, and all other stress components zero (Fig. 6.2(a)), Eq 6.28 yields (since $t_{ii} = \pm\sigma_0$),

$$\epsilon_{11} = \frac{t_{11}}{E} = \frac{\pm\sigma_0}{E} \quad \text{for } (i = j = 1) \; , \tag{6.30a}$$

$$\epsilon_{22} = -\nu\epsilon_{11} = \frac{\mp\nu\sigma_0}{E} \quad \text{for } (i = j = 2) \; , \tag{6.30b}$$

$$\epsilon_{33} = -\nu\epsilon_{11} = \frac{\mp\nu\sigma_0}{E} \quad \text{for } (i = j = 3) \; , \tag{6.30c}$$

as well as zero shear strains for $i \neq j$. Thus, E is the proportionality factor between axial (normal) stresses and strains. Geometrically, it is the slope of the one-dimensional

TABLE 6.1
Relations between elastic constants.

	λ	μ	E	ν	K
λ, μ			$\frac{\mu(3\lambda+2\mu)}{\lambda+\mu}$	$\frac{\lambda}{2(\lambda+\mu)}$	$\frac{3\lambda+2\mu}{3}$
λ, E		$(E-2\lambda)+\frac{1}{4}\sqrt{(E-3\lambda)^2+8\lambda E}$		$-(E-2\lambda)+\frac{1}{4\lambda}\sqrt{(E-3\lambda)^2+8\lambda^2}$	$(3\lambda+E)+\frac{1}{6}\sqrt{(3\lambda+E)^2-4\lambda E}$
λ, ν		$\frac{\lambda(1-2\nu)}{2\nu}$	$\frac{\lambda(1+\nu)(1-2\nu)}{\nu}$		$\frac{\lambda(1+\nu)}{3\nu}$
λ, K		$\frac{3(K-\lambda)}{2}$	$\frac{9K(K-\lambda)}{3K-\lambda}$	$\frac{\lambda}{3K-\lambda}$	
μ, E	$\frac{\mu(2\mu-E)}{E-3\mu}$			$\frac{E-2\mu}{2\mu}$	$\frac{\mu E}{3(3\mu-E)}$
μ, ν	$\frac{2\mu\nu}{1-2\nu}$		$2\mu(1+\nu)$		$\frac{2\mu(1+\nu)}{3(1-2\nu)}$
μ, K	$\frac{3K-2\mu}{3}$		$\frac{9K\mu}{3K+\mu}$	$\frac{3K-2\mu}{2(3K+\mu)}$	
E, ν	$\frac{\nu E}{(1+\nu)(1-2\nu)}$	$\frac{E}{2(1+\nu)}$			$\frac{E}{3(1-2\nu)}$
E, K	$\frac{3K(3K-E)}{(9K-E)}$	$\frac{3KE}{9K-E}$		$\frac{3K-E}{6K}$	
ν, K	$\frac{3K\nu}{1+\nu}$	$\frac{3K(1-2\nu)}{2(1+\nu)}$	$3K(1-2\nu)$		

linear stress-strain diagram (Fig. 6.1(a)). Note that $E > 0$; a specimen will elongate under tension, shorten in compression. From the second and third part of Eq 6.30 above, ν is seen to be the *ratio* of the unit lateral contraction to unit longitudinal extension for tension, and vice versa for compression. For the simple shear case shown in Fig. 6.2(b) where, say, $t_{12} = \tau_0$, all other stresses zero, we have, from Eq 6.28,

$$\epsilon_{12} = \frac{1+\nu}{E} t_{12} = \frac{\tau_0}{2G} , \tag{6.31a}$$

or for engineering strains, using Eq 4.71,

$$\gamma_{12} = \frac{t_{12}}{G} = \frac{\tau_0}{G} \tag{6.31b}$$

which casts G into the same role for simple shear as E assumes for axial tension (or compression). Hence, the name *shear modulus* for G. Finally, for the case of uniform triaxial tension (or hydrostatic compression) of Fig. 6.2(c), we take $t_{ij} = \pm p\delta_{ij}$ with $p > 0$. For this, Eq 6.28 indicates that

$$\epsilon_{ii} = \frac{1-2\nu}{E} t_{ii} = \frac{\pm 3\,(1-2\nu)}{E} p = \frac{\pm p}{K} \tag{6.32}$$

by which we infer that the bulk modulus K relates the pressure p to the volume change given by the cubical dilation ϵ_{ii} (see Eq 4.76).

It should also be pointed out that, by use of the constants G and K, Hooke's law may be expressed in terms of the spherical and deviator components of the stress and strain tensors. Thus, the pair of equations

$$S_{ij} = 2G\eta_{ij} , \tag{6.33a}$$

$$t_{ii} = 3K\epsilon_{ii} , \tag{6.33b}$$

may be shown to be equivalent to Eq 6.28 (see Problem 6.6).

6.3 Elastic Symmetry; Hooke's Law for Anisotropic Media

Hooke's law for isotropic behavior was established in Section 6.2 on the basis of **C** being a fourth-order isotropic tensor. The same result may be achieved from the concepts of *elastic symmetry*. To do so, we first define equivalent elastic directions as those specified by Cartesian axes $Ox_1x_2x_3$ and $Ox'_1x'_2x'_3$ at a point such that the elastic constants $\mathcal{C}_{\alpha\beta}$ are unchanged by a transformation between the two sets of axes. If the transformation represents a rotation about an axis, we say the material has axial elastic symmetry with respect to that axis. If the transformation is a reflection of the axes with respect to some plane, we say the material has a *plane of elastic symmetry*. Figure 6.3(a) shows the case for x_3 being the axis of elastic symmetry, whereas Fig. 6.3(b) shows the case for the x_1x_2 plane as the plane of elastic symmetry. The fact that the transformation for the reflection in Fig. 6.3(b) is an improper one (resulting in the axes being a left-handed coordinate system) does not invalidate the symmetry considerations to be used. Also, the x_3 axis in Fig. 6.3(a) is said to be of order N where $N = 2\pi/\theta$. It is also noteworthy that a *point of*

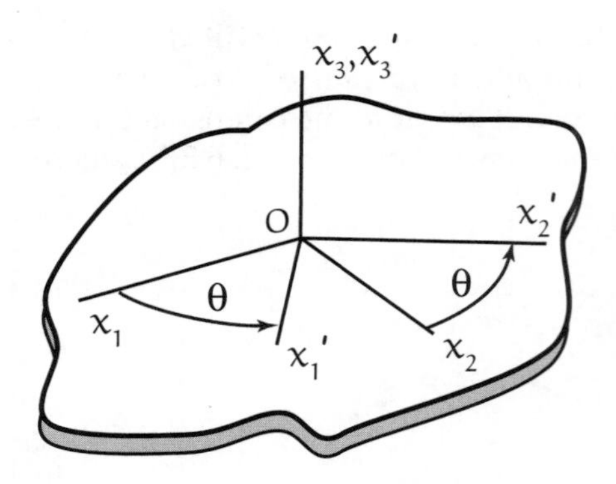

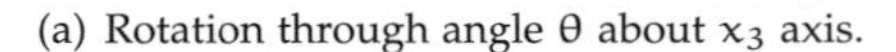

(a) Rotation through angle θ about x_3 axis.

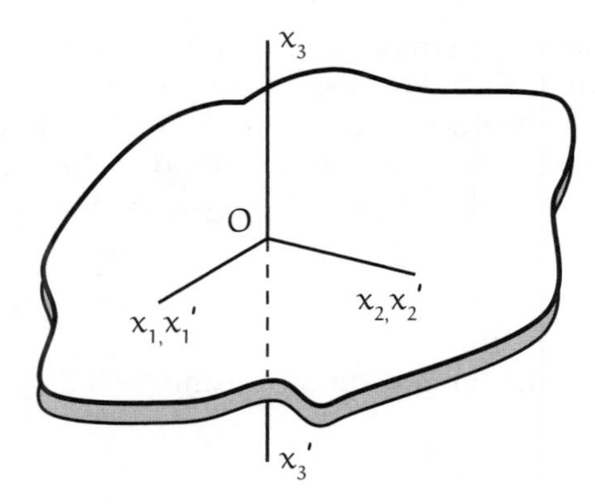

(b) Reflection in the x_1x_2 plane.

FIGURE 6.3
Axes rotations for plane stress.

elastic symmetry would imply isotropic behavior, since the elastic constants would remain unchanged for any two sets of Cartesian axes at the point.

Let us consider the consequences of the x_1-x_2 plane being a plane of elastic symmetry as shown in Fig. 6.3(b). The transformation matrix for this is clearly

$$[a_{ij}] = \begin{bmatrix} 1 & 0 & 0 \\ 0 & 1 & 0 \\ 0 & 0 & -1 \end{bmatrix}, \tag{6.34}$$

so, in the single subscript notation for stress and strain components, the transformations in matrix form are

$$\begin{bmatrix} t'_1 & t'_6 & t'_5 \\ t'_6 & t'_2 & t'_4 \\ t'_5 & t'_4 & t'_3 \end{bmatrix} = \begin{bmatrix} 1 & 0 & 0 \\ 0 & 1 & 0 \\ 0 & 0 & -1 \end{bmatrix} \begin{bmatrix} t_1 & t_6 & t_5 \\ t_6 & t_2 & t_4 \\ t_5 & t_4 & t_3 \end{bmatrix} \begin{bmatrix} 1 & 0 & 0 \\ 0 & 1 & 0 \\ 0 & 0 & -1 \end{bmatrix} \tag{6.35a}$$
$$= \begin{bmatrix} t_1 & t_6 & -t_5 \\ t_6 & t_2 & -t_4 \\ -t_5 & -t_4 & t_3 \end{bmatrix},$$

and

$$\begin{bmatrix} \epsilon'_1 & \frac{1}{2}\epsilon'_6 & \frac{1}{2}\epsilon'_5 \\ \frac{1}{2}\epsilon'_6 & \epsilon'_2 & \frac{1}{2}\epsilon'_4 \\ \frac{1}{2}\epsilon'_5 & \frac{1}{2}\epsilon'_4 & \epsilon'_3 \end{bmatrix} = \begin{bmatrix} 1 & 0 & 0 \\ 0 & 1 & 0 \\ 0 & 0 & -1 \end{bmatrix} \begin{bmatrix} \epsilon_1 & \frac{1}{2}\epsilon_6 & \frac{1}{2}\epsilon_5 \\ \frac{1}{2}\epsilon_6 & \epsilon_2 & \frac{1}{2}\epsilon_4 \\ \frac{1}{2}\epsilon_5 & \frac{1}{2}\epsilon_4 & \epsilon_3 \end{bmatrix} \begin{bmatrix} 1 & 0 & 0 \\ 0 & 1 & 0 \\ 0 & 0 & -1 \end{bmatrix} \tag{6.35b}$$
$$= \begin{bmatrix} \epsilon_1 & \frac{1}{2}\epsilon_6 & -\frac{1}{2}\epsilon_5 \\ \frac{1}{2}\epsilon_6 & \epsilon_2 & -\frac{1}{2}\epsilon_4 \\ -\frac{1}{2}\epsilon_5 & -\frac{1}{2}\epsilon_4 & \epsilon_3 \end{bmatrix}.$$

Therefore, assuming all 36 constants in Eq 6.8 are distinct, we note that for axes $Ox_1x_2x_3$

$$t_1 = \mathcal{C}_{11}\epsilon_1 + \mathcal{C}_{12}\epsilon_2 + \mathcal{C}_{13}\epsilon_3 + \mathcal{C}_{14}\epsilon_4 + \mathcal{C}_{15}\epsilon_5 + \mathcal{C}_{16}\epsilon_6\,, \tag{6.36}$$

whereas for axes $Ox'_1x'_2x'_3$, under the condition that x_1x_2 is a plane of symmetry such that the $\mathcal{C}_{\alpha\beta}$ are unchanged in this system, we have

$$t'_1 = \mathcal{C}_{11}\epsilon'_1 + \mathcal{C}_{12}\epsilon'_2 + \mathcal{C}_{13}\epsilon'_3 + \mathcal{C}_{14}\epsilon'_4 + \mathcal{C}_{15}\epsilon'_5 + \mathcal{C}_{16}\epsilon'_6\,. \tag{6.37}$$

But from Eq 6.35a, $t'_\alpha = t_\alpha$ $(a = 1,2,3,6)$ and $t'_\alpha = -t_\alpha$ $(\alpha = 4,5)$. Likewise, $\epsilon'_\alpha = \epsilon_\alpha$ $(\alpha = 1,2,3,6)$ and $\epsilon'_\alpha = \epsilon_\alpha$ $(\alpha = 4,5)$, so that Eq 6.37 becomes

$$t'_1 = t_1 = \mathcal{C}_{11}\epsilon_1 + \mathcal{C}_{12}\epsilon_2 + \mathcal{C}_{13}\epsilon_3 - \mathcal{C}_{14}\epsilon_4 - \mathcal{C}_{15}\epsilon_5 + \mathcal{C}_{16}\epsilon_6 \,. \tag{6.38}$$

Comparing Eq 6.36 with Eq 6.38, we see that, for these expressions (each representing t_1) to be equal, we must have $\mathcal{C}_{14} = \mathcal{C}_{15} = 0$. Following the same procedure, we learn that, if we compare expressions for $t'_\alpha = t_\alpha$ $(\alpha = 2,3,6)$ and for $t'_\alpha = -t_\alpha$ $(\alpha = 4,5)$, the additional elastic constants $\mathcal{C}_{24}$, $\mathcal{C}_{25}$, $\mathcal{C}_{34}$, $\mathcal{C}_{35}$, $\mathcal{C}_{41}$, $\mathcal{C}_{42}$, $\mathcal{C}_{43}$, $\mathcal{C}_{46}$, $\mathcal{C}_{51}$, $\mathcal{C}_{52}$, $\mathcal{C}_{53}$, $\mathcal{C}_{56}$, $\mathcal{C}_{64}$, and $\mathcal{C}_{65}$ must also be zero for the x_1x_2 plane to be one of elastic symmetry. Accordingly, the elastic constant matrix for this case has the form

$$[\mathcal{C}_{\alpha\beta}] = \begin{bmatrix} \mathcal{C}_{11} & \mathcal{C}_{12} & \mathcal{C}_{13} & 0 & 0 & \mathcal{C}_{16} \\ \mathcal{C}_{21} & \mathcal{C}_{22} & \mathcal{C}_{23} & 0 & 0 & \mathcal{C}_{26} \\ \mathcal{C}_{31} & \mathcal{C}_{32} & \mathcal{C}_{33} & 0 & 0 & \mathcal{C}_{36} \\ 0 & 0 & 0 & \mathcal{C}_{44} & \mathcal{C}_{45} & 0 \\ 0 & 0 & 0 & \mathcal{C}_{54} & \mathcal{C}_{55} & 0 \\ \mathcal{C}_{61} & \mathcal{C}_{62} & \mathcal{C}_{63} & 0 & 0 & \mathcal{C}_{66} \end{bmatrix}, \tag{6.39}$$

and the original 36 constants are reduced to 20. Also, if a strain energy functions exists, $\mathcal{C}_{\alpha\beta} = \mathcal{C}_{\beta\alpha}$ and these 20 nonzero constants would be further reduced to 13.

If the x_2-x_3 plane is also one of elastic symmetry at the same time as the x_1-x_2 plane at a point and we repeat the procedure outlined above, we find that $\mathcal{C}_{16}$, $\mathcal{C}_{26}$, $\mathcal{C}_{36}$, $\mathcal{C}_{45}$, $\mathcal{C}_{54}$, $\mathcal{C}_{61}$, $\mathcal{C}_{62}$, and $\mathcal{C}_{63}$ must also be zero, and the $\mathcal{C}$ matrix is further reduced to

$$[\mathcal{C}_{\alpha\beta}] = \begin{bmatrix} \mathcal{C}_{11} & \mathcal{C}_{12} & \mathcal{C}_{13} & 0 & 0 & 0 \\ \mathcal{C}_{21} & \mathcal{C}_{22} & \mathcal{C}_{23} & 0 & 0 & 0 \\ \mathcal{C}_{31} & \mathcal{C}_{32} & \mathcal{C}_{33} & 0 & 0 & 0 \\ 0 & 0 & 0 & \mathcal{C}_{44} & 0 & 0 \\ 0 & 0 & 0 & 0 & \mathcal{C}_{55} & 0 \\ 0 & 0 & 0 & 0 & 0 & \mathcal{C}_{66} \end{bmatrix} \tag{6.40}$$

having 12 nonzero coefficients, or 9 if a strain energy function exists. Interestingly enough, if x_1x_3 is also a plane of elastic symmetry along with the two considered above, no further reduction in the $\mathcal{C}_{\alpha\beta}$ matrix occurs. A material possessing three mutually perpendicular planes of elastic symmetry is called an *orthotropic material*, and its elastic constants matrix is that given in Eq 6.40.

The reduction of the orthotropic elastic matrix to that of the isotropic matrix may be completed by successive consideration of the three axes of elastic symmetry shown in Fig. 6.4 as well as their respective transformation matrices. By the rotation of 90° about the x_1 axis (Fig. 6.4(a)), we find that $\mathcal{C}_{12} = \mathcal{C}_{13}$, $\mathcal{C}_{21} = \mathcal{C}_{31}$, $\mathcal{C}_{22} = \mathcal{C}_{33}$, $\mathcal{C}_{23} = \mathcal{C}_{32}$, and $\mathcal{C}_{55} = \mathcal{C}_{66}$. For the rotation of 90° about the x_3 axis (Fig. 6.4(b)), we see that $\mathcal{C}_{12} = \mathcal{C}_{21}$, $\mathcal{C}_{11} = \mathcal{C}_{22}$, $\mathcal{C}_{12} = \mathcal{C}_{23}$, $\mathcal{C}_{31} = \mathcal{C}_{32}$, and $\mathcal{C}_{44} = \mathcal{C}_{55}$. Finally, by a rotation of 45° about the x_3 axis (Fig. 6.4(c)), we obtain $2\mathcal{C}_{44} = \mathcal{C}_{11} - \mathcal{C}_{12}$. Therefore, by setting $\mathcal{C}_{44} = \mu$ and $\mathcal{C}_{12} = \lambda$, we may identify these remaining three $\mathcal{C}_{\alpha\beta}$'s with the Lamé constants and write the elastic coefficient matrix for isotropic behavior as

$$[\mathcal{C}_{\alpha\beta}] = \begin{bmatrix} \lambda+2\mu & \lambda & \lambda & 0 & 0 & 0 \\ \lambda & \lambda+2\mu & \lambda & 0 & 0 & 0 \\ \lambda & \lambda & \lambda+2\mu & 0 & 0 & 0 \\ 0 & 0 & 0 & \mu & 0 & 0 \\ 0 & 0 & 0 & 0 & \mu & 0 \\ 0 & 0 & 0 & 0 & 0 & \mu \end{bmatrix}. \tag{6.41}$$

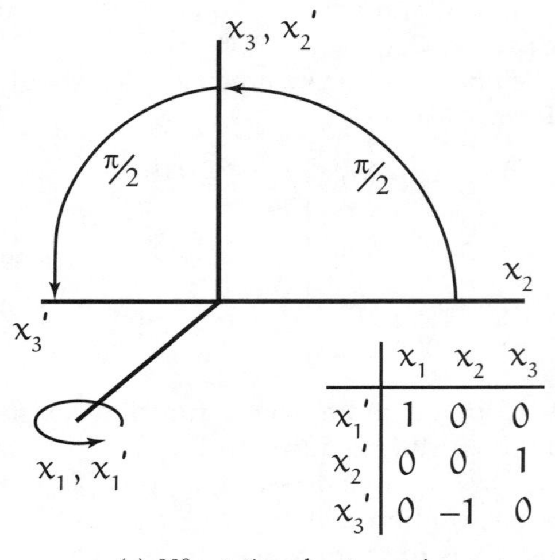

	x_1	x_2	x_3
x_1'	1	0	0
x_2'	0	0	1
x_3'	0	−1	0

(a) 90° rotation about x_1 axis

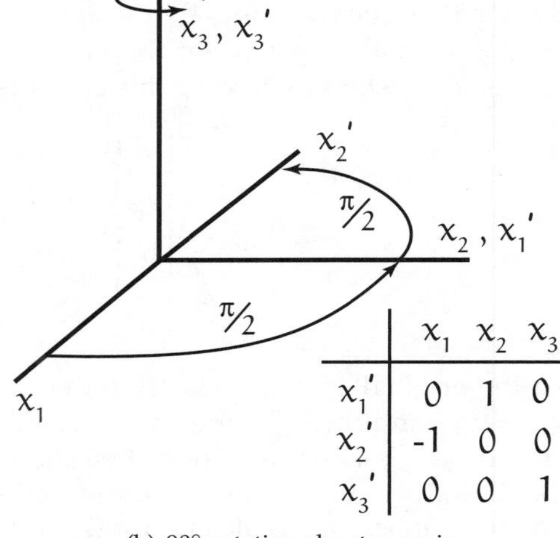

	x_1	x_2	x_3
x_1'	0	1	0
x_2'	-1	0	0
x_3'	0	0	1

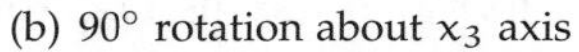

(b) 90° rotation about x_3 axis

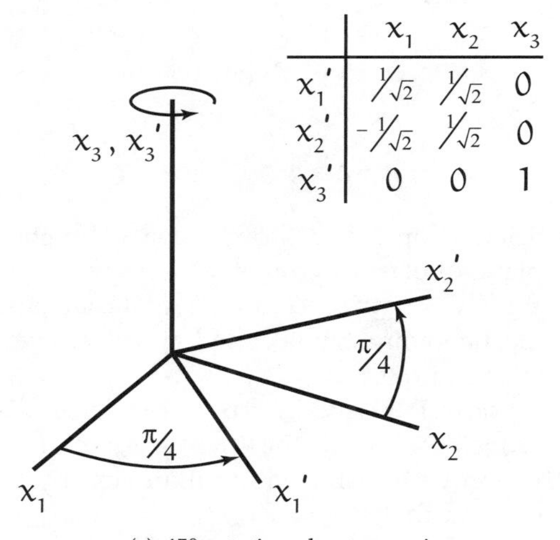

	x_1	x_2	x_3
x_1'	$1/\sqrt{2}$	$1/\sqrt{2}$	0
x_2'	$-1/\sqrt{2}$	$1/\sqrt{2}$	0
x_3'	0	0	1

(c) 45° rotation about x_3 axis

FIGURE 6.4
Geometry and transformation tables for reducing the elastic stiffness to the isotropic case.

From the definitions given in Eqs 6.27a and 6.27b, this matrix may be expressed in terms of the engineering constants E and ν so that Hooke's law for an isotropic body appears in matrix form as

$$\begin{bmatrix} t_1 \\ t_2 \\ t_3 \\ t_4 \\ t_5 \\ t_6 \end{bmatrix} = \frac{E}{(1+\nu)(1-2\nu)} \begin{bmatrix} 1-\nu & \nu & \nu & 0 & 0 & 0 \\ \nu & 1-\nu & \nu & 0 & 0 & 0 \\ \nu & \nu & 1-\nu & 0 & 0 & 0 \\ 0 & 0 & 0 & \frac{1}{2}(1-2\nu) & 0 & 0 \\ 0 & 0 & 0 & 0 & \frac{1}{2}(1-2\nu) & 0 \\ 0 & 0 & 0 & 0 & 0 & \frac{1}{2}(1-2\nu) \end{bmatrix} \begin{bmatrix} \epsilon_1 \\ \epsilon_2 \\ \epsilon_3 \\ \epsilon_4 \\ \epsilon_5 \\ \epsilon_6 \end{bmatrix} . \tag{6.42}$$

6.4 Isotropic Elastostatics and Elastodynamics, Superposition Principle

The formulation and solution of the basic problems of linear elasticity comprise the subjects we call *elastostatics* and *elastodynamics*. Elastostatics is restricted to those situations in which inertia forces may be neglected. In both elastostatics and elastodynamics, certain field equations have to be satisfied at all interior points of the elastic body under consideration, and at the same time the field variables must satisfy specific conditions on the boundary. In the case of elastodynamics problems, initial conditions on velocities and displacements must also be satisfied.

We begin our discussion with elastostatics for which the appropriate field equations are:

(a) Equilibrium equations

$$t_{ji,j} + \rho b_i = 0 \, , \tag{6.43}$$

(b) Strain-displacement relation

$$2\epsilon_{ij} = u_{i,j} + u_{j,i} \, , \tag{6.44}$$

(c) Hooke's law

$$t_{ij} = \lambda \delta_{ij} \epsilon_{kk} + 2\mu \epsilon_{ij} \, , \tag{6.45a}$$

or

$$\epsilon_{ij} = \frac{1}{E} \left[(1+\nu) t_{ij} - \nu \delta_{ij} t_{kk} \right] . \tag{6.45b}$$

It is usually assumed that the body forces b_i are known so that the solution we seek from the fifteen equations listed here is for the six stresses t_{ij}, the six strains ϵ_{ij}, and the three displacements u_i. The conditions to be satisfied on the boundary surface $\mathcal{S}$ will appear in one of the following statements:

1. displacements prescribed everywhere,

$$u_i = u_i^* (\mathbf{x}) \quad \text{on} \quad \mathcal{S} \tag{6.46}$$

where the asterisk denotes a prescribed quantity,

2. tractions prescribed everywhere,

$$t_i^{(\hat{n})} = t_i^{*(\hat{n})} \quad \text{on} \quad \mathcal{S}\,, \tag{6.47}$$

3. displacements prescribed on portion $\mathcal{S}_1$ of $\mathcal{S}$,

$$u_i = u_i^*(\mathbf{x}) \quad \text{on} \quad \mathcal{S}_1\,, \tag{6.48a}$$

with tractions prescribed on the remainder $\mathcal{S}_2$,

$$t_i^{(\hat{n})} = t_i^{*(\hat{n})} \quad \text{on} \quad \mathcal{S}_2\,. \tag{6.48b}$$

A most important feature of the field Eqs 6.43 through Eq 6.45a is that they are *linear* in the unknowns. Consequently, if $t_{ij}^{(1)}$, and $\epsilon_{ij}^{(1)}$, $u_i^{(1)}$ are a solution for body forces ${}^1b_i^*$ and surface tractions ${}^1t_i^{*(\hat{n})}$, whereas $t_{ij}^{(2)}$, $\epsilon_{ij}^{(2)}$, and $u_i^{(2)}$ are a solution for body forces ${}^2b_i^*$ and surface tractions ${}^2t_i^{*(\hat{n})}$, then

$$t_{ij} = t_{ij}^{(1)} + t_{ij}^{(2)}, \quad \epsilon_{ij} = \epsilon_{ij}^{(1)} + \epsilon_{ij}^{(2)}, \quad \text{and} \quad u_i = u_i^{(1)} + u_i^{(2)}$$

offer a solution for the situation where $b_i = {}^1b_i^* + {}^2b_i^*$ and $t_i^{(\hat{n})} = {}^1t_i^{*(\hat{n})} + {}^2t_i^{*(\hat{n})}$. This is a statement of the *principle of superposition*, which is extremely useful for the development of solutions in linear elasticity.

For those problems in which the boundary conditions are given in terms of displacements by Eq 6.46, it is convenient for us to eliminate the stress and strain unknowns from the field equations so as to state the problem solely in terms of the unknown displacement components. Thus, by substituting Eq 6.44 into Hooke's law (Eq 6.45a) and that result into the equilibrium equations (Eq 6.43), we obtain the three second-order partial differential equations

$$\mu u_{i,jj} + (\lambda + \mu)u_{j,ji} + \rho b_i = 0 \tag{6.49}$$

which are known as the *Navier equations*. If a solution can be determined for these equations that also satisfies the boundary condition Eq 6.46, that result may be substituted into Eq 6.44 to generate the strains and those in turn substituted into Eq 6.45a to obtain the stresses.

When the boundary conditions are given in terms of surface tractions (Eq 6.47), the equations of compatibility for infinitesimal strains (Eq 4.90) may be combined with Hooke's law (Eq 6.45b) and the equilibrium equations to arrive, after a certain number of algebraic manipulations, at the equations

$$t_{ij,kk} + \frac{1}{1+\nu} t_{kk,ij} + \rho\left(b_{i,j} + b_{j,i}\right) + \frac{\nu}{1-\nu}\delta_{ij}\rho b_{k,k} = 0 \tag{6.50}$$

which are known as the *Beltrami-Michell stress equations of compatibility*. In combination with the equilibrium equations, these equations comprise a system for the solution of the stress components, but it is not an especially easy system to solve. As was the case with the infinitesimal strain equation of compatibility, the body must be simply connected.

The solution for elastostatics problems is unique. Suppose that there exists two solutions to the governing equations Eq 6.43-6.45a with boundary conditions Eq 6.46-6.48b. The solutions will be denoted $\mathbf{T}^{(1)}$, $\boldsymbol{\epsilon}^{(1)}$, $\mathbf{u}^{(1)}$ and $\mathbf{T}^{(2)}$, $\boldsymbol{\epsilon}^{(2)}$, $\mathbf{u}^{(2)}$ corresponding to the same body forces and the appropriate boundary conditions. The difference of these solutions, $\mathbf{T} = \mathbf{T}^{(1)} - \mathbf{T}^{(2)}$, $\boldsymbol{\epsilon} = \boldsymbol{\epsilon}^{(1)} - \boldsymbol{\epsilon}^{(2)}$, $\mathbf{u} = \mathbf{u}^{(1)} - \mathbf{u}^{(2)}$, is also a solution to the governing equations with $\mathbf{b} = \mathbf{0}$ since the system of equations is linear and superposition applies. The solution $\mathbf{T}$, $\boldsymbol{\epsilon}$, $\mathbf{u}$ satisfies the boundary conditions

$$\mathbf{u} = 0 \quad \text{on } \mathcal{S}_1 \qquad \text{or} \qquad \mathbf{t} = 0 \quad \text{on } \mathcal{S}_2 \; .$$

Integrating over the boundary $\mathcal{S}$ of the body gives

$$\int_{\mathcal{S}} \mathbf{u} \cdot \mathbf{t} \, d\mathcal{S} = 0 \; .$$

Substituting $\mathbf{t} = \mathbf{T} \cdot \mathbf{n}$ for a symmetric stress tensor, we have

$$\int_{\mathcal{S}} \mathbf{u} \cdot \mathbf{T} \cdot \mathbf{n} \, d\mathcal{S} = 0$$

and applying the divergence theorem yields

$$\int_{\mathcal{V}} \operatorname{div}(\mathbf{u} \cdot \mathbf{T}) \, d\mathcal{V} = 0 \; . \tag{6.51}$$

Writing the integrand in Cartesian tensor notation gives

$$\operatorname{div}(\mathbf{u} \cdot \mathbf{T}) = (u_i t_{ij})_{,j} = u_{i,j} t_{ij} + u_i t_{ij,j} \; .$$

The last term is zero from the equilibrium equations $t_{ij,j} = 0$ since $\mathbf{b} = \mathbf{0}$. The first term can be written as

$$u_{i,j} t_{ij} = t_{ij} (\epsilon_{ij} + \omega_{ij}) = t_{ij} \epsilon_{ij} \tag{6.52}$$

since t_{ij} is symmetric and ω_{ij} is skew-symmetric. From Eq 6.21a and 6.52, Eq 6.51 becomes

$$\int_{\mathcal{V}} 2W(\boldsymbol{\epsilon}) \, d\mathcal{V} = 0 \; .$$

This integral is zero when $\boldsymbol{\epsilon} = 0$ provided that $W(\boldsymbol{\epsilon})$ is positive definite in Eq 6.21a. For a linear, isotropic body this requires that $E > 0$ and $-1 < \nu < \frac{1}{2}$. Since $\boldsymbol{\epsilon} = 0$, we have $\boldsymbol{\epsilon}^{(1)} = \boldsymbol{\epsilon}^{(2)}$ and $\mathbf{T}^{(1)} = \mathbf{T}^{(2)}$ from Hooke's law. There cannot be two different solutions to the linear equations of elastostatics for the same body forces and boundary conditions. A solution that satisfies the governing equations and boundary conditions is unique.

In elastodynamics, the equilibrium equations must be replaced by the equations of motion (Eq 5.27) in the system of basic field equations. Therefore, all field quantities are now considered functions of time as well as of the coordinates, so that a solution for the displacement field, for example, appears in the form $u_i = u_i(\mathbf{x}, t)$. In addition, the solution must satisfy not only boundary conditions which may be functions of time as in

$$u_i = u_i^* (\mathbf{x}, t) \quad \text{on} \quad \mathcal{S} \tag{6.53a}$$

or

$$t_i^{(\hat{\mathbf{n}})} = t_i^{*(\hat{\mathbf{n}})} (\mathbf{x}, t) \quad \text{on} \quad \mathcal{S} \; , \tag{6.53b}$$

but also *initial conditions*, which usually are taken as

$$u_i = u_i^* (\mathbf{x}, 0) \tag{6.54a}$$

and

$$\dot{u}_i = \dot{u}_i^* (\mathbf{x}, 0) \; . \tag{6.54b}$$

Analogous to Eq 6.49 for elastostatics, it is easily shown that the governing equations for displacements in elastodynamics theory are

$$\mu u_{i,jj} + (\lambda + \mu) u_{j,ji} + \rho b_i = \rho \ddot{u}_i \tag{6.55}$$

which are also called *Navier's equations*.

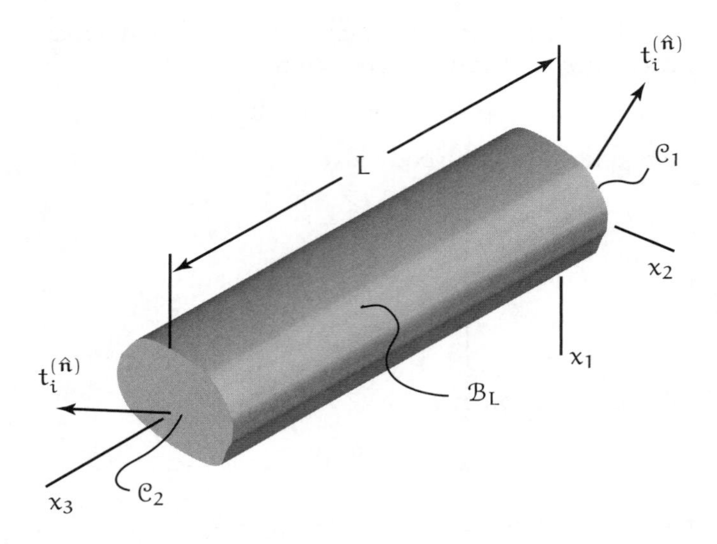

FIGURE 6.5
Beam geometry for the Saint-Venant problem.

6.5 Saint-Venant Problem

Solving an elasticity problem by satisfying Eqs 6.43, 6.44 and 6.45a along with the boundary conditions is a daunting task. This is even before the boundary conditions are considered. In the 1800s Saint-Venant studied long beams that were loaded in a variety of ways: extension, torsion, pure bending and flexure. Rather than satisfy exact boundary conditions on the lateral surface and the ends, Saint-Venant solved the problems by considering relaxed boundary conditions on the beam ends. If one considers statically equivalent force and moment systems, the solution sufficiently far away from the application of the load will be the same. This is known as the Saint-Venant principle. The relaxed boundary conditions on the beam ends requires that the resultant force and moment be considered rather than the exact traction distribution over the end.

To demonstrate Saint-Venant's solution consider a beam that is represented by a right cylinder extending in the x_3 direction as shown in Fig. 6.5. The lateral surface of the cylinder, $\mathcal{B}_L$, will be stress free. This condition is called the lateral boundary condition and can be written as

$$t_i^{(\hat{n})} = t_{i1}n_1 + t_{i2}n_2 = 0 \qquad \text{on} \quad \mathcal{B}_L \,. \tag{6.56}$$

Let the cylinder be generated by a simply connected cross section $\mathcal{C}$ which is perpendicular to the x_3 axis. For the different problems (extension, torsion, pure bending and flexure) the ends, $\mathcal{C}_1$ and $\mathcal{C}_2$, will have relaxed boundary conditions. The sum of the stress over the end's area will give a force vector T_i

$$\int_{\mathcal{C}_1} t_{3i}\, d\mathcal{S} = T_i \,. \tag{6.57}$$

Bending moments on the ends are expressed in terms of the axial stress t_{33} as follows:

$$\int_{\mathcal{C}_1} t_{33}x_2 \, d\mathcal{S} = M_1 , \qquad \int_{\mathcal{C}_1} t_{33}x_1 \, d\mathcal{S} = -M_2 . \tag{6.58}$$

Finally, the torque on the beam's end may be written as

$$\int_{\mathcal{C}_1} (t_{32}x_1 - t_{31}x_2) \, d\mathcal{S} = M_3 = M_t . \tag{6.59}$$

Similar equations hold for end $\mathcal{C}_2$.

Saint-Venant used a *semi-inverse method* in solving these problems. A reasonable assumption was made regarding the stress components, strain or displacements. From this assumption the equilibrium and compatibility equations were shown to be satisfied. Boundary conditions, exact and relaxed, were checked. Because solutions to elasticity problems are unique the solution obtained was, of course, the only solution.

In the following subsections Saint-Venant solutions for extension, torsion, pure bending and flexure are developed. The different problems are solved by setting the components of the end force T_i and moment M_i to different values:

$$\begin{array}{lll}
\text{(I)} & T_1 = T_2 = 0, \;\; T_3 = T, \;\; M_i = 0 & \text{Pure Extension} \\
\text{(II)} & T_i = 0, \;\; M_1 = M_2 = 0, \;\; M_3 = M_t & \text{Pure Torsion} \\
\text{(III)} & T_i = 0, \;\; M_1 = M_3 = 0, \;\; M_2 = M & \text{Pure Bending} \\
\text{(IV)} & T_1 = T, \;\; T_2 = T_3 = 0, \;\; M_i = 0 & \text{Pure Flexure}
\end{array}$$

More complex elasticity problems can be obtained by superposing the basic solutions.

6.5.1 Extension

For the case of pure extension the nonzero stress components are assumed to be

$$t_{11} = \frac{T}{A} \tag{6.60}$$

where the area is given by

$$A = \int_{\mathcal{C}} d\mathcal{S} ,$$

and all other components are zero. These stress components clearly satisfy the equilibrium conditions without body forces, Eq 6.43. Also, the relaxed boundary conditions are satisfied.

The strains can be easily found from Eq 6.28 to be

$$\epsilon_{33} = \frac{T}{EA} , \qquad \epsilon_{11} = \epsilon_{22} = -\frac{\nu T}{EA} \tag{6.61}$$

with all other strain components being zero. Clearly the compatibility condition, Eq 4.90 is satisfied since the strains are constant. Using the strain-displacement relationship, Eq 6.44, the displacements can be found to be

$$u_3 = \frac{T}{EA}x_3 , \qquad u_1 = -\frac{\nu T}{EA}x_1 , \qquad u_2 = -\frac{\nu T}{EA}x_2 \tag{6.62}$$

by direct integration.

6.5.2 Torsion

We begin with a brief review of the solution to the simplest of torsion problems, the case of a shaft having a constant circular cross section when subjected to equilibrating end couples, M_t as shown in Fig. 6.6(a). Let the end face at $x_3 = 0$ be fixed while the face at $x_3 = L$ is allowed to rotate about the axis of the shaft. It is assumed that plane sections perpendicular to the axis remain plane under the twisting, and that each rotates through an angle proportional to its distance from the fixed end. Accordingly, a point in the cross section at coordinate x_3 will rotate an angle of $\theta x_3$where θ is the angle of twist per unit length. Each point, say point **P**, in the cross section travels a distance $\theta x_3 R$ which is proportional to the distance R from the x_3 axis as shown in Fig. 6.6(b). The distance squared to point **P** is the square of the x_1 and x_2 coordinates (Fig. 6.6(c)). Using this distance, it is easy to define the cosine and sine of angle β. It is straightforward to write out displacements u_1 and u_2 in terms of θ since $\cos(90-\beta) = x_2/R$ and $\sin(90-\beta) = x_1/R$. Thus,

$$u_1 = -\theta x_2 x_3 \ , \quad u_2 = \theta x_1 x_3 \ , \quad u_3 = 0 \ , \tag{6.63}$$

where θ is the angle of twist per unit length of the shaft.

Recall from Eq 4.62 that $2\epsilon_{ij} = u_{i,j} + u_{j,i}$ by which we may calculate the strains from Eq 6.63. The resulting strains are then inserted into Eq 6.23 to obtain the stress components (since $\mu \equiv G$ by Eq 6.29a)

$$t_{23} = G\theta x_1 \ , \quad t_{13} = -G\theta x_2 \ , \quad t_{11} = t_{22} = t_{33} = t_{12} = 0 \ . \tag{6.64}$$

Because these stress components, as well as the strains from which they were derived, are either linear functions of the coordinates or zero, the compatibility equations Eq 4.90 are satisfied. Likewise, for zero body forces, the equilibrium equations Eq 6.43 are clearly satisfied. The lateral surface of the shaft is stress free. To verify this, consider the stress components in the direction of the normal at a point on the cross-section perimeter designated in Fig. 6.6(b). At a radius $R = a$,

$$t_{13}\frac{x_1}{a} + t_{23}\frac{x_2}{a} = \frac{G\theta}{a}\left(-x_2 x_1 + x_1 x_2\right) = 0 \ . \tag{6.65}$$

At the same time, the total shearing stress at any point of the cross section is the resultant

$$\tau = \sqrt{t_{13}^2 + t_{23}^2} = G\theta\sqrt{x_1^2 + x_2^2} = G\theta R \tag{6.66}$$

which indicates that the shear is proportional to the radius at the point, and perpendicular to that radius.

By summing the moments of the shear forces on either end face of the shaft, we find that

$$M_t = \iint \left(x_1 t_{23} + x_2 t_{13}\right) dx_1 dx_2 = G\theta \iint R^2 dx_1 dx_2 = G\theta I_P \tag{6.67}$$

where I_P is the polar moment of inertia of the cross section.

For a prismatic shaft of any cross section other than circular, shown by the schematic contour of Fig. 6.7, plane sections do not remain plane under twisting, and warping will occur. For such cases we must modify Eq 6.63 by expressing the displacements in the form

$$u_1 = -\theta x_2 x_3 \ , \quad u_2 = \theta x_1 x_3 \ , \quad u_3 = \theta\psi\left(x_1, x_2\right) \ , \tag{6.68}$$

where $\psi(x_1, x_2)$ is called the *warping function*. Note that the warping is independent of x_3 and therefore the same for all cross sections. Also, we assume that x_3 is a centroidal axis although this condition is not absolutely necessary.

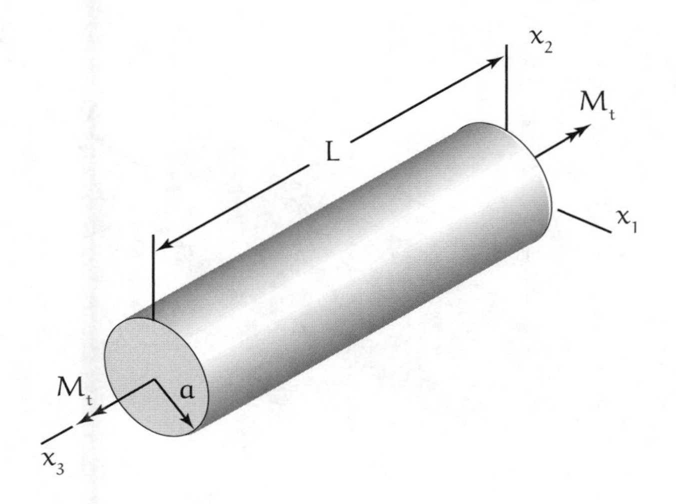

(a) Cylinder with self-equilibrating moments M_t.

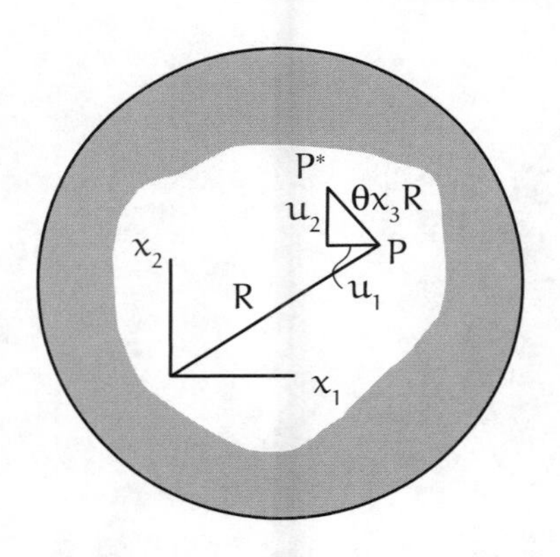

(b) Displacement of point P to P^* in cross section.

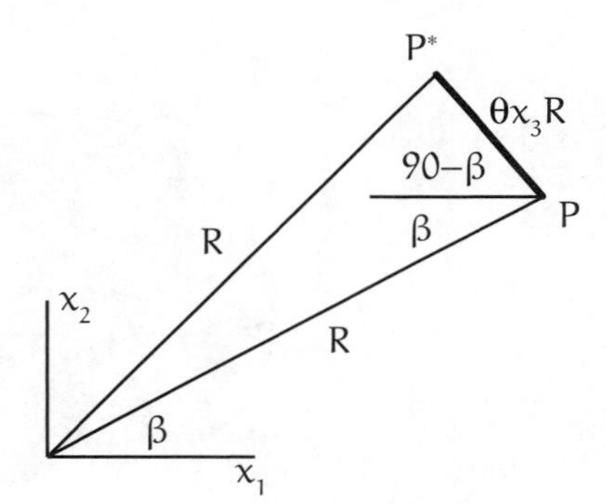

(c) Detail of cross section twist β.

FIGURE 6.6
Geometry and kinematic definitions for torsion of a circular shaft.

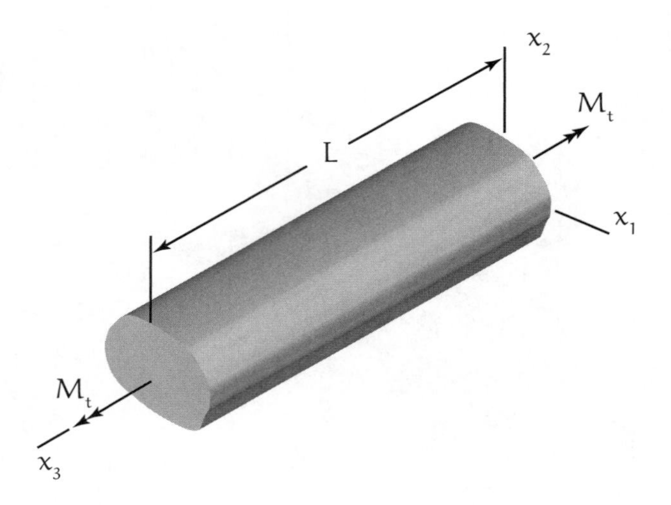

(a) Prismatic cylinder with self-equilibrating moments M_t.

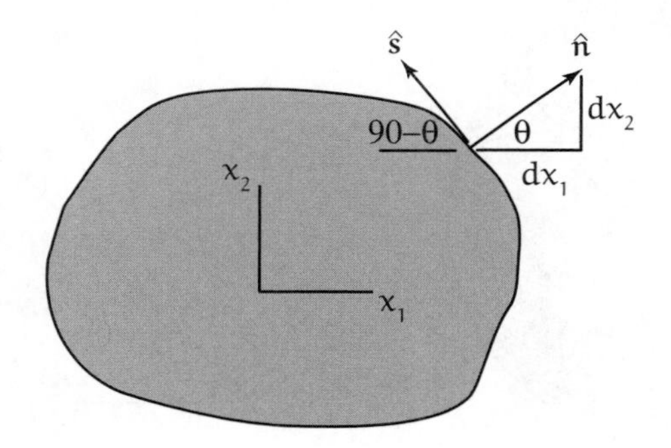

(b) Displacement of point P to P* in cross section.

FIGURE 6.7
The more general torsion case of a prismatic beam loaded by self equilibrating moments.

As in the analysis of the circular shaft we may again use Eq 4.62 along with Eq 6.23 to calculate the stress components from Eq 6.68. Thus

$$t_{11} = t_{22} = t_{33} = t_{12} = 0\,, \quad t_{13} = G\theta\,(\psi_{,1} - x_2)\,, \quad t_{23} = G\theta\,(\psi_{,2} - x_1)\,. \tag{6.69}$$

It is clear from these stress components that there are no normal stresses between the longitudinal elements of the shaft. The first two of the equilibrium equations, Eq 6.43, are satisfied identically by Eq 6.69 in the absence of body forces, and substitution into the third equilibrium equation yields

$$t_{13,1} + t_{23,2} + t_{33,3} = G\theta\,(\psi_{,11} + \psi_{,22}) = 0 \tag{6.70}$$

which indicates that ψ must be harmonic,

$$\nabla^2\psi = 0\,, \tag{6.71}$$

on the cross section of the shaft.

Boundary conditions on the surfaces of the shaft must also be satisfied. On the lateral surface, which is stress free, the following conditions based upon Eq 3.32, must prevail

$$t_{11}n_1 + t_{12}n_2 = 0\,, \quad t_{21}n_1 + t_{22}n_2 = 0\,, \quad t_{31}n_1 + t_{32}n_2 = 0\,, \tag{6.72}$$

noting that here $n_3 = 0$. The first two of these equations are satisfied identically while the third requires

$$G\theta\,(\psi_{,1} - x_2)\,n_1 + G\theta\,(\psi_{,2} - x_1)\,n_2 = 0 \tag{6.73a}$$

which reduces immediately to

$$\psi_{,1}n_1 + \psi_{,2}n_2 = \frac{d\psi}{dn} = x_2n_1 + x_1n_2\,. \tag{6.73b}$$

Therefore, $\psi\,(x_1, x_2)$ must be harmonic in the cross section of the shaft shown in Fig. 6.7, and its derivative with respect to the normal of the lateral surface must satisfy Eq 6.73b on the perimeter C of the cross section.

We note further that in order for all cross sections to be force free, that is, in simple shear over those cross sections

$$\iint t_{13}\,dx_1\,dx_2 = \iint t_{23}\,dx_1\,dx_2 = \iint t_{33}\,dx_1\,dx_2 = 0\,. \tag{6.74}$$

Since $t_{33} = 0$, the third integral here is trivial. Considering the first integral we may write

$$G\theta\iint(\psi_{,1} - x_2)\,dx_1\,dx_2 = G\theta\iint\left\{\frac{\partial}{\partial x_1}\left[x_1\,(\psi_{,1} - x_2)\right] + \frac{\partial}{\partial x_2}\left[x_1\,(\psi_{,2} - x_1)\right]\right\}dx_1\,dx_2 \tag{6.75}$$

where the condition $\nabla^2\psi = 0$ has been used. Green's theorem allows us to convert to the line integral taken around the perimeter $\mathcal{C}$

$$G\theta\int_{\mathcal{C}} x_1\left\{(\psi_{,1} - x_2)\,n_1 + (\psi_{,2} - x_1)\,n_2\right\}ds = 0 \tag{6.76}$$

which by Eq 6.73b is clearly satisfied. By an analogous calculation we find that the second integral of Eq 6.74 is also satisfied.

On the end faces of the shaft, $x_3 = 0$ or $x_3 = L$, the following conditions must be satisfied

$$\iint x_2t_{33}\,dx_1\,dx_2 = \iint x_1t_{33}\,dx_1\,dx_2 = 0\,, \quad \iint(x_1t_{23} - x_2t_{13})\,dx_1\,dx_2 = M_t\,. \tag{6.77}$$

Again, since $t_{33} = 0$, the first two of these are trivial. The third leads to

$$M_t = G\theta \iint \left(x_1^2 + x_2^2 + x_1\psi_{,2} - x_2\psi_{,1}\right) dx_1 dx_2 \,. \tag{6.78}$$

Defining the torsional rigidity as

$$K = G \iint \left(x_1^2 + x_2^2 + x_1\psi_{,2} - x_2\psi_{,1}\right) dx_1 dx_2 \tag{6.79}$$

which can be evaluated once $\psi(x_1, x_2)$ is known, we express the angle of twist as

$$\theta = \frac{M_t}{K} \,. \tag{6.80}$$

A second approach to the general torsion problem rests upon the introduction of a torsion stress function, designated here by Φ and defined so that the non-zero stresses are related to it by the definitions

$$t_{13} = \frac{\partial \Phi}{\partial x_2} \,, \qquad t_{23} = -\frac{\partial \Phi}{\partial x_1} \,. \tag{6.81}$$

Thus,

$$\frac{\partial \Phi}{\partial x_2} = G\theta\left(\psi_{,1} - x_2\right) \,, \qquad \frac{\partial \Phi}{\partial x_1} = -G\theta\left(\psi_{,2} - x_1\right) \,. \tag{6.82}$$

By eliminating ψ from this pair of equations we obtain

$$\nabla^2 \Phi = -2G\theta \,. \tag{6.83}$$

As already noted, the lateral surface of the shaft parallel to the axis must remain stress free, that is, the third of Eq 6.72 must be satisfied. However, it is advantageous to write this condition in terms of the unit vector $\hat{\mathbf{s}}$ along the boundary rather than unit normal $\hat{\mathbf{n}}$ as shown in Fig. 6.6(b). It follows directly from geometry that $n_1 = dx_2/ds$ and $n_2 = -dx_1/ds$. In terms of ds, the differential distance along the perimeter $\mathcal{C}$, the stress components in the normal direction will be given by

$$t_{13}\frac{dx_2}{ds} - t_{23}\frac{dx_1}{ds} = 0 \tag{6.84}$$

which in terms of Φ becomes

$$\frac{\partial \Phi}{\partial x_1}\frac{dx_1}{ds} + \frac{\partial \Phi}{\partial x_2}\frac{dx_2}{ds} = \frac{d\Phi}{ds} = 0 \,. \tag{6.85}$$

Thus, Φ is a constant along the perimeter of the cross section and will be assigned the value of zero here.

Finally, conditions on the end faces of the shaft must be satisfied. Beginning with the first of Eq 6.74, we have in terms of Φ

$$\iint \frac{\partial \Phi}{\partial x_2} dx_1 dx_2 = \iint \left(\frac{\partial \Phi}{\partial x_2} dx_2\right) dx_1 = \int \Phi|_a^b \, dx_1 = 0 \tag{6.86}$$

since Φ is constant on the perimeter. Likewise, by the same reasoning, the second of Eq 6.74 is satisfied, while the third is satisfied since $t_{33} = 0$. The first two conditions in Eq 6.77 are also satisfied identically and the third becomes

$$\iint \left(-x_1\frac{\partial \Phi}{\partial x_1} - x_2\frac{\partial \Phi}{\partial x_2}\right) dx_1 dx_2 = M_t \,. \tag{6.87}$$

Integrating by parts and using the fact that Φ is assumed zero on the perimeter $\mathcal{C}$ yields

$$M_t = 2\iint \Phi dx_1 dx_2 \ . \tag{6.88}$$

Thus, the solution by this approach consists of determining the stress function Φ which is zero on the cross-section perimeter, and satisfies Eq 6.83. Based upon that result we may determine θ from Eq 6.88.

Example 6.2

Determine the stresses and the angle of twist for a solid elliptical shaft of the dimensions shown when subjected to end couples M_t.

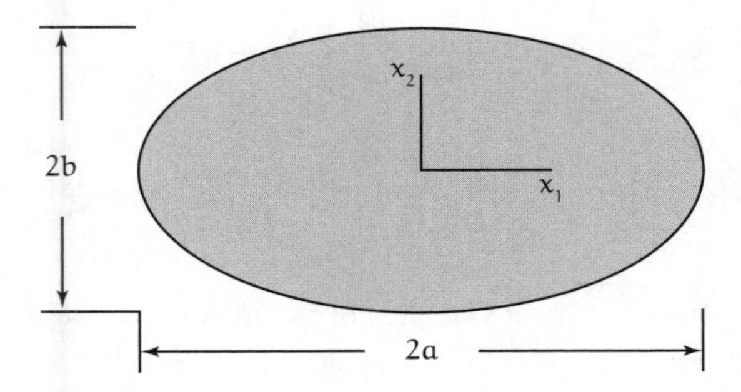

Solution

The equation of this ellipse is given by

$$\frac{x_1^2}{a^2} + \frac{x_2^2}{b^2} = 1 \ .$$

Therefore, take the stress function Φ in the form

$$\Phi = \lambda\left(\frac{x_1^2}{a^2} + \frac{x_2^2}{b^2} - 1\right) \tag{6.89}$$

where λ is a constant. Thus, Φ is zero on the cross-section perimeter. From Eq 6.83

$$2\lambda\left(\frac{1}{a^2} + \frac{1}{b^2}\right) = -2G\theta \ ,$$

so that

$$\lambda = -\frac{a^2b^2G\theta}{a^2+b^2} \ . \tag{6.90}$$

Now from Eq 6.88

$$M_t = -\frac{2a^2b^2G\theta}{a^2+b^2}\iint\left(\frac{x_1^2}{a^2} + \frac{x_2^2}{b^2} - 1\right) dx_1 dx_2 \ ,$$

and noting that

$$\iint x_1^2 dx_1 dx_2 = I_{x_2} = \frac{1}{4}\pi b a^3 \ ,$$

$$\iint x_2^2 dx_1 dx_2 = I_{x_1} = \frac{1}{4}\pi a b^3 \ ,$$

and

$$\iint dx_1 dx_2 = \pi ab$$

(the area of the cross section) we may solve for M_t which is

$$M_t = \frac{\pi a^3 b^3 G\theta}{a^2 + b^2} .$$

From this result,

$$\theta = \frac{a^2 + b^2}{\pi a^3 b^3 G} M_t$$

which when substituted into Eq 6.90 and that into Eq 6.89 gives

$$\Phi = -\frac{M_t}{\pi ab}\left(\frac{x_1^2}{a^2} + \frac{x_2^2}{b^2} - 1\right) .$$

Now, by definition

$$t_{13} = \frac{\partial \Phi}{\partial x_2} = -\frac{2M_t}{\pi ab^3} x_2 , \quad t_{23} = -\frac{\partial \Phi}{\partial x_1} = -\frac{2M_t}{\pi a^3 b} x_1 .$$

The maximum stress occurs at the ends of the minor axis, and equals

$$\tau_{max} = \pm \frac{2M_t}{\pi ab^2} ,$$

and the torsional rigidity is easily calculated to be

$$K = \frac{M_t}{\theta} = \frac{\pi a^3 b^3 G}{a^2 + b^2} = \frac{G\,(A)^4}{4\pi^2 I_P} .$$

Note also that for $a = b$ (circular cross section), the resultant stress at any point is

$$\tau = \sqrt{t_{13}^2 + t_{23}^2} = \frac{M_t}{I_P} r$$

in agreement with elementary theory.

It should be pointed out that for shafts having perimeters that are not expressible by simple equations, solutions may be obtained by using stress functions in the form of infinite series. Such analyses are beyond the scope of this introductory section.

6.5.3 Pure Bending

Consider the beam to be subject to end conditions

$$T_i = 0, \quad M_1 = M_3 = 0, \quad M_2 = M , \tag{6.91}$$

and assume the stresses to be

$$t_{33} = -\frac{Mx_1}{I} \tag{6.92}$$

with all other components zero. The compatibility equations are satisfied since the nonzero stresses are linear in the coordinates. The equilibrium equations, $t_{ji,j} = 0$, only have one equation which is not identically satisfied, and the remaining equation, $t_{33,3} = 0$, is satisfied from the assumed stress Eq 6.92.

The end boundary conditions can be shown as satisfied. First, $T_1 = T_2 = 0$ is trivially satisfied by considering Eq 6.92. Considering the torque on the end, Eq 6.59, it is clear that $M_3 = M_t = 0$ since $t_{31} = t_{32} = 0$. The axial force on the ends is given by Eq 6.57 as

$$T_3 = \int_{\mathcal{C}_1} -\frac{Mx_1}{I} d\mathcal{S} = -\frac{M}{I}\int_{\mathcal{C}_1} x_1 d\mathcal{S} = 0 \tag{6.93}$$

provided that the axes origin is chosen at the centroid of the cross section $\mathcal{C}$. A similar calculation for M_1 gives

$$M_1 = \int_{\mathcal{C}_1} -\frac{Mx_1}{I} x_2 d\mathcal{S} = -\frac{M}{I}\int_{\mathcal{C}_1} x_1 x_2 d\mathcal{S} = 0 \tag{6.94}$$

as long as the principal axes are used. Finally,

$$M_2 = -\int_{\mathcal{C}_1} \frac{Mx_1}{I} x_1 d\mathcal{S} = \frac{Mx_1}{I}\int_{\mathcal{C}_1} x_1^2 d\mathcal{S} = M \tag{6.95}$$

where

$$I = \int_{\mathcal{C}_1} x_1^2 d\mathcal{S} \, .$$

Strain components are found from Eq 6.28 giving three nonzero components

$$\epsilon_{11} = \epsilon_{22} = \frac{\nu M x_1}{EI}; \; \epsilon_{33} = -\frac{Mx_1}{EI} \, . \tag{6.96}$$

Displacements can be found using Eq 4.101. The term $u_j(\xi^0) + w_{ij}(\xi^0)\left(\xi_j - \xi_j^0\right)$ represents a rigid displacement and will be set to zero. Taking the component u_1 as an example for finding the displacements yields

$$\begin{aligned}
u_1 &= \int_0^x [\epsilon_{11} d\xi_1 + \epsilon_{12} d\xi_2 + \epsilon_{13} d\xi_3 \\
&\quad + (x_k - \xi_k)\,\epsilon_{11,k} d\xi_1 + (x_k - \xi_k)\,\epsilon_{12,k} d\xi_2 + (x_k - \xi_k)\,\epsilon_{13,k} d\xi_3 \\
&\quad - (x_k - \xi_k)\,\epsilon_{k1,1} d\xi_1 - (x_k - \xi_k)\,\epsilon_{k2,1} d\xi_2 - (x_k - \xi_k)\,\epsilon_{k3,1} d\xi_3] \\
&= \int_0^x [\epsilon_{11} d\xi_1 + (x_1 - \xi_1)\,\epsilon_{11,1} d\xi_1 \\
&\quad - (x_1 - \xi_1)\,\epsilon_{11,1} d\xi_1 - (x_2 - \xi_2)\,\epsilon_{22,1} d\xi_2 - (x_3 - \xi_3)\,\epsilon_{33,1} d\xi_3] \qquad (6.97) \\
&= \int_0^x \left[\frac{\nu M}{EI}\xi_1 d\xi_1 - (x_2 - \xi_2)\frac{\nu M}{EI} d\xi_2 + (x_3 - \xi_3)\frac{M}{EI} d\xi_3\right] \\
&= \frac{\nu M}{2EI}x_1^2 - \frac{\nu M}{EI}\left(x_2^2 - \tfrac{1}{2}x_2^2\right) + \frac{M}{EI}\left(x_3^2 - \tfrac{1}{2}x_3^2\right) \\
&= \frac{M}{2EI}\left(\nu x_1^2 - \nu x_2^2 + x_3^2\right) \, .
\end{aligned}$$

Similar calculations for u_2 and u_3 give the displacements for pure bending as

$$
\begin{aligned}
u_1 &= \frac{M}{2EI}(\nu x_1^2 - \nu x_2^2 + x_3^2) \, , \\
u_2 &= \frac{\nu M}{EI} x_1 x_2 \, , \\
u_3 &= -\frac{M}{EI} x_1 x_3 \, ,
\end{aligned}
\tag{6.98}
$$

to within an infinitesimal rigid displacement.

In the pure bending problem, $x_1 = 0$ represents the neutral surface, and the centerline displacements match elementary beam theory.

6.5.4 Flexure

As the final Saint-Venant problem, consider the beam to have end $\mathcal{C}_1$ fixed and a flexure load of T acting on $\mathcal{C}_2$. The relaxed boundary conditions are

$$T_1 = T, \; T_2 = T_3 = 0, \;\; M_i = 0 \, .$$

Assume the stresses to be

$$t_{11} = t_{22} = t_{12} = 0; \;\; t_{33} = -\frac{T}{I}(L - x_3)x_1 \tag{6.99}$$

with no assumption made on the components t_{31} and t_{32}. Equilibrium equations are

$$t_{13,3} = 0, \;\; t_{23,3} = 0, \;\; t_{31,1} + t_{32,2} = -\frac{Tx_1}{I} \tag{6.100}$$

which implies

$$t_{13} = t_{13}(x_1, x_2); \;\; t_{23} = t_{23}(x_1, x_2) \, . \tag{6.101}$$

The strain may be calculated from Eq 6.28 to obtain

$$
\begin{gathered}
\epsilon_{11} = \epsilon_{22} = \frac{\nu T}{EI}(L - x_3)x_1, \;\; \epsilon_{33} = -\frac{T}{EI}(L - x_3)x_1 \\
\epsilon_{23} = t_{23}/2\mu; \;\; \epsilon_{31} = t_{31}/2\mu; \;\; \epsilon_{12} = 0
\end{gathered}
\tag{6.102}
$$

The compatibility equations must be satisfied since the problem is posed in terms of stresses. Eq 4.91 results in two nontrivial conditions

$$\frac{\partial}{\partial x_1}(\epsilon_{23,1} - \epsilon_{13,2}) = 0 \, , \tag{6.103a}$$

$$\frac{\partial}{\partial x_2}(\epsilon_{31,2} - \epsilon_{23,1}) = \frac{\nu T}{EI} \, , \tag{6.103b}$$

Integration of Eq 6.103a gives

$$\epsilon_{23,1} - \epsilon_{31,2} = \alpha - \frac{\nu T}{EI} x_2 \tag{6.104}$$

where α is taken as a constant by virtue of Eq 6.103a. Equation 6.104 may be written as

$$\left(\epsilon_{23} - \frac{1}{2}\alpha x_1\right)_{,1} = \left(\epsilon_{31} + \frac{1}{2}\alpha x_2 - \frac{1}{2}\frac{\nu T}{EI} x_2^2\right)_{,2} \tag{6.105}$$

from which the shear strains may be found in terms of a potential function f

$$\epsilon_{32} - \frac{1}{2}\alpha x_1 = \frac{1}{2}f_{,2}; \;\; \epsilon_{31} + \frac{1}{2}\alpha x_2 - \frac{1}{2}\frac{\nu T}{EI}x_2^2 = \frac{1}{2}f_{,1} \tag{6.106}$$

The shear stresses may be written using Hooke's Law and Eq 6.106

$$\begin{aligned} t_{31} &= 2\mu\epsilon_{31} = -\mu\alpha x_2 + \frac{\nu\mu T}{EI}x_2^2 + \mu f_{,1} \, , \\ t_{32} &= 2\mu\epsilon_{32} = \mu\alpha x_1 + \mu f_{,2} \, . \end{aligned} \tag{6.107}$$

The remaining equilibrium condition, Eq 6.100 is

$$\nabla^2 f(x_1, x_2) = -\frac{Tx_1}{\mu I} = -\frac{2(1+\nu)T}{EI}x_1 \;\; \text{on } \mathcal{C} \tag{6.108}$$

subject to the lateral boundary condition $t_{i\alpha}n_\alpha = 0$ in the form

$$\frac{\partial f}{\partial n} = \alpha(x_2 n_2 - x_1 n_2) - \frac{\nu T}{EI}x_2^2 n_1 \;\; \text{on } \partial\mathcal{C} \, . \tag{6.109}$$

To solve the boundary value problem given by Eqs 6.108 and 6.109, define a function $F(x_1, x_2)$ by

$$f = \alpha\psi - \frac{T}{EI}\left\{F + \frac{\nu x_1^3}{6} + \left(1 + \frac{\nu}{2}\right)x_1 x_2^2\right\} \tag{6.110}$$

where ψ is the warping function from torsion

$$\nabla^2\psi = 0 \;\; \text{on } \mathcal{C}; \;\; \frac{\partial\psi}{\partial n} = x_2 n_1 - x_1 n_2 \;\; \text{on } \partial\mathcal{C} \, . \tag{6.111}$$

Stress components may be written in terms of ψ, α and F as

$$\begin{aligned} t_{31} &= \mu\alpha(\psi_{,1} - x_2) - \frac{T}{2(1+\nu)I}\left\{F_{,1} + \frac{\nu x_1^2}{2} + (1+\frac{\nu}{2})x_2^2\right\} \, , \\ t_{32} &= \mu\alpha(\psi_{,2} + x_1) - \frac{T}{2(1+\nu)I}\{F_{,2} + (2+\nu)x_1 x_2\} \, , \\ t_{33} &= -\frac{T}{I}(L - x_3)x_1 \, , \end{aligned} \tag{6.112}$$

subject to the lateral boundary condition

$$\frac{\partial F}{\partial n} = -\left(\frac{\nu x_1^2}{2} + \left(1 + \frac{\nu}{2}\right)x_2^2\right)n_1 - (2+\nu)\,x_1 x_2 n_2 \;\; \text{on } \partial\mathcal{C} \, . \tag{6.113}$$

Since ψ and F are harmonic it is possible to show the stresses in Eq 6.112 satisfy the equilibrium equations. Substituting the stresses from 6.112 into Eq 6.57 and using Eq 6.101 it can be shown that $T_3 = T_2 = 0$ and $T_1 = T$. Bending moments $M_1 = M_2 = 0$ since $t_{33}\,(x_3 = L) = 0$, and the torque, $M_3 = M_t$, is set to zero by expressing α in terms of F:

$$\alpha = -\frac{T}{2(1+\nu)IK}\int_{\mathcal{C}_2}\left\{x_2 F_{,1} - x_1 F_{,2} + (1+\frac{\nu}{2})x_2^3 - 2(1+\frac{\nu}{2})x_1^2 x_2\right\} dA \tag{6.114}$$

where K is the torsional rigidity. The constant α may be shown to be the average rate of rotation over a cross section.

In general, there is a twist accompanying the bending when the load T is applied at the centroid of $\mathcal{C}_2$. By superposing an appropriate torsion solution with equal and opposite α it is possible to determine a line of action parallel to T_i such that there is not twist.

The displacements are determined from the line integral Eq 4.101

$$
\begin{aligned}
u_1 &= -\alpha x_2 x_3 + \frac{T}{EI}\left\{\tfrac{1}{2}(L - x_3)\nu(x_1^2 - x_2^2) + \tfrac{1}{2}Lx_3^2 - \tfrac{1}{6}x_3^3\right\} , \\
u_2 &= \alpha x_1 x_3 + \frac{T\nu}{EI}(L - x_3)x_1 x_2 , \\
u_3 &= \alpha\psi(x_1, x_2) - \frac{T}{EI}\left\{x_1 x_3(L - \tfrac{1}{2}x_3) + x_1 x_2^2 + F(x_1, x_2)\right\} ,
\end{aligned}
\qquad (6.115)
$$

to within an infinitesimal rigid displacement. The centerline deflection is the same as that obtained in beam theory, but the cross-sections do not remain plane in the Saint-Venant solution.

6.6 Plane Elasticity

In a number of engineering applications, specific body geometry and loading patterns lead to a reduced, essentially two-dimensional form of the equations of elasticity, and the study of these situations is referred to as *plane elasticity*. Although the two basic types of problems constituting the core of this plane analysis may be defined formally by stating certain assumptions on the stresses and displacements, we introduce them here in terms of their typical physical prototypes. In *plane stress* problems, the geometry of the body is that of a thin plate with one dimension very much smaller than the other two. The loading in this case is in the plane of the plate and is assumed to be uniform across the thickness, as shown in Fig. 6.8(a). In *plane strain* problems, the geometry is that of a prismatic cylinder having one dimension very much larger than the other two and having the loads perpendicular to and distributed uniformly with respect to this large dimension (Fig. 6.8(b)). In this case, because conditions are the same at all cross sections, the analysis may be focused on a thin slice of the cylinder.

For the plane stress situation, Fig. 6.8(a), the stress components t_{33}, t_{31}, and t_{32} are taken as zero everywhere and the remaining components considered functions of only x_1 and x_2. Thus,

$$t_{ij} = t_{ij}(x_1, x_2) \quad (i, j = 1, 2) , \qquad (6.116)$$

and as a result, the equilibrium equations, Eq 6.43, reduce to the specific equations

$$t_{11,1} + t_{12,2} + \rho b_1 = 0 , \qquad (6.117a)$$

$$t_{21,1} + t_{22,2} + \rho b_2 = 0 . \qquad (6.117b)$$

The strain-displacement relations, Eq 6.44, become

$$\epsilon_{11} = u_{1,1}; \quad \epsilon_{22} = u_{2,2}; \quad 2\epsilon_{12} = u_{1,2} + u_{2,1} , \qquad (6.118)$$

and at the same time the strain compatibility equations, Eqs 4.90, take on the form given by Eq 4.92 and repeated here for convenience,

$$\epsilon_{11,22} + \epsilon_{22,11} = 2\epsilon_{12,12} . \qquad (6.119)$$

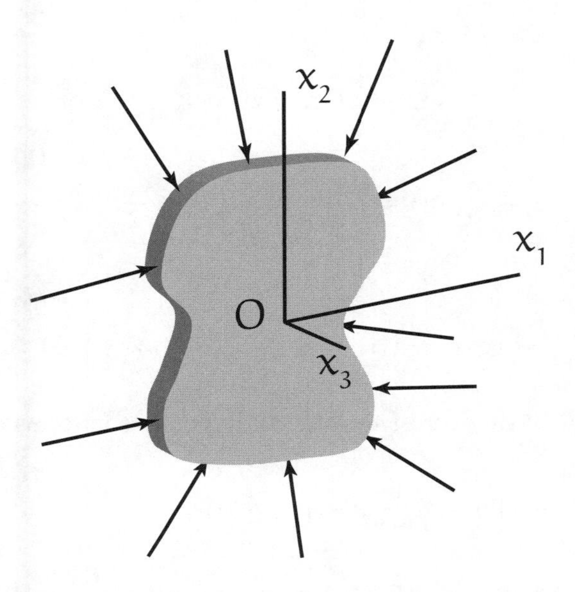

(a) Plane stress problems generally involve bodies that are thin in dimensions with loads perpendicular to that dimension.

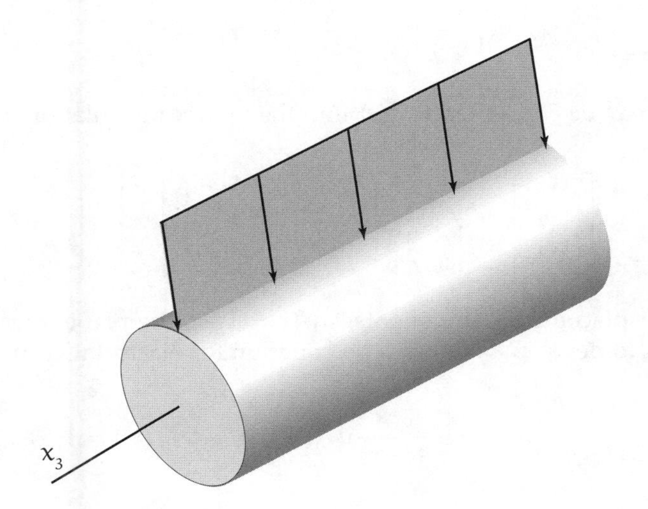

(b) Plane strain problems involve bodies that are long in on dimension with loads applied along that dimension.

FIGURE 6.8

Representative figures for plane stress and plain strain.

Hooke's law equations, Eq 6.45b, for plane stress are written

$$\epsilon_{11} = \frac{1}{E}(t_{11} - \nu t_{22}) \,, \tag{6.120a}$$

$$\epsilon_{22} = \frac{1}{E}(t_{22} - \nu t_{11}) \,, \tag{6.120b}$$

$$\epsilon_{12} = \frac{1+\nu}{E} t_{12} = \frac{t_{12}}{2G} = \frac{\gamma_{12}}{2} \,, \tag{6.120c}$$

along with

$$\epsilon_{33} = -\frac{\nu}{E}(t_{11} + t_{22}) = \frac{-\nu}{1-\nu}(\epsilon_{11} + \epsilon_{22}) \,. \tag{6.121}$$

By inverting Eq 6.120a, we express the stress components in terms of the strains as

$$t_{11} = \frac{E}{1-\nu^2}(\epsilon_{11} + \nu\epsilon_{22}) \,, \tag{6.122a}$$

$$t_{22} = \frac{E}{1-\nu^2}(\epsilon_{22} + \nu\epsilon_{11}) \,, \tag{6.122b}$$

$$t_{12} = \frac{E}{1+\nu}\epsilon_{12} = \frac{E}{2(1+\nu)}\gamma_{12} = G\gamma_{12} \,. \tag{6.122c}$$

These equations may be conveniently cast into the matrix formulation

$$\begin{bmatrix} t_{11} \\ t_{22} \\ t_{12} \end{bmatrix} = \frac{E}{1-\nu^2} \begin{bmatrix} 1 & \nu & 0 \\ \nu & 1 & 0 \\ 0 & 0 & 1-\nu \end{bmatrix} \begin{bmatrix} \epsilon_{11} \\ \epsilon_{22} \\ \epsilon_{12} \end{bmatrix} \,, \tag{6.123}$$

In terms of the displacement components, u_i $(i = 1, 2)$, the plane stress field equations may be combined to develop a Navier-type equation for elastostatics, namely,

$$\frac{E}{2(1+\nu)} u_{i,jj} + \frac{E}{2(1-\nu)} u_{j,ji} + \rho b_i = 0 \quad (i, j = 1, 2) \,. \tag{6.124}$$

For the plane strain situation (Fig. 6.8(b)) we assume that $u_3 = 0$ and that the remaining two displacement components are functions of only x_1 and x_2,

$$u_i = u_i(x_1, x_2) \quad (i = 1, 2) \,. \tag{6.125}$$

In this case, the equilibrium equations, the strain-displacement relations, and the strain compatibility equations all retain the same form as for plane stress, that is, Eqs 6.117a, 6.117b, 6.118, and 6.119, respectively. Here, Hooke's law (Eq 6.45a) may be written in terms of engineering constants as

$$t_{11} = \frac{E}{(1+\nu)(1-2\nu)} [(1-\nu)\epsilon_{11} + \nu\epsilon_{22}] \; , \tag{6.126a}$$

$$t_{22} = \frac{E}{(1+\nu)(1-2\nu)} [(1-\nu)\epsilon_{22} + \nu\epsilon_{11}] \; , \tag{6.126b}$$

$$t_{12} = \frac{E}{1+\nu}\epsilon_{12} = \frac{E}{1+\nu}\frac{\gamma_{12}}{2} = G\gamma_{12} \; , \tag{6.126c}$$

along with

$$t_{33} = \frac{E\nu}{(1+\nu)(1-2\nu)} (\epsilon_{11} + \epsilon_{22}) = \nu(t_{11} + t_{22}) \; . \tag{6.127}$$

The first three of these equations may be expressed in matrix form by

$$\begin{bmatrix} t_{11} \\ t_{22} \\ t_{12} \end{bmatrix} = \frac{E}{(1+\nu)(1-2\nu)} \begin{bmatrix} 1-\nu & \nu & 0 \\ \nu & 1-\nu & 0 \\ 0 & 0 & 1-2\nu \end{bmatrix} \begin{bmatrix} \epsilon_{11} \\ \epsilon_{22} \\ \epsilon_{12} \end{bmatrix} . \tag{6.128}$$

Furthermore, by inverting the same three equations, we may express Hooke's law for plane strain by the equations

$$\epsilon_{11} = \frac{(1+\nu)}{E} [(1-\nu)t_{11} - \nu t_{22}] \; , \tag{6.129a}$$

$$\epsilon_{22} = \frac{(1+\nu)}{E} [(1-\nu)t_{22} + \nu t_{11}] \; , \tag{6.129b}$$

$$\epsilon_{12} = \frac{1+\nu}{E} t_{12} = \frac{2(1+\nu)}{E}\frac{t_{12}}{2} = \frac{t_{12}}{2G} \; . \tag{6.129c}$$

By combining the field equations with Hooke's law for elastostatic plane strain, we obtain the appropriate Navier equation as

$$\frac{E}{2(1+\nu)} u_{i,jj} + \frac{E}{2(1+\nu)(1-2\nu)} u_{j,ji} + \rho b_i = 0 \quad (i,j = 1,2) \; . \tag{6.130}$$

It is noteworthy that Eqs 6.120a and 6.122a for plane stress become identical with the plane strain equations, Eqs 6.129 and 6.126, respectively, if in the plane stress equations we replace E with $E/(1-\nu^2)$ and ν by $\nu/(1-\nu)$. Note that, if the forces applied to the edge of the plate in Fig. 6.8(a) are not uniform across the thickness but are symmetrical with respect to the middle plane of the plate, the situation is sometimes described as a state of *generalized plane stress*. In such a case, we consider the stress and strain variables to be averaged values over the thickness. Also, a case of *generalized plane strain* is sometimes referred to in elasticity textbooks if the strain component ϵ_{33} in Fig. 6.8(b) is taken as some constant other that zero.

6.7 Airy Stress Function

As stated in Section 6.6, the underlying equations for two-dimensional problems in isotropic elasticity consist of the equilibrium relations, Eq 6.117, the compatibility condition, Eq 6.119 and Hooke's law, either in the form of Eq 6.120a (plane stress), or as Eq 6.129 (plane strain). When body forces in Eq 6.117 are conservative with a potential function $V = V(x_1, x_2)$ such that $b_i = -V_{,i}$, we may introduce the *Airy stress function*, $\phi = \phi(x_1, x_2)$ in terms of which the stresses are given by

$$t_{11} = \phi_{,22} + \rho V; \quad t_{22} = \phi_{,11} + \rho V; \quad t_{12} = -\phi_{,12} \,. \tag{6.131}$$

Note that by using this definition the equilibrium equations are satisfied identically.

For the case of plane stress we insert Eq 6.120a into Eq 6.119 to obtain

$$t_{11,22} + t_{22,11} - \nu(t_{11,11} + t_{22,22}) = 2(1+\nu)t_{12,12} \tag{6.132}$$

which in terms of ϕ becomes

$$\phi_{,1111} + 2\phi_{,1212} + \phi_{,2222} = -(1-\nu)\rho(V_{,11} + V_{,22}) \,. \tag{6.133}$$

Similarly, for the case of plane strain, when Eq 6.129 is introduced into Eq 6.119 the result is

$$(1-\nu)(t_{11,22} + t_{22,11}) - \nu(t_{11,11} + t_{22,22}) = 2t_{12,12} \,, \tag{6.134}$$

or in terms of ϕ

$$\phi_{,1111} + 2\phi_{,1212} + \phi_{,2222} = -(1-2\nu)\rho(V_{,11} - V_{,22})/(1-\nu) \,. \tag{6.135}$$

If the body forces consist of gravitational forces only, or if they are constant forces, the right-hand sides of both Eqs 6.133 and 6.135 reduce to zero and ϕ must then satisfy the bi-harmonic equation

$$\phi_{,1111} + 2\phi_{,1212} + \phi_{,2222} = \nabla^4\phi = 0 \,. \tag{6.136}$$

In each case, of course, boundary conditions on the stresses must be satisfied to complete the solution to a particular problem. For bodies having a rectangular geometry, stress functions in the form of polynomials in x_1 and x_2 are especially useful as shown by the examples that follow.

Example 6.3

For a thin rectangular plate of the dimensions shown, consider the general polynomial of the third degree as the Airy stress function and from it determine the stresses. Assume body forces are zero.

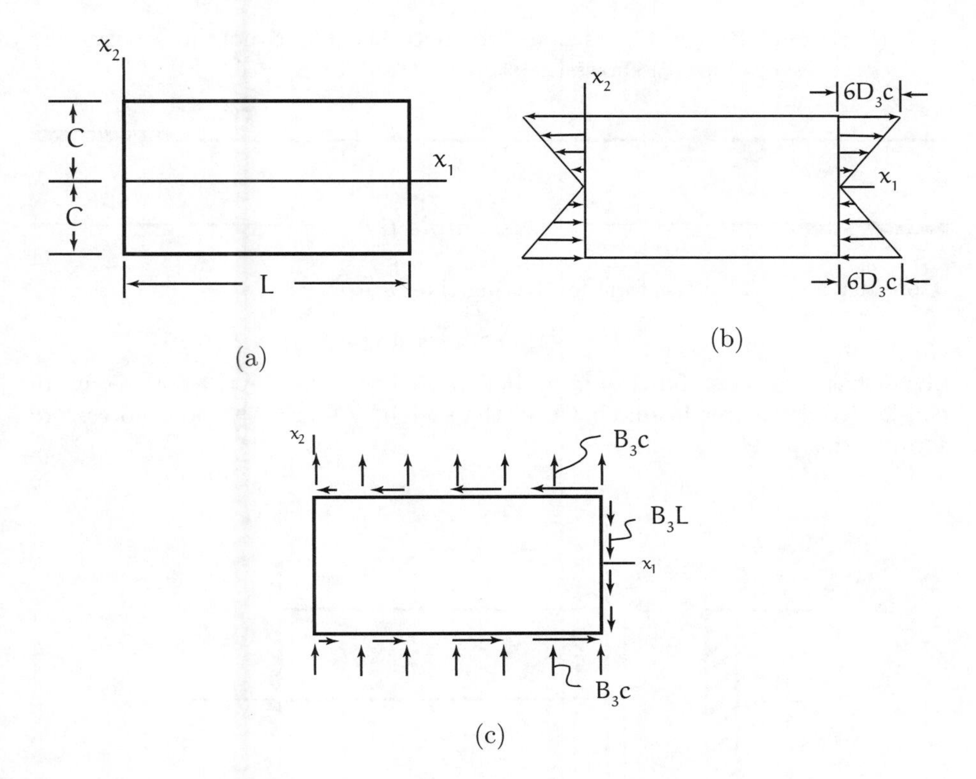

Solution

Select a polynomial stress function of the form $\phi_3 = A_3x_1^3 + B_3x_1^2x_2 + C_3x_1x_2^2 + D_3x_2^3$. Choosing this particular polynomial form for the stress function is not arbitrary; the choice is based on many trials of different order polynomials. After a certain amount of experience in observing a polynomial's effect on the stress components computed using Eq 6.131, an educated guess can be made as to what terms should be considered for a specific problem. For this reason, problems like this are often called *semi-inverse* problems.

By direct substitution into Eq 6.136 we confirm that ϕ_3 is bi-harmonic. Further, the stresses are given as

$$\begin{aligned} t_{11} &= 2C_3x_1 + 6D_3x_2 \ , \\ t_{22} &= 6A_3x_1 + 2B_3x_2 \ , \\ t_{12} &= -2B_3x_1 - 2C_3x_2 \ . \end{aligned}$$

By selecting different constants to be zero and nonzero, different physical problems may be solved. Here, two specific cases will be considered.

(a) Assume all coefficients in ϕ_3 are zero except D_3. This may be shown to solve the case of pure bending of a beam by equilibrating moments on the ends as shown by Fig. (b). Stress in the fiber direction of the beam varies linearly with the distance from the x_1 axis

$$t_{11} = 6D_3x_2; \quad t_{22} = t_{12} = 0$$

as is the case for simple bending. Similarly, by taking only A_3 as nonzero, the solution is for bending moments applied to a beam whose lengthwise direction is taken to be x_2 rather than x_1 direction.

(b) If only B_3 (or C_3) is non-zero, both shear and normal stresses are present. Fig. (c) shows the stress pattern for $B_3 \neq 0$.

Example 6.4

Consider a special stress function having the form

$$\phi^* = B_2 x_1 x_2 + D_4 x_1 x_2^3 .$$

Show that this stress function may be adapted to solve for the stresses in an end-loaded cantilever beam shown in the sketch. Assume the body forces are zero for this problem.

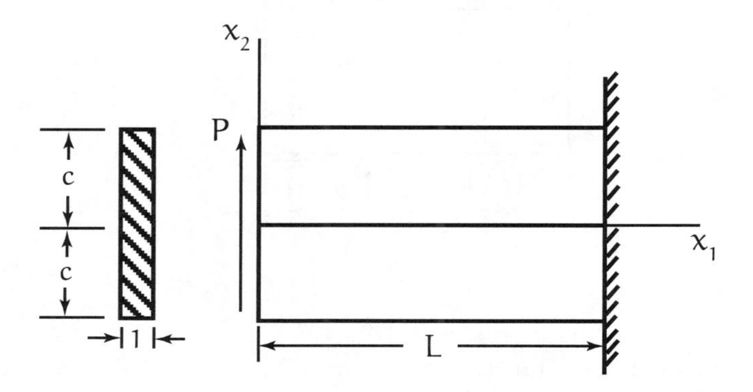

Vector $\mathbf{v}$ with respect to $Ox_1'x_2'x_3'$ and $Ox_1x_2x_3$.

Solution

It is easily verified, by direct substitution, that $\nabla^4\phi^* = 0$. The stress components are directly computed from Eq 6.131

$$t_{11} = 6D_4 x_1 x_2 ,$$
$$t_{22} = 0 ,$$
$$t_{12} = -B_2 - 3D_4 x_2^2 .$$

These stress components are consistent with an end-loaded cantilever beam, and the constants B_2 and D_4 can be determined by considering the boundary conditions. In order for the top and bottom surfaces of the beam to be stress-free, t_{12} must be zero at $x_2 = \pm c$. Using this condition B_2 is determined in terms of D_4 as $B_2 = -3D_4c^2$. The shear stress is thus given in terms of single constant B_2

$$t_{12} = -B_2 + \frac{B_2 x_2^2}{c^2} .$$

The concentrated load is modeled as the totality of the shear stress t_{12} on the free end of the beam. Thus, the result of integrating this stress over the free end of the beam at $x_1 = 0$ yields the applied force P. In equation form

$$P = -\int_{-c}^{c}\left[-B_2 + B_2\frac{x_2^2}{c^2}\right]dx_2$$

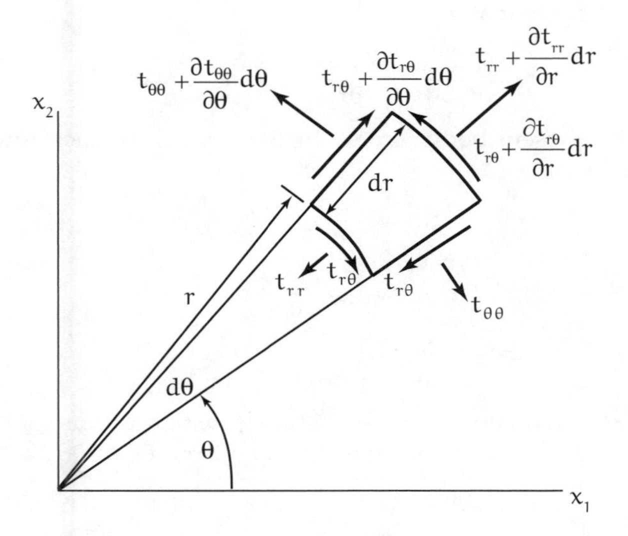

FIGURE 6.9
Differential stress element in polar coordinates.

where the minus sign is required due to the sign convention on shear stress. Carrying out the integration we have $B_2 = 3P/4c$ so that stress components may now be written as

$$t_{11} = -\frac{3P}{2c^3} x_1 x_2 ,$$
$$t_{22} = 0 ,$$
$$t_{12} = -\frac{3P}{4c}\left(1 - \frac{x_2^2}{c^2}\right) .$$

But for this beam the plane moment of inertia of the cross section is $I = 2c^3/3$ so that now

$$t_{11} = -\frac{P}{I} x_1 x_2; \quad t_{22} = 0; \quad t_{12} = -\frac{P}{2I}\left(c^2 - x_2^2\right)$$

in agreement with the results of elementary beam bending theory.

Several important solutions in plane elasticity are obtained by the Airy stress function approach when expressed in terms of *polar coordinates*. To this end we introduce here the basic material element together with the relevant stress components in terms of the coordinates r and θ as shown on Fig. 6.9. Using this element and summing forces in the radial direction results in the equilibrium equation

$$\frac{\partial t_{rr}}{\partial r} + \frac{1}{r}\frac{\partial t_{r\theta}}{\partial \theta} + \frac{t_{rr} - t_{\theta\theta}}{r} + R = 0 , \qquad (6.137a)$$

and summing forces tangentially yields

$$\frac{1}{r}\frac{\partial t_{\theta\theta}}{\partial \theta} + \frac{\partial t_{r\theta}}{\partial r} + 2\frac{t_{r\theta}}{r} + \Theta = 0 \tag{6.137b}$$

in which R and Θ represent body forces. In the absence of such forces Eq 6.137a and 6.137b are satisfied by

$$t_{rr} = \frac{1}{r}\frac{\partial \phi}{\partial r} + \frac{1}{r^2}\frac{\partial^2 \phi}{\partial \theta^2}\,, \tag{6.138a}$$

$$t_{\theta\theta} = \frac{\partial^2 \phi}{\partial r^2}\,, \tag{6.138b}$$

$$t_{r\theta} = \frac{1}{r^2}\frac{\partial \phi}{\partial \theta} - \frac{1}{r}\frac{\partial^2 \phi}{\partial r \partial \theta} = -\frac{\partial}{\partial r}\left(\frac{1}{r}\frac{\partial \phi}{\partial \phi}\right)\,, \tag{6.138c}$$

in which $\phi = \phi(r, \theta)$. To qualify as an Airy stress function ϕ must once again satisfy the condition $\nabla^4\phi = 0$ which in polar form is obtained from Eq 6.136 as

$$\nabla^4\phi = \left(\frac{\partial^2}{\partial r^2} + \frac{1}{r}\frac{\partial}{\partial r} + \frac{1}{r^2}\frac{\partial^2}{\partial \theta^2}\right)\left(\frac{\partial^2 \phi}{\partial r^2} + \frac{1}{r}\frac{\partial \phi}{\partial r} + \frac{1}{r^2}\frac{\partial^2 \phi}{\partial \theta^2}\right) = 0\,. \tag{6.139}$$

For stress fields symmetrical to the polar axis Eq 6.139 reduces to

$$\nabla^4\phi = \left(\frac{\partial^2}{\partial r^2} + \frac{1}{r}\frac{\partial}{\partial r}\right)\left(\frac{\partial^2 \phi}{\partial r^2} + \frac{1}{r}\frac{\partial \phi}{\partial r}\right) = 0\,, \tag{6.140a}$$

or

$$r^4\frac{\partial^4 \phi}{\partial r^4} + 2r^3\frac{\partial^3 \phi}{\partial r^3} - r^2\frac{\partial^2 \phi}{\partial r^2} + r\frac{\partial \phi}{\partial r} = 0\,. \tag{6.140b}$$

It may be shown that the general solution to this differential equation is given by

$$\phi = A \ln r + Br^2 \ln r + Cr^2 + D\,, \tag{6.141}$$

so that for the symmetrical case the stress components take the form

$$t_{rr} = \frac{1}{r}\frac{\partial \phi}{\partial r} = \frac{A}{r^2} + B\,(1 + 2\ln r) + 2C\,, \tag{6.142a}$$

$$t_{\theta\theta} = \frac{\partial^2 \phi}{\partial r^2} = -\frac{A}{r^2} + B\,(3 + 2\ln r) + 2C\,, \tag{6.142b}$$

$$t_{r\theta} = 0\,. \tag{6.142c}$$

When there is no hole at the origin in the elastic body under consideration, A and B must be zero since otherwise infinite stresses would result at that point. Thus, for a plate without a hole only uniform tension or compression can exist as a symmetrical case.

A geometry that does qualify as a case with a hole (absence of material) at the origin is that of a curved beam subjected to end moments as discussed in the following example.

Example 6.5

Determine the stresses in a curved beam of the dimensions shown in the figure when subjected to constant equilibrating moments.

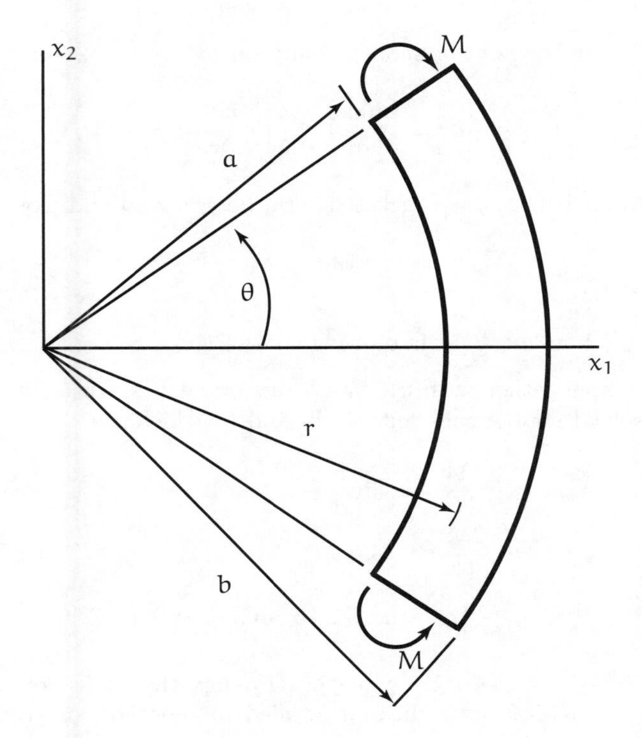

Solution
From symmetry, the stresses are as given by Eq 6.142. Boundary conditions require

(1) $t_{rr} = 0$ at $r = a$, and at $r = b$,

(2) $\int_a^b t_{\theta\theta}\,dr = 0$ on the end faces ,

(3) $\int_a^b r\, t_{rr}\,dr = -M$ on the end faces ,

(4) $t_{r\theta} = 0$ everywhere on the boundary .

These conditions result in the following equations which are used to evaluate the constants A, B, and C. The inner and outer radii are free of normal stress which can be written in terms of boundary condition (1) as

$$A/a^2 + B(1 + 2\ln a) + 2C = 0 \ ,$$

$$A/b^2 + B(1 + 2\ln b) + 2C = 0 \ .$$

No transverse loading is present on the ends of the curved beam which may be written in terms of boundary condition (2) as

$$\int_a^b t_{\theta\theta}\,dr = \int_a^b \frac{\partial^2 \phi}{\partial r^2}\,dr = \left.\frac{\partial \phi}{\partial r}\right|_a^b = 0 \ .$$

Evaluation of this integral at the limits is automatically satisfied as a consequence of boundary condition (1). Finally, the applied moments on the ends may be written in terms of boundary condition (3)

$$\int_a^b r\frac{\partial^2\phi}{\partial r^2}dr = \left[r\frac{\partial\phi}{\partial r}\right]_a^b - \int_a^b \frac{\partial\phi}{\partial r}dr = -M\ .$$

Because of condition (1) the bracketed term here is zero and from the integral term

$$\phi_b - \phi_a = M\ ,$$

or

$$A\ln b/a + B(b^2\ln b - a^2\ln a) + C(b^2 - a^2) = M\ .$$

This expression, together with the two stress equations arising from condition (1) may be solved for the constants A, B, and C, which are

$$\begin{aligned} A &= -\frac{4M}{N}a^2b^2\ln\frac{b}{a}\ , \\ B &= -\frac{2M}{N}\left(b^2 - a^2\right)\ , \\ C &= \frac{M}{N}\left[b^2 - a^2 + 2\left(b^2\ln b - a^2\ln a\right)\right]\ , \end{aligned}$$

where $N = (b^2 - a^2)^2 - 4a^2b^2[\ln(b/a)]^2$. Finally, the stress components may be written in terms of the radii and applied moment by substitution of the constants into Eq 6.142

$$\begin{aligned} t_{rr} &= -\frac{4M}{N}\left(\frac{a^2b^2}{r^2}\ln\frac{b}{a} + b^2\ln\frac{r}{b} + a^2\ln\frac{a}{r}\right)\ , \\ t_{\theta\theta} &= -\frac{4M}{N}\left(-\frac{a^2b^2}{r^2}\ln\frac{b}{a} + b^2\ln\frac{r}{b} + a^2\ln\frac{a}{r} + b^2 - a^2\right)\ , \\ t_{r\theta} &= 0\ . \end{aligned}$$

Verification of these results can be made by reference to numerous strength of materials textbooks.

If ϕ is taken as a function of both r and θ, it is useful to assume

$$\phi(r,\theta) = f(r)e^{in\theta} \tag{6.143}$$

in order to obtain a function periodic in θ. For $n = 0$, the general solution is, as expected, the same as given in Eq 6.141. As an example of the case where $n = 1$ we consider $\phi(r,\theta)$ in the form

$$\phi = (Ar^2 + B/r + Cr + Dr\ln r)\sin\theta\ . \tag{6.144}$$

Example 6.6

Show that the stress function given by Eq 6.144 may be used to solve the quarter-circle beam shown under an end load P.

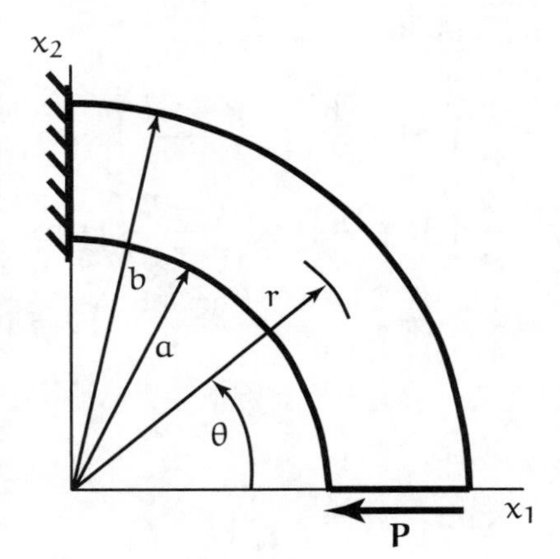

Solution
From Eq 6.138 the stress components are

$$t_{rr} = \left(2Ar - \frac{2B}{r^3} + \frac{D}{r}\right) \sin\theta \ ,$$
$$t_{\theta\theta} = \left(6Ar + \frac{2B}{r^3} + \frac{D}{r}\right) \sin\theta \ ,$$
$$t_{r\theta} = -\left(2Ar - \frac{2B}{r^3} + \frac{D}{r}\right) \cos\theta \ .$$

The inner and outer radii of the beam are stress-free surfaces leading to boundary conditions of $t_{rr} = t_{r\theta} = 0$ at $r = a$, and at $r = b$. Also, the applied force P may be taken to be the summation of the shear stress acting over the free end $\theta = 0$:

$$\int_a^b t_{r\theta}|_{\theta=0}\, dr = -P \ .$$

These conditions lead to the three equations from which constants A, B, and D may be determined

$$2Aa - \frac{2B}{a^3} + \frac{D}{a} = 0 \ ,$$
$$2Ab - \frac{2B}{b^3} + \frac{D}{b} = 0 \ ,$$
$$-A\left(b^2 - a^2\right) + B\frac{\left(b^2 - a^2\right)}{a^2b^2} - D\ln\frac{b}{a} = P \ .$$

Solving these three equations in three unknowns, the constants are determined to be

$$A = \frac{P}{2N} \ , \quad B = -\frac{Pa^2b^2}{2N} \ , \quad D = -\frac{P\left(a^2 + b^2\right)}{N} \ ,$$

where $N = a^2 - b^2 + (a^2 + b^2)\ln(b/a)$. Finally, use of these constants in the stress component equations gives

$$t_{rr} = \frac{P}{N}\left(r + \frac{a^2b^2}{r^3} - \frac{a^2+b^2}{r}\right)\sin\theta \ ,$$

$$t_{\theta\theta} = \frac{P}{N}\left(3r - \frac{a^2b^2}{r^3} + \frac{a^2+b^2}{r}\right)\sin\theta \ ,$$

$$t_{r\theta} = -\frac{P}{N}\left(r + \frac{a^2b^2}{r^3} - \frac{a^2+b^2}{r}\right)\cos\theta \ .$$

Note that, when $\theta = 0$

$$t_{rr} = t_{\theta\theta} = 0 \quad \text{and} \quad t_{r\theta} = -\frac{P}{N}\left(r + \frac{a^2b^2}{r^3} - \frac{a^2+b^2}{r}\right) \ .$$

And when $\theta = \pi/2$, $t_{r\theta} = 0$ while

$$t_{rr} = \frac{P}{N}\left(r + \frac{a^2b^2}{r^3} - \frac{a^2+b^2}{r}\right) \ ,$$

$$t_{\theta\theta} = \frac{P}{N}\left(3r - \frac{a^2b^2}{r^3} + \frac{a^2+b^2}{r}\right) \ .$$

We close this section with an example of the case when $n = 2$ in Eq 6.143.

Example 6.7

Use the stress function

$$\phi = (Ar^2 + Br^4 + C/r^2 + D)\cos 2\theta$$

to solve the stress problem of a large flat plate under a uniform axial stress T, and having a small circular hole at the origin as shown in the figure.

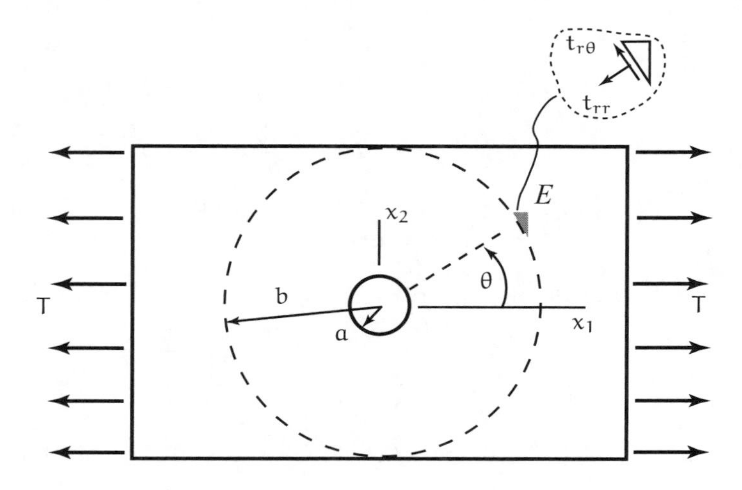

Solution

From Eq 6.138 the stress components have the form

$$t_{rr} = -\left(2A + \frac{6C}{r^4} + \frac{4D}{r^2}\right)\cos\theta\,,$$

$$t_{\theta\theta} = \left(2A + 12Br^2 + \frac{6C}{r^2}\right)\cos\theta\,,$$

$$t_{r\theta} = \left(2A + 6Br^2 - \frac{6C}{r^4} - \frac{2D}{r^2}\right)\sin 2\theta\,.$$

Assume the width of the plate b is large compared to the radius of the hole. From consideration of the small triangular element at a distance b from the origin, the following boundary conditions must hold

1. $t_{rr} = T\cos^2\theta$ at $r = b$,
2. $t_{r\theta} = -\frac{T}{2}\sin 2\theta$ at $r = b$,

and the inner surface of the hole is stress free, which may be written as

3. $t_{rr} = 0$ at $r = a$,
4. $t_{r\theta} = 0$ at $r = a$.

These conditions, when combined with the stress expressions, yield

$$2A + \frac{6C}{b^4} + \frac{4D}{b^2} = -\frac{T}{2}\,,$$

$$2A + 6Bb^2 - \frac{6C}{b^4} - \frac{2D}{b^2} = -\frac{T}{2}\,,$$

$$2A + \frac{6C}{a^4} + \frac{4D}{a^2} = 0\,,$$

$$2A + 6Ba^2 - \frac{6C}{a^4} - \frac{2D}{a^2} = 0\,.$$

Letting $b \to \infty$, the above equations may be solved to determine

$$A = -\frac{T}{4}\,, \quad B = 0\,, \quad C = -\frac{a^4 T}{4}\,, \quad D = \frac{a^2 T}{2}\,,$$

so that now the stresses are given by

$$t_{rr} = \frac{T}{2}\left(1 - \frac{a^2}{r^2}\right) + \frac{T}{2}\left(1 + \frac{3a^4}{r^4} + \frac{4a^2}{r^2}\right)\cos\theta\,,$$

$$t_{\theta\theta} = \frac{T}{2}\left(1 + \frac{a^2}{r^2}\right) - \frac{T}{2}\left(1 + \frac{3a^4}{r^4}\right)\cos\theta\,,$$

$$t_{r\theta} = -\frac{T}{2}\left(1 - \frac{3a^4}{r^4} + \frac{2a^2}{r^2}\right)\sin 2\theta\,.$$

Note that as r tends towards infinity and at $\theta = 0$ the stresses are given by $t_{rr} = T$; $t_{\theta\theta} = t_{r\theta} = 0$ (simple tension). At $r = a$, $t_{rr} = t_{r\theta} = 0$ and $t_{\theta\theta} = T - 2T\cos 2\theta$, which indicates that when $\theta = \pi/2$, or $\theta = 3\pi/2$, the stresses become $t_\theta = T - 2T(-1) = 3T$, a well-known stress concentration factor used in design. Also, when $\theta = 0$, or $\theta = \pi$, $t_{\theta\theta} = -T$ a compression factor at the centerline of the hole.

It should be pointed out that in this brief section only stresses have been determined. These, together with Hooke's law, can be used to determine the displacements for the problems considered.

6.8 Linear Thermoelasticity

When we give consideration to the effects of temperature as well as to mechanical forces on the behavior of elastic bodies, we become involved with *thermoelasticity*. Here, we address only the relatively simple *uncoupled theory* for which temperature changes brought about by elastic straining are neglected. Also, within the context of linearity we assume that the total strain is the sum

$$\epsilon_{ij} = \epsilon_{ij}^{(M)} + \epsilon_{ij}^{(T)} \tag{6.145}$$

where $\epsilon_{ij}^{(M)}$ is the contribution from the mechanical forces and $\epsilon_{ij}^{(T)}$ are the temperature-induced strains. If θ_0 is taken as a reference temperature and θ as an arbitrary temperature, the thermal strains resulting from a change in temperature of a completely unconstrained isotropic volume are given by

$$\epsilon_{ij}^{(T)} = \alpha (\theta - \theta_0)\, \delta_{ij} \tag{6.146}$$

where α is the *linear coefficient of thermal expansion*, having units of meters per meter per degree Celsius (m/m/°C). The presence of the Kronecker delta in Eq 6.146 indicates that shear strains are not induced by a temperature change in an unconstrained, homogenous, isotropic body.

By inserting Eq 6.146 into Eq 6.145 and using Hooke's law for the mechanical strains in that equation, we arrive at the thermoelastic constitutive equation

$$\epsilon_{ij} = \frac{1+\nu}{E} t_{ij} - \frac{\nu}{E} \delta_{ij} t_{kk} + \alpha (\theta - \theta_0)\, \delta_{ij} \;. \tag{6.147}$$

This equation may be easily inverted to express the stresses in terms of the strains as

$$t_{ij} = \frac{E}{(1+\nu)(1-2\nu)} \left[\nu \delta_{ij} \epsilon_{kk} + (1-2\nu)\, \epsilon_{ij} - (1+\nu)\, \alpha (\theta - \theta_o)\, \delta_{ij} \right] \;. \tag{6.148}$$

Also, in terms of the deviatoric and spherical components of stress and strain, the thermoelastic constitutive relations appear as the pair of equations

$$S_{ij} = \frac{E}{(1+\nu)} \eta_{ij} \;, \tag{6.149a}$$

$$t_{ii} = \frac{E}{(1-2\nu)} \left[\epsilon_{ii} - 3\alpha (\theta - \theta_0) \right] \;. \tag{6.149b}$$

If the heat conduction in an elastic solid is governed by the Fourier law, Eq 5.62, which we write here as

$$q_i = -\kappa\theta_{,i} \tag{6.150}$$

where κ is the *thermal conductivity* of the body (a positive constant), and if we introduce the *specific heat* constant c through the equation

$$-q_{i,i} = \rho c\dot{\theta} \tag{6.151}$$

the heat conduction equation for the uncoupled theory becomes

$$\kappa\theta_{,ii} = \rho c\dot{\theta}\ . \tag{6.152}$$

This equation, along with the thermoelastic stress-strain equations Eq 6.147 or Eq 6.148, the equilibrium equations Eq 6.43, and the strain-displacement relations Eq 6.44, constitute the basic set of field equations for uncoupled, quasi-static, thermoelastic problems. Of course, boundary conditions and the strain compatibility equations must also be satisfied.

6.9 Three-Dimensional Elasticity

Solutions of three-dimensional elasticity problems traditionally focus on two distinct formulations. First, the *displacement formulation* is based upon solutions of the Navier equations which were presented in Section 6.4, and which are developed again in the following paragraph. The second formulation, called the *stress formulation*, utilizes solutions of the equilibrium equations in association with the Beltrami-Michell stress equations previously derived in Section 6.4 and based upon the compatibility equations in terms of strains. These equations are also reviewed in the following paragraph.

Starting with the fundamental equations of elastostatics as listed in Section 6.4 and repeated here under revised numbering, we have:

Equilibrium equations, Eq 6.43

$$t_{ij,j} + \rho b_i = 0\ , \tag{6.153}$$

Strain-displacement equations, Eq 6.44

$$2\epsilon_{ij} = u_{i,j} + u_{j,i}\ , \tag{6.154}$$

Hooke's law, Eq 6.45a, or Eq 6.45b

$$t_{ij} = \lambda\delta_{ij}\epsilon_{kk} + 2\mu\epsilon_{ij}\ , \tag{6.155a}$$

or

$$\epsilon_{ij} = \frac{1}{E}\left[(1-\nu)\,t_{ij} - \nu\delta_{ij}t_{kk}\right]\ . \tag{6.155b}$$

By substituting Eq 6.154 into Hooke's law, Eq 6.155a, and that result in turn into Eq 6.153, we obtain

$$\mu u_{i,jj} + (\lambda + \mu)\, u_{j,ji} + \rho b_i = 0 \tag{6.156}$$

which comprise three second-order partial differential equations known as the *Navier equations*. For the stress formulation we convert the strain equations of compatibility introduced in Section 4.7 as Eq 4.90 and repeated here as

$$\epsilon_{ij,km} + \epsilon_{km,ij} - \epsilon_{ik,jm} - \epsilon_{jm,ik} = 0 \tag{6.157}$$

into the equivalent expression in terms of stresses using Eq 6.155a, and combine that result with Eq 6.153 to obtain

$$t_{ij,kk} + \frac{1}{1+\nu} t_{kk,ij} + \rho\,(b_{i,j} + b_{j,i}) + \frac{\nu}{1+\nu}\delta_{ij}\rho b_{k,k} = 0 \tag{6.158}$$

which are the *Beltrami-Michell equations of compatibility*. In seeking solutions by either the displacement or stress formulation, we consider only the cases for which body forces are zero. The contribution of such forces, often either gravitational or centrifugal in nature, can be appended to the homogeneous solution, usually in the form of a particular integral based upon the boundary conditions.

Let us first consider solutions developed through the displacement formulation. Rather than attempt to solve the Navier equations directly, we express the displacement field in terms of scalar and vector potentials and derive equations whose solutions result in the required potentials. Thus, by inserting an expression for u_i in terms of the proposed potentials into the Navier equations we obtain the governing equations for the appropriate potentials. Often such potentials are harmonic, or bi-harmonic functions. We shall present three separate methods for arriving at solutions of the Navier equations.

The method used in the first approach rests upon the well-known theorem of Helmholtz which states that any vector function that is continuous and finite, and which vanishes at infinity, may be resolved into a pair of components: one a rotational vector, the other an irrotational vector. Thus, if the curl of an arbitrary vector $\mathbf{a}$ is zero, then $\mathbf{a}$ is the gradient of a scalar ϕ, and $\mathbf{a}$ is *irrotational*, or as it is sometimes called, *solenoidal*. At the same time, if the divergence of the vector $\mathbf{a}$ is zero, then $\mathbf{a}$ is the curl of another vector $\boldsymbol{\psi}$, and is a rotational vector. Accordingly, in keeping with the Helmholtz theorem, we assume that the displacement field is given by

$$u_i = \phi_{,i} + \epsilon_{ipq}\psi_{q,p} \tag{6.159}$$

where $\phi_{,i}$ is representative of the irrotational portion, and curl $\boldsymbol{\psi}$ the rotational portion. Substituting this displacement vector into Eq 6.156 with b_i in that equation taken as zero, namely

$$\mu u_{i,jj} + (\lambda + \mu)\, u_{j,ji} = 0\,, \tag{6.160}$$

we obtain

$$\mu\phi_{,ijj} + \mu\varepsilon_{ipq}\psi_{q,pjj} + (\lambda + \mu)\,\phi_{.ijj} + (\lambda + \mu)\,\varepsilon_{jpq}\psi_{q,pji} = 0 \tag{6.161}$$

which reduces to

$$(\lambda + 2\mu)\,\phi_{,ijj} + \mu\varepsilon_{ipq}\psi_{q,pjj} = 0\,, \tag{6.162}$$

since $\varepsilon_{jpq}\psi_{q,pji} = 0$. In coordinate-free notation Eq 6.162 becomes

$$(\lambda + 2\mu)\,\boldsymbol{\nabla}\boldsymbol{\nabla}^2\phi + \mu\boldsymbol{\nabla} \times \boldsymbol{\nabla}^2\boldsymbol{\psi} = 0\,. \tag{6.163}$$

Any set of ϕ and ψ which satisfies Eq 6.162 provides (when substituted into Eq 6.159) a displacement field satisfying the Navier equation, Eq 6.160. Clearly, one such set is obtained by requiring ϕ and ψ to be harmonic:

$$\nabla^2\phi = 0 \,, \tag{6.164a}$$

$$\nabla^2\psi = 0 \,. \tag{6.164b}$$

It should be pointed out that while Eq 6.164 is a solution of Eq 6.160, it is not the general solution of the Navier equations. If we choose $\nabla^2\phi =$ constant, and $\psi = 0$ in Eq 6.164, the scalar function ϕ is known as the *Lamé strain potential.* By taking the divergence of Eq 6.162, and remembering that the divergence of a curl vanishes, we see that

$$\nabla^4\phi = 0 \tag{6.165}$$

is a solution of the resulting equation so that a bi-harmonic function as ϕ also yields a solution for u_i. Similarly, by taking the curl of Eq 6.162 we find that

$$\nabla^4\psi = 0 \tag{6.166}$$

also provides for a solution u_i.

The second approach for solving the Navier equations is based on the premise of expressing the displacement field in terms of the second derivatives of a vector known as the *Galerkin vector,* and designated here by $\mathbf{F} = F_i\hat{\mathbf{e}}_i$. In this approach we assume the displacement u_i is given in terms of the Galerkin vector specifically by the equation

$$u_i = 2\,(1-\nu)\,F_{i,jj} - F_{j,ji} \tag{6.167}$$

which is substituted directly into Eq 6.160. Carrying out the indicated differentiation and reducing the resulting equations with the help of the identity $\lambda = 2\nu\mu/\,(1-2\nu)$ we find that the Navier equations are satisfied if

$$\nabla^4\mathbf{F} = 0 \,. \tag{6.168}$$

Thus, any bi-harmonic vector is suitable as a Galerkin vector. As should be expected, because they are solutions to the same equation, there is a relationship between ϕ and ψ with $\mathbf{F}$. It has been shown that

$$\phi = -F_{i,i} \,, \tag{6.169a}$$

and

$$\varepsilon_{ijk}\psi_{k,j} = 2\,(1-\nu)\,F_{i,jj} \,. \tag{6.169b}$$

If F_i is not only bi-harmonic, but harmonic as well, Eq 6.169b reduces to

$$\varepsilon_{ijk}\psi_{k,j} = 0 \,, \tag{6.170a}$$

and the relationship between ϕ and F_i becomes

$$\phi_{,ii} = -F_{i,jji} \,. \tag{6.170b}$$

In this case ϕ is called the Lamé strain potential.

Example 6.8

Consider a Galerkin vector of the form $\mathbf{F} = F_i\hat{\mathbf{e}}_i$ where F_3 is a function of the coordinates, that is $F_3 = F_3\,(x_1, x_2, x_3)$. Apply this vector to obtain the solution to the problem of a concentrated force acting at the origin of coordinates in the direction of the positive x_3 axis of a very large elastic body. This is called the *Kelvin problem*.

Solution

Let $u_i = 2(1-\nu)F_{i,jj} - F_{j,ji}$ as given by Eq 6.167. Accordingly,

$$\begin{aligned} u_1 &= -F_{3,31} \ , \\ u_2 &= -F_{3,32} \ , \\ u_3 &= 2(1-\nu)(F_{3,11} + F_{3,22} + F_{3,33}) - F_{3,33} \ . \end{aligned}$$

Take F_3 to be proportional to the distance squared from the origin as defined by $F_3 = BR$ where B is a constant and $R^2 = x_1^2 + x_2^2 + x_3^2$. Thus, the displacements are

$$\begin{aligned} u_1 &= \frac{Bx_3x_1}{R^3} \ , \\ u_2 &= \frac{Bx_3x_2}{R^3} \ , \\ u_3 &= B\left[\frac{4(1-\nu)}{R} - \frac{x_1^2 + x_2^2}{R^3}\right] \ . \end{aligned}$$

From these displacement components the stresses may be computed using Hooke's law. In particular, it may be shown that

$$t_{33} = \frac{\partial}{\partial x_3}\left[(2-\nu)\left(\frac{\partial^2}{\partial x_1^2} + \frac{\partial^2}{\partial x_2^2} + \frac{\partial^2}{\partial x_3^2}\right) - \frac{\partial^2}{\partial x_3^2}\right] BR$$

which upon carrying out the indicated differentiation and combining terms becomes

$$t_{33} = -B\left[\frac{2(2-\nu)x_3}{R^3} - \frac{3(x_1^2 + x_2^2)x_3}{R^5}\right] \ .$$

This equation may be written in a more suitable form for the integration that follows by noting that $R^2 = r^2 + x_3^2$ where $r^2 = x_1^2 + x_2^2$. The modified equation is

$$t_{33} = -B\left[\frac{(1-2\nu)x_3}{R^3} + \frac{3x_3^3}{R^5}\right] \ .$$

Summing forces in the x_3 direction over the plane x_3 = constant allows us to determine B in terms of the applied force P. The required integral is

$$P = \int_0^\infty (-t_{33})\, 2\pi r\, dr \ .$$

But $r dr = R dR$ and so

$$P = 2\pi B\left[(1-2\nu)x_3\int_0^\infty \frac{dR}{R^2} + 3x_3^3\int_0^\infty \frac{dR}{R^4}\right]$$

from which we find

$$B = \frac{P}{4\pi(1-\nu)} \ .$$

The third approach for solving the Navier equations is called the *Papkovich-Neuber solution* which results in equations in terms of harmonic functions of a scalar, and a vector potential. For this we take the displacement vector to be represented by a scalar potential B, the vector potential $\mathbf{V}$, and the position vector x_i in the form

$$u_i = V_i - B_{,i} - \frac{(V_k x_k)_{,i}}{4(1-\nu)} \tag{6.171}$$

which when substituted into the homogeneous Navier equations and simplified using the identity $\lambda = 2\nu\mu/(1-2\nu)$ we obtain

$$\mu V_{i,jj} - (\lambda + 2\mu) B_{,ijj} - \frac{1}{2}(\lambda + \mu)(V_{k,ijj} + V_{i,jj}) = 0\,. \tag{6.172}$$

These equations are clearly satisfied when

$$\nabla^2 \mathbf{V} = 0\,, \tag{6.173a}$$

and

$$\nabla^2 B = 0 \tag{6.173b}$$

which indicates that any four harmonic functions, V_i with $i = 1, 2, 3$ and B, will serve to provide a displacement vector u_i from Eq 6.171 that satisfies the Navier equations. Since the displacement vector has only three components, the four scalar functions, V_i, and B are not completely independent and may be reduced to three. It can be shown that these potentials are related to the Galerkin vector through the expressions

$$\mathbf{V} = 2(1-\nu)\nabla^2 \mathbf{F}\,, \tag{6.174a}$$

and

$$B = \nabla \cdot \mathbf{F} - \frac{\mathbf{V} \cdot \mathbf{x}}{4(1-\nu)}\,. \tag{6.174b}$$

Whereas it is not usually possible to solve the Navier equations directly for problems involving a body of arbitrary geometry, in certain cases of spherical symmetry an elementary solution is available. Consider the case of a hollow spherical geometry of inner radius r_1 and outer radius r_2 that is subjected to an internal pressure p_1 and an external pressure p_2. Due to the symmetry condition here we assume a displacement field

$$u_i = \phi(r)\, x_i \tag{6.175}$$

where $r^2 = x_i x_i$ and ϕ depends solely upon r. By direct substitution of Eq 6.175 into Eq 6.160 we arrive at the ordinary differential equation

$$\frac{d^2\phi}{dr^2} + \frac{4}{r}\frac{d\phi}{dr} = 0 \tag{6.176}$$

for which the general solution may be written as

$$\phi(r) = A_1 + \frac{A_2}{r^3} \tag{6.177}$$

where A_1 and A_2 are constants of integration depending on the boundary conditions. Eq 6.155 written in terms of displacement derivatives has the form

$$t_{ij} = \lambda \delta_{ij} u_{k,k} + \mu(u_{i,j} + u_{j,i})\,. \tag{6.178}$$

It follows by making use of Eqs 6.175 and 6.177 that

$$t_{ij} = 3\lambda A_1 \delta_{ij} + 2\mu \left[\left(A_1 + \frac{A_2}{r^3} \right) \delta_{ij} - \frac{3A_2 x_i x_j}{r^3} \right] . \tag{6.179}$$

Recall that the traction vector in the radial direction (see Eq 3.51) is $\sigma_N = t_{ij} n_i n_j$ which upon substitution of Eq 6.177 becomes

$$\sigma_N = (3\lambda + 2\mu) A_1 + \frac{4\mu A_2}{r^3} \tag{6.180}$$

where the identity $x_i = r n_i$ has been used. Similarly, the tangential traction can be calculated using $\sigma_S = t_{ij} \nu_i \nu_j$ where the unit vectors ν_i are perpendicular to n_i. The result is

$$\sigma_S = (3\lambda + 2\mu) A_1 + \frac{2\mu A_2}{r^3} \tag{6.181}$$

since $\nu_i x_i = 0$.

The constants A_1 and A_2 are determined from boundary conditions on the tractions. Clearly,

$$\sigma_N = -p_1 \quad \text{at} \quad r = r_1 ,$$
$$\sigma_N = -p_2 \quad \text{at} \quad r = r_2 .$$

Carrying out the indicated algebra, we have the well-known formulas

$$\sigma_N = \frac{p_1 r_1^3 - p_2 r_2^3}{r_2^3 - r_1^3} - \frac{r_1^3 r_2^3}{r^3} \frac{p_1 - p_2}{r_2^3 - r_1^3} , \tag{6.182a}$$

$$\sigma_S = \frac{p_1 r_1^3 - p_2 r_2^3}{r_2^3 - r_1^3} + \frac{r_1^3 r_2^3}{2r^3} \frac{p_1 - p_2}{r_2^3 - r_1^3} . \tag{6.182b}$$

These equations may be easily modified to cover the case where $p_1 = 0$, or the case where $p_2 = 0$.

We conclude this section with a brief discussion of three dimensional stress functions. These functions are designed to provide solutions of the equilibrium equations. Additionally, in order for the solution to be complete it must be compatible with the Beltrami-Michell equations. Beginning with the equilibrium equations in the absence of body forces

$$t_{ij,j} = 0 \tag{6.183}$$

we propose the stress field

$$t_{ij} = \varepsilon_{ipq} \varepsilon_{jkm} \Phi_{qk,pm} \tag{6.184}$$

where Φ_{qk} is a symmetric tensor function of the coordinates. By a direct expansion of this equation the stress components may be expressed in terms of the potential Φ_{qk}. For example,

$$t_{11} = \varepsilon_{1pq} \varepsilon_{1km} \Phi_{qk,pm} \tag{6.185}$$

which when summed over the repeated indices, keeping in mind the properties of the permutation symbol, becomes

$$t_{11} = 2\Phi_{23,23} - \Phi_{22,33} - \Phi_{33,22} . \tag{6.186a}$$

Similarly,

$$t_{22} = 2\Phi_{13,13} - \Phi_{11,33} - \Phi_{33,11} \,, \tag{6.186b}$$

$$t_{33} = 2\Phi_{12,12} - \Phi_{11,22} - \Phi_{22,11} \,, \tag{6.186c}$$

$$t_{12} = \Phi_{12,33} + \Phi_{33,12} - \Phi_{23,31} - \Phi_{13,32} \,, \tag{6.186d}$$

$$t_{23} = \Phi_{23,11} + \Phi_{11,23} - \Phi_{12,13} - \Phi_{13,12} \,, \tag{6.186e}$$

$$t_{31} = \Phi_{31,22} + \Phi_{22,31} - \Phi_{12,23} - \Phi_{32,21} \,. \tag{6.186f}$$

It may be shown by direct substitution that the equilibrium equations are satisfied by these stress components.

Upon setting the off-diagonal terms of Φ_{qk} to zero, that is, if $\Phi_{12} = \Phi_{23} = \Phi_{31} = 0$, we obtain the solution proposed by Maxwell. By setting the diagonal terms of Φ_{qk} to zero, namely, $\Phi_{11} = \Phi_{22} = \Phi_{33} = 0$ we obtain the solution proposed by Morera which is known by that name. It is interesting to note that if all the components of Φ_{qk} except Φ_{33} are zero, that component is the Airy stress function introduced in Section 6.7 as can be verified by Eq 6.186. Although the potential Φ_{qk} provides us with a solution of the equilibrium equations, that solution is not compatible with the Beltrami-Michell equations except under certain conditions.

Problems

Problem 6.1
In general, the strain energy density W may be expressed in the form

$$W = \mathcal{C}^*_{\alpha\beta}\epsilon_\alpha\epsilon_\beta \quad (\alpha, \beta = 1, \ldots, 6)$$

where $\mathcal{C}^*_{\alpha\beta}$ is not necessarily symmetric. Show that this equation may be rearranged to appear in the form

$$W = \frac{1}{2}\mathcal{C}_{\alpha\beta}\epsilon_\alpha\epsilon_\beta$$

where $\mathcal{C}_{\alpha\beta}$ is symmetric, so that now

$$\frac{\partial W}{\partial \epsilon_\beta} = \mathcal{C}_{\alpha\beta}\epsilon_\beta = t_\beta$$

in agreement with Eq 6.8.

Problem 6.2
Let the stress and strain tensors be decomposed into their respective spherical and deviator components. Determine an expression for the strain energy density W as the sum of a dilatation energy density $W_{(1)}$ and a distortion energy density $W_{(2)}$.

Answer

$$W = W_{(1)} + W_{(2)} = \frac{1}{6}t_{ii}\epsilon_{jj} + \frac{1}{2}S_{ij}\eta_{ij}$$

Problem 6.3
If the strain energy density W is generalized in the sense that it is assumed to be a function of the deformation gradient components instead of the small strain components, that is, if $W = W(F_{iA})$, make use of the energy equation and the continuity equation to show that in this case Eq 6.16 is replaced by

$$Jt_{ij} = \frac{\partial W}{\partial F_{iA}}F_{jA}\ .$$

Problem 6.4
For an isotropic elastic medium as defined by Eq 6.23, express the strain energy density in terms of

(a) the components of ϵ_{ij},

(b) the components of t_{ij},

(c) the invariants of ϵ_{ij}.

Answer

$$\text{(a)}\ W = \frac{1}{2}\left(\lambda \epsilon_{ii}\epsilon_{jj} + 2\mu\epsilon_{ij}\epsilon_{ij}\right),$$

$$\text{(b)}\ W = \frac{(3\lambda + 2\mu)\, t_{ij}t_{ij} - \lambda t_{ii}t_{jj}}{4\mu\,(3\lambda + 2\mu)},$$

$$\text{(c)}\ W = \left(\tfrac{1}{2}\lambda + \mu\right)(I_\epsilon)^2 - 2\mu II_\epsilon$$

Problem 6.5

Let t_{ij} be any second-order isotropic tensor such that

$$t'_{ij} = a_{im}a_{jn}t_{mn} = t_{ij}$$

for any proper orthogonal transformation a_{ij}. Show that by successive applications of the transformations

$$[a_{ij}] = \begin{bmatrix} 0 & 0 & -1 \\ -1 & 0 & 0 \\ 0 & 1 & 0 \end{bmatrix} \quad \text{and} \quad [a_{ij}] = \begin{bmatrix} 0 & 0 & 1 \\ -1 & 0 & 0 \\ 0 & -1 & 0 \end{bmatrix}$$

every second-order isotropic tensor is a scalar multiple of the Kronecker delta, δ_{ij}.

Problem 6.6

Verify that Eqs 6.33a and 6.33b when combined result in Eq 6.28 when Eqs 6.29 are used.

Problem 6.7

For an elastic medium, use Eq 6.33 to express the result obtained in Problem 6.2 in terms of the engineering elastic constants K and G.

Answer

$$W = \tfrac{1}{2}K\epsilon_{ii}\epsilon_{jj} + G\left(\epsilon_{ij}\epsilon_{ij} - \tfrac{1}{3}\epsilon_{ii}\epsilon_{jj}\right)$$

Problem 6.8

Show that the distortion energy density $W_{(2)}$ (see Problem 6.2) for a linear elastic medium may be expressed in terms of (a) the principal stresses, $\sigma_{(1)}$, $\sigma_{(2)}$, $\sigma_{(3)}$ and (b) the principal strains, $\epsilon_{(1)}$, $\epsilon_{(2)}$, $\epsilon_{(3)}$ in the form

$$\text{(a)}\ W_{(2)} = \frac{\left(\sigma_{(1)} - \sigma_{(2)}\right)^2 + \left(\sigma_{(2)} - \sigma_{(3)}\right)^2 + \left(\sigma_{(3)} - \sigma_{(1)}\right)^2}{12G},$$

$$\text{(b)}\ W_{(2)} = \frac{1}{3}\left[\left(\epsilon_{(1)} - \epsilon_{(2)}\right)^2 + \left(\epsilon_{(2)} - \epsilon_{(3)}\right)^2 + \left(\epsilon_{(3)} - \epsilon_{(1)}\right)^2\right] G.$$

Problem 6.9

Beginning with the definition for W (Eq 6.21a), show for a linear elastic material represented by Eq 6.23 or by Eq 6.28 that $\partial W/\partial \epsilon_{ij} = t_{ij}$ and $\partial W/\partial t_{ij} = \epsilon_{ij}$ (Note that $\partial \epsilon_{ij}/\partial \epsilon_{mn} = \delta_{im}\delta_{jn}$.)

Problem 6.10

For an isotropic, linear elastic solid, the principal axes of stress and strain coincide, as was shown in Example 6.1. Show that, in terms of engineering constants E and ν this result is given by

$$\epsilon_{(q)} = \frac{(1+\nu)\,\sigma_{(q)} - \nu\left[\sigma_{(1)} + \sigma_{(2)} + \sigma_{(3)}\right]}{E}, \quad (q = 1,2,3).$$

Thus, let $E = 10^6$ psi and $\nu = 0.25$, and determine the principal strains for a body subjected to the stress field (in ksi)

$$[t_{ij}] = \begin{bmatrix} 12 & 0 & 4 \\ 0 & 0 & 0 \\ 4 & 0 & 6 \end{bmatrix}.$$

Answer

$\epsilon_{(1)} = -4.5 \times 10^{-6}$, $\epsilon_{(2)} = 0.5 \times 10^{-6}$, $\epsilon_{(3)} = 13 \times 10^{-6}$

Problem 6.11

Show for an isotropic elastic medium that

$$\text{(a)}\ \frac{1}{1+\nu} = \frac{2(\lambda+\mu)}{3\lambda+2\mu}, \quad \text{(b)}\ \frac{\nu}{1+\nu} = \frac{\lambda}{\lambda+2\mu}, \quad \text{(c)}\ \frac{2\mu\nu}{1-2\nu} = \frac{3K\nu}{1+\nu},$$

$$\text{(d)}\ 2\mu(1+\nu) = 3K(1-2\nu).$$

Problem 6.12

Let the x_1x_3 plane be a plane of elastic symmetry such that the transformation matrix between $Ox_1x_2x_3$ and $Ox'_1x'_2x'_3$ is

$$[a_{ij}] = \begin{bmatrix} 1 & 0 & 0 \\ 0 & -1 & 0 \\ 0 & 0 & 1 \end{bmatrix}.$$

Show that, as the text asserts, this additional symmetry does not result in a further reduction in the elastic constant matrix, Eq 6.40.

Problem 6.13

Let the x_1 axis be an axis of elastic symmetry of order $N = 2$. Determine the form of the elastic constant matrix $C_{\alpha\beta}$, assuming $C_{\alpha\beta} = C_{\beta\alpha}$.

Answer

$$[C_{\alpha\beta}] = \begin{bmatrix} C_{11} & C_{12} & C_{13} & C_{14} & 0 & 0 \\ C_{12} & C_{22} & C_{23} & C_{34} & 0 & 0 \\ C_{13} & C_{23} & C_{33} & C_{34} & 0 & 0 \\ C_{14} & C_{24} & C_{34} & C_{44} & 0 & 0 \\ 0 & 0 & 0 & 0 & C_{55} & C_{56} \\ 0 & 0 & 0 & 0 & C_{56} & C_{66} \end{bmatrix}$$

Problem 6.14

Assume that, by the arguments of elastic symmetry, the elastic constant matrix for an isotropic body has been reduced to the form

$$[C_{\alpha\beta}] = \begin{bmatrix} C_{11} & C_{12} & C_{12} & 0 & 0 & 0 \\ C_{12} & C_{11} & C_{12} & 0 & 0 & 0 \\ C_{12} & C_{12} & C_{11} & 0 & 0 & 0 \\ 0 & 0 & 0 & C_{44} & 0 & 0 \\ 0 & 0 & 0 & 0 & C_{44} & 0 \\ 0 & 0 & 0 & 0 & 0 & C_{44} \end{bmatrix}.$$

Show that, if the x_1 axis is taken as an axis of elastic symmetry of any order (θ is arbitrary), $C_{11} = C_{12} + 2C_{44}$. (*Hint:* Expand $t'_{23} = a_{2q}a_{3m}t_{qm}$ and $\epsilon'_{23} = a_{2q}a_{3m}\epsilon_{qm}$.)

Problem 6.15

If the axis which makes equal angles with the coordinate axes is an axis of elastic symmetry of order $N = 3$, show that there are twelve independent elastic constants and that the elastic matrix has the form

$$[C_{\alpha\beta}] = \begin{bmatrix} C_{11} & C_{12} & C_{13} & C_{14} & C_{15} & C_{16} \\ C_{13} & C_{11} & C_{12} & C_{16} & C_{14} & C_{15} \\ C_{12} & C_{13} & C_{11} & C_{15} & C_{16} & C_{14} \\ C_{41} & C_{42} & C_{43} & C_{44} & C_{45} & C_{46} \\ C_{43} & C_{41} & C_{42} & C_{46} & C_{44} & C_{45} \\ C_{42} & C_{43} & C_{41} & C_{45} & C_{46} & C_{44} \end{bmatrix}.$$

Problem 6.16

For an elastic body whose x_3 axis is an axis of elastic symmetry of order $N = 6$, show that the nonzero elastic constants are $C_{11} = C_{22}$, C_{33}, $C_{55} = C_{44}$, $C_{66} = \frac{1}{2}(C_{11} - C_{12})$, and $C_{13} = C_{23}$.

Problem 6.17

Develop a formula in terms of the strain components for the strain energy density W for the case of an orthotropic elastic medium.

Answer

$$\begin{aligned} W &= \tfrac{1}{2}(C_{11}\epsilon_1 + 2C_{12}\epsilon_2 + 2C_{13}\epsilon_3)\epsilon_1 + \tfrac{1}{2}(C_{22}\epsilon_2 + 2C_{23}\epsilon_3)\epsilon_2 \\ &+ \tfrac{1}{2}(C_{33}\epsilon_3^2 + C_{44}\epsilon_4^2 + C_{55}\epsilon_5^2 + C_{66}\epsilon_6^2) \end{aligned}$$

Problem 6.18

Show that, for an elastic continuum having x_1 as an axis of elastic symmetry of order $N = 2$, the strain energy density has the same form as for a continuum which has an x_2x_3 plane of elastic symmetry.

Answer

$$\begin{aligned} 2W &= (C_{11}\epsilon_1 + 2C_{12}\epsilon_2 + 2C_{13}\epsilon_3 + 2C_{14}\epsilon_4)\epsilon_1 + (C_{22}\epsilon_2 + 2C_{23}\epsilon_3 + 2C_{24}\epsilon_4)\epsilon_2 \\ &+ (C_{33}\epsilon_3 + 2C_{34}\epsilon_4)\epsilon_3 + C_{44}\epsilon_4^2 + (C_{55}\epsilon_5 + 2C_{56}\epsilon_6)\epsilon_5 + C_{66}\epsilon_6^2 \end{aligned}$$

Problem 6.19
Let the stress field for a continuum be given by

$$[t_{ij}] = \begin{bmatrix} x_1 + x_2 & t_{12} & 0 \\ t_{12} & x_1 - x_2 & 0 \\ 0 & 0 & x_2 \end{bmatrix}$$

where t_{12} is a function of x_1 and x_2. If the equilibrium equations are satisfied in the absence of body forces and if the stress vector on the plane $x_1 = 1$ is given by $\mathbf{t}^{(\hat{\mathbf{e}}_1)} = (1 + x_2)\,\hat{\mathbf{e}}_1 + (6 - x_2)\,\hat{\mathbf{e}}_2$, determine t_{12} as a function of x_1 and x_2.

Answer

$t_{12} = x_1 - x_2 + 5$

Problem 6.20
Invert Eq 6.45b to obtain Hooke's law in the form

$$t_{ij} = 2G\left(\epsilon_{ij} + \frac{\nu}{1-2\nu}\epsilon_{kk}\delta_{ij}\right)$$

which, upon combination with Eqs 6.44 and 6.43, leads to the Navier equation

$$G\left(u_{i,jj} + \frac{1}{1-2\nu}u_{j,ij}\right) + \rho b_i = 0\ .$$

This equation is clearly indeterminate for $\nu = 0.5$. However, show that in this case Hooke's law and the equilibrium equations yield the result

$$Gu_{i,jj} + \frac{1}{3}t_{jj,i} + \rho b_i = 0\ .$$

Problem 6.21
Let the displacement field be given in terms of some vector q_i by the equation

$$u_i = \frac{2(1-\nu)\,q_{i,jj} - q_{j,ji}}{G}\ .$$

Show that the Navier equation (Eq 6.49) is satisfied providing $b_i \equiv 0$ and q_i is bi-harmonic so that $q_{i,jjkk} = 0$. If $q_1 = x_2/r$ and $q_2 = -x_1/r$ where $r^2 = x_i x_i$, determine the resulting stress field.

Answer

$t_{11} = -t_{22} = 6QGx_1x_2/r^5\ ;\quad t_{33} = 0$
$t_{12} = t_{21} = 3QG\left(x_2^2 - x_1^2\right)/r^5$
$t_{13} = t_{31} = 3QGx_2x_3/r^5; t_{23} = t_{32} = -3QGx_1x_2/r^5;$
where $Q = 4(1-\nu)/G$

Problem 6.22

If body forces are zero, show that the elastodynamic Navier equation (Eq 6.55) will be satisfied by the displacement field

$$u_i = \phi_{,i} + \varepsilon_{ijk}\psi_{k,j}$$

provided the potential functions ϕ and ψ_k satisfy the three-dimensional wave equation.

Problem 6.23

Show that, for plane stress, Hooke's law Eq 6.120a and Eq 6.121 may be expressed in terms of the Lamé constants λ and μ by

$$\begin{aligned} \epsilon_{ij} &= \frac{1}{2\mu}\left(t_{ij} - \frac{\lambda}{3\lambda+2\mu}\delta_{ij}t_{kk}\right) \quad (i,j,k=1,2) \ , \\ \epsilon_{kk} &= -\frac{\lambda}{2\mu\,(3\lambda+2\mu)}t_{ii} \quad (i=1,2) \ . \end{aligned}$$

Problem 6.24

For the case of plane stress, let the stress components be defined in terms of the function $\phi = \phi(x_1, x_2)$, known as the Airy stress function, by the relationships,

$$t_{11} = \phi_{,22}, \quad t_{22} = \phi_{,11}, \quad t_{12} = -\phi_{,12} \ .$$

Show that ϕ must satisfy the biharmonic equation $\nabla^4\phi = 0$ and that, in the absence of body forces, the equilibrium equations are satisfied identically by these stress components. If $\phi = Ax_1^3x_2^2 - Bx_1^5$ where A and B are constants, determine the relationship between A and B for this to be a valid stress function.

Answer

$$A = 5B$$

Problem 6.25

Develop an expression for the strain energy density, W, for an elastic medium in (a) plane stress and (b) plane strain.

Answer

(a) $W = \left[t_{11}^2 + t_{22}^2 - 2\nu t_{11}t_{22} + 2\,(1+\nu)\,t_{12}^2\right]/2E$

(b) $W = \left(\mu + \frac{1}{2}\lambda\right)\left(\epsilon_{11}^2 + \epsilon_{22}^2\right) + \lambda\epsilon_{11}\epsilon_{22} + 2\mu\epsilon_{12}^2$

Problem 6.26

Show that $\phi = x_1^4x_2 + 4x_1^2x_2^3 - x_2^5$ is a valid Airy stress function, that is, that $\nabla^4\phi = 0$, and compute the stress tensor for this case assuming a state of plane strain with $\nu = 0.25$.

Answer

$$[t_{ij}] = \begin{bmatrix} 24x_1^2x_2 - 20x_2^3 & -4x_1^3 - 24x_1x_2^2 & 0 \\ -4x_1^3 - 24x_1x_2^2 & 12x_1^2x_2 + 8x_2^3 & 0 \\ 0 & 0 & \frac{1}{4}\left(9x_1^2x_2 - 3x_2^3\right) \end{bmatrix}$$

Problem 6.27

Verify the inversion of Eq 6.147 into Eq 6.148. Also, show that the two equations of Eq 6.149 may be combined to produce Eq 6.148.

Problem 6.28

Develop appropriate constitutive equations for thermoelasticity in the case of (a) plane stress and (b) plane strain.

Answer

(a) $\epsilon_{ij} = [(1+\nu)t_{ij} - \nu\delta_{ij}t_{kk}]/E + \delta_{ij}(\theta - \theta_0)\alpha \quad (i,j,k = 1,2)$
$\epsilon_{33} = \nu t_{ii}/E + \alpha(\theta - \theta_0) \quad (i = 1,2)$

(b) $t_{ij} = \lambda\delta_{ij}\epsilon_{kk} + 2\mu\epsilon_{ij} - \delta_{ij}(3\lambda + 2\mu)\alpha(\theta - \theta_0) \quad (i,j,k = 1,2)$
$t_{33} = \nu t_{ii} - \alpha E(\theta - \theta_0) = \lambda\epsilon_{ii} - (3\lambda + 2\mu)\alpha(\theta - \theta_0) \quad (i = 1,2)$

Problem 6.29

Consider the Airy stress function

$$\phi_5 = D_5 x_1^2 x_2^3 + F_5 x_2^5 \ .$$

(a) Show that for this to be valid stress function, $F_5 = -D_5/5$.

(b) Construct the composite stress function

$$\phi = \phi_5 + \phi_3 + \phi_2$$

where

$$\phi = D_5\left(x_1^2 x_2^3 - \frac{1}{5}x_2^5\right) + \frac{1}{2}B_3 x_1^2 x_2 + \frac{1}{2}A_2 x_1^2 \ .$$

For this stress function show that the stress components are

$$t_{11} = D_5\left(6x_1^2 x_2 - 4x_2^3\right) \ ,$$
$$t_{22} = 2D_5 x_2^3 + B_3 x_2 + A_2 \ ,$$
$$t_{12} = -6D_5 x_1 x_2^2 - B_3 x_1 \ .$$

Problem 6.30

A rectangular beam of width unity and length 2L carries a uniformly distributed load of q lb/ft as shown. Shear forces V support the beam at both ends. List the six boundary conditions for this beam the stresses must satisfy using stresses determined in Problem 6.29.

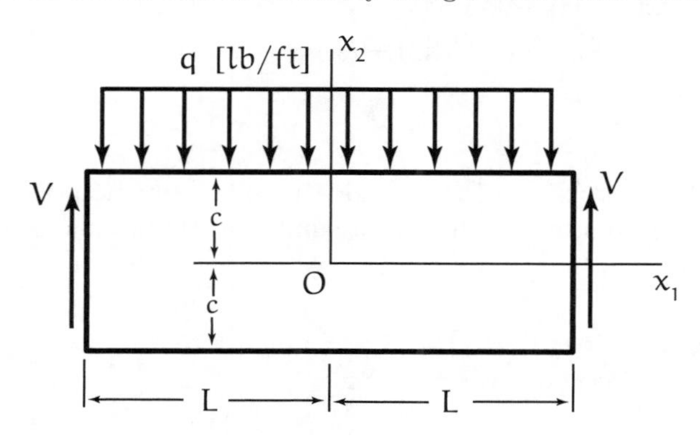

Answer

1. $t_{22} = -q$ at $x_2 = +c$
2. $t_{22} = 0$ at $x_2 = -c$
3. $t_{12} = 0$ at $x_2 = \pm c$
4. $\int_{-c}^{+c} t_{12} dx_2 = qL$ at $x_1 = \pm L$
5. $\int_{-c}^{+c} t_{11} dx_2 = 0$ at $x_1 = \pm L$
6. $\int_{-c}^{+c} t_{11} x_2 dx_2 = 0$ at $x_1 = \pm L$

Problem 6.31

Using boundary conditions 1, 2, and 3 listed in Problem 6.30, show that the stresses in Problem 6.29 require that

$$A_2 = -\frac{q}{2}, \quad B_3 = -\frac{3q}{4c}, \quad D_5 = \frac{q}{8c^3}.$$

Thus, for the beam shown the stresses are

$$t_{11} = \frac{q}{2I}\left(x_1^2 x_2 - \frac{2}{3}x_2^3\right),$$

$$t_{22} = \frac{q}{2I}\left(\frac{1}{3}x_2^3 - c^2 x_2 - \frac{2}{3}c^3\right),$$

$$t_{12} = -\frac{q}{2I}\left(x_1 x_2^2 - c^2 x_1\right),$$

where $I = \frac{2}{3}c^3$ is the plane moment of inertia of the beam cross section.

Problem 6.32

Show that, using the stresses calculated in Problem 6.31, the boundary conditions 4 and 5 are satisfied, but boundary condition 6 is not satisfied.

Problem 6.33

Continuing Problems 6.31 and 6.32, in order for boundary condition 6 to be satisfied an additional term is added to the stress function, namely

$$\phi_3 = D_3 x_2^3.$$

Show that, from boundary condition 6,

$$D_3 = \frac{3q}{4c}\left(\frac{1}{15} - \frac{L^2}{6c^2}\right),$$

so that finally

$$t_{11} = \frac{q}{2I}\left[x_1^2 - \frac{2}{3}x_2^2 + \frac{6}{15}c^2 - L^2\right] x_2.$$

Problem 6.34

Show that for the shaft having a cross section in the form of an equilateral triangle the warping function is

$$\psi(x_1, x_2) = \lambda\left(x_2^3 - 3x_1^2 x_2\right).$$

Determine

(a) the constant λ in terms of the shaft dimension,
(b) the torsional rigidity K,
(c) the maximum shearing stress.

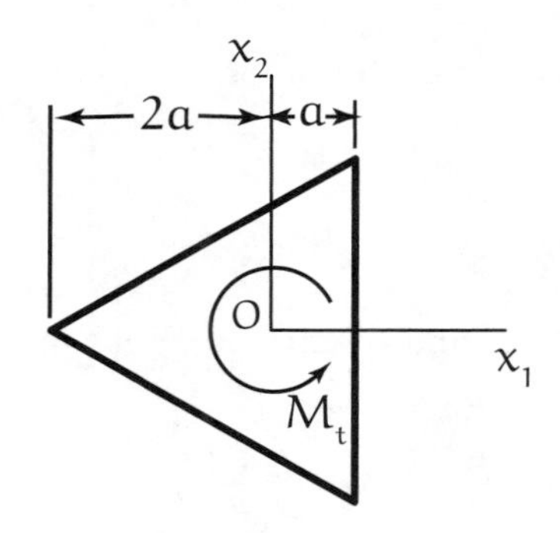

Answer

(a) $\lambda = -\dfrac{1}{6a}$

(b) $K = \dfrac{9G\sqrt{3}a^4}{5}$

(c) $t_{23}|_{max} = \dfrac{3M_t a}{2K}$ at $x_1 = a$, $x_2 = 0$

Problem 6.35
Consider the Galerkin vector that is the sum of three double forces, that is, let

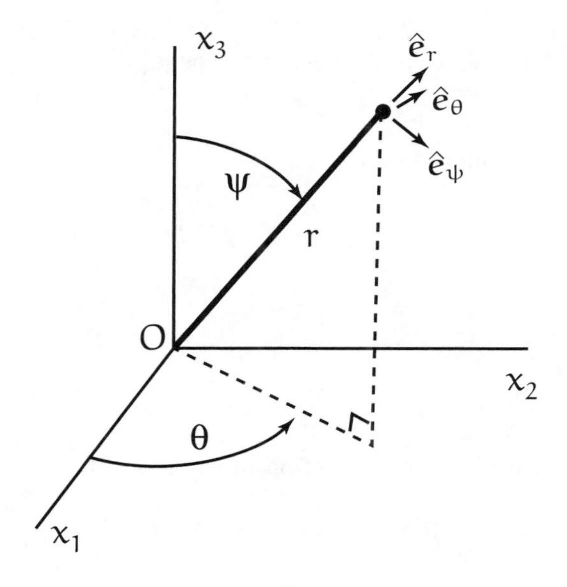

$$\mathbf{F} = B\left(\frac{x_1}{r}\hat{\mathbf{e}}_1 + \frac{x_2}{r}\hat{\mathbf{e}}_2 + \frac{x_3}{r}\hat{\mathbf{e}}_3\right)$$

where B is a constant and $r^2 = x_i x_i$. Show that the displacement components are given by

$$u_i = -\frac{2B(1-2\nu)x_i}{r^3}\ .$$

Using the sketch above, (spherical coordinates) observe that the radial displacement u_r (subscript r not a summed indice, rather indicating the radial component of displacement)

$$u_r = \frac{u_i x_i}{r}\ ,$$

and show that $u_r = -2B(1-2\nu)r^{-2}$. Also, show that $u_\psi = u_\theta = 0$. Thus

$$\epsilon_r = \frac{\partial u_r}{\partial r} = \frac{4B(1-2\nu)}{r^3} \quad \text{and} \quad \epsilon_\psi = \epsilon_\theta = -\frac{2B(1-2\nu)}{r^3}\ ,$$

so that the cubical dilatation is $\epsilon_r + \epsilon_\psi + \epsilon_\theta = 0$. From Hooke's law

$$t_{ij} = \lambda\delta_{ij}u_{k,k} + 2\mu\left(u_{i,j} + u_{j,i}\right)$$

which reduces here to $t_{ij} = 2\mu\epsilon_{ij}$ so that

$$t_{rr} = \frac{8BG(1-2\nu)}{r^3}\ ,$$

and

$$t_{\psi\psi} = t_{\theta\theta} = -\frac{4BG(1-2\nu)}{r^3}\ .$$

7

Classical Fluids

Often, continuum mechanics is taken to gain a foundation for solid mechanics like elasticity discussed in the last chapter. However, the foundation of continuum mechanics allows for the study of fluids. A fundamental characteristic of any fluid – be it a liquid or a gas – is that the action of shear stresses, no matter how small they may be, will cause the fluid to deform continuously as long as the stresses act.

7.1 Viscous Stress Tensor, Stokesian, and Newtonian Fluids

A fluid at rest (or in a state of rigid body motion) is incapable of sustaining any shear stress whatsoever. This implies that the stress vector on an arbitrary element of surface at any point in a fluid at rest is proportional to the normal n_i of that element, but independent of its direction. Thus, we write

$$t_i^{(\hat{n})} = t_{ij} n_j = -p_0 n_i \tag{7.1}$$

where the (positive) proportionality constant p_0 is the *thermostatic pressure* or, as it is frequently called, the *hydrostatic pressure*. We note from Eq 7.1 that

$$t_{ij} = -p_0 \delta_{ij} \tag{7.2}$$

which indicates that for a fluid at rest the stress is everywhere compressive, that every direction is a principal stress direction at any point, and that the hydrostatic pressure is equal to the *mean normal stress*,

$$p_0 = -\frac{1}{3} t_{ii} \,. \tag{7.3}$$

This pressure is related to the temperature θ and density ρ by an equation of state having the form

$$F(p_0, \rho, \theta) = 0 \,. \tag{7.4}$$

For a fluid in motion the shear stresses are not usually zero, and in this case we write

$$t_{ij} = -p\delta_{ij} + \tau_{ij} \tag{7.5}$$

where τ_{ij} is called the *viscous stress tensor*, which is a function of the motion and vanishes when the fluid is at rest. In this equation, the pressure p is called the *thermodynamic pressure* and is given by the same functional relationship with respect to θ and ρ as that for the static pressure p_0 in the equilibrium state, that is, by

$$F(p, \rho, \theta) = 0 \,. \tag{7.6}$$

Note from Eq 7.5 that, for a fluid in motion, p is not equal to the mean normal stress, but instead is given by

$$p = -\frac{1}{3}\left(t_{ii} - \tau_{ii}\right) , \tag{7.7}$$

so that, for a fluid at rest ($\tau_{ij} = 0$), p equates to p_0.

In developing constitutive equations for viscous fluids, we first remind ourselves that this viscous stress tensor must vanish for fluids at rest, and following the usual practice, we assume that τ_{ij} is a function of the rate of deformation tensor d_{ij}. Expressing this symbolically, we write

$$\tau_{ij} = f_{ij}\left(\mathbf{D}\right) . \tag{7.8}$$

If the functional relationship in this equation is nonlinear, the fluid is called a *Stokesian* fluid. When f_{ij} defines τ_{ij} as a linear function of d_{ij}, the fluid is known as a *Newtonian* fluid, and we represent it by the equation

$$\tau_{ij} = K_{ijpq} d_{pq} \tag{7.9}$$

in which the coefficients K_{ijpq} reflect the viscous properties of the fluid.

As may be verified experimentally, most fluids are isotropic. Therefore, K_{ijpq} in Eq 7.9 is an isotropic tensor; this, along with the symmetry properties of d_{ij} and τ_{ij}, allow us to reduce the 81 coefficients K_{ijpq} to 2. We conclude that, for a homogeneous, isotropic Newtonian fluid, the constitutive equation is

$$t_{ij} = -p\delta_{ij} + \lambda^* \delta_{ij} d_{kk} + 2\mu^* d_{ij} \tag{7.10}$$

where λ^* and μ^* are *viscosity coefficients* which denote the viscous properties of the fluid. From this equation we see that the mean normal stress for a Newtonian fluid is

$$\frac{1}{3}t_{ii} = -p + \frac{1}{3}\left(3\lambda^* + 2\mu^*\right) d_{ii} = -p + \kappa^* d_{ii} \tag{7.11}$$

where $\kappa^* = \frac{1}{3}\left(3\lambda^* + 2\mu^*\right)$ is known as the *coefficient of bulk viscosity*. The condition

$$\kappa^* = \frac{1}{3}\left(3\lambda^* + 2\mu^*\right) \tag{7.12a}$$

or, equivalently,

$$\lambda^* = -\frac{2}{3}\mu^* \tag{7.12b}$$

is known as *Stokes condition*, and we see from Eq 7.11 that this condition assures us that, for a Newtonian fluid at rest, the mean normal stress equals the (negative) pressure p.

If we introduce the deviator tensors

$$S_{ij} = t_{ij} - \frac{1}{3}\delta_{ij} t_{kk} \tag{7.13a}$$

for stress and

$$\beta_{ij} = d_{ij} - \frac{1}{3}\delta_{ij} d_{kk} \tag{7.13b}$$

for rate of deformation into Eq 7.10, we obtain

$$S_{ij} + \frac{1}{3}\delta_{ij} t_{kk} = -p\delta_{ij} + \frac{1}{3}\left(3\lambda^* + 2\mu^*\right)\delta_{ij} d_{kk} + 2\mu^* \beta_{ij} \tag{7.14}$$

which may be conveniently split into the pair of constitutive equations

$$S_{ij} = 2\mu^* \beta_{ij} \,, \tag{7.15a}$$

$$t_{ii} = -3\left(p - \kappa^* d_{kk}\right) \,. \tag{7.15b}$$

The first of this pair relates the shear effect of the motion with the stress deviator, and the second associates the mean normal stress with the thermodynamic pressure and the bulk viscosity.

7.2 Basic Equations of Viscous Flow, Navier-Stokes Equations

Inasmuch as fluids do not possess a "natural state" to which they return upon removal of applied forces, and because the viscous forces are related directly to the velocity field, it is customary to employ the Eulerian description in writing the governing equations for boundary value problems in viscous fluid theory. Thus, for the thermomechanical behavior of a Newtonian fluid, the following field equations must be satisfied:

(a) the continuity equation (Eq 5.13)

$$\dot{\rho} + \rho v_{i,i} = 0 \,, \tag{7.16}$$

(b) the equations of motion (Eq 5.22)

$$t_{ij,j} + \rho b_i = \rho \dot{v}_i \,, \tag{7.17}$$

(c) the constitutive equations (Eq 7.10)

$$t_{ij} = -p\delta_{ij} + \lambda^* \delta_{ij} d_{kk} + 2\mu^* d_{ij} \,, \tag{7.18}$$

(d) the energy equation (Eq 5.65)

$$\rho \dot{u} = t_{ij} d_{ij} - q_{i,i} + \rho r \,, \tag{7.19}$$

(e) the kinetic equation of state (Eq 7.6)

$$p = p\left(\rho, \theta\right) \,, \tag{7.20}$$

(f) the caloric equation of state (Eq 5.66)

$$u = u\left(\rho, \theta\right) \,, \tag{7.21}$$

(g) the heat conduction equation (Eq 5.62)

$$q_i = -\kappa \theta_{,i} \,. \tag{7.22}$$

This system, Eqs 7.16 through 7.22, together with the definition of the rate of deformation tensor,

$$d_{ij} = \frac{1}{2}\left(v_{i,j} + v_{j,i}\right) \tag{7.23}$$

represents 22 equations in the 22 unknowns, t_{ij}, ρ, v_i, d_{ij}, u, q_i, p, and θ. If thermal effects are neglected and a purely mechanical problem is proposed, we need only Eqs 7.16 through 7.18 as well as Eq 7.23 and a temperature independent form of Eq 7.20, which we state as

$$p = p(\rho) \ . \tag{7.24}$$

This provides a system of 17 equations in the 17 unknowns, t_{ij}, ρ, v_i, d_{ij}, and p.

Certain of the above field equations may be combined to offer a more compact formulation of viscous fluid problems. Thus, by substituting Eq 7.18 into Eq 7.17 and making use of the definition Eq 7.23, we obtain

$$\rho\dot{v}_i = \rho b_i - p_{,i} + (\lambda^* + \mu^*)\, v_{j,ji} + \mu^* v_{i,jj} \tag{7.25}$$

which are known as the *Navier-Stokes* equations for fluids. These equations, along with Eqs 7.19, 7.20, and 7.21, provide a system of seven equations for the seven unknowns, v_i, ρ, p, u, and θ. Notice that even though Eq 7.18 is a linear constitutive equation, the Navier-Stokes equations are nonlinear because in the Eulerian formulation

$$\dot{v}_i = \frac{\partial v_i}{\partial t} + v_j v_{i,j} \ .$$

If Stokes condition $\left(\lambda^* = -\frac{2}{3}\mu^*\right)$ is assumed, Eq 7.25 reduces to the form

$$\rho\dot{v}_i = \rho b_i - p_{,i} + \frac{1}{3}\mu^* \left(v_{j,ji} + 3v_{i,jj}\right) \ . \tag{7.26}$$

Also, if the kinetic equation of state has the form of Eq 7.24, the Navier-Stokes equations along with the continuity equation form a complete set of four equations in the four unknowns, v_i and ρ.

In all of the various formulations for viscous fluid problems stated above, the solutions must satisfy the appropriate field equations as well as boundary and initial conditions on both traction and velocity components. The boundary conditions at a fixed surface require not only the normal, but also the tangential component of velocity to vanish because of the "boundary layer" effect of viscous fluids. The initial and boundary conditions are

1. The velocity must be specified

$$v_i = v_i^* (\mathbf{x}, 0)$$

throughout the volume $\mathcal{V}$. For problems where the density and temperature are changing the initial conditions at $t = 0$ are

$$\rho = \rho^* (\mathbf{x}, 0) \ ,$$
$$\theta = \theta^* (\mathbf{x}, 0) \ .$$

2. The boundary conditions on the surface of the body $\mathcal{V}$ for $t > 0$ are

$$v_i = v_i^* (\mathbf{x}, t) \qquad \text{on} \quad \mathcal{S}_v \ ,$$
$$t_{ij}n_j = T_i^* (\mathbf{x}, t) \qquad \text{on} \quad \mathcal{S}_T \ ,$$

on those parts of the boundary $\mathcal{S}_v$ and $\mathcal{S}_T$ where the velocity and traction are specified. In addition, the temperature and heat flux must also be specified for problems involving temperature changes.

$$\theta = \theta^* (\mathbf{x}, t) \qquad \text{on} \quad \mathcal{S}_\theta \ .$$
$$\theta_{,i}n_i = q^* (\mathbf{x}, t) \qquad \text{on} \quad \mathcal{S}_q \ .$$

The boundary conditions on traction and velocity at solid and free surfaces occur frequently in problems and deserve special comment. On a solid surface, the velocity assumes the velocity of the surface. This is known as the *no-slip condition* for viscous fluids. The condition changes if the fluid is inviscid. Since the viscosity is zero, the shear stress is everywhere zero, and the component along the boundary will be non-zero. The normal component of the velocity will be that of the body if the surface is impermeable. Also, at free surfaces the normal traction is equal to the ambient pressure when surface tension effects are negligible. Additionally the shear stress is zero at this surface. It should also be pointed out that the formulations posed in this section are relevant only for *laminar flows*. *Turbulent flows* require additional considerations.

7.3 Specialized Fluids

Although the study of viscous fluids in the context of the equations presented in Section 7.2 occupies a major role in fluid mechanics, there is also a number of specialized situations resulting from simplifying assumptions that provide us with problems of practical interest. Here, we list some of the assumptions that are commonly made and consider briefly their meaning with respect to specific fluids.

(a) *Barotropic fluids* – If the equation of state happens to be independent of temperature as expressed by Eq 7.24, the changes of state are termed *barotropic*, and fluids which obey these conditions are called barotropic fluids. In particular, we may cite both *isothermal* changes (in which the temperature is constant) and *adiabatic* changes (for which no heat enters or leaves the fluid) as barotropic changes.

(b) *Incompressible fluids* – If the density of a fluid particle is constant, the equation of state becomes

$$\rho = \text{constant} \tag{7.27}$$

which describes *incompressibility*. This implies $\dot{\rho} = 0$ and, by the continuity equation, $v_{i,i} = 0$ for incompressible flows. Physically, incompressibility means that the elements of a fluid undergo no change in density (or volume) when subjected to a change in pressure. For incompressible flows, the Navier-Stokes equations become

$$\rho \dot{v}_i = \rho b_i - p_{,i} + \mu^* v_{i,jj} \tag{7.28}$$

due to the $v_{i,i} = 0$ condition. Water and oil, among others, are generally assumed to be incompressible, whereas most gases are highly compressible.

(c) *Inviscid (frictionless) fluids* – A fluid that cannot sustain shear stresses even when in motion is called an *inviscid*, or sometimes a *perfect* fluid. Clearly, if the coefficients λ^* and μ^* in Eq 7.10 are equal to zero, that equation describes a perfect fluid and the Navier-Stokes equations reduce to

$$\rho \dot{v}_i = \rho b_i - p_{,i} \tag{7.29}$$

which are often referred to as the *Euler equation of motion*. An ideal gas is a perfect fluid that obeys the gas law

$$p = \rho R \theta \tag{7.30}$$

where R is the gas constant for the particular gas under consideration. It should be pointed out that all real fluids are compressible and viscous to some degree.

7.4 Steady Flow, Irrotational Flow, Potential Flow

If the velocity components of a fluid are independent of time, the motion is called a *steady flow*. In such cases, the material derivative of the velocity,

$$\dot{v}_i = \frac{\partial v_i}{\partial t} + v_j v_{i,j}$$

reduces to the simpler form

$$\dot{v} = v_j v_{i,j} \ .$$

Thus, for a steady flow, the Euler equation is modified to read

$$\rho v_j v_{i,j} = \rho b_i - p_{,i} \ . \tag{7.31}$$

Furthermore, if the velocity field is constant and equal to zero everywhere, the fluid is at rest, and the theory for this condition is called *hydrostatics*. For this, the Navier-Stokes equations are simply

$$\rho b_i - p_{,i} = 0 \ . \tag{7.32}$$

Assuming a barotropic condition between ρ and p, it is possible to define a *pressure function* in the form

$$P(p) = \int_{p_0}^{p} \frac{dp}{\rho} \ . \tag{7.33}$$

In addition, if the body forces are conservative, we may express them in terms of a scalar potential function Ω by the relationship

$$b_i = -\Omega_{,i} \ . \tag{7.34}$$

From the definition Eq 7.34, it follows that

$$P_{,i} = \frac{1}{\rho} p_{,i} \quad \text{or} \quad \nabla P = \frac{\nabla p}{\rho} \ , \tag{7.35}$$

so that now Eq 7.32 may be written

$$(\Omega + P)_{,i} = 0 \tag{7.36}$$

as the governing equation for steady flow of a barotropic fluid with conservative body forces.

Example 7.1

An incompressible, Newtonian fluid maintains a steady flow under the action of gravity down an inclined plane of slope β. If the thickness of the fluid perpendicular to the plane is h and the pressure on the free surface is $p = p_0$ (a constant), determine the pressure and velocity fields for this flow.

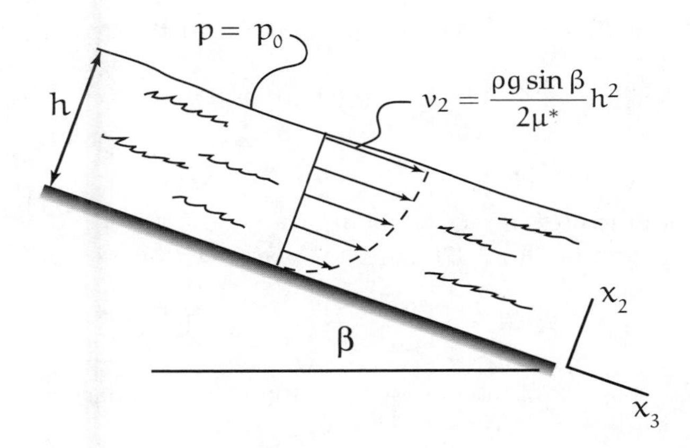

Solution

Assume $v_1 = v_3 = 0$, $v_2 = v_2(x_2, x_3)$. By the continuity equation for incompressible flow, $v_{i,i} = 0$. Hence, $v_{2,2} = 0$ and $v_2 = v_2(x_3)$. Thus, the rate of deformation tensor has components $d_{23} = d_{32} = \frac{1}{2}(\partial v_2/\partial x_3)$ and all others equal to zero. The Newtonian constitutive equation is given in this case by

$$t_{ij} = -p\delta_{ij} + 2\mu^* d_{ij}$$

from which we calculate

$$[t_{ij}] = \begin{bmatrix} -p & 0 & 0 \\ 0 & -p & \mu^* \dfrac{\partial v_2}{\partial x_3} \\ 0 & \mu^* \dfrac{\partial v_2}{\partial x_3} & -p \end{bmatrix}.$$

The equations of motion having the steady flow form

$$t_{ij,j} + \rho b_i = \rho v_j v_{i,j}$$

result in component equations

$$(\text{for } i = 1) \quad -p_{,1} = 0\,, \tag{7.37a}$$

$$(\text{for } i = 2) \quad -p_{,2} + \mu^* \frac{\partial^2 v_2}{\partial x_3^2} + \rho g \sin\beta = 0\,, \tag{7.37b}$$

$$(\text{for } i = 3) \quad -p_{,3} - \rho g \cos\beta = 0\,, \tag{7.37c}$$

since gravity is the only body force,

$$\mathbf{b} = g\,(\sin\beta\,\hat{\mathbf{e}}_2 - \cos\beta\,\hat{\mathbf{e}}_3)$$

From Eq 7.37a it is easy to see that the pressure is only a function of x_2. Namely, $p = -f(x_2)$. Integrating the Eq 7.37c gives

$$p = -\,(\rho g \cos\beta)\,x_3 + f\,(x_2)$$

where $f(x_2)$ is an arbitrary function of integration. At the free surface $(x_3 = h)$, $p = p_0$, and so

$$f(x_2) = p_0 + \rho g h \cos\beta$$

and thus

$$p = p_0 + (\rho g \cos\beta)(h - x_3)$$

which describes the pressure in the fluid.

Next, by integrating Eq 7.37b twice with respect to x_3, we obtain

$$v_2 = \frac{-\rho g \sin\beta}{2\mu^*} x_3^2 + a x_3 + b$$

with a and b constants of integration. But from the boundary conditions,

1. $v_2 = 0$ when $x_3 = 0$, therefore $b = 0$,
2. $t_{23} = 0$ when $x_3 = h$, therefore $a = \dfrac{\rho g h}{\mu^*} \sin\beta$.

Finally, therefore, from the equation for v_2 we have by the substitution of $a = \dfrac{\rho g h}{\mu^*} \sin\beta$,

$$v_2 = \frac{\rho g \sin\beta}{2\mu^*}(2h - x_3)\, x_3$$

having the profile shown in the figure.

If the velocity field of a fluid is one for which the tensor $\mathbf{W}$ vanishes identically, we say the flow is *irrotational*. In this case the vorticity vector $\mathbf{w}$, which is related to $\mathbf{W}$ by Eq 4.154, is also zero everywhere, so that for irrotational flow

$$w_i = \frac{1}{2}\varepsilon_{ijk} v_{k,j} = 0 \quad \text{or} \quad \mathbf{w} = \frac{1}{2}\nabla \times \mathbf{v} = \frac{1}{2}\operatorname{curl}\mathbf{v} = 0\,. \tag{7.38}$$

Finally, from the identity $\operatorname{curl}(\operatorname{grad}\phi) = 0$, we conclude that, for a flow satisfying Eq 7.38, the velocity field may be given in terms of a velocity potential, which we write as

$$v_i = \phi_{,i} \quad \text{or} \quad \mathbf{v} = \nabla\phi\,. \tag{7.39}$$

Indeed, it may be shown that the condition $\operatorname{curl}\mathbf{v} = 0$ is a necessary and sufficient condition for irrotational flow and the consequence expressed in Eq 7.39 accounts for the name *potential flow* often associated with this situation.

For a compressible irrotational flow, the Euler equation and the continuity equation may be linearized and combined to yield the wave equation

$$\ddot{\phi} = c^2 \phi_{,ii} \tag{7.40}$$

where c is the velocity of sound in the fluid. For a steady irrotational flow of a compressible barotropic fluid, the Euler equation and the continuity equation may be combined to give

$$\left(c^2\delta_{ij} - v_i v_j\right) v_{j,i} = 0 \tag{7.41}$$

which is called the *gas dynamics equation*. For incompressible potential flow the continuity equation reduces to a Laplace equation,

$$\phi_{,ii} = 0 \quad \text{or} \quad \nabla^2\phi = 0 \tag{7.42}$$

solutions of which may then be used to generate the velocity field using Eq 7.39. It is worthwhile to mention here that the Laplace equation is linear, so that superposition of solutions is available.

Example 7.2

Peristaltic pumping is a form of fluid transport resulting from a wave passing along a distensible fluid containing tube. This form of transport is common in biological systems. The figure below shows a two-dimensional representation of the tube. The shape of the wall is described by $x_2 = h(x_1, t) = a + b\sin\frac{2\pi}{\lambda}(x_1 - ct)$. In the x_1-x_2 frame shown, the fluid flow is unsteady. However, in a frame traveling with the wave, $\hat{x}_1$-$\hat{x}_2$, the problem becomes steady. The two frames are related by $x_1 = \hat{x}_1 + ct$, $x_2 = \hat{x}_2$ and $v_1 = \hat{v}_1 + c$. The $\hat{x}_1$-momentum equation for flows with negligible fluid inertia is

$$-\frac{dp}{d\hat{x}_1} + \mu\frac{\partial^2\hat{v}_1}{\partial\hat{x}_2^2} = 0$$

with the boundary conditions

$$\hat{v}_1 = -c \qquad \text{on } \hat{x}_2 = h \qquad \text{and} \qquad \frac{\partial\hat{v}_1}{\partial\hat{x}_2} = 0 \qquad \text{on } \hat{x}_2 = 0\ .$$

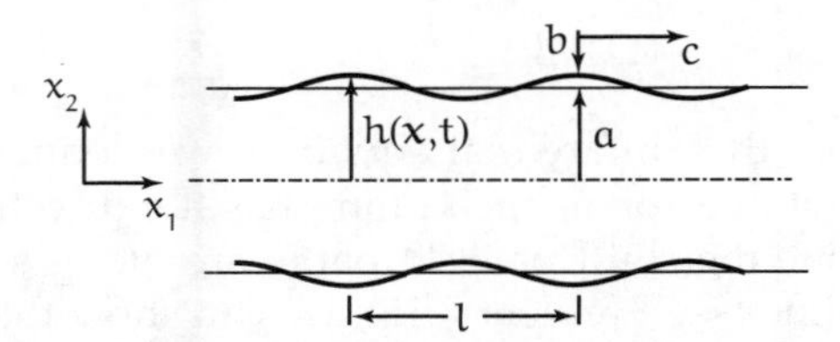

Two-dimensional figure modeling peristaltic pumping.

Determine the velocity profile $v_1(\mathbf{x}, t)$ in the x_1-x_2 frame. [1]

Solution

Integrating the governing equation twice gives

$$\hat{v}_1 = \frac{1}{2\mu}\frac{dp}{d\hat{x}_1}\hat{x}_2^2 + C\hat{x}_2 + D\ .$$

From the symmetry condition,

$$\frac{\partial\hat{v}_1}{\partial\hat{x}_2} = \frac{1}{\mu}\frac{dp}{d\hat{x}_1}(0) + C(\hat{x}_x) = 0 \qquad \text{and} \qquad C(\hat{x}_2) = 0\ .$$

On $\hat{x}_2 = h$, we have

$$\hat{v}_1(h) = -c = \frac{1}{2\mu}\frac{dp}{d\hat{x}_1}h^2 + D(\hat{x}_2) \qquad \text{and} \qquad D(\hat{x}_2) = -c - \frac{1}{2\mu}\frac{dp}{d\hat{x}_1}h^2\ .$$

[1] see A. H. Shapiro, M. Y. Jaffrin, and S. L. Weinberg (1969) Peristaltic pumping with long wavelengths at low Reynolds number, *Journal of Fluid Mechanics*, **37** (4), 799-825.

The velocity profile is

$$\hat{v}_1(\hat{\mathbf{x}}, t) = \frac{1}{2\mu}\frac{dp}{d\hat{x}_1}\left(\hat{x}_2^2 - h^2\right) - c\ ,$$

which is transformed to the stationary frame giving

$$v_1(\mathbf{x}, t) = \frac{1}{2\mu}\frac{dp}{dx_1}\left(x_2^2 - h^2\right)$$

since $v_1 = \hat{v}_1 + c$ and $dx_1 = d\hat{x}_1$. The velocity profile is a simple parabolic distribution.

7.5 The Bernoulli Equation, Kelvin's Theorem

If a fluid is barotropic with conservative body forces, Eq 7.36 may be substituted on the right-hand side of Euler's equation (Eq 7.29), giving

$$\dot{v}_i = -(\Omega + P)_{,i}\ . \tag{7.43}$$

As a step in obtaining a solution to this differential equation, we define a *streamline* as that space curve of which the tangent vector, at each point, has the direction of the fluid velocity (vector). For a steady flow, the fluid particle paths are along streamlines. By integrating Eq 7.43 along a streamline (see Problem 7.15), we can show that

$$\int_{x_1}^{x_2} \frac{\partial v_i}{\partial t} dx_i + \frac{v^2}{2} + \Omega + P = G(t) \tag{7.44}$$

where dx_i is a differential tangent vector along the streamline. This is the well-known *Bernoulli equation*. If the motion is steady, the time function $G(t)$ resulting from the integration reduces to a constant G, which may vary from one streamline to another. Furthermore, if the flow is also irrotational, a unique constant G_0 is valid throughout the flow.

When gravity is the only force acting on the body, we write $\Omega = gh$ where $g = 9.81$ m/s^2 is the gravitational constant and h is a measure of the height above a reference level in the fluid. If $h_p = P/g$ is defined as the *pressure head* and $h_v = v^2/2g$ as the *velocity head*, Bernoulli's equation for incompressible fluids becomes

$$h + h_p + h_v = h + \frac{p}{\rho g} + \frac{v^2}{2g} = G_0 \tag{7.45}$$

Recall that by Eq 2.97 in Chapter 2 we introduced Stoke's theorem, which relates the line integral around a closed curve to the surface integral over its cap. By this theorem we define the velocity circulation Γ_C around a closed path in the fluid as

$$\Gamma_C = \oint v_i dx_i = \int_S \varepsilon_{ijk} v_{k,j} n_i dS \tag{7.46}$$

where n_i is the unit normal to the surface $\mathcal{S}$ bounded by $\mathcal{C}$ and dx_i is the differential tangent element to the curve $\mathcal{C}$. Note that, when the flow is irrotational, $\operatorname{curl}\mathbf{v} = 0$ and the circulation vanishes. If we take the material derivative of the circulation by applying Eq 5.7 to Eq 7.46 we obtain

$$\dot{\Gamma}_{\mathcal{C}} = \oint (\dot{v}_i dx_i + v_i dv_i) \ . \tag{7.47}$$

For a barotropic, inviscid fluid with conservative body forces, this integral may be shown to vanish, leading to what is known as *Kelvin's theorem* of constant circulation.

Problems

Problem 7.1

Introduce the stress deviator S_{ij} and the viscous stress deviator $\tau^*_{ij} = \tau_{ij} - \frac{1}{3}\delta_{ij}\tau_{kk}$ into Eq 7.5 to prove that $S_{ij} = \tau^*_{ij}$.

Problem 7.2

Determine an expression for the stress power (a) $t_{ij}d_{ij}$ and (b) $\tau_{ij}d_{ij}$ for a Newtonian fluid. First, show that

$$\tau_{ij} = \left(\kappa^* - \frac{2}{3}\mu^*\right)\delta_{ij}d_{kk} + 2\mu^* d_{ij} \ .$$

Answer

(a) $t_{ij}d_{ij} = -pd_{ii} + \kappa^* d_{ii}d_{jj} + 2\mu^*\beta_{ij}\beta_{ij}$

(b) $\tau_{ij}d_{ij} = \kappa^*(\mathrm{tr}\,\mathbf{D})^2 + 2\mu^*\beta_{ij}\beta_{ij}$

Problem 7.3

Determine the constitutive equation for a Newtonian fluid for which Stokes condition holds, that is, for $\kappa^* = 0$.

Answer

$t_{ij} = -p\delta_{ij} + 2\mu^*\beta_{ij}$

Problem 7.4

Develop an expression of the energy equation for a Newtonian fluid assuming the heat conduction follows Fourier's law.

Answer

$\rho\dot{u} = -pv_{i,i} + \lambda^* v_{i,i}v_{j,j} + \frac{1}{2}\mu^*(v_{i,j} + v_{j,i}) + \kappa^*\theta_{,ii} + \rho r$

Problem 7.5

The dissipation potential Ψ for a Newtonian fluid is defined as a function of $\mathbf{D}$ and $\boldsymbol{\beta}$ by

$$\Psi = \frac{1}{2}\kappa^* d_{jj}d_{ii} + \mu^*\beta_{ij}\beta_{ij}, \quad \text{where } \kappa^* = \lambda^* + \frac{2}{3}\mu^* \ .$$

Show that $\partial\Psi/\partial d_{ij} = \tau_{ij}$.

Problem 7.6

Verify the derivation of the Navier-Stokes equations for a Newtonian fluid as given by Eq 7.25.

Problem 7.7
Consider a two-dimensional flow parallel to the x_2x_3 plane so that $v_1 = 0$ throughout the fluid. Assuming that an incompressible, Newtonian fluid undergoes this flow, develop a Navier-Stokes equation and a continuity equation for the fluid.

Answer

(Navier-Stokes) $\rho \dot{v}_i = \rho b_i - p_{,i} + \mu^* v_{i,jj} \quad (i,j = 2,3)$

(Continuity) $v_{i,i} = 0 \quad (i = 2,3)$

Problem 7.8
Consider a barotropic, inviscid fluid under the action of conservative body forces. Show that the material derivative of the vorticity of the fluid in the current volume $\mathcal{V}$ is

$$\frac{d}{dt}\int_{\mathcal{V}} w_i \, d\mathcal{V} = \int_{\mathcal{S}} v_i w_j n_j \, d\mathcal{S} \ .$$

Problem 7.9
Show that for an incompressible, inviscid fluid the stress power vanishes identically as one would expect.

Problem 7.10
Show that the vorticity and velocity of a barotropic fluid of constant density moving under conservative body forces are related through the equation $\dot{w}_i = w_j v_{i,j}$. Deduce that for a steady flow of this fluid $v_j w_{i,j} = w_j v_{i,j}$.

Problem 7.11
In terms of the vorticity vector $\boldsymbol{w}$, the Navier-Stokes equations for an incompressible fluid may be written as

$$\rho \dot{v}_i = \rho b_i - p_{,i} - 2\mu^* \varepsilon_{ijk} w_{k,j} \ .$$

Show that, for an irrotational motion, this equation reduces to the Euler equation

$$\rho \dot{v}_i = \rho b_i - p_{,i} \ .$$

Problem 7.12
Carry out the derivation of Eq 7.41 by combining the Euler equation with the continuity equation, as suggested in the text.

Problem 7.13
Consider the velocity potential $\phi = x_2 x_3 / r^2$ where $r^2 = x_1^2 + x_2^2$. Show that this satisfies the Laplace equation $\phi_{,ii} = 0$. Derive the velocity field and show that this flow is both incompressible and irrotational.

Problem 7.14
If the equation of state of a barotropic fluid has the form $p = \lambda \rho^k$ where k and λ are constants, the flow is termed isentropic. Show that the Bernoulli equation for a steady

motion in this case becomes

$$\Omega + \frac{kp}{(k+1)\,\rho} + \frac{1}{2}v^2 = \text{constant} .$$

Also, show that for isothermal flow the Bernoulli equation takes the form

$$\Omega + \frac{p \ln \rho}{\rho} + \frac{1}{2}v^2 = \text{constant} .$$

Problem 7.15

Derive Eq 7.44 by taking the scalar product of dx_i (the differential displacement along a streamline) with Eq 7.43 and integrating along the streamline, that is, by the integration of

$$\int_{x_1}^{x_2} (\dot{v}_i + \Omega_{,i} + P_{,i})\, dx_i .$$

Problem 7.16

Verify that Eq 7.47 is the material derivative of Eq 7.46. Also, show that for a barotropic, inviscid fluid subjected to conservative body forces the rate of change of the circulation is zero (See Eq 7.47).

Problem 7.17

Determine the circulation Γ_c around the square in the x_2x_3 plane shown in the figure if the velocity field is given by

$$\mathbf{v} = \left(x_3 - x_2^2\right)\hat{\mathbf{e}}_2 + (x_3 + x_2)\,\hat{\mathbf{e}}_3 .$$

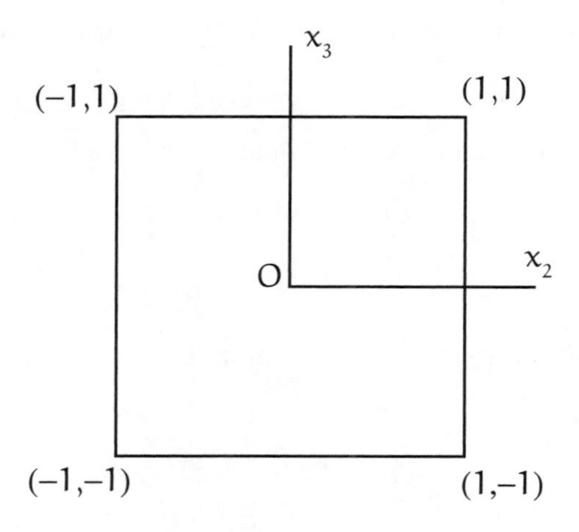

Answer

$\Gamma_c = 0$

8

Nonlinear Elasticity

Many of today's challenging design problems involve materials such as *butadiene rubber* (BR), *natural rubber* (NR), or *elastomers*. Rubber materials might be most easily characterized by the stretching and relaxing of a rubber band. The *resilience* of rubber, the ability to recover initial dimensions after large strain, was not possible with natural latex until Charles Goodyear discovered vulcanization in 1939. Vulcanization is a chemical reaction known as cross-linking which turned liquid latex into a non-meltable solid (*thermoset*) . Cross-linked rubber would also allow considerable stretching with low damping; strong and stiff at full extension, it would then retract rapidly (rebound). One of the first applications was rubber-impregnated cloth, which was used to make the sailor's "mackintosh." Tires continue to be the largest single product of rubber although there are many, many other applications. These applications exhibit some or all of rubber's four characteristics, viz. damping in motor mounts, rebound/resilience in golf ball cores, or simple stretching in a glove or bladder. While thermoset rubber remains dominant in rubber production, processing difficulties have led to the development and application of *thermoplastic elastomers* (TPEs). These materials are easier to process and are directly recyclable. While TPEs are not as rubberlike as the thermosets, they have found wide application in automotive fascia and as energy-absorbing materials.

There are several reasons why designing with plastic and rubber materials is more difficult than with metals. For starters, the stress-strain response, that is, the constitutive response, is quite different. Figure 8.1(a) shows the stress-strain curves for a mild steel specimen along with the response of a natural rubber used in an engine mount. Note that the rubber specimen strain achieves a much higher stretch value than the steel. The dashed vertical line in Fig. 8.1(b) represents the strain value of the mild steel at failure. This value is much less than the 200% strain the rubber underwent without failing. In fact, many rubber and elastomer materials can obtain 300 to 500% strain. Highly cross-linked and filled rubbers can result in materials not intended for such large strains. Golf ball cores are much stiffer than a rubber band, for instance. Each of these products is designed for different strain regimes. A golf ball's maximum strain would be on the order of 30 to 40%. Its highly crosslinked constitution is made for resilience, not for large strain. The rubber stress-strain curve exhibits nonlinear behavior from the very beginning of its deformation, whereas steel has a linear regime below the yield stress.

The reason rubber materials exhibit drastically different behavior than metals results from their sub-microscopic characteristics. Metals are crystalline lattices of atoms all being, more or less, well ordered: in contrast, rubber material molecules are made up of carbon atoms bonded into a long chain resembling a tangled collection of yarn scraps. Since the carbon-carbon (C—C) bond can rotate, it is possible for these entangled long chain polymers to rearrange themselves into an infinite number of different conformations. While the random coil can be treated as a spring, true resilience requires a cross-link to stop viscous flow. In a thermoset rubber, a chemical bond, often with sulfur, affords the tie while physical entanglements effect the same function in a TPE material. The degree of cross-linking is used to control the rubber's stiffness.

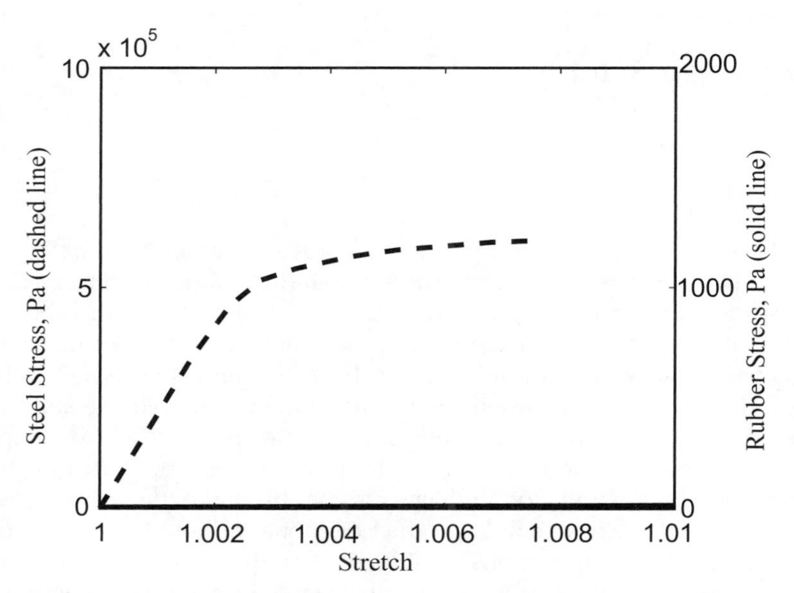

(a) Steel and rubber on strain scale appropriate for steel. The rubber stress is low and nearly coinciding with the abscissa.

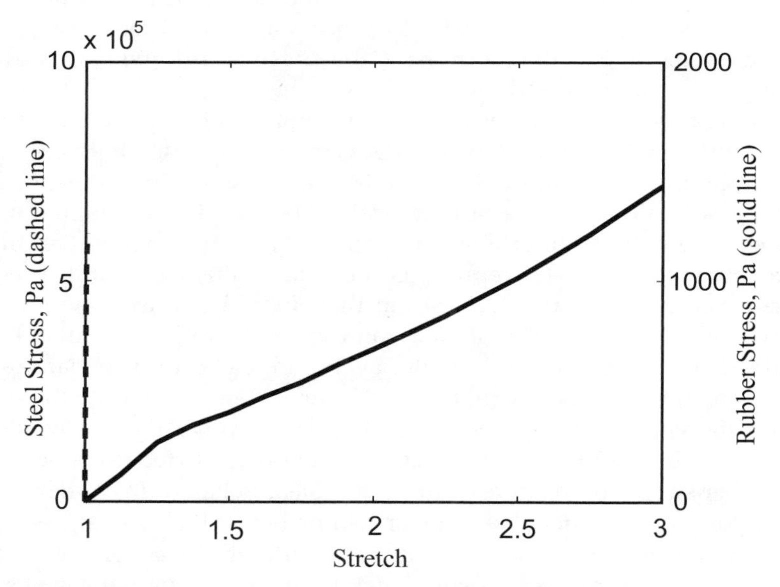

(b) Steel and rubber on strain scale appropriate for rubber. The strain values for the steel are small making the curve to be nearly coincides with the ordinate.

FIGURE 8.1
Nominal stress-stretch curves for rubber and steel. Note the same data is plotted in each figure, however, the stress axes have different scale and a different strain range is represented.

In the context of this book's coverage of continuum mechanics, material make-up on the micro-scale is inconsistent with the continuum assumption discussed in Chapter 1. However, a rubber elasticity model can be derived from the molecular level which somewhat represents material behavior at the macroscopic level. In this chapter, rubber elasticity will be developed from a first-principle basis. Following that, the traditional continuum approach is developed by assuming a form of the strain energy density and using restrictions on the constitutive response imposed by the second law of thermodynamics to obtain stress-stretch response.

8.1 Molecular Approach to Rubber Elasticity

One of the major differences between a crystalline metal and an amorphous polymer is that the polymer chains have the freedom to rearrange themselves. The term conformation is used to describe the different spatial orientations of the chain. Physically, the ease at which different conformations are achieved results from the bonding between carbon atoms. As the carbon atoms join to form the polymer chain , the bonding angle is $109.5°$, but there is also a rotational degree of freedom around the bond axis. For most macromolecules, the number of carbon atoms can range from $1,000$ to $100,000$. With each bond having a certain degree of freedom to orient itself, the number of conformations becomes quite large. Because of this substantial amount, the use of statistical thermodynamics may be used to arrive at rubber elasticity equations from first principles. In addition to the large number of conformations for a single chain, there is another reason the statistical approach is appropriate: the actual polymer has a large number of different individual chains making up the bulk of the material. For instance, a cubic meter of an amorphous polymer having 10,000 carbon atoms per molecule would have on the order of 10^{24} molecules (McCrum et al., 1997). Clearly, the sample is large enough to justify a statistical approach.

At this point, consider one particular molecule, or polymer chain, and its conformations. The number of different conformations the chain can obtain depends on the distance separating the chain's ends. If a molecule is formed of n segments each having length l, the total length would be $L = nl$. Separating the molecule's ends, the length L would mean there is only one possible conformation keeping the chain intact. As the molecule's ends get closer together, there are more possible conformations that can be obtained. Thus, a Gaussian distribution of conformations as a function of the distance between chain ends is appropriate. Figure 8.2 demonstrates how more conformations are possible as the distance between the molecule's ends is reduced.

Molecule end-to-end distance, r, is found by adding up all the segment lengths, l, as is shown in Fig. 8.3. Adding the segment lengths algebraically gives distance r from end-to-end, but it does not give an indication of the length of the chain. If the ends are relatively close and the molecule is long there will be the possibility of many conformations. Forming the magnitude squared of the end-to-end vector, $\mathbf{r}$, in terms of the vector addition of individual segments

$$r^2 = \mathbf{r} \cdot \mathbf{r} = (\mathbf{l}_1 + \mathbf{l}_2 + \cdots + \mathbf{l}_n) \cdot (\mathbf{l}_1 + \mathbf{l}_2 + \cdots + \mathbf{l}_n) \tag{8.1}$$

where $\mathbf{l}_i$ is the vector defining the ith segment of the molecule chain. Multiplying out the right-hand side of Eq 8.1 leads to

$$r^2 = nl^2 + [\mathbf{l}_1 \cdot \mathbf{l}_2 + \mathbf{l}_1 \cdot \mathbf{l}_3 + \cdots + \mathbf{l}_{n-1} \cdot \mathbf{l}_n] \ . \tag{8.2}$$

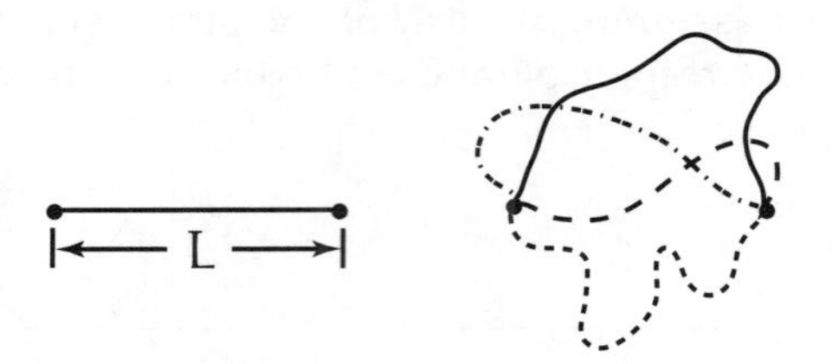

FIGURE 8.2
A schematic comparison of molecular conformations as the distance between molecule's ends varies. Dashed lines indicate other possible conformations.

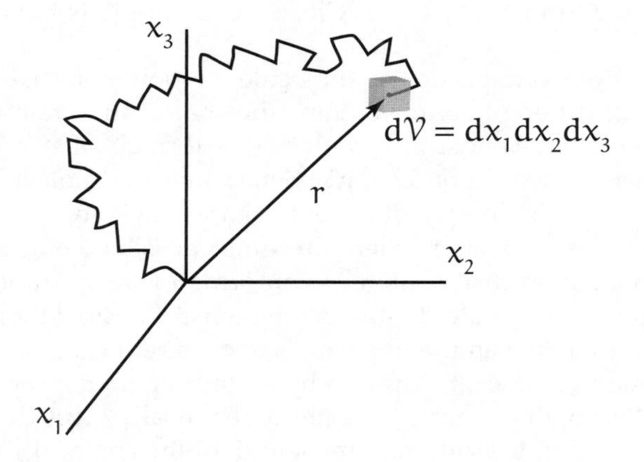

FIGURE 8.3
A freely connected chain with end-to-end vector r.

This would be the square of the end-to-end vectors for one molecule of the polymer. In a representative volume of the material there would be many chains from which we may form the mean square end-to-end distance

$$\langle r^2 \rangle = \frac{1}{N} \sum_{1}^{N} \left(nl^2 + [\mathbf{l}_1 \cdot \mathbf{l}_2 + \mathbf{l}_1 \cdot \mathbf{l}_3 + \cdots + \mathbf{l}_{n-1} \cdot \mathbf{l}_n] \right) . \tag{8.3}$$

The bracketed term in Eq 8.3 is argued to be zero from the following logic. Since a large number of molecules is taken in the sample, it is reasonable that, for every individual product $\mathbf{l}_1 \cdot \mathbf{l}_2$, there will be another segment pair product which will equal its negative. The canceling segment pair does not necessarily have to come from the same molecule. Thus, the bracketed term in Eq 8.3 is deemed to sum to zero, leaving a simple expression for the mean end-to-end distance

$$\langle r^2 \rangle = nl^2 \tag{8.4}$$

The mean end-to-end distance indicates how many segments, or carbon atoms, are in a specific chain. To address the issue of how the end-to-end distances are distributed throughout the polymer, a Gaussian distribution is assumed. Pick the coordinate's origin to be at one end of a representative chain. Figure 8.3 shows this for a single chain, with the other end of the chain in an infinitesimal volume $d\mathcal{V}$ located by the vector $\mathbf{r}$. The probability of the chain's end lying in the volume $d\mathcal{V}$ is given by

$$P(r)\, dr = \frac{e^{-\left(\frac{r}{\rho}\right)^2}}{\left(\sqrt{\pi}\rho\right)^3} dr \tag{8.5}$$

where ρ is a parameter of the distribution. Using this assumed distribution of mean end-to-end distances, it is straightforward to find

$$\langle r^2 \rangle_0 = \int_0^\infty r^2 P(r)\, dr = \frac{3}{2}\rho^2 \tag{8.6}$$

where the subscript 0 denotes that this is an intrinsic property of the chain since it was considered alone. When the chain is placed back into a crosslinked network of chains, the mean end-to-end distance is written as $\langle r^2 \rangle_i$. This latter designation takes into account the fact the chain has restrictions placed upon it by being packed into a volume with other chains. Equating Eqs 8.4 and 8.6 the distribution parameter ρ is found to be

$$\rho = \sqrt{\frac{2n}{3}}\, l \, . \tag{8.7}$$

Similar to the results of Section 5.9, the force created by stretching a uniaxial specimen is given in terms of the Helmholtz free energy

$$F = \left. \frac{\partial \psi}{\partial L} \right|_{\theta, \mathcal{V}} \tag{8.8}$$

where F is the force, L is the length, and subscripts θ and $\mathcal{V}$ designate that the change in length occurs at constant temperature and volume. Substitution of Eq 5.74b yields the force in terms of the internal energy and entropy

$$F = \frac{\partial u}{\partial L} - \theta \frac{\partial \eta}{\partial L} \tag{8.9}$$

where the constant temperature and volume subscripts have not been written for convenience.

Examination of Eq 8.9 offers an informative comparison between metals and ideal rubbers. In metals, the crystalline structure remains intact as the material is deformed. Atoms are moved closer, or further, from adjacent atoms creating a restoring force, but the relative order among the atoms remains the same. The last term of Eq 8.9 has no force contribution since the relative order of the atoms stays unchanged. For an ideal rubber, a change in length has no effect on the internal energy. Thus, the first derivative term of Eq 8.9 is zero. However, stretching of the specimen increases the mean end-to-end distance, thus reducing the possible conformations for the chains. This reduction in conformations gives rise to a negative change in entropy as the length is increased.

The entropy for a single chain will be related to the conformation through the mean end-to-end length. Noting the number of configurations is proportional to the probability per unit volume, $P(r)$, and using Boltzmann's equation, the entropy may be written as

$$\eta = \eta_0 + k \ln P(r) = \eta_0 + k \left[3 \ln \left(\sqrt{\pi} \rho \right) + \left(\frac{r}{\rho} \right)^2 \right] \tag{8.10}$$

where k is Boltzmann's constant. Use of this in Eq 8.9 for an ideal rubber gives a single chain retractive force given by

$$F = \frac{2k\theta}{\rho^2} r^2 \, . \tag{8.11}$$

Consider a polymer having forces applied resulting in stretch ratios λ_1, λ_2, and λ_3. The work done on each chain of the material is the sum of the work done in each coordinate direction x_i

$$W^{(i)} = \int_{x_i}^{\lambda^{(i)} x_i} f_i \, dx_i = \frac{2k\theta}{\rho^2} \int_{x_i}^{\lambda^{(i)} x_i} x_i \, dx_i = \frac{k\theta}{\rho^2} \left[\left(\lambda^{(i)} \right)^2 - 1 \right] x_i^2 \quad \text{(no sum on } i) \, .$$

Taking into consideration the work done on all of the chains gives total work in each coordinate direction

$$\sum_n W^{(i)} = \frac{k\theta}{\rho^2} \left[\left(\lambda^{(i)} \right)^2 - 1 \right] \sum_n x_i^2 \quad \text{(no sum on } i) \tag{8.12}$$

where the last summed term is the number of chains, n, times the initial mean end-to-end distance in the x_i direction. Assuming the rubber is initially isotropic yields

$$\sum_n x_i^2 = n \left\langle x_i^2 \right\rangle_i = \frac{n}{3} \left\langle r^2 \right\rangle_i \quad \text{(no sum on } i) \, .$$

Substituting this and ρ from Eq 8.6 into Eq 8.12 and adding all three coordinate work terms gives

$$W = \frac{nk\theta}{2} \frac{\left\langle r^2 \right\rangle_i}{\left\langle r^2 \right\rangle_0} \left[\lambda_1^2 + \lambda_2^2 + \lambda_3^2 - 3 \right] \, . \tag{8.13}$$

For convenience, this equation may be written as

$$W = \frac{\mathcal{V} G}{2} \left[\lambda_1^2 + \lambda_2^2 + \lambda_3^2 - 3 \right] \tag{8.14}$$

where $\mathcal{V}$ is the volume and G is the shear modulus which are given by

$$N = \frac{n}{\mathcal{V}} \, ,$$

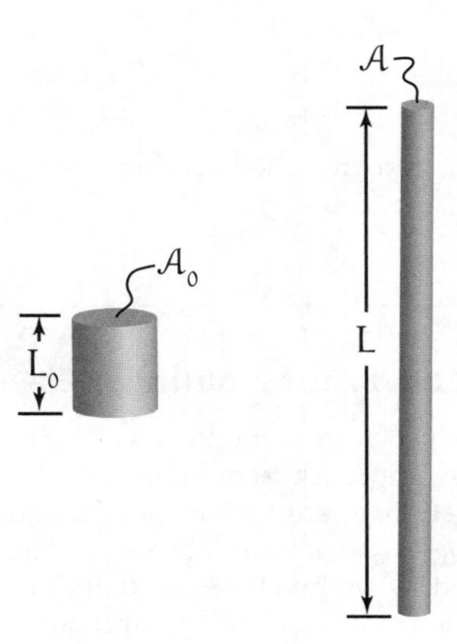

FIGURE 8.4
Rubber specimen having original length L_0 and cross-section area $\mathcal{A}_0$ stretched into deformed shape of length L and cross section area $\mathcal{A}$.

and

$$G = Nk\theta \frac{\langle r^2 \rangle_i}{\langle r^2 \rangle_0} .$$

Next, considering a uniaxial tension of a specimen (Fig. 8.4) the stretching ratios reduce to

$$\lambda_1 = \lambda, \quad \lambda_2 = \lambda_3 = \frac{1}{\sqrt{\lambda}} \tag{8.15}$$

where the nearly incompressible nature has been used in the form $\lambda_1 \lambda_2 \lambda_3 = 1$. The total work done is

$$W = \frac{\mathcal{V}G}{2} \left[\lambda^2 + \frac{2}{\lambda} - 3 \right] . \tag{8.16}$$

This is the work done on the polymer chains, but it is the same work done by external forces since an ideal rubber is assumed. Thus, the work shown in Eq 8.16 is equal to the change in Helmholtz free energy. Recalling the force from deformation as defined by Eq 8.8, the force resulting from a stretch λ is given by

$$F = \frac{\partial \psi}{\partial L} = \frac{dW}{dL} = \frac{dW}{d\lambda} \frac{d\lambda}{dL} .$$

The deformation is volume preserving, having $\mathcal{V} = \mathcal{A}_0 L_0 = \mathcal{A} L$ where the subscript 0 denotes the initial area and length. Since the stretch ratio in this case is $\lambda = L/L_0$, it is clear that $d\lambda/dL = 1/L_0$. Using this result and differentiating Eq 8.16 results in

$$F = \frac{\mathcal{V}G}{L_0} \left[\lambda - \frac{1}{\lambda^2} \right] = \mathcal{A}_0 G \left[\lambda - \frac{1}{\lambda^2} \right]$$

which may be written as

$$f = \frac{F}{\mathcal{A}_0} = G\left[\lambda - \frac{1}{\lambda^2}\right] . \tag{8.17}$$

Materials satisfying this equation are called *neo-Hookean*.

8.2 A Strain Energy Theory for Nonlinear Elasticity

The theory developed in the previous section does not represent most experimental data well at large strains. A better approach to modeling the response of rubbers comes from assuming the existence of a strain energy which is a function of the deformation gradient in the form of the *left deformation tensor* $B_{ij} = F_{i,A}F_{j,A}$. This approach, first published by Mooney (1940) and furthered by Rivlin (1948), actually predates the molecular approach discussed in Section 8.1. The basis of Mooney's and subsequent theories is the initially isotropic material has to obey certain symmetries with regards to the functional form of the strain energy function.

Assume the strain energy per unit volume to be an isotropic function of the strain in the form of the right deformation tensor invariants I_1, I_2, and I_3

$$W = W(I_1, I_2, I_3) \tag{8.18}$$

where

$$\begin{aligned} I_1 &= C_{AA} , \\ I_2 &= \frac{1}{2}\left(C_{AA}C_{BB} - C_{AB}C_{AB}\right) , \\ I_3 &= \varepsilon_{MNO}C_{1M}C_{2N}C_{3O} = \det\left[C_{AB}\right] . \end{aligned} \tag{8.19}$$

Note that if principal axes of C_{AB} are chosen the invariants of Eq 8.19 are written in terms of the stretch ratios λ_1, λ_2, and λ_3 as follows:

$$\begin{aligned} I_1 &= \lambda_1^2 + \lambda_2^2 + \lambda_3^2 , \\ I_2 &= \lambda_1^2\lambda_2^2 + \lambda_2^2\lambda_3^3 + \lambda_1^2\lambda_3^2 , \\ I_3 &= \lambda_1^2\lambda_2^2\lambda_3^2 . \end{aligned} \tag{8.20}$$

Also, it is easy to show that $I_1 = C_{AA} = B_{ii} = I_\mathbf{B} = I_\mathbf{C}$, $I_2 = II_\mathbf{B} = II_\mathbf{C}$ and $I_3 = III_\mathbf{B} = III_\mathbf{C}$ by using the definitions of the right and left deformation tensors.

An expression for the Cauchy stress in terms of the strain energy density comes from the local energy equation in the form

$$\dot{u} = \frac{1}{\rho}t_{ij}d_{ij} . \tag{8.21}$$

Since $\rho_0 u \equiv W$ and W is a function of the right deformation tensor the derivative of u may be written as

$$\dot{u} = \frac{1}{\rho_0}\frac{\partial W}{C_{AB}}\dot{C}_{AB} . \tag{8.22}$$

Noting that

$$\dot{C}_{AB} = d_{ij}\left(F_{jA}F_{iB} + F_{iA}F_{jB}\right)$$

allows us to write

$$\dot{u} = \frac{2}{\rho J} F_{iA} \frac{\partial W}{C_{AB}} F_{jB} d_{ij} \,. \tag{8.23}$$

Equating Eqs 8.21 and 8.23 for $\dot{u}$ and factoring out d_{ij} leaves

$$d_{ij} \left(t_{ij} - 2J^{-1} F_{iA} \frac{\partial W}{\partial C_{AB}} F_{jB} \right) = 0 \tag{8.24}$$

which must hold for all motions. Thus, the stress in a compressible material may be written in terms of the strain energy density as

$$t_{ij} = 2J^{-1} F_{iA} \frac{\partial W}{\partial C_{AB}} F_{jB} \,. \tag{8.25}$$

Differentiation of the strain energy, along with the chain rule, gives stress components

$$t_{ij} = 2J^{-1} F_{iA} \left[\frac{\partial W}{\partial I_1} \frac{\partial I_1}{\partial C_{AB}} + \frac{\partial W}{\partial I_2} \frac{\partial I_2}{\partial C_{AB}} + \frac{\partial W}{\partial I_3} \frac{\partial I_3}{\partial C_{AB}} \right] F_{jB} \tag{8.26}$$

where the invariants I_1, I_2 and I_3 are as given in Eq 8.19. Partial derivatives of the invariants I_1, I_2 and I_3 may be completed by noting the components are independent quantities. Thus,

$$\frac{\partial C_{AB}}{\partial C_{MN}} = \delta_{AM} \delta_{BN} \,, \tag{8.27}$$

and the partial derivatives of the invariants may be written as

$$\frac{\partial I_1}{\partial C_{AB}} = \frac{\partial C_{MM}}{\partial C_{AB}} = \delta_{MA} \delta_{MB} = \delta_{AB} \tag{8.28a}$$

$$\begin{aligned} \frac{\partial I_2}{\partial C_{AB}} &= \frac{1}{2} \delta_{AB} C_{NN} + \frac{1}{2} C_{NN} \delta_{AB} - \delta_{MA} \delta_{NB} C_{MN} \\ &= I_1 \delta_{AB} - C_{AB} \end{aligned} \tag{8.28b}$$

$$\frac{\partial I_3}{\partial C_{AB}} = \frac{1}{2} \varepsilon_{AMN} \varepsilon_{BQR} C_{MQ} C_{NR} = I_3 C^{-1}_{AB} \,. \tag{8.28c}$$

These terms are further simplified when put into Eq 8.26 since they are multiplied by $F_{iA} F_{jB}$:

$$F_{iA} \delta_{AB} F_{jB} = F_{iA} F_{jA} = B_{ij}$$

$$F_{iA} C_{AB} F_{jB} = F_{iA} F_{kA} F_{kB} F_{jB} = B_{ik} B_{kj} \,.$$

Noting that $J^{-1} = I_3^{\frac{1}{2}}$, the stress components may be written as

$$t_{ij} = \alpha_0 \delta_{ij} + \alpha_1 B_{ij} + \alpha_2 B_{ik} B_{kj} \tag{8.29}$$

where

$$\alpha_0 = 2 I_3^{\frac{1}{2}} \frac{\partial W}{\partial I_3} \,, \tag{8.30a}$$

$$\alpha_1 = 2 I_3^{-\frac{1}{2}} \left(\frac{\partial W}{\partial I_1} + I_1 \frac{\partial W}{\partial I_2} \right) \,, \tag{8.30b}$$

$$\alpha_2 = -2 I_3^{-\frac{1}{2}} \frac{\partial W}{\partial I_2} \,. \tag{8.30c}$$

A second form for the Cauchy stress can be found by use of the Cayley-Hamilton theorem:

$$t_{ij} = \beta_0 \delta_{ij} + \beta_1 B_{ij} + \beta_{-1} B_{ij}^{-1} \tag{8.31}$$

where

$$\beta_0 = 2I_3^{-\frac{1}{2}} \left(I_2 \frac{\partial W}{\partial I_2} + I_3 \frac{\partial W}{\partial I_3} \right), \tag{8.32a}$$

$$\beta_1 = 2I_3^{-\frac{1}{2}} \frac{\partial W}{\partial I_1}, \tag{8.32b}$$

$$\beta_{-1} = -2I_3^{-\frac{1}{2}} \frac{\partial W}{\partial I_2}. \tag{8.32c}$$

When a specific material model is chosen defining the specific form of the strain energy density (Eq 8.18), Eq 8.31 can be interpreted. The material parameters are represented by β_0, β_1 and β_{-1}, and the deformation is represented by δ_{ij}, B_{ij} and B_{ij}^{-1}.

Many rubber or elastomeric materials have a mechanical response that is often nearly incompressible. Even though a perfectly incompressible material is not possible, a variety of problems can be solved by assuming incompressibility. The *incompressible response* of the material can be thought of as a constraint on the deformation gradient. That is, the incompressible nature of the material is modeled by an addition functional dependence between the deformation gradient components. For an incompressible material the density remains constant. This was expressed in Section 5.2 by $v_{i,i} = 0$ (Eq 5.15) or $\rho J = \rho_0$ (Eq 5.17a). With the density remaining constant and Eq 5.17a as the expression of the continuity equation it is clear that

$$J = \det[F_{iA}] = 1. \tag{8.33}$$

At this point, we digress from specific incompressibility conditions to a more general case of a continuum with an *internal constraint*. Assume a general constraint of the form

$$\phi(F_{iA}) = 0, \tag{8.34}$$

or, since $C_{AB} = F_{iA} F_{iB}$,

$$\tilde{\phi}(C_{AB}) = 0. \tag{8.35}$$

This second form of the internal constraint has the advantage that it is invariant under superposed rigid body motions. Differentiation of ϕ results in

$$\frac{\partial \phi}{\partial C_{AB}} \dot{C}_{AB} = \frac{\partial \phi}{\partial C_{AB}} \left(\dot{F}_{iA} F_{iB} + F_{iA} \dot{F}_{iB} \right) \tag{8.36}$$

where it is understood partial differentiation with respect to a symmetric tensor results in a symmetric tensor. That is,

$$\frac{\partial \phi}{\partial C_{AB}} \equiv \frac{1}{2} \left(\frac{\partial \phi}{\partial C_{AB}} + \frac{\partial \phi}{\partial C_{BA}} \right). \tag{8.37}$$

Note that $\dot{C}_{AB}$ is essentially the same as $\dot{E}_{AB}$ which is given in Eq 4.149. Thus, Eq 8.36 may be written as

$$\phi_{ij} d_{ij} = 0 \tag{8.38}$$

where ϕ_{ij} is defined by

$$\phi_{ij} = \frac{1}{2} F_{iA} \left(\frac{\partial \phi}{\partial C_{AB}} + \frac{\partial \phi}{\partial C_{BA}} \right) F_{jB}. \tag{8.39}$$

Assume the stress is formed by adding stress components $\hat{t}_{ij}$, derivable from a constitutive response, to components of an arbitrary stress $\bar{t}_{ij}$ resulting from the internal constraint:

$$t_{ij} = \hat{t}_{ij} + \bar{t}_{ij} \ . \tag{8.40}$$

The added arbitrary stress components $\bar{t}_{ij}$ are assumed to be workless, that is,

$$\bar{t}_{ij} d_{ij} = 0 \ . \tag{8.41}$$

Comparing Eqs 8.38 and 8.41 shows that both ϕ_{ij} and $\bar{t}_{ij}$ are orthogonal to d_{ij}, so $\bar{t}_{ij}$ may be written as

$$\bar{t}_{ij} = \lambda \phi_{ij} \ . \tag{8.42}$$

For an incompressible material, $d_{ii} = 0$ which may be written as

$$\delta_{ij} d_{ij} = 0 \ , \tag{8.43}$$

or equivalently by choosing $\phi_{ij} = \delta_{ij}$ in Eq 8.38. Substituting this into Eq 8.42 yields

$$\bar{t}_{ij} = -p\delta_{ij} \tag{8.44}$$

where the scalar has been changed to reflect the pressure term it represents.

The stress in an incompressible material may now be written as

$$t_{ij} = -p\delta_{ij} + \hat{t}_{ij} \tag{8.45}$$

where the second term in Eq 8.45 is determined from the constitutive response for Cauchy stress. For an incompressible material $J = 1$, hence $I_3 = 1$, and Eq 8.18 may be written as

$$W = W(I_1, I_2) \ . \tag{8.46}$$

With the use of Eqs 8.29, the Cauchy stress components for an isotropic, incompressible material are found to be

$$t_{ij} = -p\delta_{ij} + \alpha_1 B_{ij} + \alpha_2 B_{ik} B_{kj} \tag{8.47}$$

where

$$\alpha_1 = 2\left(\frac{\partial W}{\partial I_1} + I_1 \frac{\partial W}{\partial I_2}\right) , \tag{8.48a}$$

$$\alpha_2 = -2\frac{\partial W}{\partial I_2} \ . \tag{8.48b}$$

From Eq 8.31, the stress for an isotropic, incompressible material may be written in the alternative form

$$t_{ij} = -p\delta_{ij} + \beta_1 B_{ij} + \beta_{-1} B_{ij}^{-1} \tag{8.49}$$

where

$$\beta_1 = 2\frac{\partial W}{\partial I_1} , \tag{8.50a}$$

$$\beta_{-1} = -2\frac{\partial W}{\partial I_2} \ . \tag{8.50b}$$

At this point, the strain energy has been assumed to be a function of I_1 and I_2, but the exact functional form has not been specified. Rivlin (1948) postulated the strain energy should be represented as a general polynomial in I_1 and I_2

$$W = \sum C_{\alpha\beta} (I_1 - 3)^{\alpha} (I_2 - 3)^{\beta} \ . \tag{8.51}$$

It is noted that the strain energy is written in terms of $I_1 - 3$ and $I_2 - 3$ rather than I_1 and I_2 to ensure zero strain corresponds to zero strain energy.

Depending on the type of material and deformation, that is, experimental test data, the number of terms used in Eq 8.51 is chosen. For instance, choosing $C_{10} = G$ and all other coefficients zero results in a neo-Hookean response where G is the shear modulus. Stresses are evaluated with Eq 8.49 and used in the equations of motion, Eq 5.22. This results in a set of differential equations solved with the use of the problem's appropriate boundary conditions. The indeterminate pressure is coupled with the equations of motion and is found by satisfying the boundary conditions.

8.3 Specific Forms of the Strain Energy

When confronted with the design of a specific part made of a rubber-like material, say the jacket or lumbar of an automotive impact dummy or a golf ball core, testing must be done to evaluate the various constants of the strain energy function. But before the testing can be analyzed, the specific form of the strain energy must be chosen. This is tantamount to choosing how many terms of the Mooney-Rivlin strain energy, Eq 8.51, will be assumed nonzero to effectively represent the material response. There are other common forms of the strain energy which are all somewhat equivalent to the form put forward by Mooney (see Rivlin, 1976). Since constants $C_{\alpha\beta}$ of Eq 8.51 do not represent physical quantities like, for example, modulus of elasticity, the constants are essentially curve fit parameters.

The simplest form of the strain energy for a rubber-like material is a one parameter model called a *neo-Hookean* material. The single parameter is taken to be the shear modulus, G, and the strain energy depends only on the first invariant of the deformation tensor, B_{ij}

$$W = G\,(I_1 - 3)\ . \tag{8.52}$$

Assuming principal axes for the left deformation tensor, B_{ij}, for a motion having principal stretches λ_1, λ_2, and λ_3 means Eq 8.52 is written as

$$W = G\left(\lambda_1^2 + \lambda_2^2 + \lambda_3^2 - 3\right)\ . \tag{8.53}$$

For the case of uniaxial tension the stretches are $\lambda_1^2 = \lambda^2$, $\lambda_2^2 = \lambda_3^2 = \lambda^{-1}$ and furthermore, if the material is incompressible $\lambda_1^2\lambda_2^2\lambda_3^2 = 1$. Thus,

$$W = G\left(\lambda^2 + \frac{2}{\lambda} - 3\right)\ .$$

Using this expression in Eq 8.47 for the case of uniaxial tension yields the stress per unit undeformed area as

$$f = P_{11} = \frac{\partial W}{\partial \lambda} = 2G\left(\lambda - \frac{1}{\lambda^2}\right) \tag{8.54}$$

where P_{11} is the 11-component of the first Piola-Kirchhoff stress.

The neo-Hookean material is the simplest form of the strain energy function and makes exact solutions much more tractable. A slightly more general model is a simple, or two-term, *Mooney-Rivlin* model. In this case, the strain energy function is assumed to be linear in the first and second invariants of the left deformation tensor. Again, assuming the material is isotropic and incompressible, the strain energy may be written as

$$W = C_1\left(\lambda_1^2 + \lambda_2^2 + \lambda_3^2 - 3\right) + C_2\left(\frac{1}{\lambda_1^2} + \frac{1}{\lambda_2^2} + \frac{1}{\lambda_3^2} - 3\right)\ . \tag{8.55}$$

For a uniaxial test, the principal stretches are $\lambda_1^2 = \lambda^2$, $\lambda_2^2 = \lambda_3^2 = \lambda^{-1}$. Substitution of these stretches into Eq 8.55 and differentiating with respect to λ gives the uniaxial stress per unit undeformed area

$$f = P_{11} = 2\left(\lambda - \frac{1}{\lambda^2}\right)\left(C_1 + \frac{C_2}{\lambda}\right) . \tag{8.56}$$

The Cauchy stress is easily formed by noting the area in the deformed configuration would be found by scaling dimensions in the x_2 and x_3 direction by λ_2 and λ_3, respectively. For the uniaxial case, this means that multiplying the force per undeformed area by stretch λ results in the uniaxial Cauchy stress

$$t_{11} = 2\left(\lambda^2 - \frac{1}{\lambda}\right)\left(C_1 + \frac{C_2}{\lambda}\right) . \tag{8.57}$$

Making use of Eq 4.109 while noting Λ in that equation is λ in Eq 8.57, a formula for stress in terms of strain is obtained

$$t_{11} = 2C_1\left(1 + 2E_{11} + \frac{1}{\sqrt{1+2E_{11}}}\right) + 2C_2\left(\sqrt{1+2E_{11}} + \frac{1}{1+2E_{11}}\right) . \tag{8.58}$$

Series expansion of the last three terms of Eq 8.58 followed by assuming E_{11} is small results in

$$t_{11} = 6\,(C_1 + C_2)\,E_{11} . \tag{8.59}$$

Hence, for small strain the modulus of elasticity may be written as $E = 6(C_1 + C_2)$. Furthermore, since an incompressible material is assumed, Poisson's ratio is equal to 0.5. This means, by virtue of Eq 6.29a, that the shear modulus is given by $G = 3(C_1 + C_2)$ for small strain.

The simple, two-term Mooney-Rivlin model represents material response well for small to moderate stretch values, but for large stretch higher order terms are needed. Some of these different forms for the strain energy function are associated with specific names. For instance, an *Ogden* material (Ogden, 1972) assumes a strain energy in the form

$$W = \sum_n \frac{\mu_n}{\alpha_n}\left[\lambda_1^{\alpha_n} + \lambda_2^{\alpha_n} + \lambda_3^{\alpha_n} - 3\right] . \tag{8.60}$$

This form reduces to the Mooney-Rivlin material if n takes on values of 1 and 2 with $\alpha_1 = 2$, $\alpha_2 = -2$, $\mu_1 = 2C_1$, and $\mu_2 = -2C_2$.

8.4 Exact Solution for an Incompressible, Neo-Hookean Material

Exact solutions for nonlinear elasticity problems come from using the equations of motion, Eq 5.22, or for equilibrium, Eq 5.23, along with appropriate boundary conditions. A neo-Hookean material is one whose uniaxial stress response is proportional to the combination of stretch $\lambda - 1/\lambda^2$ as was stated in Section 8.1. The proportionality constant is the shear modulus, G. The strain energy function in the neo-Hookean case is proportional to $I_1 - 3$ where I_1 is the first invariant of the left deformation tensor. A strain energy of this form leads to stress components given by

$$t_{ij} = -p\delta_{ij} + GB_{ij} \tag{8.61}$$

where p is the indeterminate pressure term, G is the shear modulus, and B_{ij} is the left deformation tensor.

The stress components must satisfy the equilibrium equations which, in the absence of body forces, are given as

$$t_{ij,j} = 0 . \tag{8.62}$$

Substitution of Eqs 4.48 and 8.61 into Eq 8.62 gives a differential equation governing equilibrium

$$\frac{\partial t_{ij}}{\partial x_j} = -\frac{\partial p}{\partial x_j} + G\left[\left(\frac{\partial^2 x_i}{\partial X_A \partial X_B}\frac{\partial X_B}{\partial x_j}\right)\frac{\partial x_j}{\partial X_A} + \left(\frac{\partial^2 x_j}{\partial X_A \partial X_B}\frac{\partial X_B}{\partial x_j}\right)\frac{\partial x_i}{\partial X_A}\right] = 0 . \tag{8.63}$$

This simplifies to

$$\frac{\partial p}{\partial x_j} = G\frac{\partial^2 x_i}{\partial X_A \partial X_A} = G\nabla^2 x_i \tag{8.64}$$

where

$$\frac{\partial X_B}{\partial x_j}\frac{\partial x_j}{\partial X_A} = \delta_{AB} ,$$

and

$$\frac{\partial^2 x_j}{\partial X_A \partial X_B}\frac{\partial X_B}{\partial x_j} = 0$$

have been used. Eq 8.64 may be made more convenient by writing the pressure term as a function of the reference configuration and using the chain rule

$$\frac{\partial P}{\partial X_A} = G\frac{\partial x_j}{\partial X_A}\nabla^2 x_i \tag{8.65}$$

where P is a function of X_A and t.

Consider the case of plane strain defined by

$$x = x(X, Y); \quad y = y(X, Y); \quad z = Z \tag{8.66}$$

from which the deformation gradient follows as

$$[F_{iA}] = \begin{bmatrix} \frac{\partial x}{\partial X} & \frac{\partial x}{\partial Y} & 0 \\ \frac{\partial y}{\partial X} & \frac{\partial y}{\partial Y} & 0 \\ 0 & 0 & 1 \end{bmatrix} . \tag{8.67}$$

Note that we have departed from the strict continuum summation convention for clarity in solving this specific problem. Substitution of the deformation gradient into the definition for the left deformation tensor and use of Eq 8.61 yields

$$t_{xx} = -p + G\left[\left(\frac{\partial x}{\partial X}\right)^2 + \left(\frac{\partial x}{\partial Y}\right)^2\right] , \tag{8.68a}$$

$$t_{yy} = -p + G\left[\left(\frac{\partial y}{\partial X}\right)^2 + \left(\frac{\partial y}{\partial Y}\right)^2\right] , \tag{8.68b}$$

$$t_{xy} = G\left[\frac{\partial x}{\partial X}\frac{\partial y}{\partial X} + \frac{\partial x}{\partial Y}\frac{\partial y}{\partial Y}\right] , \tag{8.68c}$$

where again it is noted that strict continuum notation is not used for clarity in this particular solution. The incompressibility condition, $J = \det[F_{iA}] = 1$ may be written as

$$\frac{\partial x}{\partial X}\frac{\partial y}{\partial Y} - \frac{\partial x}{\partial Y}\frac{\partial y}{\partial X} = 1 \tag{8.69}$$

The inverse of the deformation gradient can be directly calculated from Eq 8.67 to be

$$\left[F_{iA}^{-1}\right] = \begin{bmatrix} \frac{\partial X}{\partial x} & \frac{\partial X}{\partial y} & \frac{\partial X}{\partial z} \\ \frac{\partial Y}{\partial x} & \frac{\partial Y}{\partial y} & \frac{\partial Y}{\partial z} \\ \frac{\partial Z}{\partial x} & \frac{\partial Z}{\partial y} & \frac{\partial Z}{\partial z} \end{bmatrix} = \begin{bmatrix} \frac{\partial y}{\partial Y} & -\frac{\partial x}{\partial Y} & 0 \\ -\frac{\partial y}{\partial X} & \frac{\partial x}{\partial X} & 0 \\ 0 & 0 & 1 \end{bmatrix} . \tag{8.70}$$

Thus, in the case of incompressible material undergoing a plane strain motion the following relationships must hold:

$$\frac{\partial x}{\partial X} = \frac{\partial Y}{\partial y}; \quad \frac{\partial x}{\partial Y} = -\frac{\partial X}{\partial y}; \quad \frac{\partial y}{\partial X} = -\frac{\partial y}{\partial x}; \quad \frac{\partial y}{\partial Y} = \frac{\partial X}{\partial x} . \tag{8.71}$$

An alternate form of the stress components may be written by factoring out the quantity $\Pi = (t_{xx} + t_{yy})/2$ for future convenience

$$t_{xx} = \Pi + \frac{1}{2}G\left[\left(\frac{\partial X}{\partial y}\right)^2 + \left(\frac{\partial Y}{\partial y}\right)^2 - \left(\frac{\partial X}{\partial x}\right)^2 - \left(\frac{\partial Y}{\partial x}\right)^2\right] , \tag{8.72a}$$

$$t_{yy} = \Pi - \frac{1}{2}G\left[\left(\frac{\partial X}{\partial y}\right)^2 + \left(\frac{\partial Y}{\partial y}\right)^2 - \left(\frac{\partial X}{\partial x}\right)^2 - \left(\frac{\partial Y}{\partial x}\right)^2\right] , \tag{8.72b}$$

$$t_{xy} = -G\left[\frac{\partial X}{\partial x}\frac{\partial X}{\partial y} + \frac{\partial Y}{\partial x}\frac{\partial Y}{\partial x}\right] , \tag{8.72c}$$

where the results of Eq 8.71 have also been utilized. The nontrivial equilibrium equations are associated with the x and y directions. Using the stress components as defined in Eq 8.72, the equilibrium conditions are

$$\frac{\partial \Pi}{\partial x} = G\left[\frac{\partial X}{\partial x}\nabla^2 X + \frac{\partial Y}{\partial x}\nabla^2 X\right] , \tag{8.73a}$$

$$\frac{\partial \Pi}{\partial y} = G\left[\frac{\partial X}{\partial y}\nabla^2 X + \frac{\partial Y}{\partial y}\nabla^2 X\right] \tag{8.73b}$$

The incompressibility condition may also be written as

$$\frac{\partial X}{\partial x}\frac{\partial Y}{\partial y} - \frac{\partial X}{\partial y}\frac{\partial Y}{\partial x} = 1 . \tag{8.74}$$

Eqs 8.73 and 8.74 are three nonlinear partial differential equations that must be satisfied to ensure equilibrium and the constraint of incompressibility. To actually solve a problem, a solution must be found that satisfies the appropriate boundary conditions. Often the way this is done is by assuming a specific form for $X(x, y)$ and $Y(x, y)$ and demonstrating that the aforementioned equations are satisfied. Following that, the boundary conditions are defined and demonstrated, completing the solution. In a sense, this is similar to the semi-inverse method discussed in Chapter 6. The assumed "guess" of the functional form of $X(x, y)$ and $Y(x, y)$ requires some experience and or familiarity with the problem at hand. Here, a function form will be assumed in pursuit of the solution of a rectangular rubber specimen being compressed in the x direction. The faces of the specimen are perfectly attached to rigid platens. In this case, assume that $X = f(x)$. After a modest amount of work, this will be shown to lead to a solution of a neo-Hookean material compressed between rigid platens.

Substituting the assumed form of X into the incompressibility equation, Eq 8.74, yields the ordinary differential equation

$$f' \frac{\partial Y}{\partial y} = 1 \tag{8.75}$$

where the prime following f represents differentiation with respect to x. Defining $q(x) = [f'(x)]^{-1}$, Eq 8.75 may be integrated to obtain

$$Y = q(x)\, y + g(x) \tag{8.76}$$

where function $g(x)$ is an arbitrary function of integration.

Use of the given functions X and Y, the incompressibility condition is satisfied, and the equilibrium equations may be written as follows:

$$\frac{\partial \Pi}{\partial x} = G\left[f'f'' + \left(yq' + g'\right)\left(yq'' + g''\right)\right] , \tag{8.77a}$$

$$\frac{\partial \Pi}{\partial y} = G\left[q\left(yq'' + g''\right)\right] , \tag{8.77b}$$

where, again, primes after a symbol represent differentiation with respect to x. The second equation, 8.77b, is easily integrated to obtain an expression for Π

$$\frac{\Pi}{G} = \frac{1}{2} qq'' y^2 + qg'' y + M(x) \tag{8.78}$$

where $M(x)$ is a function of integration. Differentiation of Eq 8.78 with respect to x results in an equation that must be consistent with Eq 8.77a. This comparison gives rise to the following three ordinary differential equations:

$$\frac{1}{2} \frac{d}{dx}\left(qq''\right) = q'q'' , \tag{8.79a}$$

$$\frac{d}{dx}\left(qq''\right) = g'q'' + q'g'' , \tag{8.79b}$$

$$M'(x) = f'f'' + g'g'' . \tag{8.79c}$$

Eq 8.79a may be expanded and integrated to obtain $q'' = k^2 q$ from which the function $q(x)$ is determined to within the constants A and B:

$$q(x) = Ae^{kx} + Be^{-kx} . \tag{8.80}$$

Eqs 8.79a and 8.79b may be combined along with the fact that $q'' = k^2 q$ to obtain the function $g(x)$

$$g(x) = Ce^{kx} + De^{-kx} \tag{8.81}$$

where C and D are constants. With functions $q(x)$ and $g(x)$ found it is now possible to write functions $X(x, y)$ and $Y(x, y)$ as

$$X = \int \frac{dx}{Ae^{kx} + Be^{-kx}} , \tag{8.82a}$$

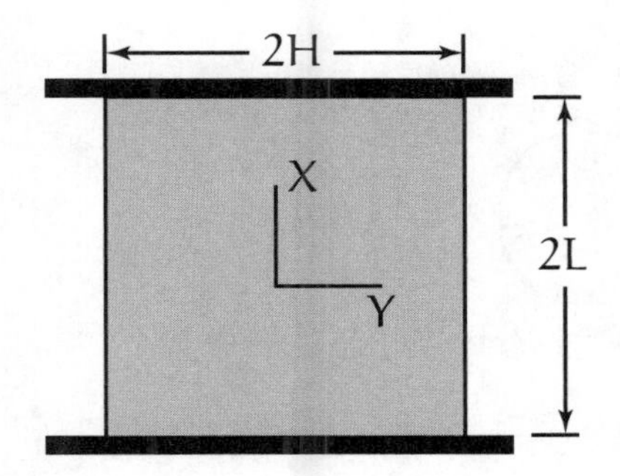

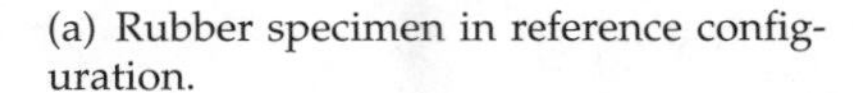

(a) Rubber specimen in reference configuration.

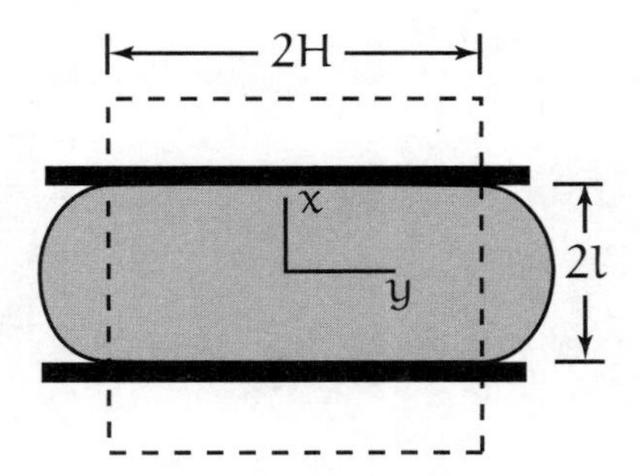

(b) Rubber specimen in deformed configuration. The dashed line represents the undeformed shape.

FIGURE 8.5
Rhomboid rubber specimen compressed by platens.

$$Y = \left(Ae^{kx} + Be^{-kx}\right) y + Ce^{kx} + De^{-kx} \ . \tag{8.82b}$$

In pursuit of the compressed rectangular rubber solution, constants C and D are taken to be identically zero and constants A and B are set equal. Thus,

$$X = (kA)^{-1} \left[\tan^{-1}\left(e^{kx}\right) + C'\right] \ , \tag{8.83a}$$

$$Y = 2Ay \cosh kx \ , \tag{8.83b}$$

where k, A, and C' are constants that must be determined. The deformation considered here has the rubber rhomboid deforming symmetrically about the origin of both the referential and deformed coordinate system (Fig. 8.5(a) and 8.5(b)). Thus, set $X = 0$ when $x = 0$ to obtain constant C', of Eq 8.83a, to have the value 0.25π. Since the platens are fixed to the specimen, $Y = y$ on planes $x = \pm l$ the constant A is found to be $\frac{1}{2}\left[\cosh(kl)\right]^{-1}$. Finally, the constant k may be evaluated in terms of the deformed and undeformed lengths l and L by

$$kL = 2\cosh(kl) \left[\tan^{-1}\left(e^{kl}\right) - \frac{\pi}{4}\right] \ . \tag{8.84}$$

To obtain an exact solution it must be shown that the barreled surfaces initially at $Y = \pm H$ are stress free in the deformed configuration. Unfortunately, this is not possible for this problem. Instead, a relaxed boundary condition may be satisfied by enforcing the resultant force on the boundary to be zero rather than have zero traction at every point.

With all the constants of Eq 8.84 determined, the stress components may be written as

$$t_{xx} = -p + \frac{1}{2} G \frac{\cosh^2(kx)}{\cosh^2(kl)} \ , \tag{8.85a}$$

$$t_{yy} = -p + G\left[k^2 y^2 \frac{\sinh^2(kx)}{\sinh^2(kl)} + 4\frac{\cosh^2(kl)}{\cosh^2(kx)}\right] \ , \tag{8.85b}$$

$$t_{xy} = -\frac{1}{2} Gky \frac{\sinh^2(2kx)}{\cosh^2(kl)} \ . \tag{8.85c}$$

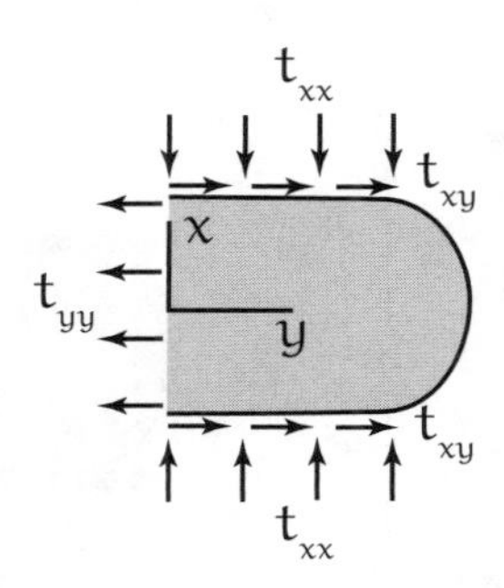

FIGURE 8.6
Rhomboid rubber specimen compressed by platens.

At this point, the stresses are known to within the additive pressure term p. A solution is sought by picking p such that the average force over the barreled edge is zero.

Symmetry of the deformation allows for considering the top half, $y \geqslant 0$, in determining the condition for zero resultant stress on the barreled top edge (Fig. 8.6). Integrating the stresses yields

$$\int_0^l t_{yy}\big|_{y=0}\, dx = \int_0^H t_{xy}\big|_{x=l}\, dy \ , \tag{8.86}$$

or

$$p = \frac{1}{4}\frac{GkH^2}{l}\frac{\sinh(2kl)}{\cosh^2(kl)} - \frac{4G}{k}\cosh^2(kl)\tanh(kl) \ . \tag{8.87}$$

Knowing the pressure term, the compressive force may be determined from

$$F = \int_0^H t_{xx}\big|_{x=l}\, dy \ . \tag{8.88}$$

Bibliography

[1] Carroll, M. M. (1988), "Finite strain solutions in compressible isotropic elasticity," *J. Elas.*, Vol. 20, No. 1, pp. 65-92

[2] Kao, B. G. and L. Razgunas (1986), "On the Determination of Strain Energy Functions of Rubbers," *Proceedings of the Sixth Intl. Conference on Vehicle Structural Dynamics* P-178, Society of Automotive Engineering, Warrendale, PA

[3] McCrum, N. G., C. P. Buckley, and C. B. Bucknall (1997), *Principles of Polymer Engineering*, Second Edition, Oxford University Press, Oxford, UK

[4] Mooney, M. (1940), "A Theory of Large Elastic Deformation," *J. Appl. Phys.*, Vol. 11, pp. 582-592

[5] Rivlin, R.S. (1948), "Large Elastic Deformations of Isotropic Materials: IV. Further Developments of the General Theory," *Phil. Trans. Roy. Soc.*, A241, pp. 379-397

[6] Rivlin, R.S. (1949), "Large Elastic Deformations of Isotropic Materials: VI. Further Results in the Theory of Torsion, Shear and Flexure," *Phil. Trans. Roy. Soc.*, A242, pp. 173-195

[7] Rivlin, R.S. and K.N. Sawyers (1976), "The Strain-Energy Function for Elastomers," *Trans. of the Society of Rheology*, 20:4, pp. 545-557

[8] Sperling, L. H. (1992), *Introduction to Physical Polymer Science*, Second Edition, Wiley & Sons, Inc., New York

[9] Ward, I. M. and D. W. Hadley (1993), *An Introduction to the Mechanical Properties of Solid Polymers*, Wiley & Sons, Inc., Chichester, UK

Problems

Problem 8.1

Referred to principal axes, the invariants of the Green deformation tensor C_{AB} are

$$I_1 = \lambda_1^2 + \lambda_2^2 + \lambda_3^2 \,,$$
$$I_2 = \lambda_1^2\lambda_2^2 + \lambda_1^2\lambda_3^2 + \lambda_2^2\lambda_3^2 \,,$$
$$I_3 = \lambda_1^2\lambda_2^2\lambda_3^2 \,.$$

For an isotropic, incompressible material, show that

$$I_1 = \lambda_1^2 + \lambda_2^2 + \frac{1}{\lambda_1^2\lambda_2^2} \,,$$
$$I_2 = \frac{1}{\lambda_1^2} + \frac{1}{\lambda_2^2} + \lambda_1^2\lambda_2^2 \,.$$

Problem 8.2

Derive the following relationships between invariants I_1, I_2, and I_3, and the deformation gradient, C_{AB}:

(a) $\dfrac{\partial I_1}{\partial C_{AB}} = \delta_{AB}$,

(b) $\dfrac{\partial I_2}{\partial C_{AB}} = C_{AB} - I_1\delta_{AB}$,

(c) $\dfrac{\partial I_3}{\partial C_{AB}} = I_3 C_{AB}^{-1}$.

Problem 8.3

Use the definitions of I_1 and I_2 in terms of the principal stretches λ_1, λ_2, and λ_3 to show

(a) $\dfrac{\partial W}{\partial \lambda_1} = \dfrac{2}{\lambda_1}\left(\lambda_1^2 - \lambda_3^2\right)\left(\dfrac{\partial W}{\partial I_1} + \lambda_2^2\dfrac{\partial W}{\partial I_2}\right)$,

(b) $\dfrac{\partial W}{\partial \lambda_2} = \dfrac{2}{\lambda_2}\left(\lambda_2^2 - \lambda_3^2\right)\left(\dfrac{\partial W}{\partial I_1} + \lambda_1^2\dfrac{\partial W}{\partial I_2}\right)$.

Problem 8.4

Let $W(I_1, I_2, I_3)$ be the strain energy per unit volume for a homogeneous, isotropic material. Show that the first Piola-Kirchhoff stress components may be written as follows:

$$P_{iA} \equiv \frac{\partial W}{\partial F_{iA}} = 2\frac{\partial W}{\partial I_1}F_{iA} + 2\frac{\partial W}{\partial I_2}\left(B_{ij} - I_1\delta_{ij}\right)F_{jA} + 2I_3\frac{\partial W}{\partial I_3}F_{iA}^{-1} \,.$$

Problem 8.5

The Cauchy stress is given by

$$t_{ij} = \frac{1}{J}F_{jA}P_{iA} \,.$$

Start with the result of Problem 8.4 to show that

$$t_{ij} = \frac{2}{J}\left[\frac{\partial W}{\partial I_1}B_{ij} + \frac{\partial W}{\partial I_2}\left(B_{ik} - I_1\delta_{ik}\right)B_{jk} + I_3\frac{\partial W}{\partial I_3}\delta_{ij}\right] .$$

Problem 8.6

Assuming a strain energy of the form

$$W = w\left(\lambda_1\right) + w\left(\lambda_2\right) + w\left(\frac{1}{\lambda_1\lambda_2}\right)$$

for an isotropic, incompressible material, show that

$$\lambda_1\frac{\partial^3 W}{\partial\lambda_1^2\partial\lambda_2} = \lambda_2\frac{\partial^3 W}{\partial\lambda_2^2\partial\lambda_1} .$$

Problem 8.7

For biaxial loading of a thin vulcanized rubber sheet the strain energy may be written as

$$W = C_1\left(I_1 - 3\right) + C_2\left(I_2 - 3\right) + C_3\left(I_2 - 3\right)^2$$

(Rivlin and Saunders, 1951).

(a) Use the definitions of invariants I_1, I_2, and I_3 in terms of stretches λ_1, λ_2, and λ_3 to show

$$\left(I_2 - 3\right)^2 = \frac{1}{\lambda_1^4} + \frac{1}{\lambda_2^4} + \frac{1}{\lambda_3^4} + 2I_1 - 6I_2 + 9 .$$

(b) Substitute the results from (a) into the strain energy above to obtain

$$w\left(\lambda_1\right) = \left(C_1 + 2C_3\right)\lambda_1^2 + \left(C_2 - 6C_3\right)\lambda_1^{-2} + C_3\lambda_3^{-4} - \left(C_1 + C_2 - 3C_3\right)$$

where

$$W = w\left(\lambda_1\right) + w\left(\lambda_2\right) + w\left(\frac{1}{\lambda_1\lambda_2}\right) .$$

Problem 8.8

Consider a material having reference configuration coordinates (R, Θ, Z) and current configuration coordinates (r, θ, z). Assume a motion defined by

$$r = \frac{R}{g\left(\Theta\right)}, \quad \theta = f\left(\Theta\right), \quad z = Z .$$

Determine F_{iA} and B_{ij} in terms of g, g', and f'.

Answer

$$F_{iA} = \begin{bmatrix} \dfrac{1}{g} & -\dfrac{g'}{g^2} & 0 \\ 0 & \dfrac{f'}{g} & 0 \\ 0 & 0 & 1 \end{bmatrix}, \qquad B_{ij} = \begin{bmatrix} \dfrac{1}{g^2} + \left(\dfrac{g'}{g^2}\right)^2 & -\dfrac{f'g'}{g^3} & 0 \\ -\dfrac{f'g'}{g^3} & \left(\dfrac{f'}{g}\right)^2 & 0 \\ 0 & 0 & 1 \end{bmatrix}$$

Problem 8.9
Show

$$t_{11} = 2\left(\lambda_1^2 - \lambda_1^{-2}\lambda_2^{-2}\right)\frac{\partial W}{\partial I_1} - 2\left(\lambda_1^{-2} - \lambda_1^2\lambda_2^2\right)\frac{\partial W}{\partial I_2},$$
$$t_{22} = 2\left(\lambda_2^2 - \lambda_1^{-2}\lambda_2^{-2}\right)\frac{\partial W}{\partial I_1} - 2\left(\lambda_2^{-2} - \lambda_1^2\lambda_2^2\right)\frac{\partial W}{\partial I_2},$$

are the nonzero Cauchy stress components for biaxial tensile loading of a homogeneous, isotropic, incompressible rubber-like material by solving for the indeterminant pressure p from the fact that $t_{33} = 0$ condition.

Problem 8.10
The following compression force-deflection data was obtained for a highly filled, polybutadiene rubber having initial gauge length of 0.490 *in* and an undeformed cross-section area of $1\ \text{in}^2$

Displ. (in)	Force (lbs)
–8.95E–04	–7.33E+00
–5.45E–03	–4.15E+01
–8.06E–03	–7.57E+01
–1.13E–02	–1.29E+02
–1.44E–02	–1.83E+02
–1.80E–02	–2.52E+02
–2.55E–02	–3.93E+02
–2.94E–02	–4.62E+02
–3.28E–02	–5.35E+02
–4.00E–02	–6.72E+02
–4.39E–02	–7.40E+02
–4.75E–02	–8.13E+02
–5.46E–02	–9.45E+02
–5.80E–02	–1.01E+03
–6.58E–02	–1.15E+03
–7.30E–02	–1.28E+03
–7.69E–02	–1.35E+03
–8.05E–02	–1.42E+03
–8.80E–02	–1.56E+03
–9.19E–02	–1.63E+03
–9.55E–02	–1.70E+03
–1.03E–01	–1.86E+03
–1.06E–01	–1.93E+03
–1.17E–01	–2.20E+03
–1.28E–01	–2.50E+03
–1.38E–01	–2.85E+03
–1.50E–01	–3.26E+03
–1.60E–01	–3.73E+03
–1.71E–01	–4.22E+03
–1.77E–01	–4.42E+03
–1.83E–01	–4.48E+03

(a) Generate a stress-stretch plot.

(b) Using a spreadsheet, or other appropriate tool, show that simple Mooney-Rivlin constants $C_1 = 1,550$ and $C_2 = -500$ represent the material for the range and type of loading given.

(c) Generate a set of significantly different constants C_1 and C_2 which might equally well model the material for this range and type of loading, showing that C_1 and C_2 are not unique.

9

Linear Viscoelasticity

The previous chapters have considered constitutive equations that deal primarily with two different types of material behavior: elastic response of solids and viscous flow of fluids. Examples of materials that behave elastically under modest loading, and at moderate temperatures, are metals such as steel, aluminum, and copper, certain polymers, and even cortical bone. Examples of viscous flow may involve a variety of fluids ranging from water to polymers under certain conditions of temperature and loading.

Polymers are especially interesting because they may behave (respond) in either elastic, viscous, or combined manners. At a relative moderate temperature and loading, a polymer such as polymethylmethacrylate (PMMA), may be effectively modeled by a linear elastic constitutive equation. However, at a somewhat elevated temperature, the same material may have to be modeled as a viscous fluid.

Polymers are by no means the only materials that exhibit different behavior under altered temperature/frequency conditions. Steel, as well as aluminum, copper, and other metals, becomes molten at high temperatures and can be poured into molds to form ingots. Additionally, at a high enough deformation rate, for example, at the 48 km/hr rate of a vehicle crash, steel will exhibit considerably altered stiffness properties.

Just as continuum mechanics is the basis for constitutive models as distinct as elastic solids (stress/strain laws) and viscous fluids (stress/strain-rate laws), it also serves as the basis for constitutive relations that describe material behavior over a range of temperature/frequency and time. One of the simplest models for this combined behavior is that of linear viscoelasticity.

9.1 Viscoelastic Constitutive Equations in Linear Differential Operator Form

One of the principal features of elastic behavior is the capacity for materials to store mechanical energy when deformed by loading, and to release this energy totally upon removal of the loads. Conversely, in viscous flow, mechanical energy is continuously dissipated with none stored. A number of important engineering materials simultaneously store and dissipate mechanical energy when subjected to applied forces. In fact, all actual materials store and dissipate energy in varying degrees during a loading/unloading cycle. This behavior is referred to as *viscoelastic*. In general, viscoelastic behavior may be imagined as a spectrum with elastic deformation as one limiting case and viscous flow the other extreme case, with varying combinations of the two spread over the range between. Thus, valid constitutive equations for viscoelastic behavior embody elastic deformation and viscous flow as special cases, and at the same time provide for response patterns that characterize behavior blends of the two. Intrinsically, such equations will involve not only stress and strain, but time-rates of both stress and strain as well.

In developing the *linear differential operator* form of constitutive equations for viscoelastic behavior as presented in Eq 5.177 we draw upon the pair of constitutive equations for elastic behavior, Eq 6.33, repeated here,

$$S_{ij} = 2G\eta_{ij} \ , \tag{9.1a}$$

$$t_{ii} = 3K\epsilon_{ii} \ , \tag{9.1b}$$

together with those for viscous flow, Eq 7.15,

$$S_{ij} = 2\mu^* \beta_{ij} \ , \tag{9.2a}$$

$$t_{ii} = -3\left(p - \kappa^* d_{ii}\right) \ , \tag{9.2b}$$

each expressed in terms of their deviatoric and dilatational responses. These equations are valid for isotropic media only. For linear viscoelastic theory we assume that displacement gradients, $u_{i,A}$, are small, and as shown by Eq 4.150, this results in

$$\dot{\epsilon}_{ij} \approx d_{ij} \tag{9.3a}$$

from which we immediately conclude that

$$\dot{\epsilon}_{ii} = d_{ii} \tag{9.3b}$$

so that now, from Eq 4.77 and Eq 7.13

$$\dot{\eta}_{ij} \approx \beta_{ij} \ . \tag{9.4}$$

If the pressure p in Eq 9.2b is relatively small and may be neglected, or if we consider the pressure as a uniform dilatational body force that may be added as required to the dilatational effect of the rate of deformation term d_{ii} when circumstances require, Eq 9.2 may be modified in view of Eq 9.3 and Eq 9.4 to read

$$S_{ij} = 2\mu^* \dot{\eta}_{ij} \ , \tag{9.5a}$$

$$t_{ii} = 3\kappa^* \dot{\epsilon}_{ii} \ . \tag{9.5b}$$

A comparison of Eqs 9.5 and Eq 9.1 indicates that they differ primarily in the physical constants listed and in the fact that in Eq 9.5 the stress tensors are expressed in terms of strain rates. Therefore, a generalization of both sets of equations is provided by introducing linear differential operators of the form given by Eq 5.178 in place of the physical constants G, K, μ^*, and κ^*. In order to make the generalization complete we add similar differential operators to the left-hand side of the equations to obtain

$$\{P\} S_{ij} = 2\{Q\}\eta_{ij} \ , \tag{9.6a}$$

$$\{M\} t_{ii} = 3\{N\}\epsilon_{ii} \ , \tag{9.6b}$$

where the numerical factors have been retained for convenience in relating to traditional elasticity and viscous flow equations. As noted, the linear differential time operators, $\{P\}$, $\{Q\}$, $\{M\}$, and $\{N\}$, are of the same form as in Eq 5.178 with the associated coefficients p_i, q_i, m_i, and n_i representing the physical properties of the material under consideration. Although these coefficients may in general be functions of temperature or other parameters, in the simple linear theory described here they are taken as constants. As stated at the outset, we verify that for the specific choices of operators $\{P\} = 1$, $\{Q\} = G$, $\{M\} = 1$,

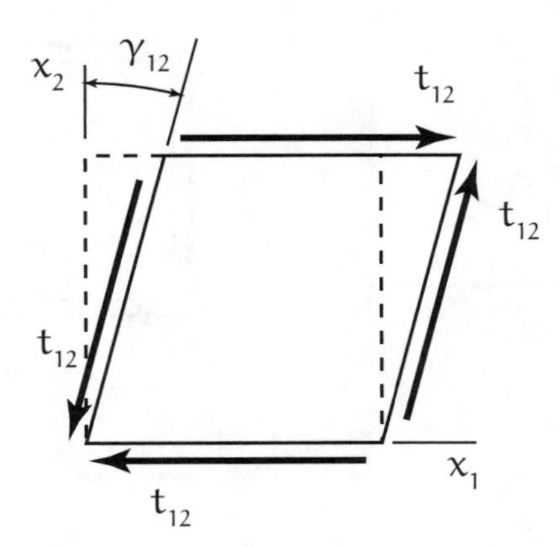

FIGURE 9.1
Simple shear element representing a material cube undergoing pure shear loading.

and $\{N\} = K$, Eqs 9.6 define elastic behavior, whereas for $\{P\} = 1$, $\{Q\} = \mu^* \partial/\partial t$, $\{M\} = 1$, and $\{N\} = \kappa^* \partial/\partial t$, linear viscous behavior is indicated.

Extensive experimental evidence has shown that practically all engineering materials behave elastically in dilatation so without serious loss of generality we may assume the fundamental constitutive equations for linear viscoelastic behavior in differential operator form to be

$$\{P\} S_{ij} = 2\{Q\}\eta_{ij} \,, \tag{9.7a}$$

$$t_{ii} = 3K\epsilon_{ii} \,, \tag{9.7b}$$

for isotropic media. For anisotropic behavior, the operators $\{P\}$ and $\{Q\}$ must be augmented by additional operators up to a total of as many as twelve as indicated by $\{P_i\}$ and $\{Q_i\}$ with the index i ranging from 1 to 6, and Eq 9.7a expanded to six separate equations.

9.2 One-Dimensional Theory, Mechanical Models

Many of the basic ideas of viscoelasticity can be introduced within the context of a one-dimensional state of stress. For this reason, and because the viscoelastic response of a material is associated directly with the deviatoric response as was pointed out in arriving at Eq 9.7, we choose the simple shear state of stress as the logical one for explaining fundamental concepts. Thus, taking a material cube subjected to simple shear, as shown by Fig. 9.1, we note that for this case Eq 9.7 reduces to the single equation

$$\{P\}t_{12} = 2\{Q\}\eta_{12} = 2\{Q\}\epsilon_{12} = \{Q\}\gamma_{12} \tag{9.8}$$

where γ_{12} is the engineering shear strain as shown in Fig. 9.1. If the deformational response of the material cube is linearly elastic, the operators $\{P\}$ and $\{Q\}$ in Eq 9.8 are constants ($P = 1$, $Q = G$) and that equation becomes the familiar

$$t_{12} = G\gamma_{12} \tag{9.9}$$

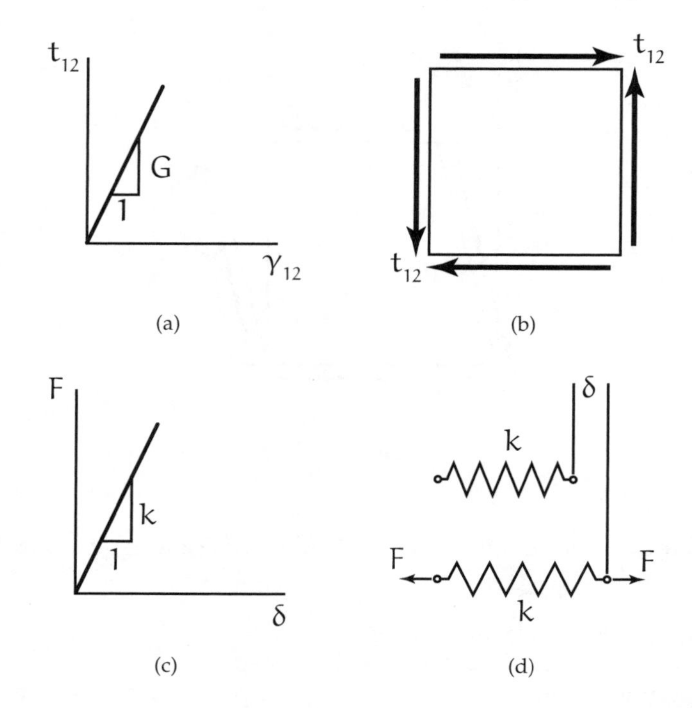

FIGURE 9.2
Mechanical analogy for simple shear.

where G is the elastic shear modulus. This equation has the same form as the one which relates the elongation (or shortening) δ of a linear mechanical spring to the applied force F as given by the equation

$$F = k\delta \tag{9.10}$$

where k is the spring constant. Because Eqs 9.9 and 9.10 are identical in form, the linear spring is adopted as the mechanical analog of simple elastic shearing with k assuming the role of G. The analogy is depicted graphically by the plots shown in Fig. 9.2.

In similar fashion, if the response of the material cube is viscous flow, Eq 9.8 is written

$$t_{12} = \mu^* \dot{\gamma}_{12} \tag{9.11a}$$

where μ^* is the coefficient of viscosity. In viscoelastic theory it is a longstanding practice that the coefficient of viscosity be represented by the symbol η, and it is in keeping with this practice that we hereafter use the scalar η for the coefficient of viscosity. Thus, Eq 9.11a becomes

$$t_{12} = \eta \dot{\gamma}_{12} \tag{9.11b}$$

in all subsequent sections of this chapter. The mechanical analog for this situation is the *dashpot* (a loose fitting piston sliding in a cylinder filled with a viscous fluid) subjected to an axial force F. Here

$$F = \eta \dot{\delta} \tag{9.12}$$

where $\dot{\delta}$ is the time-rate of extension. This analogy is illustrated in Fig. 9.3.

Based upon the two fundamental elements described above, it is easy to construct viscoelastic models by suitable combinations of this pair of elements. Two especially

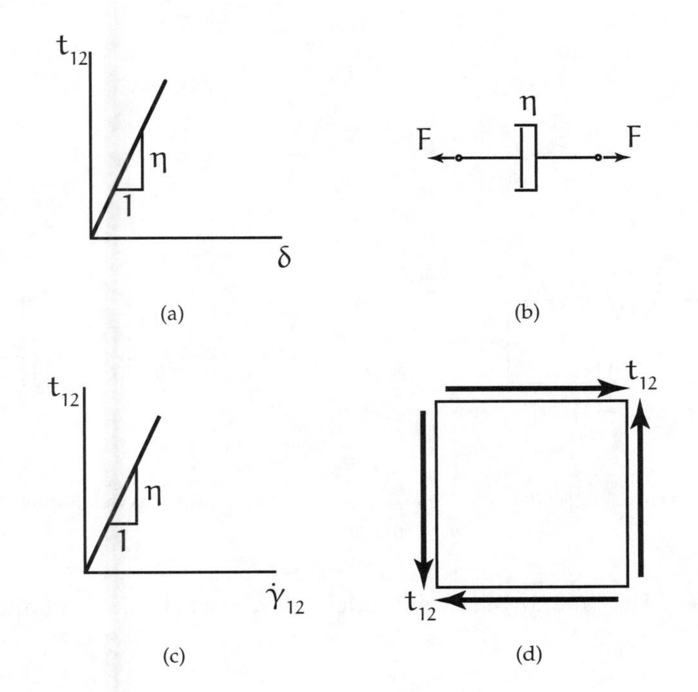

FIGURE 9.3
Viscous flow analogy.

simple combinations immediately come to mind. The first, that of the spring and dashpot in parallel, Fig. 9.4(a), portrays the *Kelvin solid* for which Eq 9.8 becomes

$$T = t_{12} = \{G + \eta\partial_t\}\gamma_{12} \tag{9.13}$$

where the partial derivative with respect to time is denoted by $\partial_t \equiv \partial/\partial t$. The second, the spring and the dashpot connected in series, Fig. 9.4(b), represents the *Maxwell fluid* having the constitutive equation

$$\left\{\partial_t + \frac{1}{\tau}\right\} t_{12} = \{G\partial_t\}\gamma_{12} \tag{9.14}$$

where $\tau = \eta/G$. In this chapter, the uppercase T is used to designate the stress applied to the model. Using this notation makes the differential equations that arise from the models easier to write than if t_{12} were used. Viscoelastic materials exhibit time dependent behavior and designating stress with a lowercase t creates confusion. Throughout this chapter T, or T(t), are used interchangeably with t_{12} to denote the stress.

Models composed of more than two elements are readily constructed. When a Kelvin unit is combined in series with the linear spring element, Fig. 9.5(a), the resulting model is said to represent the standard linear solid. If the same Kelvin unit is joined in series with a dashpot, Fig. 9.5(b), the model represents a three-parameter fluid. In general, the model of a fluid has a "free dashpot" as one of its elements. Other three-parameter models are easily imagined, for example, a Maxwell unit in parallel with a spring, or a Maxwell unit in parallel with a dashpot.

Four-parameter and higher order models may also be constructed. There are two basic patterns for systematically designing higher order models. One, leading to the *generalized*

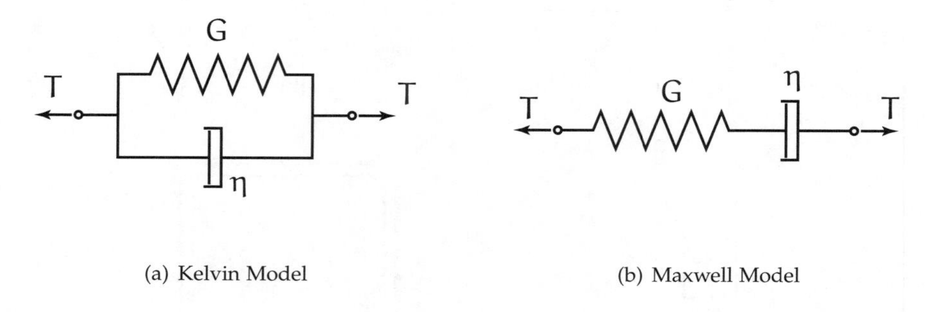

FIGURE 9.4
Representations of Kelvin and Maxwell models for a viscoelastic solid and fluid, respectively.

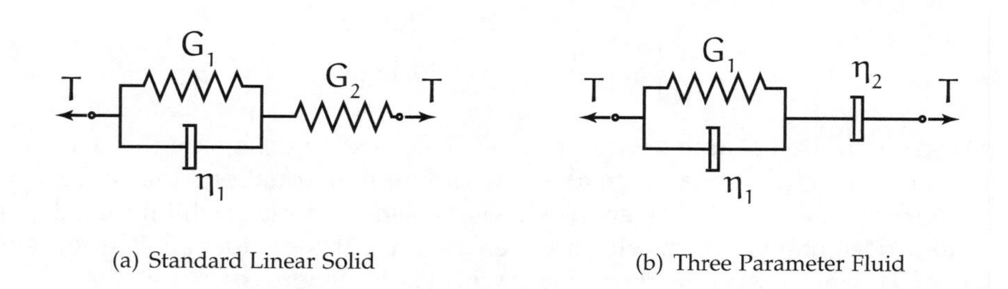

FIGURE 9.5
Three parameter standard linear solid and fluid models.

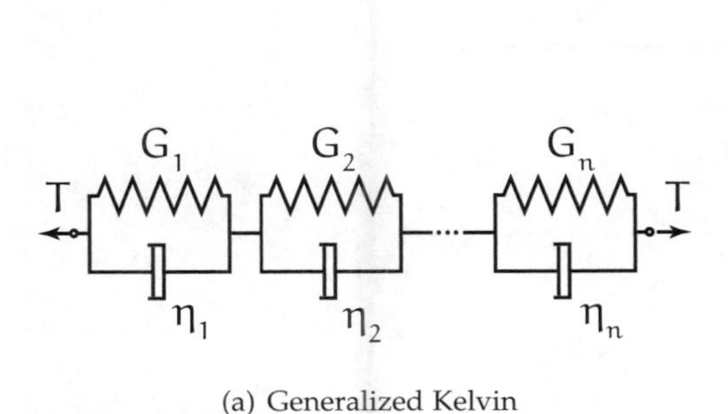

(a) Generalized Kelvin

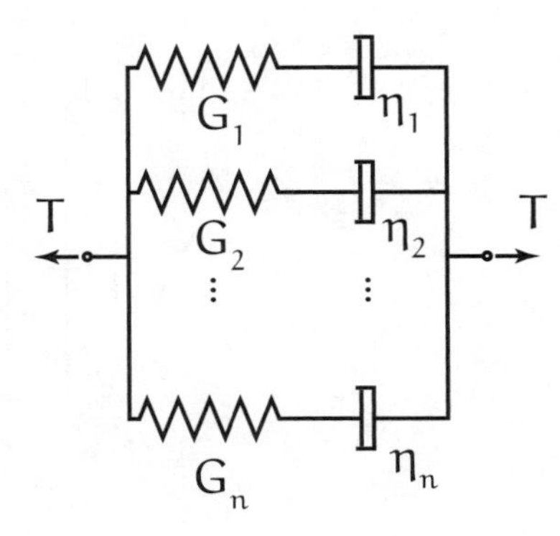

(b) Generalized Maxwell

FIGURE 9.6
Generalized Kelvin and Maxwell models constructed by combining basic models.

Kelvin model, has n-Kelvin units in series, Fig. 9.6(a). The second consists of n-Maxwell units in parallel and is called the *generalized Maxwell* model, Fig. 9.6(b).

For these models the constitutive equations (Eq 9.8) in operator form are

Kelvin $$\gamma_{12} = \frac{t_{12}}{G_1 + \eta_1 \partial_t} + \frac{t_{12}}{G_2 + \eta_2 \partial_t} + \cdots + \frac{t_{12}}{G_n + \eta_n \partial_t}, \tag{9.15a}$$

Maxwell $$t_{12} = \frac{G_1 \dot{\gamma}_{12}}{\partial_t + \frac{1}{\tau_1}} + \frac{G_2 \dot{\gamma}_{12}}{\partial_t + \frac{1}{\tau_2}} + \cdots + \frac{G_n \dot{\gamma}_{12}}{\partial_t + \frac{1}{\tau_n}}. \tag{9.15b}$$

In these generalized model equations, one or more of the constants G_i and η_i may be assigned the values 0 or ∞ in order to represent behavior for a particular material. Thus, with η_2 and all of the constants which follow it in Eq 9.15a set equal to zero, the constitutive equation for the standard solid, Fig. 9.5(a), will be given.

9.3 Creep and Relaxation

Further insight into the viscoelastic nature of the material making up the cube shown in Fig. 9.1 is provided by two basic experiments, the creep test and the stress relaxation test. The creep test consists of instantaneously subjecting the material cube to a simple shear stress t_{12} of magnitude T_0 and maintaining that stress constant thereafter while measuring the shear strain as a function of time. The resulting strain $\gamma_{12}(t)$, is called the *creep*. In the *stress relaxation* test, an instantaneous shear strain γ_{12} of magnitude γ_0 is imposed on the cube and maintained at that value while the resulting stress, $t_{12}(t)$, is recorded as a function of time. The decrease in the stress values over the duration of the test is referred to as the stress relaxation. Expressing these test loadings mathematically is accomplished by use of the unit step function, $U(t - t_1)$, defined by the equation

$$U(t - t_1) = \begin{cases} 1 & t > t_1 \\ 0 & t \leqslant t_1 \end{cases} \tag{9.16}$$

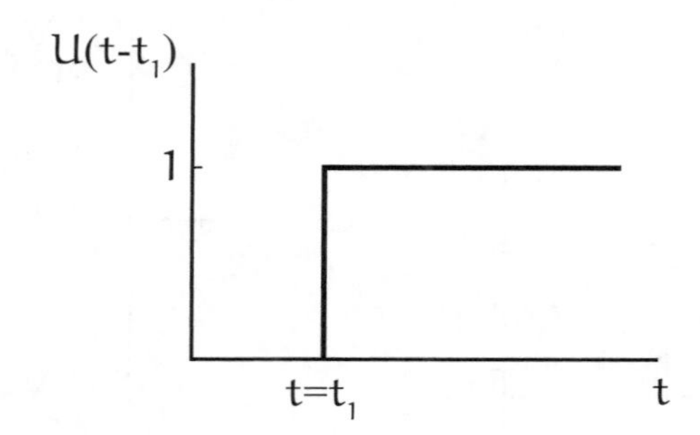

FIGURE 9.7
Graphic representation of the unit step function (often called the Heaviside step function).

which is shown by the diagram in Fig. 9.7. If the creep loading is applied at time $t = 0$, the stress is written as

$$t_{12} = T_0 U(t) . \tag{9.17}$$

By inserting this stress into the constitutive equation for a Kelvin material model, Eq 9.13, the resulting differential equation

$$T_0 U(t) = G\gamma_{12} + \eta\dot{\gamma}_{12} \tag{9.18}$$

may be integrated to yield the creep response

$$\gamma_{12}(t) = T_0 \left(1 - e^{-t/\tau}\right) \frac{U(t)}{G} \tag{9.19}$$

where e is the base of the natural logarithm system. It is interesting to note that as $t \to \infty$, the strain approaches a terminal value of T_0/G. Also, when $t = 0$, the strain rate $\dot{\gamma}_{12}$ equals T_0/η and if the creep were to continue at this rate it would reach its terminal value at time $t = \tau$. For this reason, τ is called the *retardation time*. Eq 9.19, as well as any creep response, may always be written in the general form

$$\gamma_{12}(t) = J(t)\, T_0 U(t) \tag{9.20}$$

in which $J(t)$ is called the *creep function*.[1] Thus, for the Kelvin solid, the creep function is seen to be

$$J(t) = \frac{1 - e^{-t/\tau}}{G} = J\left(1 - e^{-t/\tau}\right) \tag{9.21}$$

where the constant J, the reciprocal of the shear modulus G, is called the *shear compliance*. As a general rule, the creep function of any viscoelastic model is the sum of the creep functions of its series-connected units. Thus, for the standard linear solid of Fig. 9.5(a),

$$J(t) = J_1\left(1 - e^{-t/\tau_1}\right) + J_2 , \tag{9.22}$$

and for the generalized Kelvin model of Fig. 9.6(a)

$$J(t) = \sum_{i=1}^{n} J_i\left(1 - e^{-t/\tau_i}\right) . \tag{9.23}$$

[1]Note: The creep function, $J(t)$, is different from the Jacobian defined earlier.

With respect to creep loading for the Maxwell model, Eq 9.17 is substituted into the constitutive equation, Eq 9.14, resulting in the differential equation

$$\dot{\gamma}_{12} = T_0 \frac{\delta(t)}{G} + T_0 \frac{U(t)}{\eta} \tag{9.24}$$

where $\delta(t)$ is the *Dirac delta function*, the time derivative of the unit step function. In general,

$$\delta(t - t_1) = \frac{dU(t - t_1)}{dt}, \tag{9.25}$$

and is defined by the equations

$$\delta(t - t_1) = 0, \quad t \neq t_1, \tag{9.26a}$$

$$\int_{t_1^-}^{t_1^+} \delta(t - t_1)\, dt = 1, \tag{9.26b}$$

from which it may be shown that

$$\int_{-\infty}^{t} f(t')\,\delta(t' - t_1)\, dt' = f(t_1)\, U(t - t_1) \quad \text{for} \quad t > t_1 \tag{9.27}$$

for any continuous function, $f(t)$. Accordingly, Eq 9.24 integrates to yield the Maxwell creep response as

$$\gamma_{12} = T_0 J \left(1 + \frac{t}{\tau}\right) U(t) \tag{9.28}$$

from which the Maxwell creep function is

$$J(t) = J\left(1 + \frac{t}{\tau}\right). \tag{9.29}$$

Development of details relative to the stress relaxation test follows closely that of the creep test. With an imposed strain at time $t = 0$

$$\gamma_{12} = \gamma_0 U(t) \tag{9.30}$$

the resulting stress associated with Kelvin behavior is given directly by inserting $\dot{\gamma}_{12} = \gamma_0 \delta(t)$ into Eq 9.13 resulting in

$$t_{12} = \gamma_0 \left[G U(t) + \eta \delta(t)\right]. \tag{9.31}$$

The delta function in this equation indicates that it would require an infinite stress at time $t = 0$ to produce the instantaneous strain γ_0. For Maxwell behavior, when the instantaneous strain, Eq 9.30, is substituted into Eq 9.14, the stress relaxation is the solution to the differential equation

$$\dot{t}_{12} + \frac{1}{\tau} t_{12} = G \gamma_0 \delta(t) \tag{9.32}$$

which upon integration using Eq 9.27 yields

$$t_{12}(t) = \gamma_0 G e^{-t/\tau} U(t). \tag{9.33}$$

The initial time-rate of decay of this stress is seen to be $\gamma_0 G/\tau$, which if it were to continue would reduce the stress to zero at time $t = \tau$. Thus, τ is called the relaxation time for the Maxwell model.

Analogous to the creep function $J(t)$ associated with the creep test we define the stress relaxation function, $G(t)$, for any material by expressing $t_{12}(t)$ in its most general form

$$t_{12}(t) = G(t)\gamma_0 U(t) \tag{9.34}$$

From Eq 9.33, the stress relaxation function for the Maxwell model is

$$G(t) = Ge^{-t/\tau} \tag{9.35}$$

and for the generalized Maxwell model it is

$$G(t) = \sum_{i=1}^{n} G_i e^{-t/\tau_i} \,. \tag{9.36}$$

Creep compliance and relaxation modulus for several simple viscoelastic models are given in Table B.1. The differential equations governing the various models are also given in this table.

9.4 Superposition Principle, Hereditary Integrals

For linear viscoelasticity, the *principle of superposition* is valid just as in elasticity. In the context of stress/strain relationships under discussion here, the principle asserts that the total strain (stress) resulting from the application of a sequence of stresses (strains) is equal to the sum of the strains (stresses) caused by the individual stresses (strains). Thus, for the stepped stress history in simple shear displayed in Fig. 9.8(a) when applied to a material having a creep function $J(t)$, the resulting strain will be

$$\gamma_{12}(t) = \Delta T_0 J(t) + \Delta T_1 J(t - t_1) + \Delta T_2 J(t - t_2) = \sum_{i=0}^{2} \Delta T_i J(t - t_i) \;, \tag{9.37}$$

and by an obvious generalization to the arbitrary stress loading considered as an infinity of infinitesimal step loadings, Fig. 9.8(b), the strain is given by

$$\gamma_{12}(t) = \int_{-\infty}^{t} J(t - t') \left[\frac{dt_{12}(t')}{dt'}\right] dt' \tag{9.38}$$

which is called a *hereditary integral* since it expresses the strain at time t as a function of the entire stress history from time $t = -\infty$. If there is an initial discontinuity in the stress at time $t = 0$, and if the stress is zero up until that time (Fig. 9.9), the strain becomes

$$\gamma_{12}(t) = T_0 J(t) + \int_{0}^{t} J(t - t') \left[\frac{dt_{12}(t')}{dt'}\right] dt' \,. \tag{9.39}$$

Upon integrating the integral in this equation by parts and inserting the assigned limits of integration, the alternative form

$$\gamma_{12}(t) = J_0 t_{12}(t) + \int_{0}^{t} \left[t_{12}(t') \frac{dJ(t - t')}{d(t - t')}\right] dt' \tag{9.40}$$

is obtained where $J_0 = J(0)$.

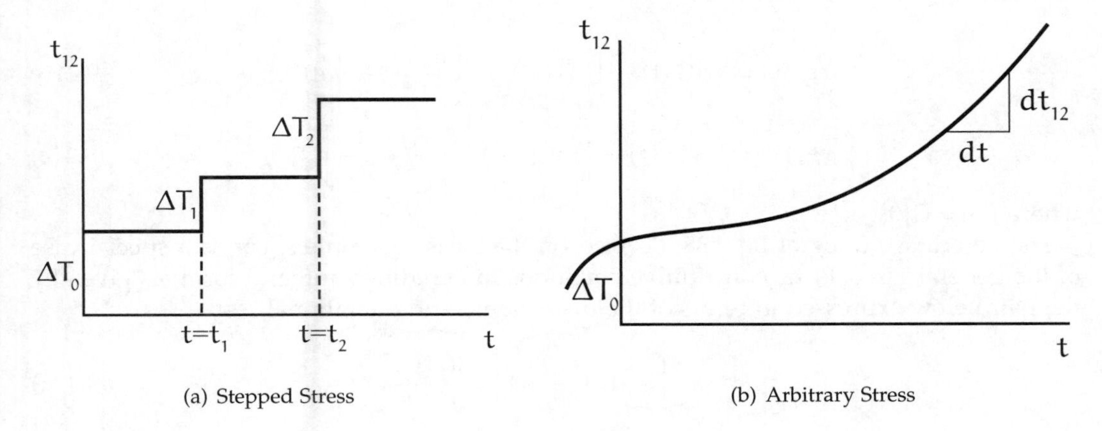

(a) Stepped Stress (b) Arbitrary Stress

FIGURE 9.8
Different types of applied stress histories.

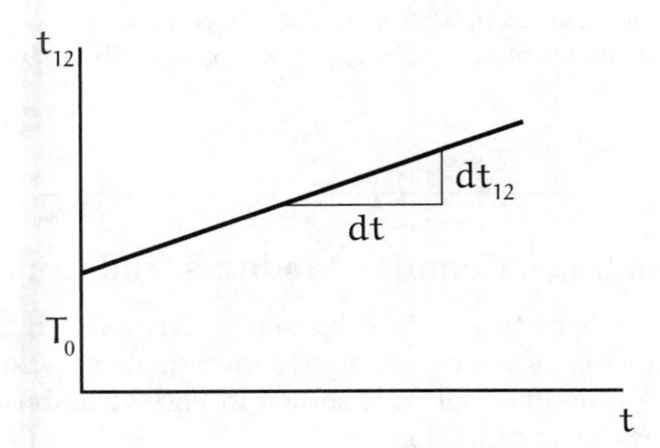

FIGURE 9.9
Stress history with an initial discontinuity.

In a completely analogous way, we may develop hereditary integrals expressing stress as the result of arbitrary strains. The basic forms are as follows:

$$t_{12}(t) = \int_{-\infty}^{t} G(t-t') \left[\frac{d\gamma_{12}(t')}{dt'}\right] dt' , \tag{9.41a}$$

$$t_{12}(t) = \gamma_0 G(t) + \int_{0}^{t} G(t-t') \left[\frac{d\gamma_{12}(t')}{dt'}\right] dt' , \tag{9.41b}$$

$$t_{12}(t) = G_0 \gamma_{12}(t) + \int_{0}^{t} \gamma_{12}(t') \left[\frac{dG(t-t')}{d(t-t')}\right] dt' , \tag{9.41c}$$

where $G_0 = G(0)$.

The hereditary integral Eq 9.38, derived on the basis of simple shear, is a special case of the general viscoelastic constitutive equations in hereditary integral form, as given by the pair below expressed in terms of the distortional and dilatational responses

$$2\eta_{ij}(t) = \int_{-\infty}^{t} J_S(t-t') \left[\frac{dS_{ij}(t')}{dt'}\right] dt' , \tag{9.42a}$$

$$3\epsilon_{kk}(t) = \int_{-\infty}^{t} J_V(t-t') \left[\frac{dt_{kk}(t')}{dt'}\right] dt' , \tag{9.42b}$$

where J_S is the shear compliance and J_V the volumetric compliance. Likewise, Eq 9.41a is the simple shear form of the general equations

$$S_{ij}(t) = \int_{-\infty}^{t} 2G_S(t-t') \left[\frac{d\eta_{ij}(t')}{dt'}\right] dt' , \tag{9.43a}$$

$$t_{kk}(t) = \int_{-\infty}^{t} 3G_V(t-t') \left[\frac{d\epsilon_{kk}(t')}{dt'}\right] dt' , \tag{9.43b}$$

where G_S is the relaxation modulus in shear and G_V the relaxation modulus in dilatation. Because we have assumed elastic behavior in dilatation, $J_V = 1/G_V = K$ and both Eq 9.42b and Eq 9.43b reduce to the form $t_{kk} = 3K\varepsilon_{kk}$ in keeping with Eq 9.7b.

9.5 Harmonic Loadings, Complex Modulus, and Complex Compliance

The behavior of viscoelastic bodies when subjected to harmonic stress or strain is another important part of the theory of viscoelasticity. To investigate this aspect of the theory, we consider the response of the material cube shown in Fig. 9.1 under an applied harmonic shear strain of frequency ω as expressed by

$$\gamma_{12}(t) = \gamma_0 \sin \omega t , \tag{9.44a}$$

or by

$$\gamma_{12}(t) = \gamma_0 \cos \omega t . \tag{9.44b}$$

Mathematically, it is advantageous to combine these two by assuming the strain in the complex form

$$\gamma_{12}(t) = \gamma_0 (\cos \omega t + i \sin \omega t) = \gamma_0 e^{i\omega t} \tag{9.45}$$

where $i = \sqrt{-1}$. It is understood that physically the real part of the resulting stress corresponds to the real part of the applied strain, and likewise, the imaginary parts of each are directly related. The stress resulting from the excitation prescribed by Eq 9.45 will have the same frequency, ω, as the imposed strain. Therefore, expressing the stress as

$$t_{12}(t) = T^* e^{i\omega t} \tag{9.46}$$

where T^* is complex, the response will consist of two parts: a steady-state response which will be a function of the frequency ω, and a transient response that decays exponentially with time. It is solely with the steady-state response that we concern ourselves in the remainder of this section.

Substituting Eq 9.45 and Eq 9.46 into the fundamental viscoelastic constitutive equation given by Eq 9.8, and keeping in mind the form of the operators $\{P\}$ and $\{Q\}$ as listed by Eq 5.178 we obtain

$$\sum_{k=0}^{N} T^* p_k (i\omega)^k e^{i\omega t} = \sum_{k=0}^{M} \gamma_0 q_k (i\omega)^k e^{i\omega t} . \tag{9.47}$$

Canceling the common factor $e^{i\omega t}$ we solve for the ratio

$$\frac{T^*}{\gamma_0} = \frac{\sum_{k=0}^{N} q_k (i\omega)^k}{\sum_{k=0}^{M} p_k (i\omega)^k} \tag{9.48}$$

which we define as the complex modulus, $G^*(i\omega)$, and write it in the form

$$\frac{T^*}{\gamma_0} = G^*(i\omega) = G'(i\omega) + iG''(i\omega) . \tag{9.49}$$

The real part, $G'(i\omega)$, of this modulus is associated with the amount of energy stored in the cube during a complete loading cycle and is called the *storage modulus*. The imaginary part, $G''(i\omega)$, relates to the energy dissipated per cycle and is called the *loss modulus*.

In terms of the complex modulus, the stress t_{12} as assumed in Eq 9.46 may now be written

$$t_{12} = G^*(i\omega)\gamma_0 e^{i\omega t} = [G'(\omega) + iG''(\omega)]\gamma_0 e^{i\omega t} , \tag{9.50}$$

and by defining the absolute modulus, $\tilde{G}(\omega)$ as the magnitude of $G^*(i\omega)$ according to

$$\tilde{G}(\omega) = \sqrt{[G'(\omega)]^2 + [G''(\omega)]^2} \tag{9.51}$$

together with the loss angle, δ between $\tilde{G}(\omega)$ and $G'(\omega)$ as given by its tangent

$$\tan\delta = \frac{G''(\omega)}{G'(\omega)} \tag{9.52}$$

the stress t_{12} (Eq 9.50), may now be written

$$t_{12} = \tilde{G}(\omega) e^{i\delta}\gamma_0 e^{i\omega t} = \tilde{G}(\omega)\gamma_0 e^{i(\omega t+\delta)} . \tag{9.53}$$

From this equation we see that the peak value of the stress is

$$T_0 = \tilde{G}(\omega)\gamma_0 , \tag{9.54}$$

and that the strain lags behind the stress by the loss angle δ. Figure 9.10 provides a graphical interpretation of this phenomenon. In Fig. 9.10(a), the two constant magnitude

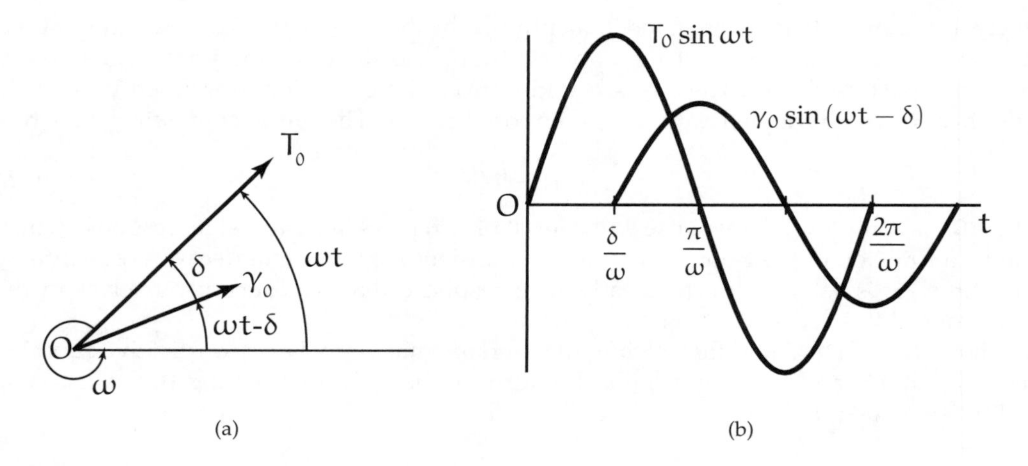

FIGURE 9.10
Different types of applied stress histories.

stress and strain vectors, separated by the constant angle δ, rotate about a fixed origin with a constant angular velocity ω. The vertical projections of these vectors, representing the physical values of the stress and strain, are plotted against time in Fig. 9.10(b). From Fig. 9.10(a), the portion of the stress in phase with the strain is $T_0 \cos\delta$ and by Eq 9.54 together with Eq 9.52, the storage modulus may be expressed as

$$G' = \frac{T_0 \cos\delta}{\gamma_0} . \tag{9.55a}$$

Similarly, the loss modulus is written as

$$G'' = \frac{T_0 \sin\delta}{\gamma_0} \tag{9.55b}$$

Consistent with the duality present in all of viscoelastic theory we reverse the roles of stress and strain in the preceding portion of this section to define the complex compliance, $J^*(i\omega)$ along with its associated real and imaginary parts. Briefly, we assume an applied stress

$$t_{12} = T_0 e^{i\omega t} \tag{9.56}$$

together with the resulting strain

$$\gamma_{12} = \gamma^* e^{i\omega t} \tag{9.57}$$

which when substituted into Eq 9.8 leads to

$$\frac{\gamma^*}{T_0} = J^*(i\omega) = J'(\omega) - iJ''(\omega) \tag{9.58}$$

where the minus sign reflects the fact that the strain lags the stress by the loss angle δ, defined in this case by

$$\tan\delta = \frac{J''(\omega)}{J'(\omega)} . \tag{9.59}$$

In analogy with the complex modulus components, $J'(\omega)$ is called the *storage compliance*, $J''(\omega)$ the *loss compliance*, and

$$\tilde{J}(\omega) = \sqrt{[J'(\omega)]^2 + [J''(\omega)]^2} = \frac{\gamma_0}{T_0} \tag{9.60}$$

the *absolute compliance,* in which γ_0 is the peak value of the strain as given by

$$\gamma_0 = \tilde{J}(\omega)\, T_0 \tag{9.61}$$

Based upon the definitions of G^* and J^*, it is clear that these complex quantities are reciprocals of one another. Thus

$$G^*(i\omega)\, J^*(i\omega) = 1\,. \tag{9.62}$$

A simple procedure for calculating G^* (or J^*) of a specific model or material is to replace the partial differential operator ∂_t in the material's constitutive equation by $i\omega$ and solve the resulting algebraic equation for the ratio $t_{12}/\gamma_{12} = G^*$, (or $\gamma_{12}/t_{12} = J^*$ if that is the quantity required). Accordingly, from Eq 9.13 for the Kelvin solid, the constitutive equation $t_{12} = \{G + \eta\partial_t\}\gamma_{12}$ becomes $t_{12} = (G + i\eta\omega)\gamma_{12}$ which yields $t_{12}/\gamma_{12} = G(1 + i\tau\omega) = G^*$. Likewise, from Eq 9.14 for the Maxwell fluid, $\{\partial_t + 1/\tau\}t_{12} = \{G\partial_t\}\gamma_{12}$ becomes $(i\omega+1/\tau)t_{12} = (Gi\omega)\gamma_{12}$ from which $t_{12}/\gamma_{12} = G(\tau^2\omega^2+i\tau\omega)/(1+\tau^2\omega^2) = G^*$.

In developing the formulas for the complex modulus and the complex compliance we have used the differential operator form of the fundamental viscoelastic constitutive equations. Equivalent expressions for these complex quantities may also be derived using the hereditary integral form of constitutive equations. To this end we substitute Eq 9.45 into Eq 9.41a. However, before making this substitution it is necessary to decompose the stress relaxation function $G(t)$ into two parts as follows,

$$G(t) = G_0[1 - \phi(t)] \tag{9.63}$$

where $G_0 = G(0)$, the value of $G(t)$ at time $t = 0$. Following this decomposition and the indicated substitution, Eq 9.41a becomes

$$t_{12}(t) = i\omega\gamma_0 G_0 \int_{-\infty}^{t} e^{i\omega t'}\,[1 - \phi(t - t')]\,dt'\,. \tag{9.64}$$

In this equation, let $t - t' = \xi$ so that $dt' = -d\xi$ and such that when $t' = t$, $\xi = 0$, and when $t' = -\infty$, $\xi = \infty$. Now

$$t_{12}(t) = i\omega\gamma_0 G_0 \left\{ \frac{e^{i\omega t}}{i\omega} - \int_0^\infty e^{i\omega\xi} e^{i\omega t}\phi(\xi)\,d\xi \right\} \tag{9.65}$$

which reduces to

$$t_{12}(t) = \gamma_0 \left\{ G_0 - G_0 \int_0^\infty (i\omega\cos\omega\xi + \omega\sin\omega\xi)\,\phi(\xi)\,d\xi \right\} e^{i\omega t}\,. \tag{9.66}$$

But by Eq 9.50, $t_{12}(t) = \gamma_0[G'(\omega) + iG''(\omega)]e^{i\omega t}$ so that from Eq 9.66

$$G'(\omega) = G_0 - \omega G_0 \int_0^\infty \sin\omega\xi\,\phi(\xi)\,d\xi\,, \tag{9.67a}$$

$$G''(\omega) = -\omega G_0 \int_0^\infty \cos\omega\xi\,\phi(\xi)\,d\xi\,. \tag{9.67b}$$

For a Kelvin material, $G(t) = G[1 + \tau\delta(t)]$ so that $G_0 = G$ and $\phi(t) = -\tau\delta(t)$. Thus, for a Kelvin solid, Eqs 9.67 yield

$$G'(\omega) = G - G\omega \int_0^\infty -\tau\delta(\xi)\sin\omega\xi d\xi = G\,,$$

$$G''(\omega) = -G\omega \int_0^\infty -\tau\delta(\xi)\cos\omega\xi d\xi = G\omega\tau = \omega\eta \; .$$

For a Maxwell material, $G(t) = G[1 - (1 - e^{-t/\tau})]$ so that $G_0 = G$ and $\phi(t) = 1 - e^{-t/\tau}$. Thus, for a Maxwell fluid, Eqs 9.67 yield

$$G'(\omega) = G - G\omega \int_0^\infty \left(1 - e^{-\xi/\tau}\right)\sin\omega\xi d\xi$$
$$= G - G\omega\left[\frac{1}{\omega} - \frac{\omega\tau^2}{1+\tau^2\omega^2}\right] = \frac{G\omega^2\tau^2}{1+\tau^2\omega^2} \, ,$$

$$G''(\omega) = -G\omega \int_0^\infty \left(1 - e^{-\xi/\tau}\right)\cos\omega\xi d\xi = \frac{G\omega\tau}{1+\tau^2\omega^2} \; .$$

These values for G' and G'' for the Kelvin and Maxwell models agree with those calculated in the previous paragraph.

In a completely analogous fashion, if we adopt the hereditary form Eq 9.38 as the constitutive equation of choice, and substitute into that equation $t_{12} = T_0 e^{i\omega t}$, together with the decomposition of $J(t)$ in the form $J(t) = J_0[1 + \psi(t)]$, we obtain

$$\gamma_{12}(t) = \int_{-\infty}^{t} J_0\left[1 - \psi\left(t - t'\right)\right] i\omega T_0 e^{i\omega t'} dt' \; . \tag{9.68}$$

Upon making the same change in variable of integration, $t - t' = \xi$, this equation becomes

$$\gamma_{12}(t) = T_0\left[J_0 + J_0\omega\int_0^\infty \sin\omega\xi\,\psi(\xi)\,d\xi + iJ_0\omega\int_0^\infty \cos\omega\xi\,\psi(\xi)\,d\xi\right] e^{i\omega t} \tag{9.69}$$

from which we extract

$$J'(\omega) = J_0 + J_0\omega\int_0^\infty \sin\omega\xi\,\psi(\xi)\,d\xi \; , \tag{9.70a}$$

$$J''(\omega) = -J_0\omega\int_0^\infty \cos\omega\xi\,\psi(\xi)\,d\xi \; . \tag{9.70b}$$

These expressions may be specialized to obtain $J'(\omega)$ and $J''(\omega)$ for any particular model for which the creep function $J(t)$ is known.

9.6 Three-Dimensional Problems, The Correspondence Principle

The fundamental viscoelastic constitutive equations in differential operator form as expressed by Eqs 9.6 distinguish between the distortional response (a change in shape at constant volume due to the deviatoric portion of the applied stress, Eq 9.6a), and the dilatational response (a change in volume without a change in shape due to the spherical portion of the applied stress, Eq 9.6b). Extensive experimental evidence indicates that practically all materials of engineering importance behave elastically in the dilatational mode, and for this reason the reduced form of Eq 9.6 as given by Eq 9.7 is used in this book. In expanded component notation these equations appear as

$$\{P\}\left(t_{11} - \tfrac{1}{3}t_{ii}\right) = 2\{Q\}\left(\epsilon_{11} - \tfrac{1}{3}\epsilon_{ii}\right) \; , \tag{9.71a}$$

$$\{P\}\left(t_{22} - \tfrac{1}{3}t_{ii}\right) = 2\{Q\}\left(\epsilon_{22} - \tfrac{1}{3}\epsilon_{ii}\right) , \quad (9.71b)$$

$$\{P\}\left(t_{33} - \tfrac{1}{3}t_{ii}\right) = 2\{Q\}\left(\epsilon_{33} - \tfrac{1}{3}\epsilon_{ii}\right) , \quad (9.71c)$$

$$\{P\}t_{12} = 2\{Q\}\epsilon_{12} , \quad (9.71d)$$

$$\{P\}t_{23} = 2\{Q\}\epsilon_{23} , \quad (9.71e)$$

$$\{P\}t_{31} = 2\{Q\}\epsilon_{31} , \quad (9.71f)$$

$$t_{ii} = 3K\epsilon_{ii} . \quad (9.71g)$$

Depending upon the particular state of applied stress, some of Eq 9.71 may be satisfied identically. For example, the state of simple shear in the x_1-x_2 plane, introduced earlier to develop the basic concepts of viscoelastic behavior, results in Eq 9.71 being reduced to a single equation, Eq 9.71d as expressed by Eq 9.8. Similarly, for a hydrostatic state of stress with $t_{11} = t_{22} = t_{33} = p_0$, and $t_{12} = t_{23} = t_{31} = 0$ (or for a uniform triaxial tension having $t_{11} = t_{22} = t_{33} = T_0$, with $t_{12} = t_{23} = t_{31} = 0$), the behavior is simply elastic. On the other hand, for a simple one-dimensional tension or compression in one of the coordinate directions, several of Eq 9.71 enter into the analysis as discussed in the following paragraph.

Let an instantaneously applied constant stress T_0 be imposed uniformly in the x_1 direction on a member having a constant cross section perpendicular to that direction. Thus, let $t_{11} = T_0U(t)$ with all other components zero which results in $t_{ii} = T_0U(t)$ so that from Eq 9.71g, $\epsilon_{ii} = T_0U(t)/3K$. For this situation the first three of Eq 9.71 become

$$\{P\}\left(T_0U(t) - \tfrac{1}{3}T_0U(t)\right) = 2\{Q\}\left(\epsilon_{11} - \frac{T_0U(t)}{9K}\right) , \quad (9.72a)$$

$$\{P\}\left(-\tfrac{1}{3}T_0U(t)\right) = 2\{Q\}\left(\epsilon_{22} - \frac{T_0U(t)}{9K}\right) , \quad (9.72b)$$

$$\{P\}\left(-\tfrac{1}{3}T_0U(t)\right) = 2\{Q\}\left(\epsilon_{33} - \frac{T_0U(t)}{9K}\right) , \quad (9.72c)$$

and the next three of the set indicate zero shear strains. Clearly, from Eq 9.72b and Eq 9.72c we see that $\epsilon_{22} = \epsilon_{33}$. Furthermore, Eq 9.72a may be solved directly for ϵ_{11} to yield

$$\epsilon_{11} = T_0U(t)\frac{3K\{P\} + \{Q\}}{9K\{Q\}} = \frac{T_0U(t)}{\{E\}} \quad (9.73)$$

where $\{E\} = 9K\{Q\}/(3K\{P\} + \{Q\})$ is the operator form of E, Young's modulus. Similarly, from Eq 9.72 using $\epsilon_{22} = \epsilon_{33}$ we find that

$$\frac{\epsilon_{22}}{\epsilon_{11}} = \frac{\epsilon_{33}}{\epsilon_{11}} = -\frac{3K\{P\} - 2\{Q\}}{6K\{P\} + 2\{Q\}} = -\{\nu\} \quad (9.74)$$

which designates the operator form of Poisson's ratio, ν.

In order to compute a detailed solution for a particular material we need the specific form of the operators $\{P\}$ and $\{Q\}$ for that material as illustrated by the following example.

Example 9.1

Let the stress $t_{11} = T_0U(t)$ be applied uniformly to a bar of constant cross section made of a Kelvin material and situated along the x_1 axis. Determine ϵ_{11} and ϵ_{22} as functions of time.

Solution

From the constitutive relation for a Kelvin material, Eq 9.13, we note that $\{P\} = 1$ and $\{Q\} = \{G + \eta\partial_t\}$, which when inserted into Eq 9.72a results (after some algebraic manipulations) in the differential equation

$$\dot{\epsilon}_{11} + \frac{\epsilon_{11}}{\tau} = T_0 U(t)\left[\frac{3K+G}{9KG}\right] + \frac{T_0\delta(t)}{9K}\ . \tag{9.75}$$

This differential equation may be solved by standard procedures to yield the solution

$$\epsilon_{11}(t) = T_0 U(t)\left\{\frac{3K+G}{9KG}\left(1-e^{-t/\tau}\right) + \frac{e^{-t/\tau}}{9K}\right\}\ . \tag{9.76}$$

When $t = 0$, $\epsilon_{11} = T_0 U(t)/9K$ which is the result of elastic behavior in bulk. As $t \to \infty$, $\epsilon_{11} \to T_0[(3K+G)/9KG] = T_0/E$, the terminal elastic response.

From Eq 9.72b the governing differential equation for determining ϵ_{22} is (when expressed in its standard form)

$$\dot{\epsilon}_{22} + \frac{\epsilon_{22}}{\tau} = T_0 U(t)\left[\frac{G}{9K} - \frac{1}{6}\right]\frac{1}{\eta} + \frac{T_0\delta(t)}{9K} \tag{9.77}$$

which upon integration and simplification yields

$$\epsilon_{22}(t) = -T_0 U(t)\frac{3K-2G}{18KG}\left(1-e^{-t/\tau}\right) - T_0 U(t)\frac{e^{-t/\tau}}{9K}\ . \tag{9.78}$$

When $t = 0$, $\epsilon_{22} = T_0/9K$ which, due to the elastic dilatation effect, is identical with the initial value of ϵ_{11}. As $t \to \infty$, $\epsilon_{22} \to (2G-3K)/18KG$.

Up until now in this section we have discussed three-dimensional problems from the point of view of constitutive equations in differential operator form, but our analysis can be developed equally well on the basis of the hereditary integral form of constitutive equations as given by Eq 9.42 or Eq 9.43. With respect to the uniaxial stress loading analyzed above, Eq 9.42a (assuming elastic behavior in dilatation with $t_{kk} = 3K\epsilon_{kk}$ and $\epsilon_{ii} = T_0 U(t)/3K$, along with zero stress at time $t = 0$) results in the equations

$$2\left[\epsilon_{11} - \frac{T_0 U(t)}{9K}\right] = \int_0^t \left(T_0 - \frac{T_0}{3}\right)\delta(t')\, J_S(t-t')\, dt'\ , \tag{9.79a}$$

$$2\left[\epsilon_{22} - \frac{T_0 U(t)}{9K}\right] = \int_0^t \left(-\frac{T_0}{3}\right)\delta(t')\, J_S(t-t')\, dt'\ , \tag{9.79b}$$

$$2\left[\epsilon_{33} - \frac{T_0 U(t)}{9K}\right] = \int_0^t \left(-\frac{T_0}{3}\right)\delta(t')\, J_S(t-t')\, dt'\ . \tag{9.79c}$$

From these equations it is again apparent that $\epsilon_{22} = \epsilon_{33}$, and that in order to develop the solution details for a particular material we need the expression for J_S, the shear compliance of that material as shown by the example that follows.

Example 9.2

Develop the solution for the problem of Example 9.7-1 using the hereditary integral form of constitutive equations for a bar that is Kelvin in distortion, elastic in dilatation.

Solution

For a Kelvin material, the shear (creep) compliance, Eq 9.21, is $J_S = (1 - e^{-t/\tau})/G$ so that Eq 9.79a becomes for $t_{11} = T_0 U(t)$

$$2\left[\epsilon_{11} - \frac{T_0 U(t)}{9K}\right] = \int_0^t \frac{2T_0 \delta(t')}{3G}\left(1 - e^{-\frac{t-t'}{\tau}}\right) dt' \tag{9.80}$$

which may be integrated directly using Eq 9.27 to yield

$$\epsilon_{11}(t) = T_0 U(t)\left[\frac{1 - e^{-\frac{t}{\tau}}}{3G} + \frac{1}{9K}\right], \tag{9.81}$$

or by a simple rearrangement

$$\epsilon_{11}(t) = T_0 U(t)\left[\frac{3K + G}{9KG} - \frac{e^{-\frac{t}{\tau}}}{3G}\right]$$

in agreement with Eq 9.76. Likewise, from Eq 9.79b we obtain for this loading

$$2\left[\epsilon_{22} - \frac{T_0 U(t)}{9K}\right] = \int_0^t -\frac{T_0 \delta(t')}{3G}\left(1 - e^{-\frac{t-t'}{\tau}}\right) dt' \tag{9.82}$$

which also integrates directly using Eq 9.27 to yield

$$\epsilon_{22}(t) = -T_0 U(t)\left[\frac{1 - e^{-\frac{t}{\tau}}}{6G} + \frac{1}{9K}\right] \tag{9.83}$$

in agreement with Eq 9.78.

The number of problems in viscoelasticity that may be solved by direct integration as in the examples above is certainly quite limited. For situations involving more general stress fields, or for bodies of a more complicated geometry, the *correspondence principle* may be used to advantage. This approach rests upon the analogy between the basic equations of an associated problem in elasticity and those of the Laplace transforms of the fundamental equations of the viscoelastic problem under consideration. For the case of quasi-static viscoelastic problems in which inertia forces due to displacements may be neglected, and for which the elastic and viscoelastic bodies have the same geometry, the correspondence method is relatively straightforward as described below. In an elastic body under constant load, the stresses, strains, and displacements are independent of time, whereas in the associated viscoelastic problem, even though the loading is constant or a slowly varying function of time, the governing equations are time dependent and may be subjected to the Laplace transformation. By definition, the Laplace transform of an arbitrary continuous time-dependent function, say the stress $t_{ij}(\mathbf{x}, t)$ for example, is given by

$$\bar{t}_{ij}(\mathbf{x}, s) = \int_0^t t_{ij}(\mathbf{x}, t)\, e^{-st} dt \tag{9.84}$$

in which s is the transform variable, and barred quantities indicate transform. Standard textbooks on the Laplace transform list tables giving the transforms of a wide variety of time-dependent functions. Of primary importance to the following discussion, the Laplace transforms of the derivatives of a given function are essential. Thus, for the first two derivatives of stress

$$\int_0^t \frac{dt_{ij}(\mathbf{x}, t)}{dt} e^{-st} dt = -t_{ij}(0) + s\bar{t}_{ij}(\mathbf{x}, s)\ , \tag{9.85a}$$

$$\int_0^t \frac{d^2 t_{ij}(\mathbf{x}, t)}{dt^2} e^{-st} dt = -t_{ij}(0) - st_{ij}(0) + s^2\bar{t}_{ij}(\mathbf{x}, s)\ , \tag{9.85b}$$

and so on for higher derivatives. Note that in these equations the values of the stress and its derivatives at time $t = 0$ become part of the transform so that initial conditions are built into the solution.

We may now examine the details of the correspondence method by considering the correspondence between the basic elasticity equations (as listed in Chapter 6 and repeated here)

Equilibrium	(Eq 6.43)	$t_{ij,j}(\mathbf{x}) + \rho b_i(\mathbf{x}) = 0\ ,$
Strain – displacement	(Eq 6.44)	$2\epsilon_{ij}(\mathbf{x}) = u_{i,j}(\mathbf{x}) + u_{j,i}(\mathbf{x})\ ,$
Constitutive relations	(Eq 6.33a)	$S_{ij}(\mathbf{x}) = 2G\eta_{ij}(\mathbf{x})\ ,$
	(Eq 6.33b)	$t_{ii}(\mathbf{x}) = 3K\epsilon_{ii}(\mathbf{x})\ ,$

and the Laplace transforms of the associated time-dependent viscoelastic equations

Equilibrium	$\bar{t}_{ij,j}(\mathbf{x}, s) + \rho\bar{b}_i(\mathbf{x}, s) = 0\ ,$	(9.86a)
Strain – displacement	$2\bar{\epsilon}_{ij}(\mathbf{x}, s) = \bar{u}_{i,j}(\mathbf{x}, s) + \bar{u}_{j,i}(\mathbf{x}, s)\ ,$	(9.86b)
Constitutive relations	$\bar{P}(s)\,\bar{S}_{ij}(\mathbf{x}, s) = 2\bar{Q}(s)\,\bar{\eta}_{ij}(\mathbf{x}, s)\ ,$	(9.86c)
	$\bar{t}_{ii}(\mathbf{x}, s) = 3K\bar{\epsilon}_{ii}(\mathbf{x}, s)\ ,$	(9.86d)

where barred quantities are transforms, and in which $\bar{P}(s)$ and $\bar{Q}(s)$ are polynomials in the transform variable s in accordance with Eq 9.85. A comparison of Eqs 9.86, which are algebraic, time-independent relations, with the elasticity equations above indicates a complete analogy between the barred and unbarred entities if we assign the equivalency of the ratio $\bar{Q}(s)/\bar{P}(s)$ to the shear modulus G. This allows us to state the correspondence principle as follows:

> If the solution of a problem in elasticity is known, the Laplace transform of the solution of the associated viscoelastic problem is constructed by substituting the quotient $\bar{Q}(s)/\bar{P}(s)$ of the transformed operator polynomials in place of the shear modulus G, and the actual time-dependent loads by their Laplace transforms.

Since many elasticity solutions are written in terms of Young's modulus, E and Poisson's ratio, ν, it is useful to extract from Eqs 9.73 and 9.74 the transform replacements for these constants which are

$$\bar{E}(s) \rightarrow \frac{9K\bar{Q}}{3K\bar{P} + \bar{Q}}\ , \tag{9.87}$$

$$\bar{\nu}(s) \rightarrow \frac{3K\bar{P} - 2\bar{Q}}{6K\bar{P} + 2\bar{Q}} . \tag{9.88}$$

As an illustration of how the correspondence method works, we consider the following problem.

Example 9.3

The radial stress t_{rr} in an elastic half-space under the action of a concentrated constant force F_0 acting at the origin as shown in the figure below is

$$t_{rr}(r, z) = \frac{F_0}{2\pi}\left[(1 - 2\nu) A(r, z) - B(r, z)\right] .$$

where $A(r, z)$ and $B(r, z)$ are known functions of the coordinates. Determine the time-dependent viscoelastic stress $t_{rr}(r, z, t)$ for a half-space that is Kelvin in shear and elastic in dilatation if the force at the origin is given by the step loading $F(t) = F_0 U(t)$.

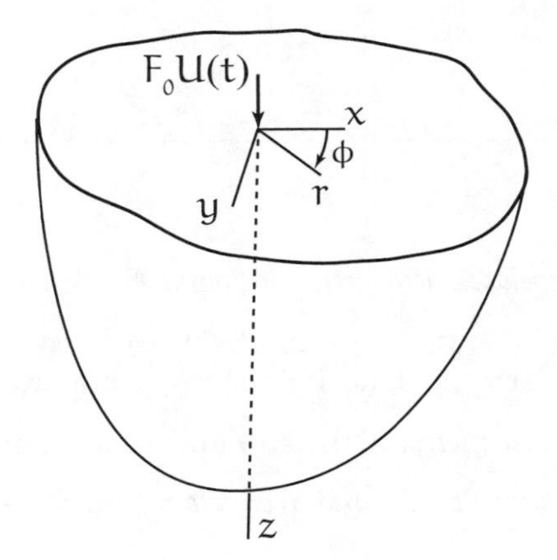

Solution

From Eq 9.88 for $\bar{\nu}(s)$ we may directly calculate the Laplace transform of the expression for $1 - 2\nu$ appearing in the elastic solution as $3\bar{Q}/\left(3K\bar{P} + \bar{Q}\right)$. The Laplace transform of the load function $F_0 U(t)$ is given by F_0/s. Thus, the Laplace transform of the associated viscoelastic solution is

$$\bar{t}_{rr}(r, z, s) = \frac{F_0}{2\pi s}\left[\frac{3\bar{Q}}{3K\bar{P} + \bar{Q}} A(r, z) - B(r, z)\right] .$$

From the Kelvin model constitutive equation, Eq 9.13, we have $\{Q\}/\{P\} = \{G + \eta\partial_t\}$ for which $Q(s)/P(s) = G + \eta s$ so that

$$\bar{t}_{rr}(r, z, s) = \frac{F_0}{2\pi s}\left[\frac{3(G + \eta s)}{3K + G + \eta} A(r, z) - B(r, z)\right] .$$

This expression may be inverted with the help of a table of transforms from any standard text on Laplace transforms to give the time-dependent viscoelastic solution

$$t_{rr}(r,z,t) = \frac{F_0}{2\pi}\left\{\left[\frac{\frac{3G}{3K+G} + 9Ke^{-\frac{(3K+G)t}{\eta}}}{3K+G}\right] A(r,z) - B(r,z)\right\} .$$

Notice that when $t = 0$

$$t_{rr}(r,z,0) = \frac{F_0}{2\pi}\left[3A(r,z) - B(r,z)\right] ,$$

and as $t \to \infty$

$$t_{rr}(r,z,t \to \infty) = \frac{F_0}{2\pi}\left[\frac{3G}{3K+G} A(r,z) - B(r,z)\right]$$

which is the elastic solution.

References

[1] Ferry, J. D. (1961), *Viscoelastic Properties of Polymers*, Wiley and Sons, New York

[2] Findley, W. N., Lai, J. S., and Onanran, O. (1976), *Creep and Relaxation of Nonlinear Viscoelastic Materials*, North-Holland Publishing Company, London

[3] Flugge, W. (1967), *Viscoelasticity*, Blaisdell Publishing Company, Waltham, MA

[4] Fried, J.R. (1995), *Polymer Science and Technology*, Prentice Hall PTR, Upper Saddle River, NJ

[5] McCrum, N. G., Buckley, C. P., and Bucknall, C. B. (1997), *Principles of Polymer Engineering*, Second Edition, Oxford University Press, New York

[6] Pipkin, A. C. (1972), *Lectures on Viscoelasticity Theory*, Springer-Verlag, New York

Problems

Problem 9.1

By substituting $S_{ij} = t_{ij} - \frac{1}{3}\delta_{ij}t_{kk}$ and $\eta_{ij} = \epsilon_{ij} - \frac{1}{3}\delta_{ij}\epsilon_{kk}$ into Eq 9.7 and combining those two equations, determine expressions in operator form for

(a) the Lamé constant, λ,

(b) Young's modulus, E,

(c) Poisson's ratio, ν.

Answer

(a) $\{\lambda\} = K - 2\{Q\}/3\{P\}$

(b) $\{E\} = 9K\{Q\}/(3K\{P\} + \{Q\})$

(c) $\{\nu\} = (3K\{P\} - 2\{Q\})/(6K\{P\} + 2\{Q\})$

Problem 9.2

Compliances are reciprocals of moduli. Thus, in elasticity theory $D = 1/E$, $J = 1/G$, and $B = 1/K$. Show from the stress-strain equations of a simple one-dimensional tension that

$$D = \frac{1}{3}J + \frac{1}{9}B\ .$$

Problem 9.3

The four-parameter model shown consists of a Kelvin unit in series with a Maxwell unit. Knowing that $\gamma_{MODEL} = \gamma_{KELVIN} + \gamma_{MAXWELL}$, together with the operator equations Eqs 9.13 and 9.14, determine the constitutive equation for this model.

Answer

$$G_2\eta_1\ddot{\gamma} + G_1G_2\dot{\gamma} = \eta_1\ddot{T} + (G_1 + G_2 + \eta_1/\tau_2)\dot{T} + (G_1/\tau_2)T$$

Problem 9.4

Develop the constitutive equations for the three-parameter models shown.

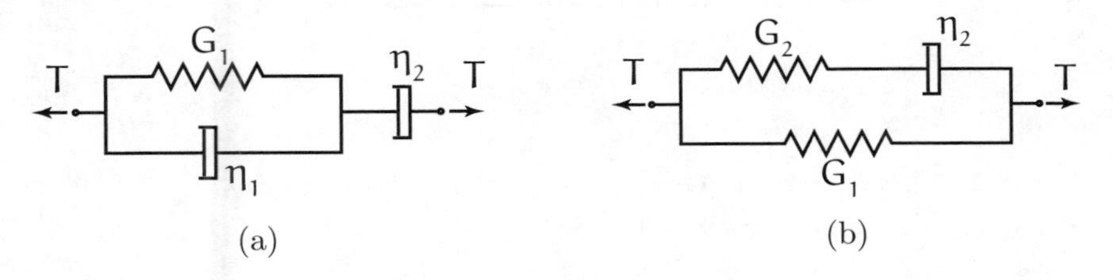

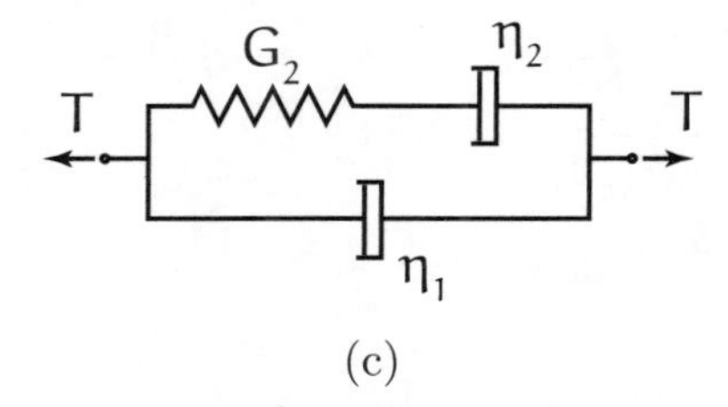

(c)

Answer

(a) $\ddot{\gamma} + \dot{\gamma}/\tau_1 = [(\eta_1 + \eta_2)/\eta_1\eta_2]\dot{T} + (1/\tau_1\eta_2)T$
(b) $\dot{T} + T/\tau_2 = (G_2 + G_1)\dot{\gamma} + (G_1/\tau_2)\gamma$
(c) $\dot{T} + T/\tau_2 = \eta_1\ddot{\gamma} + (G_2 + \eta_1/\tau_2)\dot{\gamma}$

Problem 9.5
A proposed model consists of a Kelvin unit in parallel with a Maxwell unit. Determine the constitutive equation for this model.

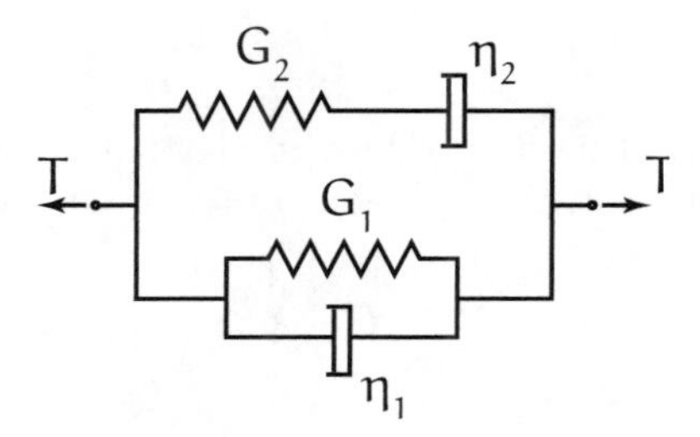

Answer

$$\dot{T} + T/\tau_2 = \eta_1\ddot{\gamma} + (G_1 + G_2 + \eta_1/\tau_2)\dot{\gamma} + (G_1/\tau_2)\gamma$$

Problem 9.6
For the four-parameter model shown, determine

(a) the constitutive equation,
(b) the relaxation function, $G(t)$. [Note that $G(t)$ is the sum of the $G(t)$'s of the parallel joined units.]

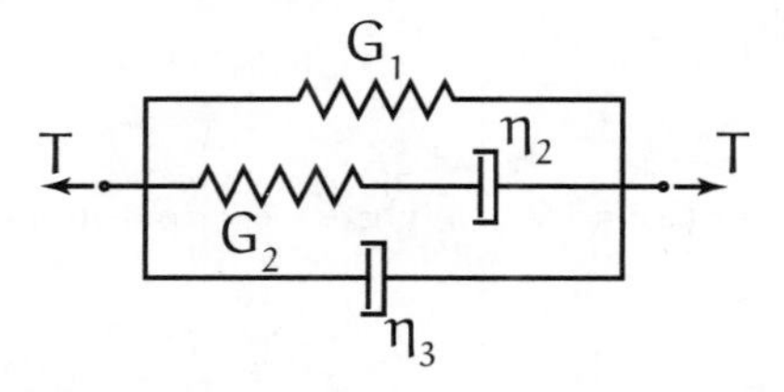

Answer

(a) $\dot{T} + T/\tau_2 = \eta_3\ddot{\gamma} + (G_1 + G_2 + \eta_3/\tau_2)\dot{\gamma} + (G_1/\tau_2)\gamma$
(b) $G(t) = G_1 + G_2 e^{-t/\tau_2} + \eta_3\delta(t)$

Problem 9.7

For the model shown the stress history is given by the accompanying diagram. Determine the strain $\gamma(t)$ for this loading during the intervals

(a) $0 \leqslant t/\tau \leqslant 2$,
(b) $0 \leqslant t/\tau \leqslant 4$.

Use superposition to obtain answer (b).

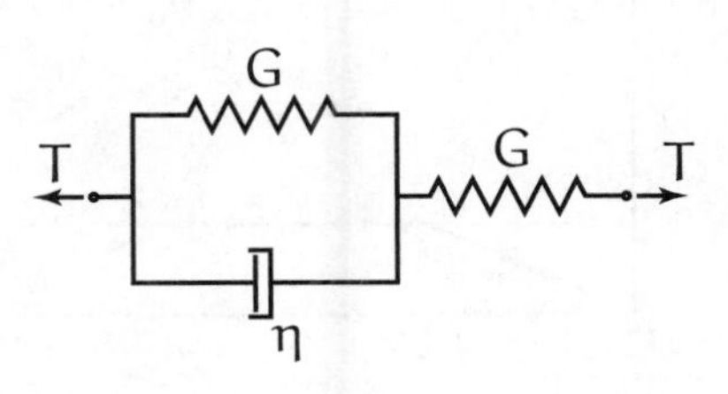

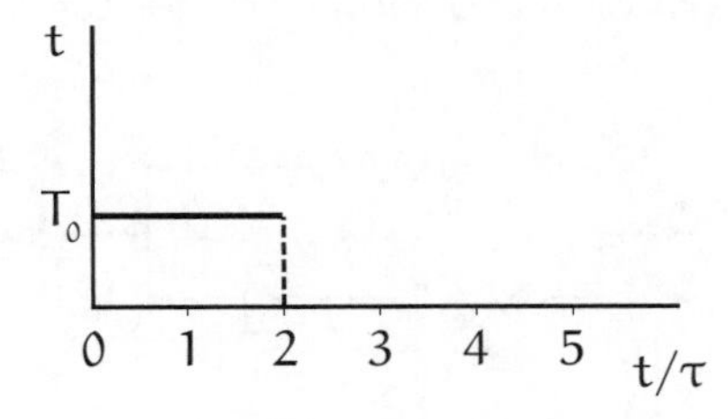

Answer

(a) $\gamma(t) = T_0 J\left(2 - e^{-t/\tau}\right) U(t)$
(b) $\gamma(t) = T_0 J\left(2 - e^{-t/\tau}\right) U(t) - T_0 J\left(2 - e^{-(t-2\tau)/\tau}\right) U(t - 2\tau)$

Problem 9.8

For the model shown determine

(a) the constitutive equation,
(b) the relaxation function, $G(t)$,
(c) the stress, $T(t)$ for $0 \leqslant t \leqslant t_1$, when the strain is given by the accompanying graph.

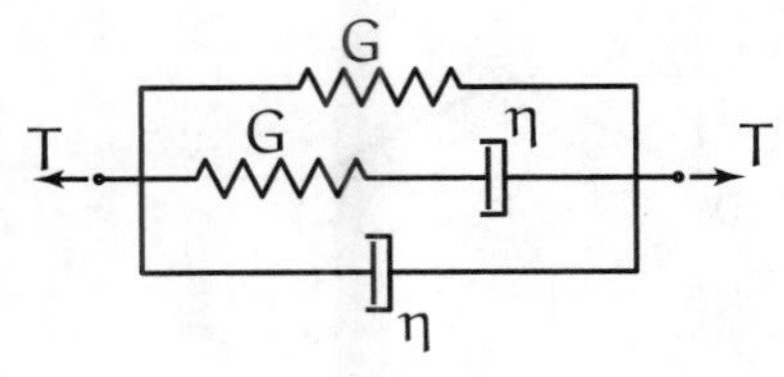

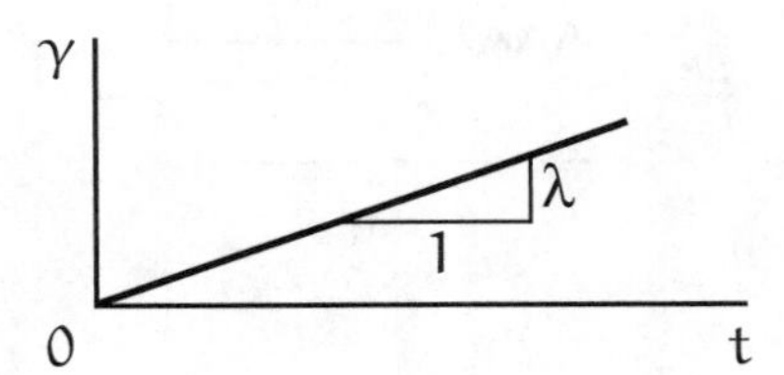

Answer

(a) $\dot{T} + T/\tau = \eta\ddot{\gamma} + 3G\dot{\gamma} + (G/\tau)\gamma$
(b) $G(t) = \eta\delta(t) + G\left(1 + e^{-t/\tau_2}\right)$
(c) $T(t) = \lambda\left(2\eta - 2e^{-t/\tau}Gt\right) U(t)$

Problem 9.9

For the model shown in Problem 9.8, determine $T(t)$ when $\gamma(t)$ is given by the diagram shown here.

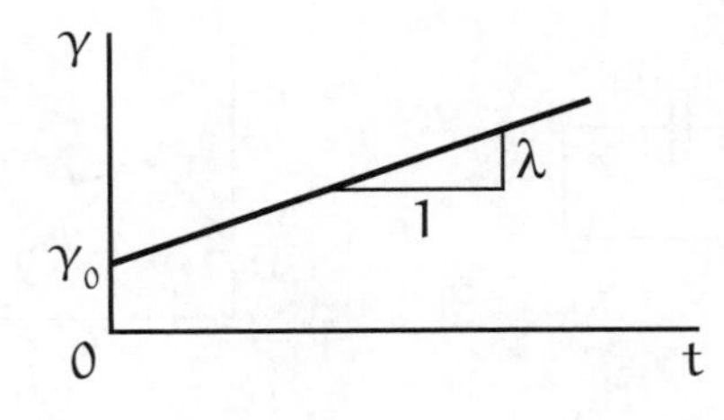

Answer

$$T(t) = \gamma_0 \left[\eta\delta(t) + G\left(1 + e^{-t/\tau}\right)\right] U(t) + \lambda \left[\eta\left(2 - e^{-t/\tau}\right) + Gt\right] U(t)$$

Problem 9.10

The three-parameter model shown is subjected to the strain history pictured in the graph. Use superposition to obtain $T(t)$ for the $t \geqslant t_1$, from $T(t)$ for $t \leqslant t_1$. Let $\gamma_0/t_1 = \lambda$.

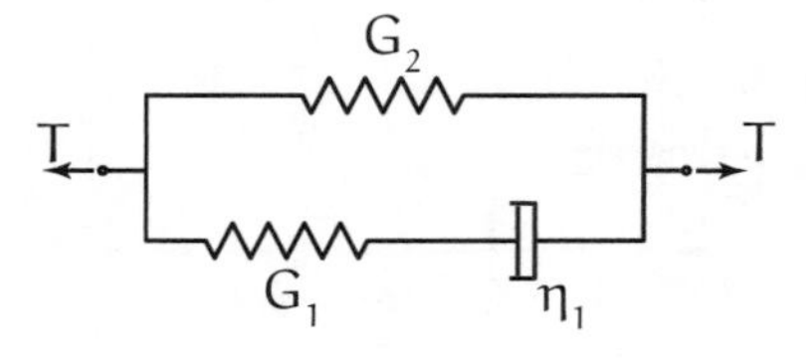

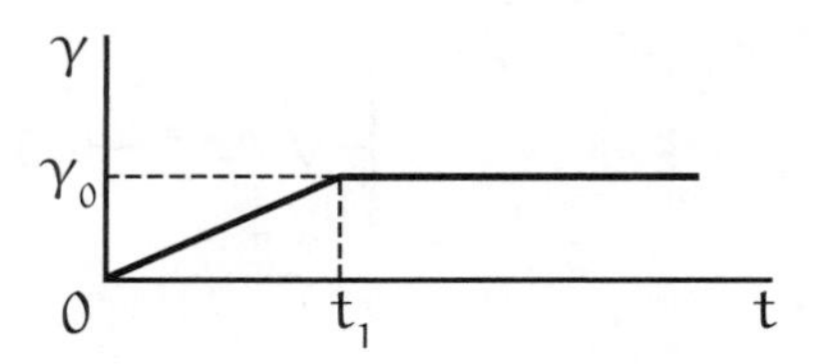

Answer

$$\text{For } t \leqslant t_1;\ T(t) = \lambda\left[\eta_1\left(1 - e^{-t/\tau_1}\right) + G_2 t\right] U(t)$$
$$\text{For } t \geqslant t_1;\ T(t) = \lambda\left[\eta_1\left(1 - e^{-t/\tau_1}\right) + G_2 t\right] U(t)$$
$$-\lambda\left[\eta_1\left(1 - e^{-(t-t_1)/\tau_1}\right) + G_2(t - t_1)\right] U(t - t_1)$$

Problem 9.11

For the model shown determine the stress, $T(t)$ at (a) $t = t_1$; (b) $t = 2t_1$; and (c) $t = 3t_1$, if the applied strain is given by the diagram. Use superposition for (b) and (c).

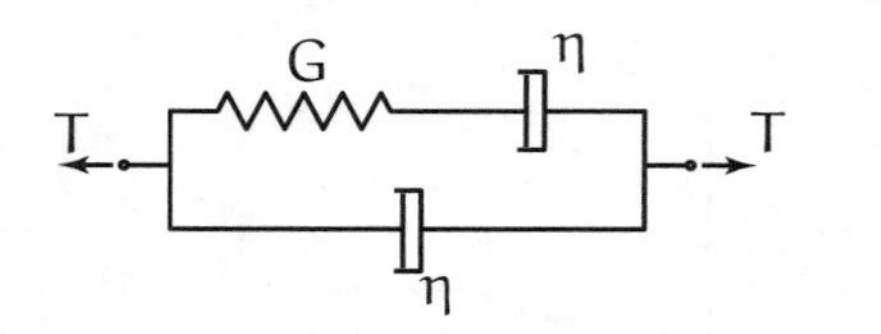

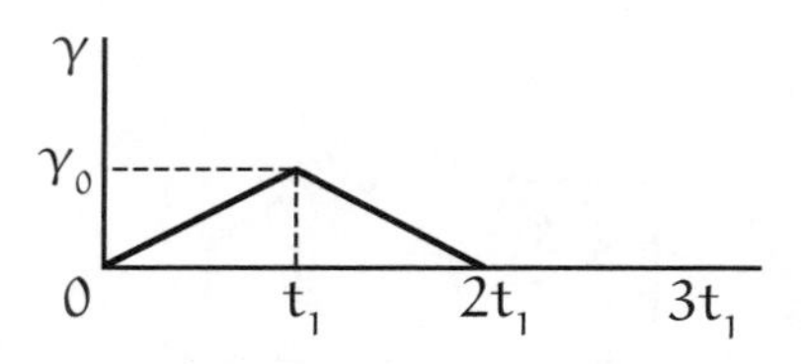

Answer

(a) $T(t_1) = (\gamma_0\eta/t_1)\left[2 - e^{-t_1/\tau}\right]$

(b) $T(2t_1) = (\gamma_0\eta/t_1)\left[-2 + 2e^{-t_1/\tau} - e^{-2t_1/\tau}\right]$

(c) $T(3t_1) = (\gamma_0\eta/t_1)\left[-e^{-t_1/\tau} + 2e^{-2t_1/\tau} - 3e^{-3t_1/\tau}\right]$

Problem 9.12

If the model shown in the sketch is subjected to the strain history $\gamma(t) = (\sigma_0/2G)\left[2 - e^{-t/2\tau}\right] U($ as pictured in the time diagram, determine the stress, $T(t)$.

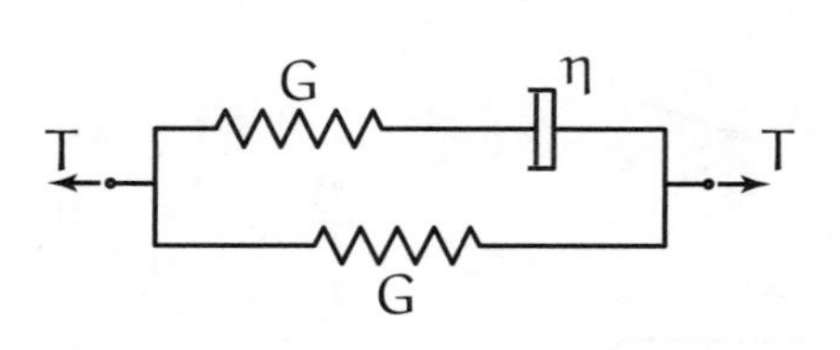

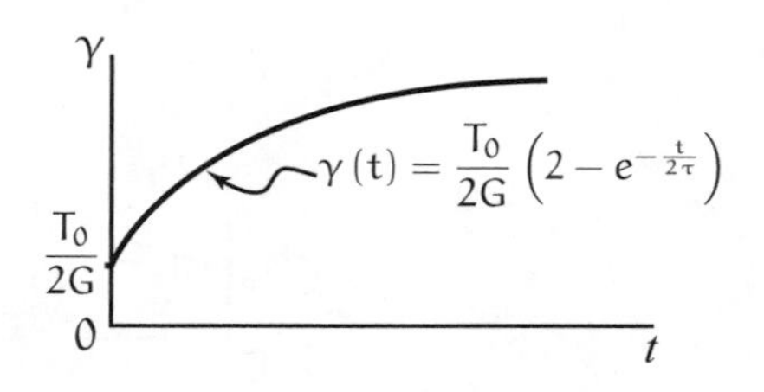

Answer

$$T(t) = T_0\left(e^{-t/\tau} + U(t)\right)$$

Problem 9.13

For the hereditary integral, Eq 9.38

$$\gamma(t) = \int_{-\infty}^{t} J(t - t')\left(dT(t')/dt'\right)dt'$$

assume $T(t) = e^{st}$ where s is a constant. Let $\tau = t - t'$ be the "elapsed time" of the load application and show that $\gamma(t) = se^{st}\bar{J}(s)$ where $\bar{J}(s)$ is the Laplace transform of $J(t)$.

Problem 9.14

Using $T(t) = e^{st}$ as in Problem 9.13, together with the hereditary integral Eq 9.41a

$$T(t) = \int_{-\infty}^{t} G(t - t')\left(d\gamma(t')/dt'\right)dt'\ ,$$

and the result of Problem 9.13 show that $\bar{G}(s)\bar{J}(s) = 1/s^2$ where $\bar{G}(s)$ is the Laplace transform of $G(t)$. Assume s is real.

Problem 9.15

Taking the hereditary integrals for viscoelastic behavior in the form Eq 9.40

$$\gamma(t) = J_0 T(t) + \int_0^t T(t')\left[dJ(t - t')/d(t - t')\right]dt'$$

and Eq 9.41c

$$T(t) = G_0\gamma(t) + \int_0^t \gamma(t')\left[dG(t - t')/d(t - t')\right]dt'\ ,$$

show that for the stress loading $T(t) = e^{st}$ and with $\tau = t - t'$, the expression

$$G_0\bar{A}(s) + J_0\bar{B}(s) + \bar{A}(s)\bar{B}(s) = 0$$

results, where here

$$\bar{A}(s) = \int_0^{\infty} e^{-s\tau}\left(dJ/d\tau\right)d\tau$$

and

$$\bar{B}(s) = \int_0^{\infty} e^{-s\tau}\left(dG/d\tau\right)d\tau\ .$$

Problem 9.16

Let the stress relaxation function be given as $G(t) = a(b/t)^m$ where a, b, and m are constants and t is time. Show that the creep function for this material is $J(t) = \frac{1}{am\pi}\sin mx\left(\frac{t}{b}\right)^m$ with $m < 1$. Use the identity $\bar{G}(s)\bar{J}(s) = 1/s^2$ where barred quantities are Laplace transforms.

Problem 9.17

A three-parameter solid has the model shown. Derive the constitutive equation for this model and from it determine (a) the relaxation function, and (b) the creep function for the model.

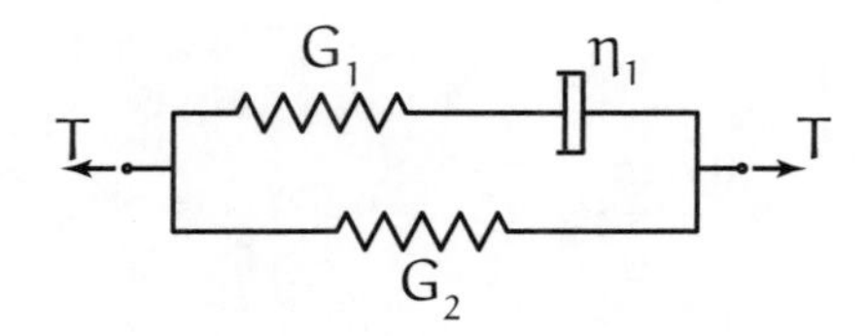

Answer

(a) $G(t) = G_2 + G_1 e^{-t/\tau_1}$
(b) $J(t) = (1/(G_1 + G_2))\, e^{-t/\tau_1^*} + (1/G_2)\left(1 - e^{-t/\tau_1^*}\right)$
where $\tau_1^* = (G_1 + G_2)\,\tau_1/G_2$

Problem 9.18
A material is modeled as shown by the sketch. (a) For this model determine the relaxation function, $G(t)$. (b) If a ramp function strain as shown by the diagram is imposed on the model, determine the stress, using the appropriate hereditary integral involving $G(t)$.

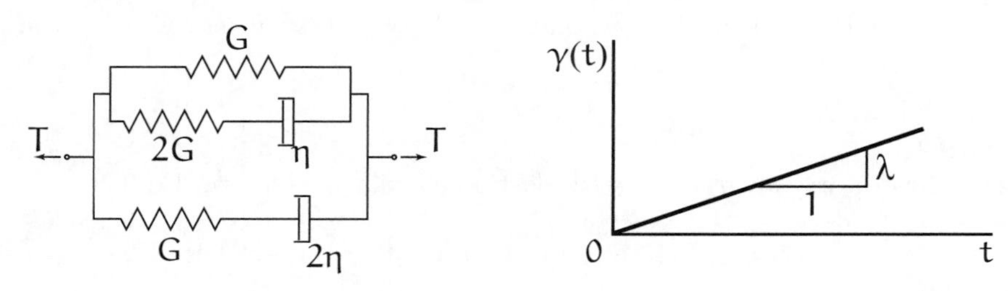

Answer

(a) $G(t) = G + 2Ge^{-2t/\tau} + Ge^{-t/2\tau}$
(b) $T(t) = G\lambda\left[t + 3\tau - \tau e^{-2t/\tau} - 2\tau e^{-t/2\tau}\right] U(t)$

Problem 9.19
Determine the complex modulus, $G^*(i\omega)$ for the model shown using the substitution $i\omega$ for ∂_t in the constitutive equation.

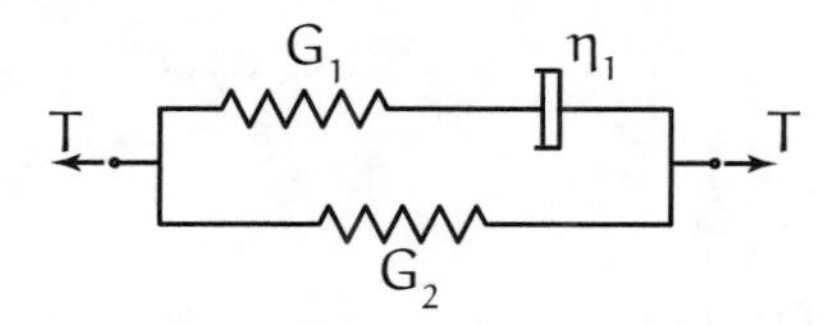

Answer

$$G^*(i\omega) = \left[G_2 + (G_1 + G_2)\,\tau_1^2\omega^2 + iG_1\tau_1\omega\right] / \left(1 + \tau_1^2\omega^2\right)$$

Problem 9.20
Show that, in general, $J' = 1/G'(1 + \tan 2\delta)$ and verify that G' and J' for the Kelvin model satisfies this identity. (Hint: Begin with $G^* J^* = 1$.)

Problem 9.21

Let the complex viscosity (denoted here by $\eta^*(i\omega)$) be defined through the equation

$$T_0 e^{i\omega t} = \eta^* \left[i\omega\gamma_0 e^{i\omega t} \right] .$$

Determine $\eta^*(i\omega)$ in terms of $G^*(i\omega)$ (see Eq 9.49) and calculate $\eta^*(i\omega)$ for the model shown below.

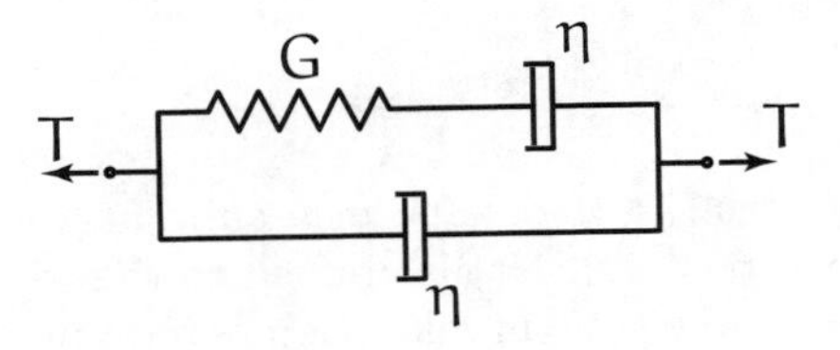

Answer

$$\eta^*(i\omega) = \left[\eta\left(2 + \omega^2\tau^2\right) - i\tau\eta\omega\right] / \left(1 + \omega^2\tau^2\right)$$

Problem 9.22

From Eq 9.55a in which $G' = T_0 \cos\delta/\gamma_0$ and 9.55b in which $G'' = (T_0 \sin\delta)/\gamma_0$ show that $J' = (\gamma_0 \cos\delta)/T_0$ and that $J'' = (\gamma_0 \sin\delta)/T_0$. Use $G^*J^* = 1$.

Problem 9.23

Show that the energy dissipated per cycle is related directly to the loss compliance, J'' by evaluating the integral $\int T d\gamma$ over one complete cycle assuming $T(t) = T_0 \sin\omega t$. (See Fig. 9.10.)

Answer

$$\int T d\gamma = T_0^2 \pi J''$$

Problem 9.24

For the rather complicated model shown here, determine the constitutive equation and from it $G^*(i\omega)$. Sketch a few points on the curve G'' vs. $\ln(\omega\tau)$.

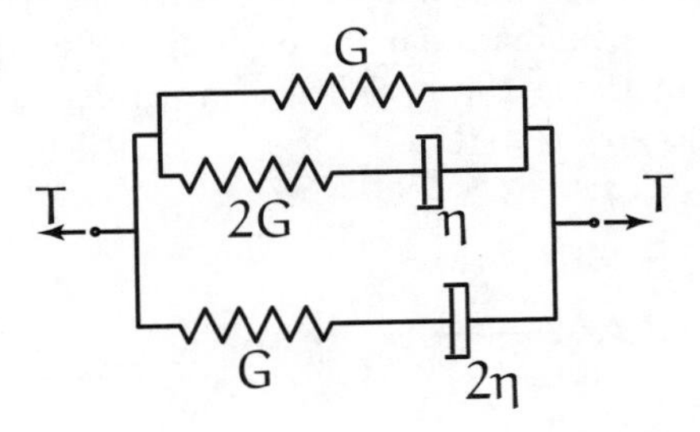

Answer

$$\ddot{T} + (5/2\tau)\dot{T} + \left(1/\tau^2\right)T = 4G\ddot{\gamma} + (11G/2\tau)\dot{\gamma} + \left(G/\tau^2\right)\gamma$$

$$G^*(i\omega) = G' + iG'' \quad \text{where}$$

$$G' = G\left[1 + (35/4)\tau^2\omega^2 + 4\tau^4\omega^4\right] / \left[1 + (17/4)\tau^2\omega^2 + \tau^4\omega^4\right]$$

$$G'' = G\left[3\tau\omega + (9/2)\,\tau^3\omega^3\right] / \left[1 + (17/4)\,\tau^2\omega^2 + \tau^4\omega^4\right]$$

For $\ln(\omega\tau) = 0,\ G'' = 1.2G$

For $\ln(\omega\tau) = 1,\ G'' = 1.13G$

For $\ln(\omega\tau) = 2,\ G'' = 0.572G$

For $\ln(\omega\tau) = \infty,\ G'' = 0$

Problem 9.25

A block of viscoelastic material in the shape of a cube fits snugly into a rigid container. A uniformly distributed load $p = -p_0 U(t)$ is applied to the top surface of the cube. If the material is Maxwell in shear and elastic in dilatation determine the stress component $t_{11}(t)$ using Eq 9.71. Evaluate $t_{11}(0)$ and $t_{11}(\infty)$.

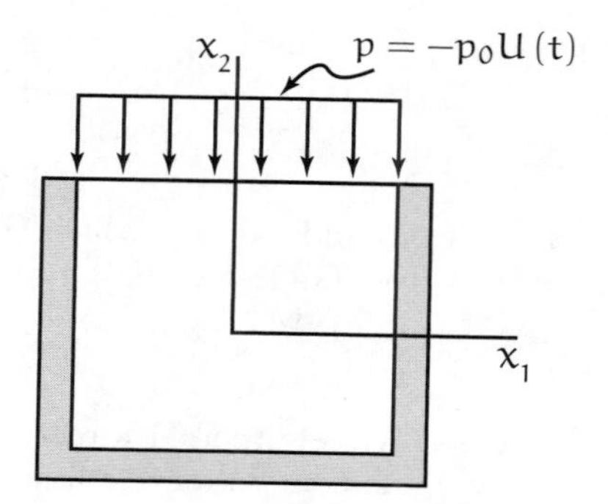

Answer

$$t_{11}(t) = -p_0[1 - (6G/(3K + 4G))]e^{-(3K/(3K+4G)\tau)t}]U(t)$$

$$t_{11}(0) = -p_0[(3K - 2G)/(3K + 4G)]$$

$$t_{11}(\infty) = -p_0$$

Problem 9.26

A slender viscoelastic bar is loaded in simple tension with the stress $T(t) = t_{11}(t) = T_0 U(t)$. The material may be modeled as a standard linear solid in shear having the model shown, and as elastic in dilatation. Using the hereditary integrals, Eq 9.42, determine the axial strain $\epsilon_{11}(t)$ and the lateral strain $\epsilon_{22}(t)$.

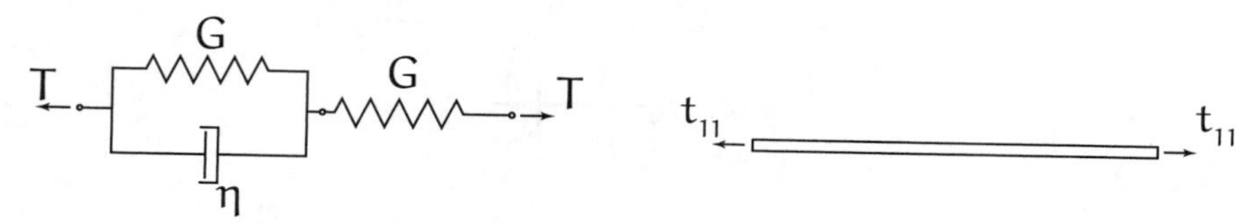

Answer

$$\epsilon_{11}(t) = T_0\{[(6K + G)/3K - e^{-t/\tau}]/3G\}U(t)$$

$$\epsilon_{22}(t) = T_0[(1/9K) - (2 - e^{-t/\tau})/6G)]U(t)$$

Problem 9.27

A cylinder of viscoelastic material fits snugly into a rigid container so that $\epsilon_{11} = \epsilon_{22} = \epsilon_{rr} = 0$ (no radial strain). The body is elastic in dilatation and has a creep compliance $J_S = J_0(1+t)$ with J_0 a constant. Determine $t_{33}(t)$ if $\dot{\epsilon}_{33} = A$ (a constant).

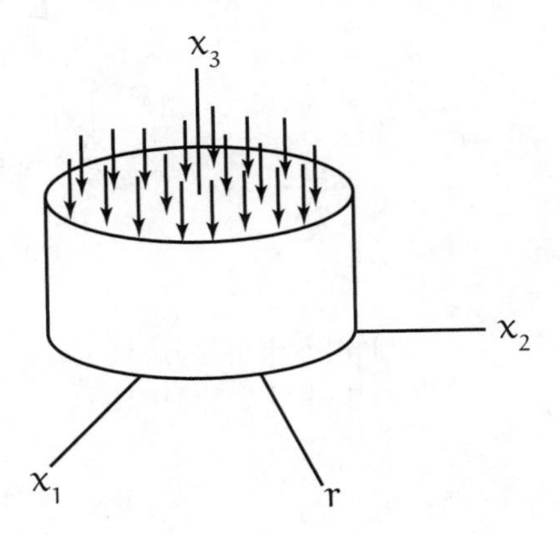

Answer

$$t_{33}(t) = \{A[Kt + 4(1 - e^{-t})/3J_0]\}U(t)$$

Problem 9.28

A viscoelastic body in the form of a block is elastic in dilatation and obeys the Maxwell law in distortion. The block is subjected to a pressure impulse $t_{11} = -p_0\delta(t)$ distributed uniformly over the x_1 face. If the block is constrained so that $\epsilon_{22} = \epsilon_{33} = 0$, determine $\epsilon_{11}(t)$ and $t_{22}(t)$.

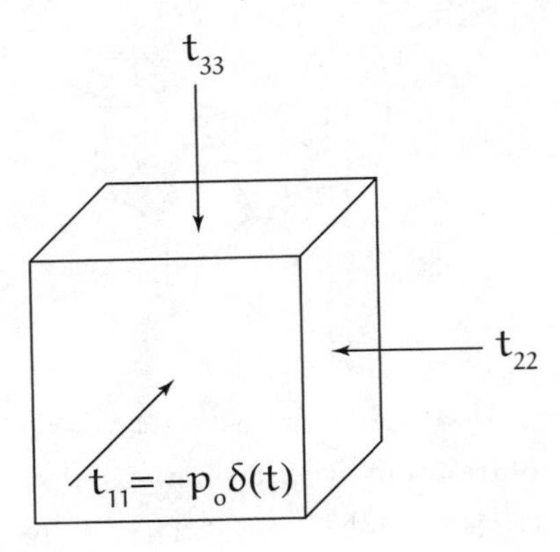

Answer

$$t_{22}(t) = p_0[(2G - 3K)\delta(t)/(3K + 4G) - (6G/(3K + 4G))e^{[-3K/(3K+4G)\tau]t}]U(t)$$
$$\epsilon_{11}(t) = p_0[3\delta(t)/(3K + 4G) - [4G/(3K + 4G)K]e^{[-3K/(3K+4G)\tau]t}]U(t)$$

Problem 9.29

A viscoelastic cylinder is inserted into a snug fitting cavity of a rigid container. A flat, smooth plunger is applied to the surface $x_1 = 0$ of the cylinder and forced downward at a constant strain rate $\dot{\epsilon}_{11} = \epsilon_0$. If the material is modeled as the three-parameter solid shown in shear and as elastic in dilatation, determine $t_{11}(t)$ and $t_{22}(t)$ during the downward motion of the plunger.

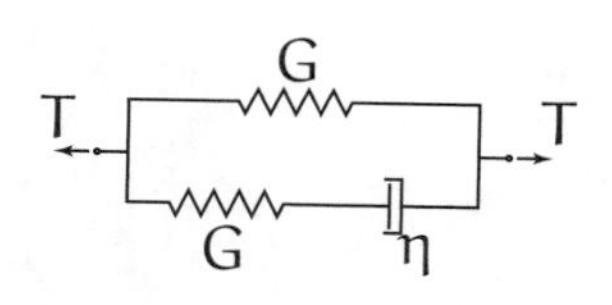

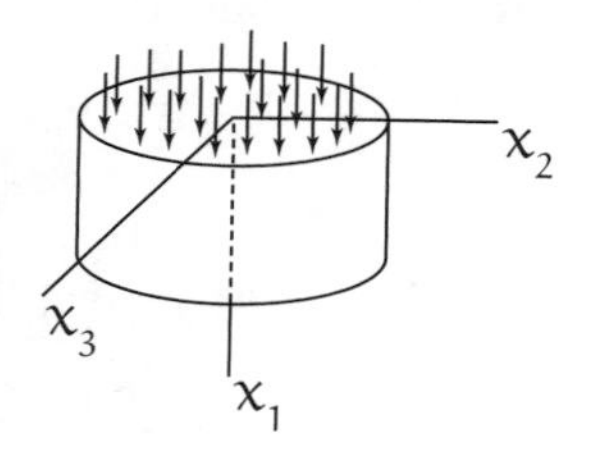

Answer

$$t_{11}(t) = -\epsilon_0[(4G\tau/3)(1 - e^{-t/\tau}) + (K + 4G/3)t]U(t)$$
$$t_{22}(t) = \epsilon_0[(2G\tau/3)(1 - e^{-t/\tau}) + (-K + 2G/3)t]U(t)$$

Problem 9.30

For a thick-walled elastic cylinder under internal pressure p_0, the stresses are $t_{rr} = A - B/r^2$; $t_{\theta\theta} = A + B/r^2$ and the radial displacement is given by $u_r = \frac{1+\nu}{E}[A(1-2\nu)r + B/r]$ where A and B are constants involving p_0, E is Young's modulus, and ν is Poisson's ratio. Determine t_{rr}, $t_{\theta\theta}$, and u_r for a viscoelastic cylinder of the same dimensions that is Kelvin in shear and elastic dilatation if $p = p_0U(t)$.

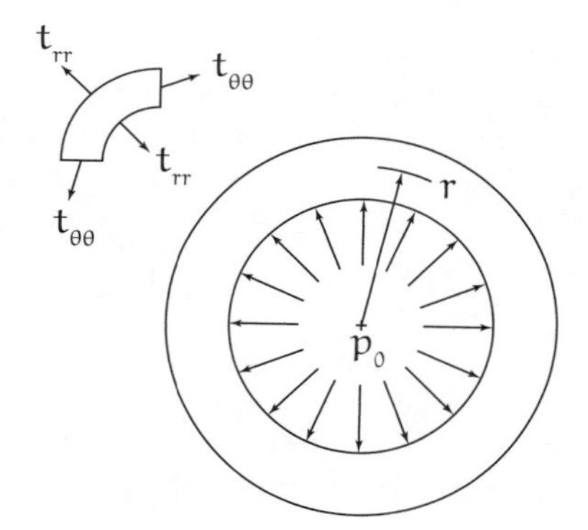

Answer

$$t_{rr} = \text{same as elastic solution but with } p_0 \text{ now } p_0U(t)$$
$$t_{\theta\theta} = \text{same as elastic solution but with } p_0 \text{ now } p_0U(t)$$
$$u_r(t) = (3Ar/(6K + 2G))(1 - e^{-(3K+G)t/G\tau})U(t) + (B/2Gr)(1 - e^{-t/\tau})U(t)$$

Problem 9.31

A viscoelastic half-space is modeled as Kelvin in shear, elastic in dilatation. If the point force $P = P_0e^{-t}$ is applied at the origin of a stress free material at time $t = 0$, determine $t_{rr}(t)$ knowing that for an elastic half-space the radial stress is

$$t_{rr} = P_0[(1 - 2\nu)A - B]/2\pi$$

where A and B are functions of the coordinates only.

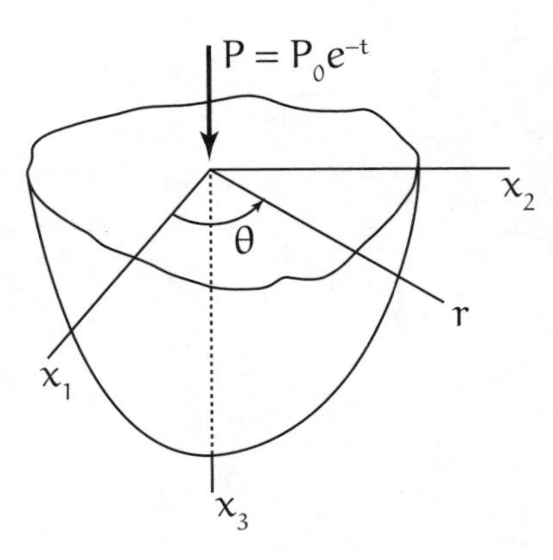

Answer

$$t_{rr} = (P_0/2\pi)[(3(G-\eta)e^{-t} + 9Ke^{-(3K+G)t/\eta}]A/(3K+G-\eta) - e^{-t}B$$

Problem 9.32

The deflection at $x = L$ for an end-loaded cantilever elastic beam is $w = P_0L^3/3EI$. Determine the deflection $w(L,t)$ for a viscoelastic beam of the same dimensions if $P = P_0U(t)$ assuming (a) one-dimensional analysis based on Kelvin material, and (b) three-dimensional analysis with the beam material Kelvin in shear, elastic in dilatation. Check $w(L,\infty)$ in each case.

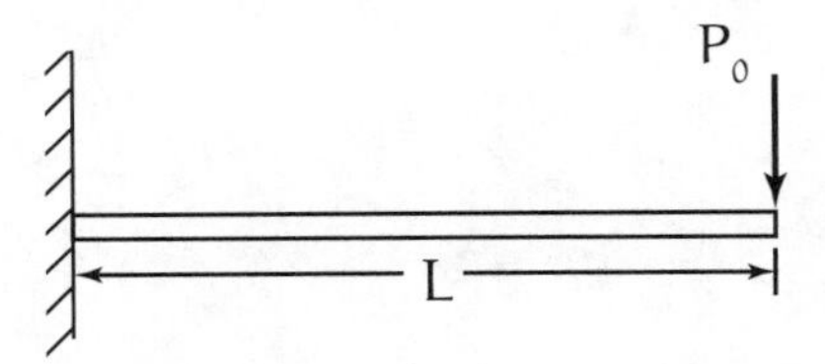

Answer

(a) $w(L,t) = (P_0L^3/3EI)(1 - e^{-t/\tau_E})U(t), \tau_E = \eta/E$
$w(L,\infty) = (P_0L^3/3EI)$, the elastic deflection.
(b) $w(L,t) = (P_0L^3/3EI)[(1 - e^{-t/\tau}) + (1/9K)e^{-t/\tau}]U(t)$
$w(L,\infty) = (P_0L^3/3EI)$, the elastic deflection.

Problem 9.33

A simply-supported viscoelastic beam is subjected to the time-dependent loading $f(\mathbf{x},t) = q_0t$ where q_0 is a constant and t is time. Determine the beam deflection $w(\mathbf{x},t)$ in terms of the elastic beam shape $X(\mathbf{x})$ if the beam material is assumed to be (a) one-dimensional Kelvin, and (b) three-dimensional Kelvin in shear, and elastic in dilatation. Compare the results.

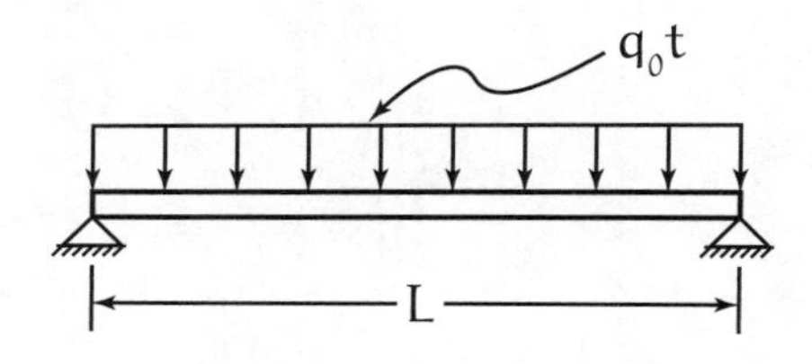

Answer

(a) $w(\mathbf{x}, t) = X(\mathbf{x}) \left[t - \tau_E(1 - e^{-t/\tau_E})\right] U(t)$

(b) $w(\mathbf{x}, t) = X(\mathbf{x}) \left[t - (3\tau_E/(3K + G))(1 - e^{-t/\tau_E})\right] U(t)$
where $\tau_E = \eta/E$

Appendix A: General Tensors

The balance laws and the constitutive relations describing the behavior of continuous media were formulated using direct or coordinate free notation. This is a recognition of the invariance of the physical principles. These laws were written using indicial notation in a rectangular-Cartesian coordinate frame. The study of tensors in other coordinate systems may be better suited for solving particular boundary value problems. These coordinate frames might provide ways to describe vectors and tensors that better describe the geometry of the body. The study of general tensors also aids in understanding much of the literature that uses general tensors.

It would be convenient if there was a way to convert the Cartesian equations to general coordinate systems without a great deal of effort. It is the purpose of this appendix to give such a scheme. The following section begins by developing ideas about the general representation of vectors. Next methods are given for taking derivatives in general coordinate systems. Then a set of rules is introduced that will allow one to take an expression in Cartesian coordinates and convert it to an expression in a general coordinate system.

A.1 Representation of Vectors in General Bases

Vectors are quantities that do not depend on a coordinate system. The representation of vectors does depend on the basis for computing the components of a vector. Any set of three non-coplanar vectors can be used to represent a vector $\mathbf{v}$. For a Cartesian coordinate system $\{\hat{\mathbf{e}}_1, \hat{\mathbf{e}}_2, \hat{\mathbf{e}}_3\}$ forms a basis that allows the vector $\mathbf{v}$ to be written as

$$\mathbf{v} = v_1 \hat{\mathbf{e}}_1 + v_2 \hat{\mathbf{e}}_2 + v_3 \hat{\mathbf{e}}_3 = v_i \hat{\mathbf{e}}_i \ .$$

The basis $\{\hat{\mathbf{e}}_1, \hat{\mathbf{e}}_2, \hat{\mathbf{e}}_3\}$ is the same for every point in the space defined in the Cartesian coordinate system. The coordinate system is said to be homogenous. It is also possible to choose a set of base vectors $\{\mathbf{g}_1, \mathbf{g}_2, \mathbf{g}_3\}$ that is not orthonormal, Fig A.1. The vectors $\mathbf{g}_1$ and $\mathbf{g}_2$ lie in a plane, and the vector $\mathbf{g}_3$ is out of the plane. These base vectors may also vary from point to point in the space. Such a system is called nonhomogeneous.

A vector can be represented by

$$\mathbf{v} = v^1 \mathbf{g}_1 + v^2 \mathbf{g}_2 + v^3 \mathbf{g}_3 = v^i \mathbf{g}_i \ .$$

The summation is over a subscript and superscript. The superscripts are indices not exponents. An exponent would be represented by $v^i \times v^i = \left(v^i\right)^2$. The components of the vector $\mathbf{v}$ are found by projecting the vector onto the planes formed by $\mathbf{g}_1$-$\mathbf{g}_2$, $\mathbf{g}_2$-$\mathbf{g}_3$, and $\mathbf{g}_1$-$\mathbf{g}_3$. This gives the unique scalar components v^1, v^2, and v^3 with respect to the basis $\{\mathbf{g}_i\}$. The calculation of the components v^i are not as straight forward as in the Cartesian system. The basis vectors are not orthonormal and

$$\mathbf{g}_i \cdot \mathbf{g}_j \neq \begin{cases} 0 & \text{for } i \neq j \\ 1 & \text{for } i = j \end{cases}$$

as in the Cartesian system.

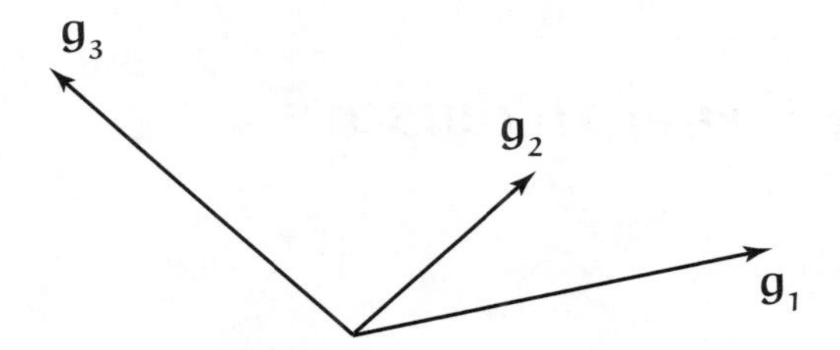

FIGURE A.1
A set of non-orthonormal base vectors.

Example A.1

Given the set of base vectors $\{g_i\}$ with Cartesian components

$$g_1 = 3\hat{e}_1 + 2\hat{e}_2 + \hat{e}_3$$
$$g_2 = \hat{e}_1 - 2\hat{e}_3$$
$$g_3 = \hat{e}_1 + \hat{e}_3$$

and a vector

$$v = 4\hat{e}_1 + \hat{e}_2 + 8\hat{e}_3 \ .$$

Find v^1, v^2, and v^3.

Solution
We know that

$$v = v^1 g_1 + v^2 g_2 + v^3 g_3 \ .$$

Taking the dot product $v \cdot \hat{e}_i$, gives

$$\begin{array}{ll} v \cdot \hat{e}_1 & 4 = 3v^1 + v^2 + v^3 \\ v \cdot \hat{e}_2 & 1 = 2v^1 + v^3 \\ v \cdot \hat{e}_3 & 8 = v^1 - 2v^3 \end{array}$$

Solving yields

$$v^1 = 2, \qquad v^2 = 1, \qquad v^3 = -3 \ .$$

The vector is

$$v = 2g_1 + g_2 - 3g_3 \ .$$

Determining whether a set of vectors forms a basis is not a simple task. A simple test for linear independence is to form a matrix, $\mathcal{G}$, from the basis vectors expressed in the Cartesian components.

$$\mathcal{G} = [g_1, g_2, g_3]$$

The determinate of $\mathcal{G}$ being non-zero assures that the basis is linearly independent. The mathematical condition for this is

$$\det \mathcal{G} \neq 0$$

and is called the Jacobian matrix of the basis $\{g_i\}$.

Example A.2

From the previous example, show that the basis $\mathbf{g}_i$ is linearly independent.

Solution

The matrix $\mathcal{G}$ is

$$\mathcal{G} = \begin{bmatrix} 3 & 1 & 1 \\ 2 & 0 & 1 \\ 1 & 0 & -2 \end{bmatrix}$$

and

$$\det \mathcal{G} = \begin{vmatrix} 3 & 1 & 1 \\ 2 & 0 & 1 \\ 1 & 0 & -2 \end{vmatrix} = 5 \neq 0 \,.$$

The basis is linearly independent. Note that the determinate of $\mathcal{G}$ being non-zero is just the condition that the equations for the components v^i has a solution.

A.2 The Dot Product and the Reciprocal Basis

The dot product between two vectors $\mathbf{u}$ and $\mathbf{v}$ is

$$\mathbf{u} \cdot \mathbf{v} = u^i \mathbf{g}_i \cdot v^j \mathbf{g}_j = u^i v^j \mathbf{g}_i \cdot \mathbf{g}_j \,.$$

If the basis is Cartesian, the dot product $\mathbf{g}_i \cdot \mathbf{g}_j$ is just the Kronecker delta. However, since the basis vectors are not orthonormal, the dot product becomes

$$\begin{aligned} \mathbf{u} \cdot \mathbf{v} =& u^1 v^1 \mathbf{g}_1 \cdot \mathbf{g}_1 + u^1 v^2 \mathbf{g}_1 \cdot \mathbf{g}_2 + u^1 v^3 \mathbf{g}_1 \cdot \mathbf{g}_3 + \cdots \\ & + u^3 v^1 \mathbf{g}_3 \cdot \mathbf{g}_1 + u^3 v^2 \mathbf{g}_3 \cdot \mathbf{g}_2 + u^3 v^3 \mathbf{g}_3 \cdot \mathbf{g}_3 \,. \end{aligned}$$

The dot product of the base vectors is a set of scalars and possesses the symmetry

$$\mathbf{g}_i \cdot \mathbf{g}_j = g_{ij} = \mathbf{g}_j \cdot \mathbf{g}_i = g_{ji} \,.$$

The dot product would be simplified considerably if the dot product of the basis vectors were orthogonal. This can be accomplished by computing a reciprocal or dual basis. The reciprocal vectors can be computed in a number of ways but have the property that

$$\mathbf{g}_i \cdot \mathbf{g}^j = \delta_i^j = \begin{cases} 1 & \text{if } i = j \\ 0 & \text{if } i \neq j \end{cases} \,.$$

δ_i^j is the Kronecker delta. One method for the computation of the reciprocal basis is given in Problem 2.2. The reciprocal basis allows us to write a vector as

$$\mathbf{v} = v^i \mathbf{g}_i = v_i \mathbf{g}^i \,.$$

Note that the implied summation is always across a subscript and superscript pair. With the reciprocal basis, the dot product becomes

$$\mathbf{u} \cdot \mathbf{v} = u^i v_j \mathbf{g}_i \cdot \mathbf{g}^j = u^i v_j \delta_i^j = u^i v_i \, .$$

It is also possible to form the dot product of the base vectors with themselves. The dot products of the basis vectors give

$$\begin{aligned}
\mathbf{g}_i \cdot \mathbf{g}_j &= g_{ij} \\
\mathbf{g}^i \cdot \mathbf{g}^j &= g^{ij} \\
\mathbf{g}_i \cdot \mathbf{g}^j &= g_i^j = \delta_i^j \\
\mathbf{g}^i \cdot \mathbf{g}_j &= g^i_j = \delta^i_j
\end{aligned}$$

These equations give the metric tensors of the space.

The metric tensor is associated with distances or lengths in the space. For a vector $\mathbf{v} = v^i \mathbf{g}_i$, we have

$$\mathbf{v} \cdot \mathbf{v} = |\mathbf{v}|^2 = v^i v^j \mathbf{g}_i \cdot \mathbf{g}_j = g_{ij} v^i v^j$$

or

$$\mathbf{v} \cdot \mathbf{v} = |\mathbf{v}|^2 = v_i v_j \mathbf{g}^i \cdot \mathbf{g}^j = g^{ij} v_i v_j \, .$$

The magnitude of the vector is related to the metric tensors g_{ij} and g^{ij}.

The products

$$g^{ij} = \mathbf{g}^i \cdot \mathbf{g}^j$$

are the dual, reciprocal or inverse metric tensor. This latter designation is seen from the following calculation. The bases are related by

$$\mathbf{g}_i = g_{ij} \mathbf{g}^j \qquad \text{and} \qquad \mathbf{g}^i = g^{ij} \mathbf{g}_j \, .$$

Now, this gives

$$\mathbf{g}_i \cdot \mathbf{g}^j = \delta_i^j = g_{il} \mathbf{g}^l \cdot g^{jk} \mathbf{g}_k = g_{il} g^{jk} \delta_k^l = g_{il} g^{jl}$$

and

$$g_{il} g^{jl} = g_{il} g^{lj} = \delta_i^j \, .$$

Hence the designation inverse metric.

The components of a vector are related by

$$v_i = \mathbf{v} \cdot \mathbf{g}_i = v^k \mathbf{g}_k \cdot \mathbf{g}_i = g_{ki} v^k = g_{ik} v^k$$

from the symmetry of the metric tensor. This can also be repeated for the $\mathbf{g}^i$ base vectors. The results are

$$v_i = g_{ij} v^j \qquad \text{and} \qquad v^i = g^{ij} v_j \, .$$

This is termed *raising and lowering the indices* and will be useful when we describe the rules for generating general tensor expressions from Cartesian tensor expression.

A.3 Components of a Tensor

The components of a tensor in an orthonormal basis are defined by the operation of the tensor on the base vectors $\hat{\mathbf{e}}_i$. The components are

$$T_{ij} = \hat{\mathbf{e}}_i \cdot \mathbf{T} \cdot \hat{\mathbf{e}}_j \, .$$

This process can be extended to the general basis. However, unlike the orthonormal basis, the general basis gives four different component representations. Two of these tensor components are

$$T_{ij} = \mathbf{g}_i \cdot \mathbf{T} \cdot \mathbf{g}_j$$
$$T^{ij} = \mathbf{g}^i \cdot \mathbf{T} \cdot \mathbf{g}^j \, .$$

The components are formed in the standard way. In addition, there are two sets of mixed components. These are[2]

$$T^i_{\ j} = \mathbf{g}^i \cdot \mathbf{T} \cdot \mathbf{g}_j$$
$$T_i^{\ j} = \mathbf{g}_i \cdot \mathbf{T} \cdot \mathbf{g}^j$$

Thus, the four representations of a tensor are

$$\mathbf{T} = T^{ij}\mathbf{g}_i\mathbf{g}_j = T_{ij}\mathbf{g}^i\mathbf{g}^j = T^i_{\ j}\mathbf{g}_i\mathbf{g}^j = T_i^{\ j}\mathbf{g}^i\mathbf{g}_j \, .$$

The mixed components of a tensor are not related. In general

$$T_i^{\ j} \neq T^i_{\ j} \, .$$

However, for a symmetric tensor

$$\mathbf{T} = \mathbf{T}^T$$

the result is

$$T_{ij} = T_{ji}$$
$$T^{ij} = T^{ji}$$
$$T_i^{\ j} = T_j^{\ i}$$
$$T^i_{\ j} = T^j_{\ i} \, .$$

Just as with orthonormal bases, transformation rules exist between two sets of general basis vectors $\{\mathbf{g}_i\}$ and $\{\mathbf{g}'_i\}$. The transformation follows the same patterns established previously. The only change being that summation occurs over subscripts and superscripts. The transformation of a vector is

$$v'_i = a_i^j v_j$$

and

$$v_i = \left(a^{-1}\right)_i^j v'_j \, .$$

A second order tensor transforms as

$$T'_{ij} = a_i^p a_j^k T_{pk} \, .$$

For the other components, we need to have the transformation rules for the reciprocal basis

$$\mathbf{g}'^i = \left(a^{-1}\right)_j^i \mathbf{g}^j$$

and

$$\mathbf{g}^i = a_j^i \mathbf{g}'^j \, .$$

These can be used to determine the transformation laws for the other tensor components.

[2] Sometimes dots are used to hold the positions of the indices, $T_i^{\ j} = T_{i}^{\cdot j}$. This is most commonly done when writing the mixed components. Typesetting offers the opportunity to clearly separate the superscript and subscript.

A.4 Determination of the Base Vectors

We have seen that vectors can be represented in terms of two sets of base vectors $\{\mathbf{g}_i\}$ or $\{\mathbf{g}^i\}$.

$$\mathbf{v} = v^i \mathbf{g}_i = v_i \mathbf{g}^i .$$

The base vectors $\{\mathbf{g}_i\}$ are called the covariant basis vectors, and $\{\mathbf{g}^i\}$ is called a set of contravariant basis vectors. The covariant basis vectors are defined by the derivatives of a coordinate transformation

$$\mathbf{x} = \tilde{\mathbf{x}}\left(u^j\right) = \tilde{x}^i\left(u^j\right)\hat{\mathbf{e}}_i .$$

The vectors $\{\hat{\mathbf{e}}_i\}$ are the orthonormal base vectors of the rectangular system. The functions $\tilde{x}^i\left(u^j\right)$ are the coordinate functions relating the rectangular coordinates, x^i, to the new coordinate variables, u^j. The covariant base vectors are

$$\mathbf{g}_j = \frac{\partial \tilde{\mathbf{x}}}{\partial u^j} = \frac{\partial \tilde{x}^i}{\partial u^j}\hat{\mathbf{e}}_i .$$

The base vectors are the derivatives along the coordinate curves. They vary along the coordinate curves. The dual or reciprocal basis is the contravariant basis vectors. These vectors have the property that

$$\mathbf{g}_i \cdot \mathbf{g}^j = \delta_i^j = \begin{cases} 1 & \text{if } i = j \\ 0 & \text{if } i \neq j \end{cases}$$

δ_i^j is the Kronecker delta.

Example A.3

Determine the base vectors for the circular-cylindrical coordinate system.

Solution

Figure A.3 shows the circular-cylindrical coordinate system for $x^3 = 0$. The position of the point $\mathbf{p}$ is

$$\begin{aligned} \mathbf{x} &= x^1\hat{\mathbf{e}}_1 + x^2\hat{\mathbf{e}}_2 + x^3\hat{\mathbf{e}}_3 \\ &= r\cos\theta\hat{\mathbf{e}}_1 + r\sin\theta\hat{\mathbf{e}}_2 + z\hat{\mathbf{e}}_3 . \end{aligned}$$

This shows that $\mathbf{x} = \tilde{\mathbf{x}}(r, \theta, z)$. Now

$$\mathbf{g}_r = \frac{\partial \tilde{\mathbf{x}}}{\partial r} = \cos\theta\hat{\mathbf{e}}_1 + \sin\theta\hat{\mathbf{e}}_2$$

$$\mathbf{g}_\theta = \frac{\partial \tilde{\mathbf{x}}}{\partial \theta} = -r\sin\theta\hat{\mathbf{e}}_1 + r\cos\theta\hat{\mathbf{e}}_2$$

and

$$\mathbf{g}_z = \frac{\partial \tilde{\mathbf{x}}}{\partial z} = \hat{\mathbf{e}}_3 .$$

The basis vectors $\{\mathbf{g}_i\}$ can be shown to be a basis by forming the Jacobian, $\mathcal{G}(\mathbf{g}_r, \mathbf{g}_\theta, \mathbf{g}_z)$. This gives

$$\mathcal{G} = \det\begin{vmatrix} \cos\theta & \sin\theta & 0 \\ -r\sin\theta & r\cos\theta & 0 \\ 0 & 0 & 1 \end{vmatrix} = r\left(\cos^2\theta + \sin^2\theta\right) = r .$$

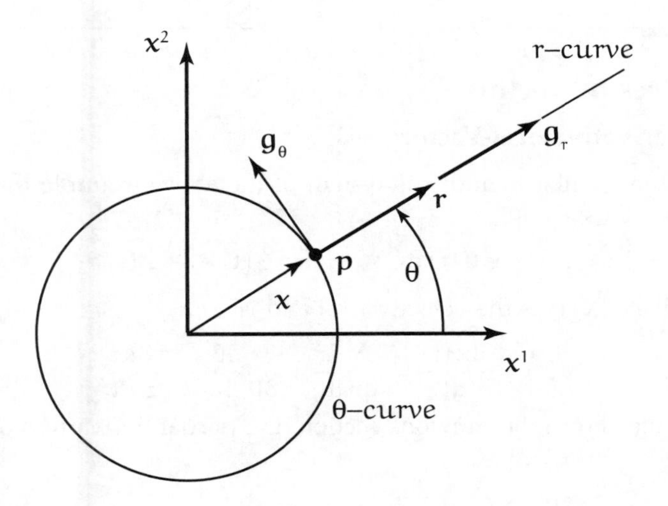

FIGURE A.2
Circular-cylindrical coordinate system for $x^3 = 0$.

The vectors form a basis except for the point $r = 0$. The vector

$$\mathbf{g}_r = \frac{\partial \tilde{x}}{\partial r} = \cos\theta\,\widehat{\mathbf{e}}_1 + \sin\theta\,\widehat{\mathbf{e}}_2$$

lies along the r-coordinate. Similarly, $\mathbf{g}_\theta$ is a vector that is perpendicular to $\mathbf{g}_r$. The basis vectors form a non-homogeneous basis. The two vectors are tangent to the r and θ curves. This is consistent with the definition of the covariant basis vectors.

The dual basis is found from the computation in Problem 2.2. These vectors are

$$\mathbf{g}^r = \cos\theta\,\widehat{\mathbf{e}}_1 + \sin\theta\,\widehat{\mathbf{e}}_2$$

$$\mathbf{g}^\theta = \frac{1}{r}\left(-\sin\theta\,\widehat{\mathbf{e}}_1 + \cos\theta\,\widehat{\mathbf{e}}_2\right)$$

and

$$\mathbf{g}^z = \widehat{\mathbf{e}}_3 \,.$$

The properties of the bases can be verified.

A.5 Derivatives of Vectors

A.5.1 Time Derivative of a Vector

Suppose that in the circular cylindrical system of the above example that $r = r(t)$, $\theta = \theta(t)$, and $z = z(t)$. This gives

$$\mathbf{x} = r(t)\cos\theta(t)\,\hat{\mathbf{e}}_1 + r(t)\sin\theta(t)\,\hat{\mathbf{e}}_2 + z(t)\,\hat{\mathbf{e}}_3 .$$

The time derivative of $\mathbf{x}(t)$ is the velocity $\mathbf{v}(t)$ and is

$$\mathbf{v}(t) = \frac{d\mathbf{x}(t)}{dt} = \frac{\partial \mathbf{x}}{\partial r}\frac{dr}{dt} + \frac{\partial \mathbf{x}}{\partial \theta}\frac{d\theta}{dt} + \frac{\partial \mathbf{x}}{\partial z}\frac{dz}{dt}$$

using the chain rule. From the previous section, the partial derivatives of $\mathbf{x}$ are the base vectors $\mathbf{g}_i$ and

$$\mathbf{v}(t) = \frac{d\mathbf{x}(t)}{dt} = \frac{dr}{dt}\mathbf{g}_r + \frac{d\theta}{dt}\mathbf{g}_\theta + \frac{dz}{dt}\mathbf{g}_z = v^r\mathbf{g}_r + v^\theta\mathbf{g}_\theta + v^z\mathbf{g}_z .$$

The result is

$$v^r = \frac{dr}{dt} \qquad v^\theta = \frac{d\theta}{dt} \qquad \text{and} \qquad v^z = \frac{dz}{dt} .$$

The first and third terms have the dimensions of a velocity. However, the second term does not. It only has units of s^{-1}. This raises the need to develop the physical components of a vector.

The physical components of a vector $\mathbf{w}$ in the direction $\mathbf{u}$ are

$$\mathbf{w}\cdot\hat{\mathbf{u}} .$$

Here $\hat{\mathbf{u}}$ is a unit vector in the direction of $\mathbf{u}$.

$$\hat{\mathbf{u}} = \frac{\mathbf{u}}{|\mathbf{u}|}$$

The physical components of $\mathbf{v}$ denoted by $v^{<i>}$ are the components of the vector along the covariant unit vectors. The unit vectors $\hat{\mathbf{g}}_i$ are

$$\hat{\mathbf{g}}_i = \frac{\mathbf{g}_i}{\sqrt{g_{ii}}} \qquad \text{no sum on } i .$$

Now for a vector $\mathbf{v} = v^i\mathbf{g}_i$,

$$\mathbf{v} = v^i\mathbf{g}_i = v^1\sqrt{g_{11}}\hat{\mathbf{g}}_1 + v^2\sqrt{g_{22}}\hat{\mathbf{g}}_2 + v^3\sqrt{g_{33}}\hat{\mathbf{g}}_3 .$$

The physical components of $\mathbf{v}$ are

$$v^{<i>} = \sqrt{g_{ii}}v^i \qquad \text{no sum on } i$$

and

$$\mathbf{v} = v^{<k>}\hat{\mathbf{g}}_k .$$

For the circular cylindrical system

$$v^{<r>} = \mathbf{v}\cdot\hat{\mathbf{g}}^r = \frac{dr}{dt}$$

and

$$v^{<\theta>} = \mathbf{v}\cdot\hat{\mathbf{g}}^\theta = \mathbf{v}\cdot r\mathbf{g}^\theta = r\frac{d\theta}{dt} .$$

These are the familiar components of the velocity that appear in elementary dynamics.

A.5.2 Covariant Derivative of a Vector

The spatial derivative of a vector field frequently arises in applications. This is the case for gradients of displacement fields defining strain fields or the divergence of a velocity field in a fluid. The derivative of a vector $\mathbf{v}$ with respect to a curvilinear coordinate u^i is

$$\frac{\partial \mathbf{v}}{\partial u^j} = \frac{\partial v^i}{\partial u^j}\mathbf{g}_i + v^i \frac{\partial \mathbf{g}_i}{\partial u^j} .$$

The latter term arises because the base vectors, $\mathbf{g}_i$, can vary from point to point in the space.

It is desirable to express the derivative of the base vector in terms of components along the base vector. This allows the writing of an expression analogous to that in a Cartesian coordinate system.

$$\frac{\partial \mathbf{v}}{\partial x_j} = \frac{\partial v_i}{\partial x_j}\hat{\mathbf{e}}_i = v_{i,j}\hat{\mathbf{e}}_i .$$

Here, the comma notation denotes partial differentiation.

The derivative of the base vector is

$$\frac{\partial \mathbf{g}_i}{\partial u^j} = \frac{\partial}{\partial u^j}\left(\frac{\partial \tilde{x}^k}{\partial u^i}\hat{\mathbf{e}}_k\right) = \frac{\partial^2 \tilde{x}^k}{\partial u^j \partial u^i}\hat{\mathbf{e}}_k .$$

The Cartesian base vectors are expressible in terms of the covariant base vectors.

$$\hat{\mathbf{e}}_i = \frac{\partial u^j}{\partial \tilde{x}^i}\mathbf{g}_j .$$

This is the inverse of the original expression defining $\mathbf{g}_i$. Inserting this expression for $\hat{\mathbf{e}}_i$ gives

$$\frac{\partial \mathbf{g}_i}{\partial u^j} = \frac{\partial^2 \tilde{x}^k}{\partial u^i \partial u^j}\frac{\partial u^l}{\partial \tilde{x}^k}\mathbf{g}_l .$$

The coefficients in the above expression are known as the *Christoffel symbols of the second kind.* These are given a special notation

$$\frac{\partial^2 \tilde{x}^k}{\partial u^i \partial u^j}\frac{\partial u^l}{\partial \tilde{x}^k} = \Gamma^l_{ij} = \left\{ \begin{matrix} l \\ i\,j \end{matrix} \right\}$$

Using this notation gives,

$$\frac{\partial \mathbf{g}_i}{\partial u^j} = \Gamma^l_{ij}\mathbf{g}_l = \left\{ \begin{matrix} l \\ i\,j \end{matrix} \right\}\mathbf{g}_l .$$

The different notations for the Christoffel symbols appear in the literature and are presented to aid in understanding the literature on general tensors.

This expression gives one way to compute the Christoffel symbols. Taking the dot product of both sides with $\mathbf{g}^p$ gives

$$\Gamma^p_{ij} = \frac{\partial \mathbf{g}_i}{\partial u^j}\cdot \mathbf{g}^p$$

after rearranging the indices. The process for computing the Christoffel symbols is

1. From the coordinate transformation $\mathbf{x} = \tilde{x}^i\left(u^j\right)\hat{\mathbf{e}}_i$, find $\mathbf{g}_k = \frac{\partial \tilde{x}^i}{\partial u^k}\hat{\mathbf{e}}_i$ and compute

$$\mathbf{g}_{i,j} = \frac{\partial^2 \tilde{x}^k}{\partial u^j \partial u^i}\hat{\mathbf{e}}_k .$$

2. Determine the contravariant base vectors $\mathbf{g}^k$.
3. Form the dot product to find the Christoffel symbol.

$$\Gamma^l_{ij} = \frac{\partial \mathbf{g}_i}{\partial u^j} \cdot \mathbf{g}^l \,.$$

Another method is available for finding the Christoffel symbols that is not dependent on the coordinate transformation. It will be presented subsequently.

Example A.4

Compute the Christoffel symbols for the circular cylindrical coordinate system.

Solution

The covariant base vectors are

$$\mathbf{g}_r = \frac{\partial \tilde{\mathbf{x}}}{\partial r} = \cos\theta\hat{\mathbf{e}}_1 + \sin\theta\hat{\mathbf{e}}_2$$

$$\mathbf{g}_\theta = \frac{\partial \tilde{\mathbf{x}}}{\partial \theta} = -r\sin\theta\hat{\mathbf{e}}_1 + r\cos\theta\hat{\mathbf{e}}_2$$

and

$$\mathbf{g}_z = \frac{\partial \tilde{\mathbf{x}}}{\partial z} = \hat{\mathbf{e}}_3 \,.$$

The computation is not as daunting as it might first appear. Many of the Christoffel symbols are zero. Differentiating the base vectors we have

$$\frac{\partial \mathbf{g}_r}{\partial r} = \frac{\partial \mathbf{g}_r}{\partial z} = 0, \qquad \frac{\partial \mathbf{g}_r}{\partial \theta} = \frac{\partial \mathbf{g}_\theta}{\partial r} = -\sin\theta\hat{\mathbf{e}}_1 + \cos\theta\hat{\mathbf{e}}_2$$

$$\frac{\partial \mathbf{g}_\theta}{\partial z} = 0, \qquad \frac{\partial \mathbf{g}_\theta}{\partial \theta} = -r\cos\theta\hat{\mathbf{e}}_1 - r\sin\theta\hat{\mathbf{e}}_2$$

and

$$\frac{\partial \mathbf{g}_z}{\partial r} = \frac{\partial \mathbf{g}_z}{\partial \theta} = \frac{\partial \mathbf{g}_z}{\partial z} = 0 \,.$$

The reciprocal base vectors are

$$\mathbf{g}^r = \cos\theta\hat{\mathbf{e}}_1 + \sin\theta\hat{\mathbf{e}}_2$$

$$\mathbf{g}^\theta = \frac{1}{r}\left(-\sin\theta\hat{\mathbf{e}}_1 + \cos\theta\hat{\mathbf{e}}_2\right)$$

and

$$\mathbf{g}^z = \hat{\mathbf{e}}_3 \,.$$

The Christoffel symbols found from $\Gamma^l_{ij} = \frac{\partial \mathbf{g}_i}{\partial u^j} \cdot \mathbf{g}^l$ give

$$\Gamma^r_{rr} = \Gamma^\theta_{rr} = \Gamma^z_{rr} = \Gamma^r_{rz} = \Gamma^\theta_{rz} = \Gamma^z_{rz} = 0$$

$$\Gamma^r_{r\theta} = \Gamma^r_{\theta r} = \Gamma^z_{r\theta} = \Gamma^z_{\theta r} = 0 \qquad \Gamma^\theta_{r\theta} = \Gamma^\theta_{\theta r} = \frac{1}{r}$$

$$\Gamma^\theta_{\theta\theta} = 0 \qquad \Gamma^r_{\theta\theta} = -r \,.$$

The remaining components are zero.

The derivative of the vector can now be written as

$$\frac{\partial \mathbf{v}}{\partial u^j} = \frac{\partial v^i}{\partial u^j}\mathbf{g}_i + v^i \frac{\partial \mathbf{g}_i}{\partial u^j} = \frac{\partial v^i}{\partial u^j}\mathbf{g}_i + v^i \Gamma^l_{ij}\mathbf{g}_l \, .$$

This can be rearranged to give

$$\frac{\partial \mathbf{v}}{\partial u^j} = \frac{\partial v^i}{\partial u^j}\mathbf{g}_i + v^k \Gamma^i_{kj}\mathbf{g}_i = \left(\frac{\partial v^i}{\partial u^j} + v^k \Gamma^i_{kj}\right)\mathbf{g}_i \, .$$

The last form is the result of changing the dummy indices. The coefficients of the covariant base vectors are

$$v^i|_j = \frac{\partial v^i}{\partial u^j} + v^k \Gamma^i_{kj} \, .$$

This is the *covariant derivative* of the contravariant components of a vector. It is a natural generalization of the result for partial differentiation in Cartesian systems.

To this point, all calculation was done using the contravariant components of the vector. If $\mathbf{v} = v_i \mathbf{g}^i$ represents the vector, then

$$\frac{\partial \mathbf{v}}{\partial u^j} = \frac{\partial v_i}{\partial u^j}\mathbf{g}^i + v_i \frac{\partial \mathbf{g}^i}{\partial u^j} \, .$$

This presents the problem of computing the derivative of the second term. To do this, note that

$$\mathbf{g}_i \cdot \mathbf{g}^j = \delta^j_i$$

and

$$\frac{\partial}{\partial u^k}\left(\mathbf{g}_i \cdot \mathbf{g}^j\right) = \frac{\partial \mathbf{g}_i}{\partial u^k} \cdot \mathbf{g}^j + \mathbf{g}_i \cdot \frac{\partial \mathbf{g}^j}{\partial u^k} = \frac{\partial \delta^j_i}{\partial u^k} = 0 \, .$$

This gives

$$\frac{\partial \mathbf{g}^j}{\partial u^k} \cdot \mathbf{g}_i = -\frac{\partial \mathbf{g}_i}{\partial u^k} \cdot \mathbf{g}^j = -\Gamma^l_{ik}\mathbf{g}_l \cdot \mathbf{g}^j = -\Gamma^j_{ik} \, .$$

The right hand side is the covariant component of a vector and

$$\frac{\partial \mathbf{g}^j}{\partial u^k} = -\Gamma^j_{ik}\mathbf{g}^i \, .$$

This result gives

$$\frac{\partial \mathbf{v}}{\partial u^j} = \left(\frac{\partial v_i}{\partial u^j} - v_k \Gamma^k_{ij}\right)\mathbf{g}^i \, .$$

The covariant derivative of the covariant components is

$$v_i|_j = \frac{\partial v_i}{\partial u^j} - v_k \Gamma^k_{ij} \, .$$

Compare this to the previous result.

A.6 Christoffel Symbols

A.6.1 Types of Christoffel Symbols

The definition of the Christoffel symbols of the second kind indicates that there is symmetry in the indices i and j.

$$\frac{\partial^2 \tilde{x}^k}{\partial u^i \partial u^j}\frac{\partial u^l}{\partial \tilde{x}^k} = \frac{\partial^2 \tilde{x}^k}{\partial u^j \partial u^i}\frac{\partial u^l}{\partial \tilde{x}^k}$$

$$\Gamma^l_{ij} = \Gamma^l_{ji} = \left\{ \begin{matrix} l \\ i\,j \end{matrix} \right\} = \left\{ \begin{matrix} l \\ j\,i \end{matrix} \right\}$$

The Christoffel symbols of the second kind are the partial derivatives of the basis vectors in a curvilinear coordinate system with respect to the coordinate variables. Also, it is useful to look at the placement of the indices. The indices are all free indices. The upper index corresponds to the term in the numerator and the lower indices correspond to the terms in the denominator.

The above discussion indicates that the Christoffel symbols are of two types. The *Christoffel symbols of the first kind* are denoted by $[ij, l]$. These are computed from the following formula

$$[ij, l] = g_{lm}\Gamma^m_{ij}$$

Similarly, using the metric g_{lp}, we find

$$\Gamma^p_{ij} = g^{lp}\,[ij, l] \ .$$

The Christoffel symbols of the first kind have a representation similar to those of the second kind. To find this representation, note that

$$g_{lm} = \mathbf{g}_l \cdot \mathbf{g}_m = \frac{\partial \tilde{x}^q}{\partial u^l}\widehat{\mathbf{e}}_q \cdot \frac{\partial \tilde{x}^k}{\partial u^m}\widehat{\mathbf{e}}_k = \frac{\partial \tilde{x}^q}{\partial u^l}\frac{\partial \tilde{x}^k}{\partial u^m}\delta_{qk} \ .$$

The Christoffel symbols of the first kind are

$$[ij, l] = g_{lm}\Gamma^m_{ij} = \frac{\partial \tilde{x}^q}{\partial u^l}\frac{\partial \tilde{x}^k}{\partial u^m}\delta_{qk}\frac{\partial^2 \tilde{x}^p}{\partial u^i \partial u^j}\frac{\partial u^m}{\partial \tilde{x}^p}$$

but

$$\frac{\partial \tilde{x}^k}{\partial u^m}\frac{\partial u^m}{\partial \tilde{x}^p} = \frac{\partial \tilde{x}^k}{\partial \tilde{x}^p} = \delta_{kp} \qquad \text{and} \qquad \delta_{kp}\delta_{qk} = \delta_{pq} \ .$$

This gives

$$[ij, l] = \frac{\partial^2 \tilde{x}^p}{\partial u^i \partial u^j}\frac{\partial \tilde{x}^p}{\partial u^l} \ .$$

Notice that all the indices are in the denominator. Also, the grouping appears in a logical sequence.

A.6.2 Calculation of the Christoffel Symbols

From the above discussion, it would appear that the Christoffel symbols are connected to a particular rectangular Cartesian coordinate system. However, the Christoffel symbols can be computed directly from the metric and the inverse metric for the space. This is shown in the following.

The metric of the space is

$$g_{ij} = \mathbf{g}_i \cdot \mathbf{g}_j \ .$$

We differentiate this with respect to the coordinate u^k and obtain

$$\frac{\partial g_{ij}}{\partial u^k} = \frac{\partial \mathbf{g}_i}{\partial u^k} \cdot \mathbf{g}_j + \mathbf{g}_i \cdot \frac{\partial \mathbf{g}_j}{\partial u^k} \ .$$

This can be written as

$$\frac{\partial g_{ij}}{\partial u^k} = \Gamma^l_{ik}\mathbf{g}_l \cdot \mathbf{g}_j + \mathbf{g}_i \cdot \mathbf{g}_p\Gamma^p_{jk} = \Gamma^l_{ik}g_{lj} + g_{ip}\Gamma^p_{jk} \ .$$

This is

$$\frac{\partial g_{ij}}{\partial u^k} = [ik, j] + [jk, i] \ .$$

Cyclically permuting the indices generates two more formulae.

$$\frac{\partial g_{ki}}{\partial u^j} = [kj, i] + [ij, k]$$

$$\frac{\partial g_{jk}}{\partial u^i} = [ji, k] + [ki, j]$$

Adding the first 2 expressions and subtracting the last expression gives

$$[jk, i] = \frac{1}{2}\left(\frac{\partial g_{ij}}{\partial u^k} + \frac{\partial g_{ki}}{\partial u^j} - \frac{\partial g_{jk}}{\partial u^i}\right) .$$

The symmetry of the Christoffel symbols was exploited in obtaining this result.

The Christoffel symbols of the second kind can be obtained from the above expression and the relation $\Gamma^p_{jk} = g^{ip}\,[jk, i]$. The result is

$$\Gamma^p_{jk} = g^{ip}\,[jk, i] = \frac{1}{2} g^{ip}\left(\frac{\partial g_{ij}}{\partial u^k} + \frac{\partial g_{ki}}{\partial u^j} - \frac{\partial g_{jk}}{\partial u^i}\right) .$$

This demonstrates that the Christoffel symbols are not dependent on the Cartesian basis that was used to motivate their introduction.

The use of the metric tensor is especially useful in determining the Christoffel symbols when the general coordinate bases are orthogonal. In this case, we have $g_{mn} = 0$ if $m \neq n$. This immediately shows that many of the Christoffel symbols will be zero in an orthogonal basis. If i, j, and k are all distinct, then $\Gamma^k_{ij} = 0$. The only non-zero values will occur when two of the indices are equal in a general, orthogonal coordinate system.

A.7 Covariant Derivatives of Tensors

In a manner similar to that used to find the covariant derivative of a vector, the covariant derivative of a tensor can be found. To calculate $\frac{\partial \mathbf{T}}{\partial u^k}$, begin with the representation of a tensor in the basis $\{\mathbf{g}_i\}$, $\mathbf{T} = T^{ij}\mathbf{g}_i\mathbf{g}_j$. Then

$$\begin{aligned}
\frac{\partial \mathbf{T}}{\partial u^k} &= \frac{\partial}{\partial u^k}\left(T^{ij}\mathbf{g}_i\mathbf{g}_j\right) \\
&= \frac{\partial T^{ij}}{\partial u^k}\mathbf{g}_i\mathbf{g}_j + T^{ij}\frac{\partial \mathbf{g}_i}{\partial u^k}\mathbf{g}_j + T^{ij}\mathbf{g}_i\frac{\partial \mathbf{g}_j}{\partial u^k} \ .
\end{aligned}$$

Using the Christoffel symbols, this can be written as

$$\frac{\partial \mathbf{T}}{\partial u^k} = \frac{\partial T^{ij}}{\partial u^k}\mathbf{g}_i\mathbf{g}_j + T^{ij}\Gamma^l_{ik}\mathbf{g}_l\mathbf{g}_j + T^{ij}\Gamma^l_{jk}\mathbf{g}_i\mathbf{g}_l \ .$$

Rearranging the dummy indices gives

$$\begin{aligned}
\frac{\partial \mathbf{T}}{\partial u^k} &= \left(\frac{\partial T^{ij}}{\partial u^k} + T^{lj}\Gamma^i_{lk} + T^{il}\Gamma^j_{lk}\right)\mathbf{g}_i\mathbf{g}_j \\
&= T^{ij}|_k\,\mathbf{g}_i\mathbf{g}_j \ .
\end{aligned}$$

The term in parentheses is the covariant derivative of the contravariant components of the tensor.

$$T^{ij}|_k = \frac{\partial T^{ij}}{\partial u^k} + T^{lj}\Gamma^i_{lk} + T^{il}\Gamma^j_{lk} .$$

Other covariant derivatives can be formed. These are

$$T_{ij}|_k = \frac{\partial T_{ij}}{\partial u^k} - T_{lj}\Gamma^l_{ik} - T_{il}\Gamma^l_{jk}$$

$$T^i{}_j|_k = \frac{\partial T^i{}_j}{\partial u^k} + T^l{}_j\Gamma^i_{lk} - T^i{}_l\Gamma^l_{jk} .$$

They are computed using the same techniques as were employed previously.

A.8 General Tensor Equations

Tensors and tensor equations describing physical phenomena are invariant with respect to coordinate systems. A number of relations in direct and Cartesian tensor notation were developed. These relations span the spectrum from inner products of vectors to local forms of the First Law of Thermodynamics.

The following presents a way to convert these relations from Cartesian tensor notation to general tensor notation. Vectors and tensors are represented in terms of covariant and contravariant base vectors. The techniques presented allow us to compute the derivatives of vector fields and tensor fields. Three rules allow the conversion of Cartesian tensor expressions to expressions in general coordinate systems.

Rule 1: *Rewrite summations over dummy indices in Cartesian expressions so that the summation is only over raised and lowered indices.*

This rule is a direct consequence of the way that summation was introduced for vector and tensor representation in a general coordinate system. We defined vectors and tensors in terms of covariant and contravariant bases. This led to the convention that implied summation occurs only over a subscript and a superscript pair.

Summation across pairs of indices may be accomplished in a number of ways. We may simply change an index in the Cartesian expression from a subscript to a superscript. This is possible since we have Kronecker deltas δ^{ij} and δ_{ij} in the Cartesian system. We may raise or lower indices by noting that the metric or inverse metric acts as a generalized Kronecker delta. Table A.1 gives examples of Cartesian expressions and the corresponding general tensor expressions.

Rule 2: *Replace partial differentiation in Cartesian tensor expressions with covariant differentiation.*

This rule is useful in converting derivatives and differential operators. An example is the divergence of a vector field. In Cartesian tensor notation, we have

$$\operatorname{div} \mathbf{v} = v_{i,i} .$$

This becomes

$$\operatorname{div} \mathbf{v} = v^i|_i$$

TABLE A.1
Converting from Cartesian tensor notation to general tensor notation. Summation over only subscript and superscript pairs.

Cartesian Tensor Expression	General Tensor Expression
d_{kk}	$d_k{}^k = g_{km} d^{km} = g^{km} d_{km}$
$a_i b_i$	$a_i b^i = g_{ik} a^k b^i = g^{ik} a_i b_k$
$d_i d_i$	$d_i d^i = g_{ij} d^i d^j = g^{ij} d_i d_j$
$v_i = w_{ikk}$	$v_i = w_{ik}{}^k = g_{km} w_i{}^{km} = g^{km} w_{ikm}$

in general tensor notation. Note that the summation is over a raised and lowered index. Another example is the equilibrium equations. In Cartesian notation, they read

$$T_{ij,j} + b_i = \rho a_i \ .$$

In general tensor notation, this becomes

$$T^{ij}|_j + b^i = \rho a^i \ .$$

Are there other forms that could be used to represent these equations?

These general tensor forms appear to be quite simple. However, it must be remembered that the covariant derivative involves Christoffel symbols. For general coordinate systems, these are quite complicated. In orthogonal systems, things improve since the non-zero Christoffel symbols have a pair of matching indices. The distinction disappears in a Cartesian system where the Christoffel symbols are all zero. Things are not always as simple as they first appear!

Rule 3: *Replace the Cartesian tensor permutation symbol, ε_{ijk}, by the general tensor permutation symbol, e_{ijk} or e^{ijk}, and rewrite summations over raised and lowered indices.*

This rule is useful when considering the cross product of vectors or vector operators. The cross product of two vectors in a general coordinate system is

$$\mathbf{u} \times \mathbf{v} = u^i v^j \mathbf{g}_i \times \mathbf{g}_j$$

but

$$\mathbf{g}_i \times \mathbf{g}_j = e_{ijk} \mathbf{g}^k \ .$$

This gives

$$\mathbf{u} \times \mathbf{v} = e_{ijk} u^i v^j \mathbf{g}^k \ .$$

Also note that

$$e_{ijk} = (\mathbf{g}_i \times \mathbf{g}_j) \cdot \mathbf{g}_k = \varepsilon_{ijk} \sqrt{g}$$

where

$$g = \det |g_{ij}| \ .$$

These three rules allow one to conveniently change Cartesian tensor expressions into general tensor expressions. Now the laws and equations that are valid in Cartesian reference frames can be rewritten so that they have forms appropriate for general coordinate systems. This gives us a very powerful methodology for exploring the mechanics of continuous media.

A.9 General Tensors and Physical Components

A tensor has the following representations

$$\mathsf{T} = T^{ii}\mathbf{g}_i\mathbf{g}_j = T_{ij}\mathbf{g}^i\mathbf{g}^j = T^{<ij>}\hat{\mathbf{g}}_i\hat{\mathbf{g}}_j \,.$$

The equilibrium equations are given by the derivatives of the physical components $T^{<ij>}$. For cylindrical-polar coordinates, these are

$\hat{\mathbf{g}}_1 = \hat{\mathbf{e}}_r$

$$\frac{\partial T^{<rr>}}{\partial r} + \frac{1}{r}\frac{\partial T^{<r\theta>}}{\partial \theta} + \frac{\partial T^{<rz>}}{\partial z} + \frac{T^{<rr>} - T^{<\theta\theta>}}{r} = 0$$

$\hat{\mathbf{g}}_2 = \hat{\mathbf{e}}_\theta$

$$\frac{1}{r^2}\frac{\partial \left(r^2 T^{<\theta r>}\right)}{\partial r} + \frac{1}{r}\frac{\partial T^{<\theta\theta>}}{\partial \theta} + \frac{\partial T^{<\theta z>}}{\partial z} = 0$$

$\hat{\mathbf{g}}_3 = \hat{\mathbf{e}}_z$

$$\frac{1}{r}\frac{\partial \left(r T^{<zr>}\right)}{\partial r} + \frac{1}{r}\frac{\partial T^{<\theta z>}}{\partial \theta} + \frac{\partial T^{<zz>}}{\partial z} = 0 \,.$$

The covariant derivative of the tensors for computing the equilibrium equation can be written in two forms.

Form 1:

$$T^{ji}\Big|_i = \frac{\partial T^{ji}}{\partial u^i} + T^{jl}\Gamma^i_{li} + T^{li}\Gamma^j_{li} = 0$$

This is an acceptable form for the equilibrium equations since the tensor components are the contravariant components. These are similar to the contravariant physical components used above.

$\mathbf{g}_1$ $(j = 1)$

$$\frac{\partial T^{1i}}{\partial u^i} + T^{1l}\Gamma^i_{li} + T^{li}\Gamma^1_{li} = 0$$

This equation contains many terms. However, all but two of the Christoffel symbols are zero. The non-zero Christoffel symbols are

$$\Gamma^2_{12} = \Gamma^2_{21} = \frac{1}{r} \qquad \Gamma^1_{22} = -r \,.$$

This gives

$$\frac{\partial T^{1i}}{\partial u^i} + T^{11}\Gamma^2_{12} + T^{22}\Gamma^1_{22} = 0 \,;$$

$$\frac{\partial T^{11}}{\partial u^1} + \frac{\partial T^{12}}{\partial u^2} + \frac{\partial T^{13}}{\partial u^3} + \frac{1}{r}T^{11} - rT^{22} = 0 \,.$$

The contravariant components in this expression are not the physical components of the tensor. What is the relationship between the two? The representation of a tensor above gives

$$T^{ii}\mathbf{g}_i\mathbf{g}_j = T^{<ij>}\hat{\mathbf{g}}_i\hat{\mathbf{g}}_j = T^{<ij>}\frac{\mathbf{g}_i\,\mathbf{g}_j}{g_i\;g_j} \,.$$

g_i and g_j are the magnitudes of the vectors. These are

$$g_i = \sqrt{\mathbf{g}_i \cdot \mathbf{g}_i} = \sqrt{g_{ii}} \quad \text{no sum on } i \,.$$

g_{ij} is the metric tensor. For cylindrical coordinates, this is

$$[g_{ij}] = \begin{bmatrix} 1 & 0 & 0 \\ 0 & r^2 & 0 \\ 0 & 0 & 1 \end{bmatrix} .$$

The connection between the components are

$$T^{<ij>} = \sqrt{g_{ii}}\sqrt{g_{jj}}\, T^{ij} \quad \text{no sum on i or j} .$$

$$\begin{bmatrix} T^{<rr>} & T^{<r\theta>} & T^{<rz>} \\ T^{<\theta r>} & T^{<\theta\theta>} & T^{<\theta z>} \\ T^{<zr>} & T^{<z\theta>} & T^{<zz>} \end{bmatrix} = \begin{bmatrix} T^{11} & rT^{12} & T^{13} \\ rT^{21} & r^2T^{22} & rT^{23} \\ T^{31} & rT^{32} & T^{33} \end{bmatrix}$$

This gives

$$\frac{\partial T^{<rr>}}{\partial u^1} + \frac{\partial}{\partial u^2}\left(\frac{1}{r}T^{<r\theta>}\right) + \frac{\partial T^{<rz>}}{\partial u^3} + \frac{1}{r}T^{<rr>} - r\left(\frac{1}{r^2}T^{<\theta\theta>}\right) = 0 .$$

This is the $\hat{\mathbf{e}}_r$ equilibrium equation where (u^1, u^2, u^3) is (r, θ, z).

$\mathbf{g}_2$ $(j = 2)$

$$\frac{\partial T^{2i}}{\partial u^i} + T^{2l}\Gamma^i_{li} + T^{li}\Gamma^2_{li} = 0$$

or

$$\frac{\partial T^{2i}}{\partial u^i} + T^{21}\Gamma^2_{12} + T^{12}\Gamma^2_{12} + T^{21}\Gamma^2_{21} = \frac{\partial T^{2i}}{\partial u^i} + \frac{3}{r}T^{21} = 0 .$$

This leads to

$$\frac{\partial T^{21}}{\partial u^1} + \frac{\partial T^{22}}{\partial u^2} + \frac{\partial T^{23}}{\partial u^3} + \frac{3}{r}T^{21} = 0 .$$

Writing this in terms of the physical components and the coordinates (r, θ, z) gives

$$\frac{\partial}{\partial r}\left(\frac{1}{r}T^{<\theta r>}\right) + \frac{\partial}{\partial \theta}\left(\frac{1}{r^2}T^{<\theta\theta>}\right) + \frac{\partial}{\partial z}\left(\frac{1}{r}T^{<\theta z>}\right) + \frac{3}{r}\left(\frac{1}{r}T^{<\theta r>}\right) = 0$$

and

$$\frac{1}{r^2}\frac{\partial}{\partial r}\left(r^2 T^{<\theta r>}\right) + \frac{\partial}{\partial \theta}\left(\frac{1}{r}T^{<\theta\theta>}\right) + \frac{\partial}{\partial z}\left(T^{<\theta z>}\right) = 0$$

$\mathbf{g}_3$ $(j = 3)$

$$\frac{\partial T^{3i}}{\partial u^i} + T^{3l}\Gamma^i_{li} + T^{li}\Gamma^3_{li} = 0$$

or

$$\frac{\partial T^{3i}}{\partial u^i} + T^{31}\Gamma^2_{12} + T^{li}(0) = 0 .$$

This yields

$$\frac{\partial T^{31}}{\partial u^1} + \frac{\partial T^{32}}{\partial u^2} + \frac{\partial T^{33}}{\partial u^3} + \frac{1}{r}T^{31} = 0 .$$

Converting to physical components gives

$$\frac{1}{r}\frac{\partial\left(rT^{<zr>}\right)}{\partial r} + \frac{\partial}{\partial \theta}\left(\frac{1}{r}T^{<z\theta>}\right) + \frac{\partial T^{<zz>}}{\partial z} = 0$$

Form 2:

This form uses the covariant tensor components

$$T_{ij}|_i = \frac{\partial T_{ij}}{\partial u^j} - T_{mj}\Gamma^m_{ij} + T_{im}\Gamma^m_{jj} = 0$$

This is not an acceptable form for the equilibrium equations for the physical components since the tensor components are the covariant components rather than the contravariant components. These are not the same as the contravariant physical components used in the equilibrium equations. While these expressions will give partial differential equations for the covariant components, they will not give an equivalent set of equations to those already presented. The contravariant and covariant components and associated equations become equivalent in an orthonormal rectangular coordinate system.

References

[1] A. C. Eringen (1962) *Nonlinear Theory of Continuous Media*, McGraw-Hill Book Co., New York.

[2] Y. C. Fung (1965) *Foundations of Solid Mechanics*, Prentice Hall, Englewood Cliffs.

[3] L. E. Malvern (1969) *Introduction to the Mechanics of a Continuous Media*, Prentice Hall, Englewood Cliffs.

[4] M. N. L. Narasimhan (1993) *Principles of Continuum Mechanics*, John Wiley & Sons, Inc., New York.

[5] J. G. Simmonds (1994) *A Brief on Tensor Analysis*, Springer, New York.

Appendix B: Viscoelastic Creep and Relaxation

TABLE B.1: Creep and relaxation responses for various viscoelastic models.

Model/Name	Constitutive Equation	Creep Compliance, J(t)	Relaxation Modulus, G(t)
Elastic Solid  (G)	$T = G\gamma$	$\frac{1}{G} = J = \frac{U(t)}{G}$	$G(t) = GU(t)$
Viscous Fluid 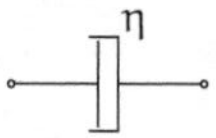 (η)	$T = \eta\dot{\gamma}$ $T = \eta\dot{\epsilon}$	$\frac{t}{\eta} = \frac{t}{G\tau}$	$\eta\delta(t)$
Maxwell (Fluid)  (G, η)	$\dot{T} + \frac{1}{\tau}T = G\dot{\gamma}$ $\left\{\partial_t + \frac{1}{t}\right\}T = \{G\partial_t\}\gamma$	$\frac{1}{G} + \frac{t}{\eta} = \frac{\tau + t}{\eta}$	$Ge^{-t/\tau}$
Kelvin (Solid) 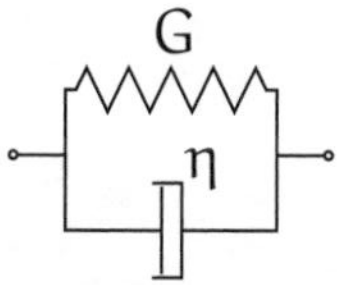 (G, η)	$T = G\gamma + \eta\dot{\gamma}$ $T = \{G + \eta\partial_t\}\gamma$ $\frac{1}{\eta}T = \left\{\frac{1}{\tau} + \partial_t\right\}\gamma$	$\frac{1}{G}(1 - e^{-t/\tau})$	$G + \eta\delta(t)$
Three Parameter Solid 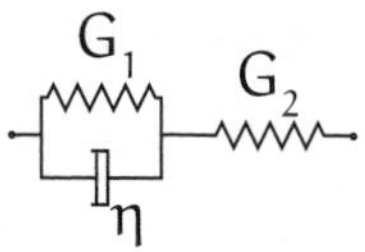 (G_1, G_2, η)	$G_1G_2\gamma + G_1\eta_2\dot{\gamma} =$ $(G_1 + G_2)T + \eta_2\dot{T}$	$\frac{e^{-t/\tau}}{G_1} + \frac{G_1 + G_2}{G_1G_2}(1 - e^{t/\tau_2})$ or $\frac{1}{G_1} + \frac{1}{G_2}(1 - e^{-t/\tau_2})$	$G_1e^{-t/\tau'} + \frac{G_1G_2}{G_1 + G_2}(1 - e^{-t/\tau'})$ or $\frac{G_1}{G_1 + G_2}(G_2 + G_1e^{-t/\tau'})$ where $\tau' = \frac{\eta_2}{G_1 + G_2}$

Continued on next page

TABLE B.1 – continued from previous page

Model/Name	Constitutive Equation	Creep Compliance, J(t)	Relaxation Modulus, G(t)
G_1, η, G_2	$\dot{T}+\dfrac{T}{\tau_2}=(G_1+G_2)\dot{\gamma}+\dfrac{G_1}{\tau_2}\gamma$	$\dfrac{e^{-t/\tau'}}{G_1+G_2}+\dfrac{1}{G_1}\left(1-e^{-t/\tau'}\right)$ or $\dfrac{1}{G_1}-\dfrac{G_2}{G_1(G_1+G_2)}e^{-t/\tau'}$ where $\tau'=\dfrac{G_1+G_2}{G_1}\tau_2$	$G_1+G_2e^{-t/\tau_2}$
G, η_1, η_2	$G_2\eta_1\dot{\gamma}+\eta_1\eta_2\ddot{\gamma}=G_2T+(\eta_1+\eta_2)\dot{T}$ $\ddot{\gamma}+\dfrac{\dot{\gamma}}{\tau_2}=\dfrac{\eta_1+\eta_2}{\eta_1\eta_2}\dot{T}+\dfrac{T}{\tau_2\eta_1}$	$\dfrac{1}{G_1}\left(1-e^{-t/\tau'}\right)+\dfrac{t}{\eta_2}$	$\dfrac{\eta_1\eta_2}{\eta_1+\eta_2}\delta(t)+\dfrac{G_2}{\eta_1+\eta_2}\left(\eta_1+\dfrac{\eta_1\eta_2}{\eta_1+\eta_2}\right)e^{\tau'}$ where $\tau'=\dfrac{\eta_1+\eta_2}{G_2}$
G, η_1, η_2	$\dot{T}+\dfrac{T}{\tau_2}=\left(G_2+\dfrac{\eta_1}{\tau_2}\right)\dot{\gamma}+\eta_1\ddot{\gamma}$	$\dfrac{t}{\eta_1+\eta_2}+\dfrac{1}{G_2}\left(\dfrac{\eta_2}{\eta_1+\eta_2}\right)^2\left(1-e^{-\lambda t}\right)$ where $\lambda=\dfrac{G_2(\eta_1+\eta_2)}{\eta_1\eta_2}$	$\eta_1\delta(t)+G_2e^{-t/\tau_2}$
G_1, G_2, η_1, η_2	$\ddot{\gamma}+\dfrac{\dot{\gamma}}{\tau_2}=\dfrac{\ddot{T}}{G_1}+\left(\dfrac{1}{\eta_2}+\dfrac{1}{G_1\tau_2}+\dfrac{1}{\eta_1}\right)\dot{T}+\dfrac{1}{\eta_1\tau_2}T$	$\dfrac{1}{G_1}+\dfrac{t}{\eta_1}+\dfrac{1}{G_2}\left(1-e^{-t/\tau_2}\right)$	

Continued on next page

TABLE B.1 – continued from previous page

Model/Name	Constitutive Equation	Creep Compliance, J(t)	Relaxation Modulus, G(t)
G_1, G_2, η_1, η_2	$(\eta_1 + \eta_2)\dot{T} + (G_1 + G_2)T = G_1 G_2 \gamma + (\eta_2 G_1 + \eta_1 G_2)\dot{\gamma} + \eta_1 \eta_2 \ddot{\gamma}$		$\frac{1}{\eta_1 + \eta_2}\left\{\eta_1\eta_2\left(\delta(t) - \frac{e^{-t/\tau'}}{\tau'}\right)\right\} + G_1 G_2 \tau' \left(1 - e^{-t/\tau'}\right)$

Index